# Geographic Information Systems for Transportation

SPATIAL INFORMATION SYSTEMS

*General Editors*
M. F. Goodchild
P. A. Burrough
R. A. McDonnell
P. Switzer

# GEOGRAPHIC INFORMATION SYSTEMS FOR TRANSPORTATION

## Principles and Applications

Harvey J. Miller
Shih-Lung Shaw

UNIVERSITY PRESS

2001

# OXFORD
UNIVERSITY PRESS

Oxford   New York
Athens   Auckland   Bangkok   Bogotá   Buenos Aires   Cape Town
Chennai   Dar es Salaam   Delhi   Florence   Hong Kong   Istanbul   Karachi
Kolkata   Kuala Lumpur   Madrid   Melbourne   Mexico City   Mumbai   Nairobi
Paris   São Paulo   Shanghai   Singapore   Taipei   Tokyo   Toronto   Warsaw

with associated companies in
Berlin   Ibadan

Copyright © 2001 by Oxford University Press, Inc.

Published by Oxford University Press, Inc.
198 Madison Avenue, New York, New York 10016

Oxford is a registered trademark of Oxford University Press.

All rights reserved. No part of this publication may be reproduced,
stored in a retrieval system, or transmitted, in any form or by any means,
electronic, mechanical, photocopying, recording, or otherwise,
without the prior permission of Oxford University Press.

Library of Congress Cataloging-in-Publication Data
Miller, Harvey J.
    Geographic information systems for transportation : principles and applications / Harvey
J. Miller, Shih-Lung Shaw.
        p. cm.—(Spatial information systems)
    Includes bibliographical references and index.
    ISBN 0-19-512394-8
    1. Transportation—United States—Planning.   2. Transportation—Europe—Planning.
3. Transportation—Japan—Planning.   4. Geographic information systems—United States.
5. Geographic information systems—Europe.   6. Geographic information systems—Japan.
I. Shaw, Shih-Lung.   II. Title.   III. Series.
HE206.2 .M55   2001
388'.028'5—dc21          00-063674

9 8 7 6 5

Printed in the United States of America
on acid-free paper

# Preface

To paraphrase Mark Twain, reports of the "death of distance" are greatly exaggerated. The rise of the information economy triggered an even greater explosion in the private transportation industry: every movement of a bit implies movement of a far greater number of atoms. Despite the hype about telecommuting, teleshopping, and virtual entertainment, we still sit in traffic jams and watch our urban airsheds turn opaque. Like squeezing a balloon, we have traded congestion in the urban core for congestion at the urban periphery in many cities of the world. Transportation systems serve some neighborhoods (and cities and regions) well. Some continue to be neglected. Some suffer so that others can be better connected. As Waldo Tobler recently pointed out in remarks to the 1999 ESRI user conference, while the world may be "shrinking," it is also "shriveling": transportation cost differentials are increasing and creating a new, complex geography of transportation.

It is likely that transportation-related problems will be at the forefront of private sector decision making and public sector debate as the world navigates through the twenty-first century. Our connected and crowded world requires efficient, responsive and environmentally friendly transportation systems. We have also come to the realization (too late in some cases) that transportation decisions have enormous impacts on land, lives, and life.

Geographic Information Systems for Transportation (GIS-T) refers to the application of information technology and related knowledge to transportation problems. As we point out in chapter 1, GIS-T has "arrived" and now is one of the most important and rapidly growing applications for GIS. GIS-T offers a new toolkit to transportation professionals just when it is needed. GIS-T encourages a holistic approach to transportation analysis, supporting integrated analysis of all components of a transportation system within its geographic context.

In *Geographic Information Systems for Transportation: Principles and Applications*, we emphasize the role of GIS-T as a toolkit for transportation research and decision making. We provide a thorough but accessible discussion of the relevant geographic information science (GISci), spatial analytical and computer science principles at the core of GIS-T as well as their application to major transportation problems. This provides an introduction and reference for students, researchers and professionals interested in GIS-T software tool development. It also provides a foundation so that the application-oriented student, researcher and professional can be an intelligent consumer of these tools.

We conceived this project during the Ft. Worth meeting of the Association of American Geographers. Shaw was a panelist in a session discussing GIS-T education needs; Miller was an audience member. We discovered that our views and perspectives on GIS-T education coincided: the dearth of good material on transportation analysis and modeling from a GIS perspective was hampering our ability to teach these subjects to students (and occasionally, professionals) with eclectic backgrounds and needs.

As is typical in a project of this magnitude, we have benefited from support, encouragement, and feedback from numerous individuals. Professor Bruce Ralston (University of Tennessee) wrote chapter 11 ("GIS and Logistics") of this text; we greatly appreciate his contribution. We received very helpful comments on material in this text from Mace Bowen, Tom Cova, Bertrand Granberg, George Hepner, and Yi-Hwa Wu (University of Utah), Cheng Liu (Oak Ridge National Laboratory), and students in Dr. Shaw's GIS-T courses at the University of Tennessee ("GEOG549: Topics in Transportation Geography", Fall 1999 and "GEOG649: Seminar in Transportation Geography", Spring 2000). Al Butler (GIS Director, Hamilton County, Tennessee Government) provided figure 3-15 and commented on the GIS-T data model material in chapter 3. David O'Sullivan (Centre for Advanced Spatial Analysis, University College London) provided figure 8-5. Will Fontanez and Tom Wallin of the Cartographic Services Laboratory at the University of Tennessee provided professional help with some of the figure production for this text. Finally, Joyce Berry at Oxford University Press was very patient in working with us to realize our vision for this book.

Harvey Miller thanks the College of Social and Behavioral Sciences, University of Utah, for a sabbatical leave that greatly facilitated the completion of this book. He also thanks his colleagues in the Department of Geography at the University of Utah for their support, encouragement, and tolerance. Special thanks go to his wife and friend Susanne, who now knows more about GIS-T than she thought was possible (or necessary).

Shih-Lung Shaw thanks the College of Arts and Sciences and the Department of Geography at the University of Tennessee for providing the opportunity and the resources for his completion of this project. He extends special appreciation to Bruce Ralston for his constant support and encouragement. He is most indebted to his parents and family for their support and understanding. There was many nights and weekends not spent with his family. Thank you—Mei, Chris, and Chrissy.

# Acknowledgments

We are grateful for permission to reprint the following figures and tables:

Figures 2-3 and 2-4 are reprinted from Worboys, M.F. (1995). *GIS: A Computing Perspective*, (Figures 2-16 and 5-10, respectively), Taylor and Francis. Used with permission of Taylor and Francis, Ltd.

Figure 2-9 is reprinted from Cova, T.J. and Goodchild, M.F. (1994). "Spatially distributed navigable databases for intelligent vehicle highway systems," *GIS/LIS '94 Proceedings*, 191-2001. Used with permission of American Congress on Surveying and Mapping.

Figures 3-1, 3-2, and 3-3 are reprinted from Sheffi, Y. (1985). Urban Transportation Networks: Equilibrium Analysis with Mathematical Programming Methods (Figures 1.5, 1.6, and 1.7, respectively). Englewood Cliffs, NJ: Prentice-Hall.

Figure 3-6 is reprinted from Nyerges, T.L. (1990). "Locational referenceing and highway segmentation in a geographic information system," *ITE Journal* 60(3): 27–31, Figure 1. © 1990 Institute of Transportation Engineers. Used by permission.

Figure 3-11 is reprinted from *Transportation Research C: Emerging Technologies*, 4C, Miller, H.J. and Storm, J.D. (1996). "Geographic information system design for network equilibrium-based travel demand models," 373–389. © 1996, with permission from Elsevier Science.

Figure 3-14 is reprinted with permission from the Urban and Regional Information Systems Association, from Dueker, K.J. and Butler, J.A. (1997). *GIS-T Enterprise Data Model with Suggested Implementation Choices*, Discussion paper, Center for Urban Studies, Portland State University, Figure 7.

Figure 3-15 is reprinted from Fohl, P., Curtin, K.M., Goodchild, M.F., and Church, R.L. (1996). "A non-planar, lane-based navigable data model for ITS." *7$^{th}$ International Symposium on Spatial Data Handling*, Delft, The Netherlands. Used with permission of Anthony Yeh, GIS/LIS Research Center, University of Hong Kong.

Figure 4-4 is reprinted by kind permission of Ordnance Survey © Crown Copyright NC/00/1232.

Figure 5-2 is reprinted from Phillips, D.T. and Garcia-Diaz, A. (1981). *Fundamentals of Network Analysis*, Englewood Cliffs, NJ: Prentice-Hall. Used with permission of Prentice-Hall.

Figure 7-7 is reprinted from Open GIS Consortium Technical Committee (1998). *Introduction to Interoperable Geoprocessing and the OpenGIS Specification*, 3$^{rd}$ Edition, available at http://www.opengis.org. Open GIS Consortium, Inc., Wayland, MA. Used with permission of the Open GIS Consortium, Inc.

Figure 8-2 is reprinted from You, J., Nedovic-Budic, Z., and Kim, T.J. (1997). "A GIS-based traffic analysis zone design: Implementation and evaluation," *Transportation Planning and Technology*, 21, 69–91, Figure 1. Used with the permission of Overseas Publishers Association and Gordon and Breach Publishers.

Figure 8-5 is reprinted from O'Sullivan, D., Morrison, A., and Shearer, J. (2000). "Using desktop GIS for the investigation of accessibility by public transport: An isochrone approach," *International Journal of Geographical Information Science*, 14, 85–104, Figure 2. Used with permission of Taylor and Francis, Ltd.

Figure 8-6 is reprinted from Miller, H.J. and Wu, Y.-H. (2000). "GIS software for measuring space-time accessibility in transportation planning and analysis," *GeoInformatica* 4, 141–159, Figure 10. Used with permission of Kluwer Academic Publishers.

Figure 8-7 is reprinted from Koninger, A. and Bartel, S. (1998), "3D-GIS for urban purposes," *GeoInformatica* 2, 79–103, Figure 11. Used with permission of Kluwer Academic Publishers.

Figure 8-8 is reprinted from Jankowski, P. (1995). "Integrating geographical information systems and multiple criteria decision-making methods," *International Journal of Geographical Information Science* 9, 251–273, Figure 1. Used with permission of Taylor and Francis, Ltd.

Figures 8-9 and 8-10 are reprinted from Nielsen, O.A. (1995). "Using GIS in Denmark for traffic planning and decision support," *Journal of Advanced Transportation* 29, 335–354 (Figures 1 and 2, respectively). Used with permission of Institute for Transportation, Inc.

Figure 9-1 is reprinted from Keller, H. (ed.) (1999). *Telematics Applications Programme: Concertation and Achievements Report of the Transport Sector (CARTS)*, Co-ordinated Dissemination in Europe (CODE) Deliverable 2.2, Figure 1-1. Used with permission of European Commission, DG Information Society.

Figure 9-3 is reprinted from the Japanese Ministry of Construction (1999). *ITS Handbook 1999–2000*, Section 5.8.

Figure 9-4 is reprinted from the Japanese Ministry of Construction (1999). *System Architecture for ITS in Japan—Summary*, Figure 4.2-1.

Figures 9-7 and 9-8 are reprinted from Vehicle Intelligence & Transportation Analysis Laboratory (1998). *The Cross Streets Profile with Coordinates—Technical Evaluation*, Report prepared for US Department of Transportation, Federal Highway Administration (Figures 1 and 3, respectively). Used with permission of the Vehicle Intelligence & Transportation Analysis Laboratory, National Center for Geographic Information and Analysis, University of California, Santa Barbara.

Figure 9-9 is reprinted from Plewe, B. (1997). *GIS Online: Information, Retrieval, Mapping, and the Internet*, from the figure on p. 280. Santa Fe, NM: OnWord Press, Used with permission of Thomson Learning.

Figure 10-1 is reprinted from Davis, T.J. and Keller, C.P. (1997). "Modeling uncertainty in natural resource analysis using fuzzy sets and Monte Carlo simulation: Slope stability prediction," *International Journal of Geographical Information Science* 11, 409–434, Figure 1. Used with permission of Taylor and Francis, Ltd.

Figure 10-3 is reprinted from Stern, E. and Sinuany-Stern, Z. (1989). "A behavioural-based simulation model for urban evacuation," *Papers of the Regional Science Association* 66, 87–103, Figure 1. Used with permission of the Regional Science Association International.

Table 2-1 is reprinted from Fletcher, D. (1987). "Modeling GIS transportation networks," *Proceedings*, 25th annual meeting of the Urban and Regional Information Systems Association, 84–92, Table 1. Used with permission from the Urban and Regional Information Systems Association.

Table 4-1 is reprinted from Dana, P.H. (2000). "Geodetic datum overview," *The Geographer's Craft Project*, Department of Geography, University of Colorado at Boulder, available at http://www.colorado.edu/geography/gcraft/notes/datum/datum_f.html Used with permission of Peter Dana, Geographer's Craft Project.

Table 4-2 is reprinted with permission from Vonderohe, A.P., Travis, L., Smith, R.L., and Tsai, V. (1993), *Adaptation of Geographic Information Systems for Transportation*, Washington, DC: Report 359, National Cooperative Highway Research Program, National Academy Press. © 1993 by the National Academy of Sciences, Courtesy of the National Academy Press, Washington, DC.

Table 5-1 is reprinted from *Computers and Intractability: A Guide to the Theory of NP-Completeness* by Michael R. Garey and David S. Johnson. © 1979 Bell Laboratories, Inc. Used with permission of W.H. Freeman and Company.

Table 9-2 is reprinted from the Japanese Ministry of Construction (1996). *Comprehensive Plan for ITS in Japan*.

Table 9-5 is reprinted from Vehicle Intelligence and Transportation Analysis Laboratory (1999). *The LRMS Linear Referencing Profile—Technical Evaluation*. Report prepared for US Department of Transportation, Federal Highway Administration, Table 2. Used with permission of the Vehicle Intelligence and Transportation Analysis Laboratory, National Center for Geographic Information and Analysis, University of California, Santa Barbara.

# Contents

1. **Introduction, 3**
   Geographic information systems for transportation, 3
   Geographic information *science* for transportation, 6
   The scope of this book, 7
   The outline of this book, 8

2. **Data Modeling and Database Design, 14**
   Data domains and data modeling in GIS-T, 15
   Data modeling techniques, 16
   Other data modeling and design issues, 38
   Conclusion, 52

3. **GIS-T Data Models, 53**
   Mathematical foundations: Graph theory and network analysis, 54
   Network representation of a transportation system, 55
   Linear referencing methods and linear referencing systems, 62
   Transportation data models for ITS and related applications, 77
   Conclusion, 84

4. Transportation Data Sources and Integration, 85
   Basic mapping concepts, 86
   Transportation data capture and data products, 93
   Transportation data integration, 111
   Spatial data quality, 120
   Spatial and network aggregation, 125
   Conclusion, 129

5. Shortest Paths and Routing, 130
   Fundamental network properties, 131
   Fundamental properties of algorithms, 134
   Shortest path algorithms, 139
   Routing vehicles within networks, 158
   Conclusion, 170

6. Network Flows and Facility Location, 172
   Flow through uncongested networks, 173
   Flow through congested networks, 178
   Facility location within networks, 199
   Spatial aggregation in network routing and location problems, 212
   Conclusion, 213

7. GIS-Based Spatial Analysis and Modeling, 214
   GIS and spatial analysis, 215
   GIS analytical functions, 218
   Coupling transportation analysis and modeling with GIS, 229
   Customizing GIS, 232
   Supporting advanced transportation analysis in GIS, 236
   Geographic visualization, 238
   Conclusion, 245

8. Transportation Planning, 247
   Transportation analysis zone design, 248
   Travel demand analysis, 252
   Land-use/transportation modeling, 276
   Route planning, 282
   Decision support for transportation planning, 290
   Conclusion, 294

9. **Intelligent Transportation Systems**, 295
   ITS development, 296
   ITS applications, 303
   Marketing ITS: What services do travelers want?, 307
   ITS architectures, 307
   ITS architectures and geographic information, 317
   Integrating GIS and ITS: Some examples, 332
   Conclusion, 340

10. **Transportation, Environment, and Hazards**, 341
    Transportation and the environment, 342
    Transportation and hazards, 360
    Conclusion, 378

11. **Logistics**, 380
    Supply chains and logistics, 382
    GIS and logistics management, 383
    Some complex aspects of logistics problems, 396
    Conclusion, 398

**References**, 401

**Index**, 448

# Geographic Information Systems for Transportation

# 1

# Introduction

## Geographic information systems for transportation

*Geographic information systems for transportation* (GIS-T) are interconnected hardware, software, data, people, organizations, and institutional arrangements for collecting, storing, analyzing, and communicating particular types of information about the Earth. The particular types of information are transportation systems and geographic regions that affect or are affected by these systems (Fletcher 2000).

GIS-T have "arrived" and represent one of the most important applications of GIS. The GIS-T community has its own, widely recognized moniker (namely, "GIS-T"). There are dedicated conferences and well-attended session tracks at mainstream conferences. Papers and articles about GIS-T can be found in a wide range of general and specialized GIS journals and trade publications. GIS-T consultants abound. There are career opportunities in the public and private sectors (Maquire et al. 1993; Waters 1999). Indeed, some have recognized the recent emergence of a "second GIS-T renaissance" as GIS data and services continue to improve by leaps and bounds (Fletcher 2000; Wiggins et al. 2000)

GIS-T applications cover much of the broad scope of transportation. Transportation analysts and decision makers are using GIS tools in infrastructure planning, design and management, public transit planning and operations, traffic analysis and control, transportation safety analysis, environmental impacts assessment, hazards mitigation, and configuring and managing complex logistics systems, just to name a few application domains. Intelligent transportation systems, including services such as intelligent vehicle highway systems and automatic vehicle location systems, are a particularly ambitious integration of GIS and communication technologies to a wide variety of transportation services (see Souleyrette and Strauss 1999; Waters 1999).

What is the fuss about GIS-T? Part of the excitement is certainly spillover from the rise of geographic information science and systems in general. GIS are profoundly changing geographic analysis and decision-making across a wide range of domains (Longley et al. 1999). This is only at the beginning of the geographic information revolution. We will continue to see vast, perhaps even accelerating, increases in the volume, scope and spectrum of digital geographic data, capabilities for processing geographic data into geographic information, and information technology (IT) for moving these data and information to where it is needed. We may soon enter the era of the *geographically enabled* scientist, engineer and citizen who can access, process and communicate digital geographic information anywhere at anytime (Fletcher 2000; Mark 1999).

In addition to the supply-side "push" of increasingly pervasive and powerful tools, there are several transportation trends that are increasing the demand for GIS-T. First is a widening recognition that land and transportation are complex, highly interrelated systems. Transportation exists to overcome geographic discrepancies in resources, goods and services by moving material, people or information between where things are and where things are wanted (or vice versa). By overcoming these geographic discrepancies for certain locations and regions, transportation systems dramatically alter accessibility. This in turn influences travel demands and eventually land-use patterns, creating new and sometimes unintended outcomes.

There is also widening recognition that transportation is a major component of the quality-of-life and sustainability (Wiggins et al. 2000). In this respect, transportation trends are not encouraging. Population growth, urbanization, and suburbanization and continued intensive use of the internal combustion engine as a transport technology are creating negative consequences at many geographic scales. The transition to an information and service-oriented economy with activities that are less spatially and temporally focused than machine-based economies is propagating congestion across entire communities and regions for many hours of the weekly clock (Cervero 1986; Hanson 1995). This restricts accessibility to opportunities and has major negative impacts on the environment, including high demands for nonrenewable energy sources, declining air quality and the consequent impact on human health and well-being.

The heightened visibility and changing perspectives on transportation in our public arena is occurring at the same time that dramatic transformations are occurring in the institutional frameworks for public sector planning and decision making. Governments around the world are outsourcing, privatizing, and decentralizing services and operations due to political pressures for smaller bureaucracies and increased competition from the private sector for services traditionally considered public (e.g., education, transportation, and data provision). This is not the place to debate the merit of these trends but only to recognize their existence and likely continuation. This is decentralizing public sector decision making to local levels. Many public sector transportation and metropolitan planning organizations are also evolving from engineering and design agencies to multipurpose information providers operating in a context of increasing public scrutiny (Fletcher 2000).

GIS-T can play a central role in the new environment for public land-use and transportation decision making. By allowing a wide range of information to be integrated based on location, GIS-T fosters (but certainly does not guarantee) a holistic perspective on complex land-use and transportation problems. GIS-T allows analytical and computational tools to be used in conjunction with detailed representations of the local geography, allowing analysis and problem-solving to be tailored to the local context. GIS-T can also greatly reduce the gulf between analysis and communication, allowing greater public input into analytical decisions such as choice of data, modeling assumptions and scenario development. This could lead to greater public "buy-in" to transportation decisions, particularly important in an era of pervasive "not-in-my-backyard" attitudes. GIS-T can also make transportation information more accessible, potentially enhancing location and transportation decision-making by the public-at-large and encouraging wider participation in the transportation planning process.

GIS-T in the private sector is also expanding and showing few signs of abating. Dramatically improving information technologies (particularly communication technologies) are creating an increasingly hypercompetitive, global-scale economy. This creates new requirements for efficiency and customer-responsiveness. Meeting these objectives requires an effective logistics system for managing the flow and storage of material, information and services from their points of origin to their place of final consumption. Since many organizations are dispersed across geographic space, sometimes at national and global-scales, the "supply chains" from origin to consumption are geographically dispersed. GIS-T is being increasingly used to configure and manage geographical supply chains for maximum efficiency and responsiveness.

GIS-T is also diffusing to the public at large. Wireline and wireless access to the World Wide Web (WWW) and the Internet is increasing and will soon achieve penetration rates similar to telephony. Wireless GIS services will become increasingly available, both from the private sector as well as from public and quasi-public organizations. Many of the queries posed by casual user will be transportation and travel-related (e.g., "Is there a good Mexican restaurant within ten minutes of my current location?"; see Smyth 2001). Not only will this create direct demands for GIS-T services such as intelligent transportation systems, but will also quite likely improve the geographic and transportation literacy of the citizenry at large. This could create additional, indirect demands for GIS-T services, including greater sophistication in citizen scrutiny of transportation decisions in the public and private arenas.

All of these trends point to an enhanced role for GIS-T in public and private sector decision making. While current trends are encouraging, the actual outcomes may be less than rosy. Increasing the quantity or scope of transportation information in society does not necessarily imply an increase in the quality of analysis and decision making. As Fotheringham (2000) and Longley (2000) warn, GIS may result in a step backward for scientific research and applied decision making if these systems only make bad tools more accessible or poor analysis appear more sophisticated. This is where geographic information science plays a role.

## Geographic information science for transportation

As Longley et al. (1999) point out, the continuing existence of GIS relies on a widely shared view that there is something "special about spatial," in other words, geographic information is unique and requires a particular type of processing. Two unique properties of geographic information are *spatial dependency* and *spatial heterogeneity*. Spatial dependency is the tendency for things closer in geographic space to be more related. Spatial heterogeneity is the tendency of each location in geographic space to exhibit some degree of uniqueness (see Anselin 1989). GIS would not be worthwhile if geographic information did not exhibit spatial dependency and heterogeneity. Spatial dependency means that it is meaningful to record, organize and analyze data by geographic location. Spatial heterogeneity means that it is worthwhile to consider local geographic context rather than just global generalities.

From an information science perspective, geographic information is unique. Geographic information exhibits a high degree of interrelation among two, three, or four of its dimensions (the two or three dimensions of geographic space, possibly with time). In contrast, many of the other information spaces processed by computers have dimensions that can be isolated and therefore easily unraveled into separate one-dimensional structures. This is convenient since computers are basically one-dimensional "information pipelines" (although we can sometimes run these pipelines in parallel). A challenge in GIS software development is determining the best way to arrange and process geographic information through a 1-D pipeline.

Spatial dependency and heterogeneity also have implications for analysis and modeling. Spatial dependency implies that data recording in the real world should occur at different locations depending on whether one wants to control or capture these relationships. Analytical models and statistical estimation techniques that do not consider spatial dependencies in geographic data ignore valuable information, are not properly specified and can give misleading results. Spatial heterogeneity implies that spatially aggregated measures can be artificial and arbitrary. Heterogeneity also means that most boundaries are lies and create measurement artifacts that may not correspond with the geographic reality.

Spatial dependency and heterogeneity imply that the science and tools to support geographic information processing and decision making must be tailored to recognize and exploit the unique nature of geographic information. *Geographic information science* (GISci) refers to the theory and methods at the foundation of GIS (Goodchild 1992). *Geographic information science for transportation* (GISci-T) refers to the theory and methods that underlie GIS-T. This is a subset of GISci that develops theory and method for capturing, processing, analyzing, and communicating digital transportation information.

As with GIS in general, the proper use of GIS-T requires knowledge not only of the operational level (i.e., software commands, programming languages) but also what happens "behind-the-screen" when invoking procedures. This is most obviously true for software engineers and application developers working on new GIS-T tools. However, deeper knowledge beyond commands and languages is also required if one is to be an intelligent consumer of these tools. A lack of deeper GISci-T knowledge is particularly noticeable when we encounter unusual situations where the tools generate unexpected results. The GIS-T analyst who understands the deeper GISci-T principles can recover and respond gracefully in these difficult situations (that always seem to happen at the worst times).

The software environment for GIS-T is also changing and placing greater responsibilities on the transportation analyst. GIS vendors recognize that they cannot provide all the tools that every user wants for different types of analyses. Even if they tried, the outcome would be less than desirable, with GIS software resembling a multifunction "Swiss army knife" rather than a precision toolkit. Interoperable and customizable software environments can prevent this unfortunate situation by allowing an analyst who knows a particular domain well to build appropriate tools. Componentware or "plug-and-play" software modules increasingly blur the lines between software developer and user, allowing the user to develop effective GIS-T tools tailored for a particular problem. IT will also allow individuals to share these tools freely, meaning propagation of both good and bad tools. A foundation in GISci-T is necessary for distinguishing between good and bad tools and building new tools when none are available.

### The scope of this book

This book is about principles and applications of geographic information science (GISci) and geographic information systems (GIS) to transportation. Chapters 2–7 discuss the fundamental scientific principles of capturing, representing, integrating, processing and analyzing digital geographic information about transportation infrastructure and systems. Chapters 8–11 build on this foundation and discuss applications of GIS-T software, tools and related technologies to transportation planning, intelligent transportation systems, environmental and hazards analysis, and logistics.

This book focuses mainly on ground transportation. While we recognize the applicability of GIS to other modes such as water, air and space transport, we restrict our application focus to ground transportation for two reasons. One is that this is the major focus of GIS-T activity. A second reason is that ground transportation forms a coherent core for GIS-T. If GIS-T have truly "arrived," then there should be some core knowledge that all GIS-T analysts should know, particularly if adopting an integrated perspective on transportation problems. We have attempted to identify a core for GIS-T by focusing on ground transportation.

We assume that the reader has only a basic understanding of GIS comparable with a college-level "Principles of GIS" course. This includes basic cartographic data models (raster, vector, and terrain data models), digitizing and editing geographic data, the basic analytical functions of a GIS (such as buffering, and overlay operations) and principles of cartographic output. We will occasionally revisit these topics, sometimes only briefly and other times at a higher level than a "principles" course. Readers who are unfamiliar with these introductory GIS topics may wish to consult a good text such as Burrough and McDonnell (1998) or DeMers (1997). We also assume a comfort with college-level algebra and (to a much lesser extent) calculus, although we explain mathematical equations and other formalisms from an intuitive perspective.

## The outline of this book

At its heart, GIS are geographic database management systems. Consequently, data are central to GIS. We begin the "principles" chapters of this book (chapters 2–7) with three chapters on data modeling, sources, and integration. To provide a solid foundation in database design, chapter 2 addresses the fundamental data requirements and data modeling techniques with special reference to GIS-T. It may seem that developing data models for GIS-T is relatively easy since transportation has a central object of study, namely, the *transportation network* in a study area. However, transportation networks have complex properties associated with their multimodal nature, different logical views and "one-to-many" relationships among transportation features. Data modeling techniques are tools for translating information about some real-world domain into effective digital representations.

After reviewing data modeling techniques, chapter 2 discusses three general database design issues that are relevant to GIS-T. The first is handling time in a GIS database. This is a complex issue due to the increase in dimensionality and multiple types of "time" that might be represented in a GIS. A second issue is *metadata* or "data about data," that is, ancillary data that describe the database. Metadata are critical for determining the fitness of a database for particular uses. This is important in transportation applications since data are often integrated from multiple sources with varying degrees of accuracy, scale, and resolution. Finally, we will discuss *data warehousing*, an increasingly important issue as the volume, scope and spectrum of digital transportation data continues to increase. A data warehouse allows the analyst to store read-only, historical copies of the operational databases for querying, exploration and decision support. A data warehouse involves challenging design issues, particularly for geographic data.

Chapter 3 discusses data models and design issues that are specifically oriented to GIS-T. We begin with graph theory, the mathematical basis for representing networks (particularly for modeling and analysis). We discuss the traditional node-arc network model for representing a transportation system as well as major extensions such as intersection "turn-tables" and maintaining one-to-many relationships between the transportation network and events such as accidents or pavement condition. We also review some advanced, cutting-edge data models for intelligent transportation systems and related applications, including data models that support navigation.

We conclude chapter 3 by discussing distributed and interoperable GIS-T data models. Distributed and interoperable databases are critical for intelligent transportation systems (ITS). ITS components are likely to be distributed over many locations both within a single jurisdiction and among different jurisdictions. These concepts can also be applied across all GIS-T domains: distributed network computing and enterprise systems are applied to many types of information processing activities.

After designing a GIS-T database, we must populate it with data. Chapter 4 discusses issues surrounding GIS data sources and integration for transportation-related applications. Although a GIS transcends the traditional analog map, it shares many of its fundamental representational properties. Therefore, using geographic data effectively requires a solid foundation in basic mapping concepts. After reviewing basic mapping concepts, chapter 4 discusses primary and secondary sources for GIS-T data. We will discuss methods for collecting geographic data for transportation, including the Global Positioning System, remote sensing and traffic recording devices. We also briefly review some major secondary data sources and products in the United States and Europe, including spatial data infrastructures or information networks for making geographic data accessible to researchers, analysts and the public at large.

Transportation data are often acquired from different sources with different formats and varying quality. Although spatial data standards are emerging to facilitate GIS data exchanges, data integration remains as a challenge. Chapter 4 also discusses some of the most frequently encountered GIS-T data integration issues. In addition to spatial data standards, we discuss integration methods such as areal interpolation and network conflation. We will also discuss the closely related issues of spatial data aggregation and the "modifiable areal unit problem" or the troubling phenomenon that changing the configuration of aggregate spatial units changes the results of data measurement and analysis. Data integration often requires aggregation; as we will see, aggregating spatial data can create confounding effects on modeling and analysis. This is a long-recognized and serious problem that is beginning to yield (albeit slowly) due to increasing capabilities for handling and reconfiguring spatial data through GIS.

The next two "principles" chapters concern network algorithms. At the core of many analytical procedures in GIS-T software are procedures or algorithms for conducting analyses and solving routing and location problems within a network. Chapter 5 focuses on methods for calculating shortest paths and solving vehicle and fleet routing problems. We begin chapter 5 by discussing some crucial properties of networks and algorithms. This includes precise definitions of important network connectivity properties and fundamental properties of algorithms, including the difference between "exact" and "heuristic" algorithms and methods for predicting an algorithm's run time for a given problem size. Even if one is not involved in constructing new algorithms, knowing these basic properties is important for evaluating and choosing "off-the shelf" algorithms (and, by extension, software).

We continue chapter 5 by discussing the shortest path algorithm, perhaps the central algorithm in transportation network analysis. The discussion includes a generic shortest path algorithm and different methods for implementation of this algorithm within a computational platform such as a computer. As we will see, different implementations of the same algorithm can vary widely in performance, particularly when applied to real-world transportation networks. Knowing these principles is not only important for choosing GIS-T software but is also a good basis for understanding physical processing structures used in computers. We also discuss methods for computing shortest paths very quickly, i.e., in real-time or near-real-time. This will include a review of parallel processing and heuristic implementations.

We conclude chapter 5 by discussing two closely related vehicle routing problems, namely, the traveling salesman problem (TSP) and the vehicle routing problem (VRP). The TSP attempts to find the least cost route for a vehicle that tours a set of locations that returns to the starting location. The VRP attempts to determine the simultaneous tours of multiple vehicles through a set of locations where each vehicle starts and stops at a central location (the "depot"). These problems are very important due to their direct application in transportation as well for a surprising range of non-transportation problems. They are also fascinating in their own right since they are easy to state but notoriously difficult to solve.

Chapter 6 completes the survey of basic GIS-T network algorithms by discussing methods for solving network flow problems and finding optimal or near-optimal facility locations within a network. In contrast with the shortest path, traveling salesman and vehicle routing problems, network flow modeling concerns the case where multiple entities may travel through a given network arc. There are two broad classes of flow problems, namely, uncongested and congested. Uncongested means that each unit of flow does not affect other flows; this case often corresponds to "abstract" networks that represent *logistics* problems (the optimal distribution of materials, services information and control within an organization such as manufacturing, warehousing and retailing). In the congested case, the network is saturated and flows affect each other by creating delays and other congestion-related costs. This case corresponds to physical networks, in particular, flow through street and other transportation networks. Both sets of methods are being increasingly incorporated into GIS software, either as "turn-key" systems or through software integration and customization by researchers, users or third-party vendors.

We conclude chapter 6 by discussing methods for locating facilities within a network. Facility location methods are important for configuring logistics systems. In addition, the new transportation demands created by facilities also should be considered when evaluating new land use proposals; facility location models can be helpful for predicting induced travel demands. We can also use facility location methods to solve flow interception and sampling problems such as locating vehicle inspection stations and traffic counters.

Chapter 7 concludes the "principles" section of the book by discussing GIS spatial analysis and modeling capabilities with special reference to transportation. We will begin with a discussion of general issues surrounding GIS and *spatial analysis* (the scientific study of properties that vary with geographic location). Transportation and land-use systems are examples of phenomena whose properties vary by geographic location. We also discuss GIS analytical functions, including basic tools common to most GIS software and transportation-specific tools that are being integrated into many commercial GIS products. Since it is unlikely that many of the sophisticated tools required for transportation analysis will be built into GIS software, chapter 7 also discusses coupling GIS and transportation analysis software, including recent developments in GIS interoperability and componentware that support tight software integration. We also review software tools such as macrolanguages and object-oriented programming that allow users to customize GIS software.

Advanced transportation applications, particularly disaggregate travel demand modeling approaches and intelligent transportation systems, require representation of complex and abstract transportation features that are not well supported by the node-arc data model. Toward the end of chapter 7, we expand on a similar discussion in chapter 3 to set a context for the use of GIS in some of the advanced transportation applications discussed in the second part of this book.

We conclude chapter 7 by discussing geographic visualization, a blending of scientific visualization and cartography. We will discuss several modes and strategies for geographic visualization, including virtual reality. These information exploration and communication tools will become increasingly important in the emerging data-rich environment for GIS-T.

Chapter 8 is the first of several on applications of geographic information science and systems to transportation problems. In chapter 8, we discuss the role of GISci and GIS in transportation planning. The objective of transportation planning is to guide development of a land-use/transportation system to achieve beneficial economic, social and environmental outcomes. This includes tactical decisions such as planning new right-of-ways or public transit routes. This also includes long-term, strategic planning of the entire land-use transportation system. (Issues concerning operational or day-to-day transportation management will be discussed in the next chapter on intelligent transportation systems.)

GISci and GIS have much to offer to transportation planning. Many transportation analysts are familiar with GIS support at the "front-end" as a spatial database management system and at the "back-end" to produce graphic and cartographic visualizations of present and future scenarios. These are very valuable in transportation planning. In the middle are tools for processing geographic data into geographic information. This requires consideration of spatial dependency and spatial heterogeneity. Modeling techniques that do not take these properties into account do not exploit the full range of information in geographic data and in fact may lead to violations of the assumptions of "standard" techniques (particularly statistical and parameter estimation methods).

GIS software tools as well as techniques from the underlying GISci support processing of very large quantities of land-use/transportation data and the appropriate analysis of these data to produce reliable geographic information. This can help determine effective solutions to transportation problems, identify the factors that influence the land-use/transportation system, determine desirable future states of the system and illuminate the paths to achieve the desirable states. GIS can also greatly enhance the quantity and quality of information flows among all components of transportation planning process, substantially influencing decision-making. Chapter 8 discusses spatial decision support systems for helping analysts, decision makers and stakeholders choose among future land-use/transportation scenarios.

Chapter 9 discusses an important special case of a distributed and interoperable GIS, namely, intelligent transportation systems (ITS). Urbanization, suburbanization, and population growth combined with desire for dispersed locations and continued reliance on the internal combustion engine as a transport technology are creating undesirable negative impacts on economies and individual quality of life in many communities of the world. Conventional approaches to tackling these problems increase transportation supply by building new highways and widening existing roads. Unfortunately, as many (but sadly not all) political leaders, decision-makers and stakeholders are realizing, we cannot build our way out of congestion. Rather than increasing the physical supply of transportation infrastructure, ITS try to improve system efficiency through the use of advanced computing, real-time data sensor and communication technologies. The objective is to make transportation systems more efficient, safer, and environmentally friendly.

ITS are integrated information technologies for monitoring and influencing a land-use/transportation system through direct control (e.g., traffic signals) or indirect persuasion (e.g., variable message signage and WWW). Chapter 9 provides an overview of ITS, particularly as they relate to geographic data, information, and GIS. We begin with a discussion of ITS development in three different national and pan-nation settings (Japan, Europe, and the United States). We then identify and discuss the range of possible ITS user services. This sets the stage for a review of ITS architectures that provide the logical configuration of geographic information, communication technologies and ITS services. Chapter 9 compares and contrasts ITS architectures in the three international settings mentioned above. We also discuss the role of geographic information in ITS architectures, particularly with respect to location referencing and geographic data error. Chapter 9 concludes by providing examples of GIS-linked ITS services for in-vehicle navigation systems and Internet-based transportation information systems.

As mentioned above, transportation systems are major sources of environmental degradation and risk in modern societies. Building transportation facilities requires large alterations to the physical environment. Automobile-oriented transportation systems are a major source of energy consumption and air pollution. Transportation systems also have an inherent degree of risk due to accidents, potentially causing harm to individuals and property, particularly if hazardous materials (HazMats) are being transported when an accident occurs. Transportation systems are also central to emergency response and management. Under extreme circumstances, we may need to rapidly evacuate the population within an affected area, often using personal transportation.

In chapter 10, we discuss GIScience principles and GIS tools for assessing the environmental impacts of transportation and mitigating the effects of transportation hazards. Chapter 10 discusses the impacts of transportation infrastructure on sensitive environmental features, hydrological systems and scenic views. We also discuss the impact of transportation systems on air quality. With respect to hazards, chapter 10 discusses accident and safety analysis, routing hazardous materials and evacuation planning. All of these applications require tight integration of transportation and geography, particularly with respect to high resolution and accurate representations of the physical environment and detailed estimates of flow and congestion over geographic space and real time.

Bruce Ralston (University of Tennessee) contributed chapter 11, which concerns applications of GIS in logistics or the planning and management of material, service, information or capital flows within an organization. Since many organizations are geographically dispersed, sometimes at global scales, logistics involves planning and managing transportation networks and/or services. Logistics increasingly involves sophisticated information and communication technologies to coordinate these dispersed and intricate systems. A recurring theme in chapter 11 will be the mismatch between narrow spectrum of commercial GIS software and the complexity of most logistics problems. This simplification is due to many factors, including the complexity of many logistics problems, the biases of GIS software vendors, and the limits of the current state of the art in transportation theory and methods. There is great opportunity for transportation analysts and the GIS-T community to contribute to reducing this mismatch by building more-sophisticated logistics tools for GIS.

# 2

# Data Modeling and Database Design

This chapter addresses the fundamental data requirements and data modeling techniques with special reference to GIS-T. This will set the stage for chapter 3 that focuses on data models developed specifically for GIS-T applications. At first glance, developing data models to meet GIS-T requirements may not appear to be a formidable problem. In contrast with many other GIS application domains, GIS-T has a central object of study, namely, the *transportation network* in a study area. However, as we will soon see, digital representation of these networks is not trivial. Transportation networks have complex properties associated with their multimodal nature, varying legal jurisdictions and logical domains. There is also a need to reference events (e.g., accidents and pavement quality) within the network. A network can have varying representations depending on the map scale of interest. There is often need to represent the relationships between the network and other non-network data. Finally, advanced GIS-T applications require the ability to track conditions or objects over time as well as solve navigation problems.

This chapter assumes an understanding of the geographic data coding and data modeling principles embodied in standard GIS-based approaches (e.g., the raster and vector spatial data models). We refer readers unfamiliar with these GIS data models to Burrough and McDonnell (1998) or DeMers (1997) for basis reviews or Worboys (1995) for a more advanced treatment.

The majority of this chapter discusses data modeling techniques. These are tools for translating information about some real-world domain into effective digital representations. As we will see, this must occur at several levels within a GIS or database management system. We will also see that this must be a careful well-reasoned process: populating a poor database design can lead to substantial problems, particularly over time as the database is updated.

After reviewing data modeling techniques, we will discuss three general database design issues. The first is handling *spatiotemporal* data in a GIS. As we will see, this is a complex issue due to the increase in dimensionality and the varying types of "time" that are relevant to GIS database design. The second issue is *metadata* or "data about data," that is, ancillary data that describes the database. Metadata is critical for determining the fitness of a database for particular uses. Finally, we will discuss *data warehousing*, an increasingly important issue as the volume, scope and spectrum of digital transportation data continues to explode. A data warehouse allows the analyst to store read-only, historical copies of the heterogeneous database systems across an enterprise. The data warehouse can be explored using "data mining" and visualization tools. This wealth of data can also support directed decision making through "shallow" queries. This requires meeting some challenging design considerations, particularly for geographic data.

## Data domains and data modeling in GIS-T

A complexity associated with GIS-T data is the multifaceted nature of transportation data entities. Table 2-1 highlights several aspects (Fletcher 1987). First, transportation entities have obvious *physical* descriptions but can also have *logical* relationships with other transportation entities. Second, these entities exist both in the *real* world and in the database or *virtual* world. This results in the four possible existence modes illustrated in table 2-1. A one-to-one correspondence should exist between entities in the physical and logical realms if the database is an unambiguous representation of the real world. However, the relationships between the physical and logical realms are often one-to-many (i.e., one entity in the domain corresponds to many entities in the co-domain). This introduces database design complexities.

Table 2-1: GIS-T Modeling Transformations

|  | Logical | Physical |
|---|---|---|
| Real | Legal definitions | Actual facilities |
|  | Route | Highways |
|  | State trunk network | Roads |
|  | County trunk network | Interchanges |
|  | Street network | Intersections |
|  | Political boundary |  |
| Virtual | Data structures | Data values |
|  | Networks | Lines |
|  | Chains | Points |
|  | Links | Polylines |
|  | Nodes | Polygons |
|  | Lattices | Attributes |

*Source*: After Fletcher 1987.

The real/physical mode is the most familiar: this corresponds to transportation facilities as constructed and used in the real world (e.g., physical facilities such as highways, intersections and interchanges). Also in the real world are real/logical or legally defined transportation entities such as state and federal routes. The relationship between real/physical entities and real/logical entities are often one-to-many. These one-to-many relationships occur in both directions. For example, two state routes may share the same physical highway. Conversely, a state route can (and often will) traverse several physical streets in an urban area.

One-to-many relationships also exist among logical and physical entities in the virtual domain. Virtual/logical entities correspond to data structures such as nodes, links, networks and polygons. Virtual/physical entities correspond to geometric and attribute data associated with the transportation entity. These latter data are often the information displayed graphically by the GIS. A one-to-many relationship occurs when two or more network links correspond to the same graphical line when displaying the network at a given map scale (e.g., displaying a two-way street represented logically by two directed arcs as a single cartographic line at small map scales). Also, several cartographic lines can represent one link (e.g., displaying modal-specific flow in a network link).

Fletcher (1987) discusses the general database design implications of physical/logical and real/virtual data in transportation. One obvious implication is that a GIS must maintain both logical and physical views of the real world. The GIS must support transformations from the real/logical to the virtual/logical realms. Data input procedures must support transformations from the real/physical to the virtual/physical realms. Finally, some data redundancies are expected due to the one-to-many relationships between the physical and logical views. As will be seen below, database design strategies attempt to avoid redundancies due to potential losses of integrity after database updates. Thus, a GIS-T database design issue is reconciling the need to maintain one-to-many relationships with the need to minimize or eliminate database redundancies.

### Data modeling techniques

A *data model* is an abstract representation of some real-world situation or domain of interest about which information will be stored in a database (Illingworth and Pyle 1996). Data models are key to database systems in general and GIS in particular. The data models supported by the system determines the analytical and query operations available to the user (Goodchild 1998; Worboys 1995).

Data models consist of three major components (Date 1995). The first component is a collection of data objects or *entity types* that are the basic building blocks for the database. The second component is a set of general *integrity rules* that constrain the occurrences of entities to those that can legally appear in the database. The final component is a set of *operators* that can be applied to entities in the database.

Data modeling can occur at several levels, typically the *conceptual, logical* and *physical* levels. (Note: These terms are distinct from Fletcher's 1987 terminology.) The data modeling process typically involves stepping through these levels in sequence from the "highest" (conceptual) to "lowest" (physical). However, a particular data modeling exercise may not involve the lower levels, particularly the physical level. Physical data models are imposed by the particular "off-the-shelf" database management system (DBMS) or GIS. These schemas are hidden from the user, both for usability and proprietary reasons.

The *conceptual data model* is a concise description of the users' data requirements. These models attempt to represent the application domain or *miniworld* in a manner that is understandable to users and independent of the system implementation (Date 1995; Worboys 1995). A conceptual model should contain detailed information about the data types, relationships among data types and constraints on relationships and data values. These are expressed using the concepts and language provided by the particular conceptual data modeling technique (e.g., entity-relationship diagrams, discussed below). Because the conceptual modeling language does not concern system-dependent implementation details, the conceptual model can communicate the fundamental data structure to non-technical users. This allows the data analyst to ensure that users' data requirements are met and no conflicts exist among these requirements (Elmasri and Navathe 1994). In fact, conceptual data models are often *semantic* models, that is, concentrating on the *meaning* of the data (Date 1995).

The next level is *logical data modeling*. This is a system-dependent translation of the conceptual data model into a database schema (Elmasri and Navathe 1994). The logical level refers to how the user perceives the data when interacting with the system (Date 1995). A common logical data model (discussed in detail below) is the *relational data model*, that is, data organized as "tables" with specified domain and integrity constraints. The raster and vector spatial representations are common logical data models for handling digital geographic data. Also, these are often linked to attribute data stored in a relational format (e.g., the so-called *georelational* model; see below). A logical data model is an abstraction of the physical data model, i.e., how the data is physically stored in *primary memory* (e.g., RAM), *secondary memory* (e.g., the computer's hard disk) or *tertiary memory* (e.g., "external" storage devices such as CD-ROMs, mass storage disks, or tape). The logical data model intends to hide these physical implementation details from the user (Date 1995).

At the lowest level is the *physical data model*. As implied above, the physical data model concerns the actual physical storage of the data within the system. This concerns storage details such as stored record placement and access structures such as indexing systems (Date 1995). Physical data models are usually not a concern of most system users and will not be discussed in detail in this chapter. For a general discussion of physical data structures and access methods; see Date (1995, pp. 712–760). For specific discussions related to spatial data storage and GIS, see Nievergelt and Widmayer (1997), van Oosterom (1999), or Worboys (1995).

## Conceptual data modeling

Although many conceptual data modeling techniques exist (see Date 1995), by far the most common techniques are *entity-relationship* (E-R) and *extended entity-relationship* (EER) models. These are also referred to as *E-R* and *EER diagrams*. As the term "diagram" implies, these techniques are predominately graphical devices. This allows easy communication of fundamental data properties to users.

### Entity-relationship model

The E-R model treats data as consisting of *entities* and *relationships* among entities. In addition, entities also exhibit properties or *attributes*. Entities correspond to "things" in the real world; these can be physical (e.g., a person, a house, or a city) or conceptual (e.g., a job or a project). An entity *type* is an aggregation of particular entities; this refers to a set of entities with the same attributes. Each entity's attributes describe the relevant properties of that entity type with respect to the database. A relationship describes a set of associations among different entity types (Elmasri and Navathe 1994).

Figure 2-1 summarizes a simplified graphical notation set for E-R modeling. (Also refer to figures 2-2 and 2-4 for example E/R diagrams. These will be discussed in detail below.) A box labeled with the entity name represents each entity type. Each entity type can also have several attributes. E/R diagrams indicate the number of entities involved in the relationship by graphical notations at each end of the line connecting the entity types. The number of entities involved in a relationship can be exactly one, one or more (i.e., never zero), zero or one, or zero or more. We can also qualify relationships by specifying the type of relationship. Two methods are available. One method uses a diamond-shaped polygon containing a label describing the relationship type. This method allows us to list attributes associated with the relationship type. The second, simpler method labels the line connecting the two entity types with the relation type. This does not allow attributes to be associated with relationship type.

E/R diagrams can have much greater detail; for a more-complete graphical notation set, see Elmasri and Navathe (1994).

Figure 2-2 provides a simple example of an E-R diagram (Butler 1998). The E-R diagram illustrates the entities Land, Parcel, and Street Address, presumably with the intention of building a cadastral database. (Note that, by convention, we capitalize the name of entity types. References to entity instances and to attributes are in lower case. We also use a typewriter font to distinguish these labels from other terms within the text.) The E-R diagram captures the following relationships:

i. Parcel may have one Land.
ii. Land must have one Parcel.
iii. Parcel may have zero or more Street Address, for example, over time or if a parcel is a corner lot.
iv. Street Address may relate to only one Parcel.

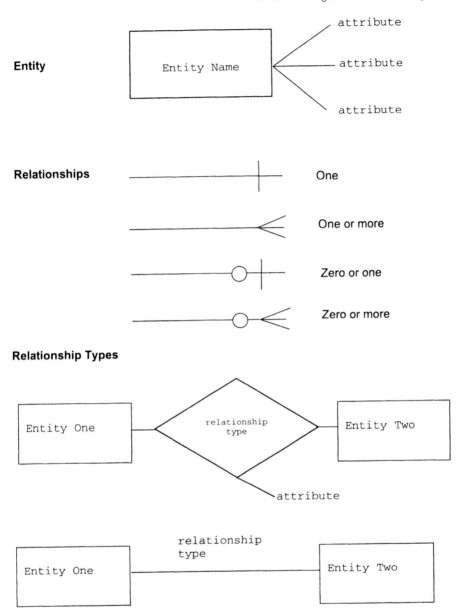

Figure 2-1: Simplified E-R diagram notation

These entities and relationships lead to the implications:
   i. Not all parcels include land; for example, a condominium may not include ownership of land. This is why Street Address is related to Parcel rather than to Land.
   ii. A parcel record must exist before a land record can be attached to it.

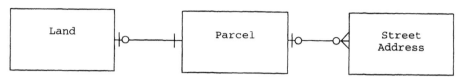

Figure 2-2: E-R diagram example (after Butler 1998)

iii. Not all parcels have a street address. This allows a parcel to be created before assigning an address.
iv. Multiple street addresses may be assigned to a single parcel; for example, multiple buildings on a single parcel may have different addresses.

As this example illustrates, an E-R diagram can summarize a large amount of information in a clear and easily communicated manner.

*Extended entity-relationship models*

EER models extend the basic language in E-R models to include the concepts of *generalization* and *specialization* among entity types. Figure 2-3 provides a simple example of these additional concepts (based on Worboys 1995). Subtypes are formed through specialization from an entity type. In figure 2-3, Automobile and Bus are subtypes formed through specialization of the entity type Travel Mode. Generalization is the inverse process: We can generalize the entity types Driver and Bus Passenger by forming the supertype Traveler. Similarly, Travel Mode is considered a *supertype* of the entity types Automobile and Bus. An important aspect of specialization and generalization is *inheritance*: the subtypes inherit the attributes of its supertype. For example, Automobile and Bus inherit the attributes of Travel Mode. Each adds its own attributes as part of the specialization process.

The symbols *d* and *o* in figure 2-3 refer to the relationship among the subtypes. If the subtypes are *disjoint*, an entity can only be in one subtype, e.g., a transportation mode can be an Automobile or a Bus but not both. *Overlapping* subtypes means that an entity can be in more than one subtype. For example, a Traveler can be both a Driver and a Bus Passenger within the same trip (e.g., a "park-and-ride" system configuration) (Elmasri and Navathe 1994; Worboys 1995).

*Conceptual data models for spatial data: An example*

The conceptual data model that underlies many basic vector GIS databases is the *node-arc-area* (NAA) model (Worboys 1995). The core entities in this model are Node (a zero-dimensional extent or "point"), Directed Arc (a 1-D extent or "line segment" with a specified orientation) and Area (a 2-D extent without holes enclosed by arcs, except for a designated "universe" or external area). The following integrity rules characterize this data model:

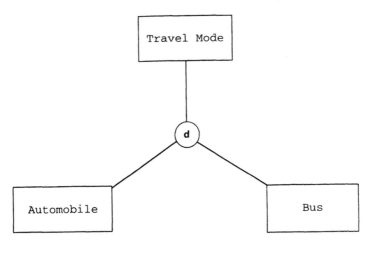

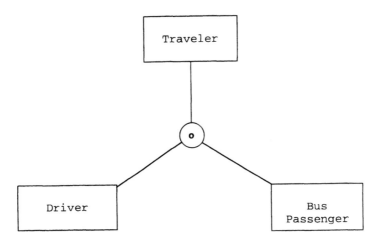

**Figure 2-3:** Additional notation for EER modeling (after Worboys 1995)

i. Each Directed Arc has exactly one start and one end Node.
ii. Each Node must be the start Node, end Node, or both of at least one Directed Arc.
iii. One or more Directed Arcs bound each Area.
iv. Directed Arcs intersect only at Nodes.
v. Each Directed Arc has exactly one Area on its left and one Area on its right.
vi. Each Area must be the left Area, right Area, or both of at least one Directed Arc.

A vector-based GIS database that satisfies these rules is *topologically-consistent*, that is, the connectivity and inclusion relationships are logically consistent. Figure

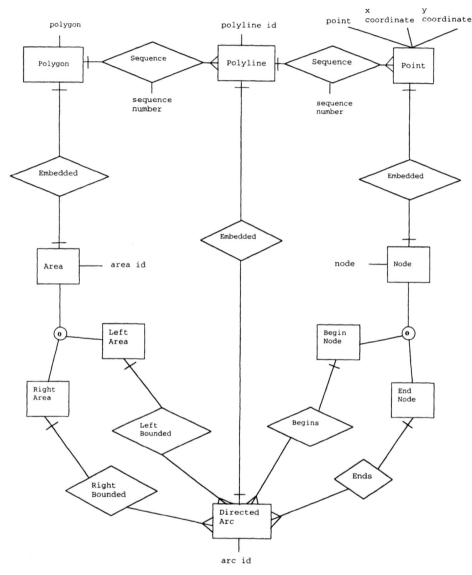

Figure 2-4: ER diagram for node-arc-area representation (after Worboys 1995)

2-4 provides the corresponding E-R diagram for the NAA representation. The core entities of Area, Directed Arc, and Node are in the lower half of the figure. Corresponding to these are the entities Polygon, Polyline, and Point. These represent the *planar embedding* of the core entities, that is, the spatial objects referenced within the 2-D plane. Polyline is a sequence of directed arcs; each Directed Arc belongs to exactly one Polyline. A closed Polyline defines a Polygon; hence, the planar embedding of an Area is a closed sequence of directed arcs that demarcates its boundary. The planar embedding of a Node is a Point with an x coordinate and a y coordinate. Note that these coordinates are the central spatial referencing entities of the planar embedded NAA model.

## Logical data modeling

Logical data modeling translates the conceptual data model into a system-specific data schema. This section reviews an important and popular logical data schema, the *relational data model*. A prime reason for the popularity of the relational data model is its robustness. Relational data modeling has a very strong formal theory at its foundation, namely *relational algebra*. This theory integrates algebraic theory and set theory as applied to *relations* or data "tables." Because of its strong theoretical foundation, the state of the database after updates and the results of database query operations follow well-known and predictable properties. Thus, we can be assured that database inconsistencies and anomalies will be minimized. We discuss the major rules of this theory below; for more detail, see Date (1995).

### *The relational data model*

Relational data modeling typically involves two phases. First, translating the conceptual data model into a set of *relations* and *functional dependencies* generates an initial relational data model. Next, the *normalization* process refines the initial relational model by eliminating to the greatest degree possible any data *redundancies* among the relations. Eliminating data redundancies reduces the likelihood of data integrity problems after database updates, such as disagreement between two occurrences of the same attribute for the same entity.

*Basic relational data model concepts*   The fundamental object in the relational data model is the *relation*. Loosely speaking, each relation is a table that maintains observations about some generic "thing" or fact represented by the database. Since the facts represented in our database are entities and relationships, in general there is a relation corresponding to each entity type and relationship type identified in our E-R or EER model, at least in the initially derived relational model. This may change as we refine the initial model during the normalization process.

Each row or *tuple* in a relation corresponds to an instance or observation of the entity type or relation type. The ordering of tuples in the relation is irrelevant: No information is gained by considering the tuple ordering (Date 1995). The columns of the relation correspond to the attributes of the entity type or relation type. Therefore, each field of a tuple maintains information on the attribute corresponding to that column of the relation. Each column corresponds to a particular *data domain*. A data domain specifies a set of *atomic* (non-reducible, nondivisible) values that meet user-designated restrictions; examples include integers, real numbers, real numbers between 0 and 100,000 and so on. The values in any field must be drawn from the data domain corresponding to that column. If not, the value violates the *domain constraint* of that field (Elmasri and Navathe 1994).

For some tuples we may not have the attribute data that belongs in a particular field. In this case, we can indicate the missing data through a special value known as *NULL*. NULL indicates the state of "no data" or "unknown" for that attribute. This condition does *not* indicate the lack of the value, e.g., zero flow on a network link. The use of NULLs in the relational data model is controversial. Some authors (in particular, Date 1995) argue strenuously that NULL should never appear in a relation; instead, the offending tuple should be eliminated from the database. The specific reasons are technical; in brief, NULLs negates the ability for a relational algebra operation to be evaluated (theoretically) as *TRUE* or *FALSE*. Instead, NULLs require *three-value logic*; that is, results can be (theoretically) *TRUE, FALSE*, or *UNKNOWN*. The latter state can lead to anomalous results from some relational algebra operations. More pragmatically, relational *database management system* (DBMS) software that allow NULLs often require auxiliary rules on how to handle these values in database operations (e.g., are NULLs simply ignored, or does the operation return NULL if one is encountered?). Neglecting or forgetting these rules can lead to misleading or erroneous results.

An important relational database rule is that each tuple within a relation must be unique. A relation's *primary key* maintains (or, perhaps more correctly, "identifies") the uniqueness among its tuples. A primary key is a unique identifier for each tuple formed from one or more of its fields. Ideally, we should form the primary key by combining a subset of the relation's attributes (columns). If the database is designed well, it should be easy to identify the subset of attributes that make each tuple unique. It is also possible to have more than one *candidate key* for the primary key for a relation, although this is unusual and may indicate poor database design. In this case, we designate one key as the primary key and the other as an *alternative key* (Date 1995). Since key fields uniquely identify a tuple, the *entity integrity constraint* dictates that no field in a primary key can have a NULL value (Elmasri and Navathe 1994). Some authors also extend this rule to include alternative keys, since they cannot serve as a primary key if they contain a NULL (Date 1995).

A primary key can also be as simple as a unique index number added to the set of attributes in the relation. There are often good reasons to use a unique index number rather than the intrinsic primary key (e.g., query support). However, a relation that *relies* on an index for tuple uniqueness may be designed poorly or inappropriate for the application. This means that duplicate tuples can exist in the relation, indicating the potential for database redundancies and inconsistencies after updates (Date 1995). If the only unique distinguishing property among tuples is an index number generated when entering the record, it is possible to generate duplicate tuples within the database, undermining the primary key restriction. That being said, there are often good reasons for violating these principles in practice. For example, in a transportation database we may want two or more arcs between the same nodes; these arcs may appear identical with respect to candidate keys such as the "from-node" and "to-node." However, the user should be aware of the potential consequences.

As an example of a basic relational schema, consider the NAA conceptual data model (figure 2-4). The NAA EER diagram implies the following relations (Worboys 1995):

Arc (<u>arc ID</u>, from node, to node, left area, right area)

Polygon (<u>area ID</u>, <u>arc ID</u>, sequence number)

Polyline (<u>arc ID</u>, <u>point ID</u>, sequence number)

Point (<u>point ID</u>, x coordinate, y coordinate)

Node (<u>node ID</u>, point ID)

The relation names are in capital letters to the left of the parentheses. Each name within a parenthesis corresponds to a field name of the relation. An underline indicates the primary key for each relation.

*Functional dependencies and the normalization process*  Keys reflect a deeper concept in relational database theory, namely the idea of *functional dependencies*. A functional dependency is a constraint between two sets of attributes (fields) in a relational database. More formally, if field (or set of fields) Y is functionally dependent on field (or set of fields) X, then the values that can appear in field(s) Y are entirely determined by the values in field(s) X. (Note that a functional dependency can involve more than one field on both "sides.") In this case, we say that X *functionally determines* Y or X → Y. The classic (although not entirely accurate) example is a person's name (Name) and his/her social security number (SSN). The functional dependency SSN → NAME implies that a given name can only be associated with a single social security number. If two tuples with the same SSN have two different names, then the database is in an anomalous state.

As might be guessed, the concepts of key fields and functional dependencies are closely related. A key is a subset of its fields that functionally determines the remaining fields. This follows from the desired property of keys; namely, that it uniquely determines the tuple. Functional dependencies can also exist between fields in different relations. For example, besides indexing the fields in its own relation, SSN may also index information stored in other relations. In this case, SSN may be both a primary key in its relation as well as a *foreign key* linking to tuples in another relation. However, this coincidence is not required (i.e., another field may be the foreign key instead of the relation's primary key). The use of foreign keys will become clearer as we discuss the normalization process.

Functional dependencies and key fields are at the basis of normalization. Normalization is a process of refining the initial relational data model so that it achieves certain desirable properties. These properties are (Date 1995):

i. The database requires the *minimum amount of storage space*. This means that, with the exception of foreign keys, data items are not stored in more than one place.
ii. *The risk of database inconsistencies is minimized.* Since we store data only once (if possible), we minimize the risk of data values being inconsistent.
iii. *The potential for update and delete anomalies is minimized.* Storing multiple copies of the same data value can lead to anomalies after delete or updates; for example, one copy is updated or deleted while the other remains unchanged. Normalization reduces this possibility.

iv. *The stability of the data structure is maximized.* Normalization assists in associating data attributes based on the inherent data properties rather than the particular application. Thus, a new application is less likely to require changes in the database design.

These properties are important since designing, populating and maintaining a database can require substantial expenditures of resources and time. Therefore, it is worth considering the normalization process in some detail.

As mentioned above, normalization is a process by which an initial or "first-cut" relational database design is refined into a more-elegant and stable design. Functional dependencies are central to this process: Much of normalization occurs by "decomposing" relations according to their functional dependencies. Normalization consists of a sequence of steps or stages that must be achieved in order. Each stage is inclusive and assumes the previous stages are achieved. Also, the normalization process is *information preserving*; that is, it can be reversed since no information is lost.

Although six normal forms exist, we can restrict our attention to the first three forms since, except in rare cases, we achieve the desirable properties after reaching third form (Date 1995). These forms are: (i) *first normal form* (1NF), (ii) *second normal form* (2NF), and (iii) *third normal form* (3NF). Obtaining 3NF achieves the desirable database properties by making the semantics (meaning) of each field explicit and clear (1NF) and by eliminating database redundancies (2NF and 3NF) (Elmasri and Navathe 1994). In the discussion that follows, we will use intuitive but very informal definitions of the three normal forms; we encourage the reader to examine Date (1995) or Elmasri and Navathe (1994) for the more-precise, formal definitions.

1NF is embedded within the widely accepted formal definition of a relation. 1NF requires each field in a relation to consist of atomic (nonreducible, simple, indivisible) values. This is achieved by requiring the domains corresponding to the relation's fields to consist of atomic values and each field to contain only one value from its corresponding domain. Thus, 1NF does not allow "relations within relations" or "relations as attributes of tuples." By disallowing the "mixing" of information within a field, the interpretation of each field is explicit and clear (Elmasri and Navathe 1994).

2NF and 3NF eliminate redundancies within relations. 2NF is based on the idea of a *full functional dependency* among key and non-key fields in the relational schema. We say that a functional dependency $X \rightarrow Y$ is a full-functional dependency if the removal of any field from $X$ eliminates the functional dependency (Elamasri and Navathe 1994). A relation is in 2NF if and only if it is in 1NF and every non-key field is irreducibly dependent on any key field (primary or alternative); that is, the relationship between any key and the non-key fields is a full functional dependency (Date 1995). To achieve 2NF, we must remove any field from a relation that does not depend on the *entire* primary key of that relation. We can achieve this by decomposing the relation into smaller relations.

To illustrate a 2NF violation, consider the following simple example in figure 2-5. The top half of the figure (figure 2-5a) shows a relation named Road. The relation name is at the top. The primary keys are underlined. The functional dependencies in this relation are as follows:

## Road

| road ID | segment ID | road name | speed limit | functional class |
|---|---|---|---|---|
| 0001 | 0001 | Emerson | 55 | A |
| 0001 | 0002 | Emerson | 45 | B |
| 0001 | 0003 | Emerson | 35 | C |
| 0002 | 0001 | Redondo | 35 | C |
| 0002 | 0002 | Redondo | 25 | D |

a) Relation in violation of 2NF

## Road Name

| road ID | road name |
|---|---|
| 0001 | Emerson |
| 0002 | Redondo |

## Segment Speeds

| road ID | segment ID | Speed limit | functional class |
|---|---|---|---|
| 0001 | 0001 | 55 | A |
| 0001 | 0002 | 45 | B |
| 0001 | 0003 | 35 | C |
| 0002 | 0001 | 35 | C |
| 0002 | 0002 | 25 | D |

b) Decomposed relations that satisfy 2NF

**Figure 2-5:** Second normal form example

road ID → road name

road ID, segment ID → speed limit

road ID, segment ID → functional class

speed limit → functional class

Note that road name is only partially dependent on the primary key (road ID and segment ID). This leads to unnecessary repeated information in the relation and potential update anomalies. For example, a name change of "Emerson" to "Kensington" must be repeated for every occurrence of "Emerson" in Road; if not, the database is in an anomalous state. To resolve this violation, we decompose Road along its functional dependencies. The new relations are shown in figure 2-5b. The field road ID is now a foreign key linking the relations Road Name and Segment Speeds. Note that a road name change is easily accomplished by changing only one record in Road Name.

3NF is based on the idea of a *transitive dependency*. A transitive dependency occurs when a non-key field functionally depends on another non-key field in addition to the key fields. More precisely, assume X and Y are two fields in a relation. A transitive dependency X → Y exists if there is a set of non-key fields Z and the functional dependencies X → Z and Z → Y both hold (Elamsri and Navathe 1994). In other words, the functional dependency X → Y is implied by the functional dependencies X → Z and Z → Y since functional dependencies are transitive (Date 1995). At first glance, this may not seem to be a problem: if X is the primary key then the functional dependency X → Y would seem desirable. Although this is desirable in this case, the problem is that the functional dependency occurs in an indirect and redundant manner through a non-key field. This means that some non-key fields in the relation (e.g., Y and Z) are mutually dependent. This can cause database update anomalies.

The Road relation discussed above violates 3NF. This violation was translated to the Segment Speed relation. Specifically, we have the redundant transitive dependency (road ID, segment ID) → functional class implied by a dependency based on a non-key field, speed limit → functional class. Again, we can see unnecessary information redundancy and potential update anomalies. For example, if the relationship between specific speed limits and functional classes changes (i.e., if the functional class "C" is no longer associated with a speed limit of "35") we must update more than one record of the database. If all changes do not occur, the database is in an anomalous state. To resolve this difficulty, we can decompose Segment Speed into two relations. Figure 2-6 shows the new relations that satisfy 3NF. The relationship between each functional class and speed limit is stored only once, increasing the likelihood of database consistency after updates. The database specified by the relations Road Name, Segment, and Speed is now in 3NF.

Since 3NF is inclusive (i.e., it assumes that 1NF and 2NF have been obtained), our final relational schema consists only of full and nontransitive functional dependencies on the key field(s). Therefore, a good mnemonic device for remembering 3NF is that each attribute (non-key field) represents a fact about the primary key, the whole primary key and nothing but the primary key (Healy 1991).

Although the properties obtained by achieving 3NF are desirable, we should note that there might be good reasons for violating these guidelines in practice (Healey 1991). One reason is performance: "joining" relations (see below) to extract information fragmented during the normalization process can require considerable computational effort, especially for very large databases. Relational database theory is a guideline and we can deviate from these guidelines as long as if we keep the potential consequences in mind.

*Relational algebra*   One of the most powerful features of the relational database model is the ability to manipulate and extract information using relational algebra. The traditional relational algebraic operations are (Date 1995):

**Segment**

| road ID | segment ID | functional class |
|---------|------------|------------------|
| 0001    | 0001       | A                |
| 0001    | 0002       | B                |
| 0001    | 0003       | C                |
| 0002    | 0001       | C                |
| 0002    | 0002       | D                |

**Speed**

| functional class | speed limit |
|------------------|-------------|
| A                | 55          |
| B                | 45          |
| C                | 35          |
| D                | 25          |

Figure 2-6: Third normal form example

i. *Select*: Select certain tuples within a relation based on a stated criterion
ii. *Project*: Select certain fields within a relation
iii. *Join*: Logical linking of tuples across different relations based on the matching of specified fields
iv. *Union*: Combination of all tuples of two relations into a single relation
v. *Intersection*: Find the tuples that are identical across two relations
vi. *Difference*: Generate all tuples that are not also in another relation
vii. *Product*: Generate all possible combinations of tuples from two relations (more precisely, generate a set of tuples that correspond to the ordered pairs that result from taking each tuple in the first relation and combining it with every tuple from the second relation)

The last four operations are extensions of traditional set theoretic operations while the first three are special relational algebraic operations.

A critical feature of relational algebra is the property of *closure* (Date 1995). Closure means that the result of any operation is the same object as the input. In other words, both the inputs to and the output from a relational algebraic operation are relations. This implies that the output from a relational algebraic operation can be used as the input to another operation. This also means we can create arbitrarily complex operations by nesting relational algebraic operations. This allows us to formulate very powerful algebraic operations and procedures from the existing set of operators. The ability to derive sophisticated database operations from a very small initial set of commands is perhaps the major reason why the relational data model is so widespread.

The most common method of implementing relational algebra within a DBMS software is through *structured query language* (SQL; pronounced "ess-kew-el" or "sequel"). Although few (if any) existing SQLs are completely faithful implementations of relational algebra (Date 1995), most retain the major powerful features of this system (including closure). A discussion of SQL is beyond the scope of this text; see Emerson, Darnovsky and Bowman (1989) for a general discussion or Worboys (1995) for a discussion within a GIS context.

*The relational data model and spatial data* The relational data model was developed to handle aspatial (textual, numeric) data; consequently, it is not well suited for spatial data. In particular, the atomicity requirement of 1NF creates problems for storing spatial data. Typically, finite (computable) spatial objects consist of multiple, related components; e.g., we often represent polylines and polygons as a sequence of node coordinates. Therefore, it is desirable to represent these elements as related components in a single field; e.g., a node sequence in a field labeled `polyline` or `polygon`. However, this straightforward solution violates the atomicity requirement of 1NF since these data are divisible.

Despite the difficulties, there are often good reasons to maintain spatial data through the relational model. Perhaps the strongest reason is that many legacy databases are in relational form. Therefore, for pragmatic reasons, it is often easier to bring the geographic data to the legacy relational database rather than develop a new schema and repopulate the database. Another reason relates to the benefits of a common, enterprise-wide database. Rather than develop and maintain separate databases, a common database shared across multiple users and applications is often more cost effective and certainly has greater likelihood of maintaining integrity over time. The relational model offers good robustness for supporting multiple users and multiple applications (Scarponcini 1997).

Scarponcini (1997) identifies four strategies for incorporating spatial data in a relational database. The first is through *fully normalized tables*. In this approach, we maintain the standard properties and functionality of the relational database approach and "fit" spatial data into the relational schema. The label "fully normalized tables" refers to maintaining the atomicity requirement of 1NF; it does not necessarily indicate obtaining third (or higher) normal forms. One example is the GEOVIEW design of Waugh and Healey (1987). *Base tables* contain the geographic database. *Symbolism tables* contain information on how to graphically display entities. Finally, *system tables* maintain *metadata* and information for interfacing with external devices. Spatial queries are formed through standard SQL scripts. Although effective, GEOVIEW and related designs are not elegant. Unnecessary data redundancy is often required, as are extensive data dictionaries to explain the complex relational structure to users.

Another approach to storing spatial data using fully normalized tables is the *georelational* model (Healy 1991; Morehouse 1984). This approach uses a hybrid strategy to sidestep problems associated with storing spatial data in a normalized table. The georelational design separates spatial and attribute data into different data models. A logical spatial data model such as the vector or raster models maintain spatial information while a relational database maintains attribute data. Unique identifiers associated with each spatial entity (e.g., a `polygon ID`) provide links to records in the relational model and its data on the entity's attributes. This is not an elegant nor robust solution since it artificially separates spatial and attribute data. Also, it does not allow the relationships between a spatial object and its attributes have their own attributes (Goodchild 1998). Nevertheless, it is effective and the georelational approach is widely present in GIS software.

A second approach is extending the relational data model to allow *abstract data types*. In this case, the user has freedom to define arbitrarily the data domain for each field, even if the resulting fields are nonatomic. For example, a user could define the data domains of `point`, `line`, and `polygon` for three fields of a `Spatial Object` relation. Besides specifying the characteristics of these data, the user must also specify associated functionality, including methods for creating, destroying and manipulating instances (see Osborn and Heaven 1986). This is the strategy being adopted for newer versions of SQL and by some DBMS vendors. For example, the Informix RDBMS implements the Open GIS Consortium's specifications of abstract data types for geographic data in SQL3. We should note that strict relational theorists (e.g., Date 1995) argue against this approach since it reduces the generality of the relational data model and can compromise valuable properties of the underlying theory (such as integrity constraints, recovery, and security).

A third approach is to support *binary large object* (BLOB) encoding within a field. In this case, spatial data are encoded into bit strings (strings of zeros and ones) and stored as an atomic object within a relational field. Relational advantages are now maintained since there is no violation of the underlying theory. Unfortunately, there are no standards for encoding BLOBs. Also, any access to spatial data must occur through the BLOB encoding/decoding software rather than through the relational database system.

A fourth approach is to provide *spatially extended relational database systems*, i.e., adding explicit (but nonarbitrary) spatial data domains and functionality as an extension to traditional relational database model. This is a halfway approach between the strategies of fully normalized tables and abstract data types. The user is not required to "fit" spatial data into the restrictive standard relational data but is not free to define arbitrary data domains and functions. This approach relaxes the relational data requirements in a controlled manner that (hopefully) does not negate its straightforward and robust structure. Unfortunately, the spatial data types and functionality can be vendor-specific, resulting in potential compatibility and data transfer problems.

Another problem with the relational data model and spatial data concerns support for spatial querying. As mentioned previously, a powerful feature of the relational data model is a strong underlying formal theory (relational algebra) for manipulating relations and a standard language for implementing the formal theory (SQL). However, relational algebra does not support the proper set of operations to support geometric and graphic operations. Therefore, unless specific support for spatial operations is provided, relational algebra and SQL are very limited for querying and data manipulation within a GIS environment (Egenhofer 1992).

Several authors have attempted to reconcile the limitations of SQL by developing extensions for supporting geometric and graphic queries (e.g., Egenhofer 1991; Frank 1982; Herring, Larsen, and Shivakumar 1988). Egenhofer (1992) reviews these attempts and concludes that any SQL-based spatial query language will have serious deficiencies. These deficiencies include:

i. *Object identity queries.* Object identity is a property that distinguishes an object from all other objects independent of their attribute values. This is critical in many GIS applications, since the properties of spatial objects can change over time (e.g., road alignment, lanes, surface composition, and name). SQL only supports equality comparisons based on attributes; this prevents queries that relate its state at different points in time.

ii. *Metadata queries.* SQL queries require knowledge of the relations' structures; it does not allow the user to ask questions about the data itself, that is, the metadata (data about the data). Metadata queries can be important for spatial databases. For example, the user may wish to point to a symbol on the cartographic display and ask, "What is this?" The answer is not the value of a particular attribute; it is the name of a relation or attribute. It is possible to circumvent this difficulty by storing metadata as a relation, but this introduces potential anomalies associated with redundant data storage.

iii. *Knowledge queries.* These are closely related to metadata queries: These explain the reasoning process that underlies a particular representation or query language. Computed spatial relationships (e.g., proximity and adjacency) are strongly influenced by the spatial data model. Users should be able to inquire about the rules used to compute a spatial relationship. Similarly, users may also wish to determine the geometric consistency constraints in the spatial data model (e.g., "a road cannot cross a river unless a bridge is present"). These capabilities are beyond current SQL implementations.

iv. *Qualitative queries.* SQL requires quantitative measures in its queries. For example, SQL allows the user to formulate the query "return all road segments that are wider than 300 feet." However, SQL does not support qualitative queries such as "What is the relationship between the widths of Emerson Street and Redondo Street?" (with an expected answer such as "Emerson Street is wider than Redondo Street."). This is because SQL is a *first-order language*; that is, it only allows constant values for predicate names. *Second-order languages* that allow both constants and variables as predicates can support qualitative queries such as the example above.

v. *Decoupled retrieval and display.* Standard SQL disregards the presentation of the query result to the user. The single default display consists of the results in tabular (relational) form. The retrieval language is separate from display of query results: SQL is neither "aware" of what is displayed nor does it allow manipulation of query results based on what is displayed. This is an obvious shortcoming with respect to spatial data and spatial queries.

These flaws suggest that alternate data models may be more appropriate for spatial and transportation data.

## Object-oriented data modeling

*Object-orientation* (OO) is a departure from the conceptual-logical-physical data modeling progression discussed above. A problem with the multi-level sequence is the *impedance mismatch* between the different levels: Ideas and concepts are unique to each level and do not translate cleanly or completely. OO is an attempt to reduce impedance mismatch by "collapsing" the conceptual and logical levels to a single level. This allows users' interactions with the logical data structures within the computer to more closely match their conceptual or semantic understanding of the data. There are many types of OO, including *object-oriented analysis*, *object-oriented programming*, *object-oriented user interfaces* (e.g., the Apple and Windows operating systems), and *object-oriented database systems* (OODBMS). Our concern in this chapter is OODBMS, although the general concepts discussed below are common across all types of OO.

### *Major OO concepts*

Object-orientation involves three major concepts, namely, *abstract data types*, *inheritance* and *object identity* (Khoshafian and Abnous 1995). Closely related are the concepts of *encapsulation* and *overriding* (Graham 1993). We will review these ideas briefly; for more details, see Graham (1993) or Khoshafian and Abnous (1995).

Most DBMS such as relational systems only support prespecified or *builtin* data types. Although the relational database designer can specify the data domain for each field of a relation, these domains must typically be drawn from the system-supported builtin data types (e.g., integer, real, alphanumeric, and time). In contrast, object-oriented database management systems (OODBMS) allow the database designer to specify completely arbitrary or *abstract* data types. For example, the designer can specify an abstract data type Highway; this is now a data domain that can support observations or *objects* from the real world. The attributes that define the domain are entirely user specified; the highway domain can consist of the attributes number of lanes, designated name, flow capacity, and free flow travel time. These in turn can be defined using other, user-specified attributes or "standard" domains such as real numbers, integers or alphanumeric characters. Note the difference with the relational data model. In the relational model, each tuple is the union of attributes drawn from standard, builtin data domains. Thus, from the relational perspective a "highway" is a loose amalgamation of generic attributes (integer, real, alphanumeric) rather than a domain with intuitive properties that closely related to characteristics in the real world. In the OO model, the user can define the domain of each object at will. A specific object such as a highway is an example of a general class "highway" with intuitive, highway-specific properties. This subtle but critical difference is more in tune with our intuition about things in the real world.

Abstract data types in OO extend the concept of data domains beyond the representation of object attributes. Abstract data types also support the representation of object *behaviors*. We may wish to have a `Highway` calculate its current travel cost. We can add the behavior `calculate travel cost` to the abstract data type. The actual method for calculating this cost is hidden or *encapsulated* within any `Highway` object. The user need not worry about how travel cost is calculated. Instead, the user sends a *message* to the `Highway` object telling it to calculate its current travel cost. *Public declarations* within the abstract data type specify the input and output formats for messages to and from the object. *Private declarations* are attributes and behaviors that are internal to the object and are not accessible to other objects. We sometimes refer to this strategy as the *object-message paradigm*.

The OO concept of *inheritance* allows hierarchical extension of abstract data types. Inheritance allows the user to declare attributes and behaviors only once even if used in other objects. It also forces the user to organize explicitly the abstract data types. This can expose gaps, logical contradictions and redundancies in the user's data model.

Figure 2-7 provides an example abstract data hierarchy for links in a transportation network. Similar to EER diagrams, the data types lower on the tree are subtypes of the linked data types higher in the tree. The subtypes inherit both the attributes and behaviors of its supertypes. In figure 2-7, `road` inherits the attributes and behavior of `link`. `Addressed` inherits the attributes and behaviors of `Road` (which also includes the attributes and behaviors of `Link`). Thus, there is no reason to redefine many of the attributes and behaviors of abstract data types lower in the inheritance tree; definitions of subtypes can be restricted to additional attributes and behaviors. Some abstract data types may not have any objects in the OO database. In figure 2-7, `link`, `road`, `addressed`, `local`, and `arterial` will not have corresponding objects in the database. These are conceptual placeholders in the data hierarchy. Objects in the database are the leaves of the inheritance tree.

Related to the concept of inheritance is behavior *overriding*. Behavior overriding allows implementation of new methods for a declared behavior at a lower level of the inheritance hierarchy. For example, the precise method used for the behavior `calculate flow cost` can change to reflect domain-specific properties as we move down the hierarchy from `road` to `addressed` to `arterial` to `major`. These functional differences are encapsulated; the user simply sends the message `calculate flow cost` to the object and "let it worry" about how to do the precise calculation. This has two advantages. First, the user need not worry about the precise algorithm or function between levels; he or she simply needs to recall the input and output formats. Second, in some database or programming procedures the user may not know precisely what object will be present at a given stage of the process. The user can simply send the message `calculate flow cost`; this will apply to any object in the inheritance succession. Thus, code and procedures are more robust and flexible.

The final OO concept is *object identity*. In OO systems, each object has a permanent identity that distinguishes it from all other objects regardless of changes in its structure or state. This property is not supported by conventional programming languages or database systems. In conventional programming languages, object identity derives from its location in memory. If the memory

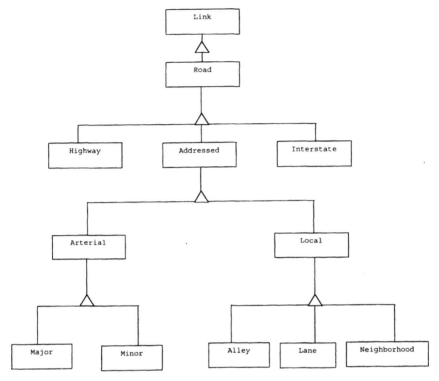

**Figure 2-7**: Example of a transportation network link hierarchy (after Kwan, Golledge, and Speigle 1996)

location changes, identity is lost. The relational model defines object identity through some combination of "special" or unique attributes (such as a primary key). If these attributes change, object identity is lost. Also, the relational model requires "join" operations to link explicitly an object with (non-key) attributes in other relations. In contrast, in OO systems the object identifier is inherently and intimately associated with the object regardless of changes in its attributes (or behaviors) (Khoshafian and Abnous 1995).

*Object-oriented analysis*

As mentioned, OO attempts to reduce the impedance mismatch problem between conceptual and logical data modeling by collapsing these into a single, integrated model. This philosophy is particularly clear when analyzing a system using the techniques of *object-oriented analysis* (OOA). Similar to E-R and EER diagrams, OOA typically involves graphical tools for capturing a concise and meaningful description of *what* a real-world system must do, not necessarily *how* it operates (Rumbaugh et al. 1991). However, OOA techniques are more encompassing than E-R or EER diagrams. Instead of capturing the data requirements only, OOA attempts to capture three dimensions of any complex system (Graham 1994):

i. required *data*
ii. system *processes*, typically in terms of data or activity flow among components of the system
iii. system *dynamics*, typically in terms of allowable transitions among system states

One or more of these three dimensions may be emphasized depending on the particular application. The OOA provides the initial blueprint for system design and implementation, often using other OO techniques.

Several OOA techniques exit. A fairly comprehensive OOA system is the *object modeling technique* (OMT) developed by Rumbaugh and associates (Graham 1994; Rumbaugh et al. 1991). Figure 2-8 provides the major graphical components of the OMT system; see Rumbaugh et al. (1991) for more details. (Booch 1994 provides another competitive and widely used technique.) Note that several of the concepts in E-R and EER diagrams are available in OOA. A unique concept in OOA is *class aggregation*. This can be viewed roughly as a "has-a" relationship; in other words, an aggregate class "has-a" component corresponding to each of its disaggregate classes. This is distinct from inheritance or "is-a" relationships (e.g., an Interstate "is-a" Road in figure 2-7). We will see an example of OMT in the next chapter when we discuss enterprise linear referencing system data models.

### *Advantages and disadvantages of object-orientation*

OO has several advantages over traditional data modeling and database methods. The major advantage is that an OO approach allows a more natural fit between real-world concepts and their representation in the database. Although this sounds trivial at first glance, this is a critical issue (perhaps the critical issue) with respect to database usability.

OO also supports effectively other important database requirements. These include:

i. *versioning* (tools and constructs to support multiple versions or configurations of data)
ii. *concurrency control* (allow simultaneous interactions with the database by multiple users without conflicts)
iii. *transaction management* (ensuring that user interactions take the database results from a consistent state to another)
iv. *integrity* (rules that define consistent database states)
v. *security* (protecting data from unauthorized users)

Although most DBMS can accomplish these requirements, an OO-DBMS supports these at a much finer level of granularity (and therefore with greater flexibility) than many traditional approaches. See Khoshafian and Abnous (1995) for a detailed discussion.

Kwan, Golledge, and Speigle (1996) discuss the benefits of OO approaches in GIS-T. The natural and intuitive data models supported by the OO approach are appropriate for GIS-T applications that require complex, real-time information such as intelligent transportation systems (ITS). ITS applications require very detailed transportation system representations, including detailed representa-

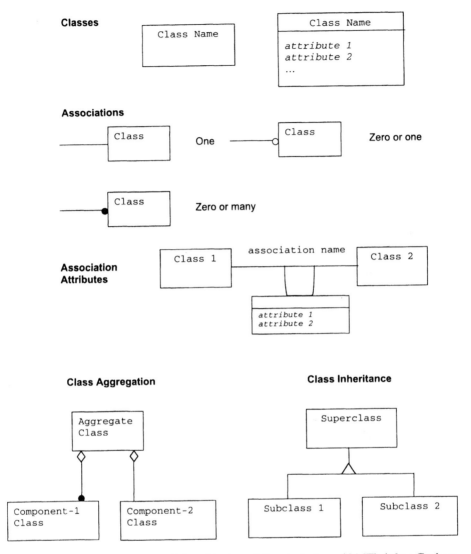

Figure 2-8: Major components of the object modeling technique (OMT) (after Graham 1994)

tions of intersections, lanes and lane changes, highway entrances and exits, and even road curvatures and geometry. A second ITS requirement is representation at varying map scales. This dictates a multilevel representation of the transportation system, from details at the local level (e.g., city or county) to less-detailed, general representations of state or regional level highways. In addition, linkages must exist between the representations at different map scales. Finally, ITS requires real-time updating of traffic information and the transmission of this information to multiple system elements (e.g., traffic controllers, in-vehicle navigation systems) across several jurisdictions. This implies a strong need for system *interoperability*. These requirements are well beyond the needs of traditional transportation analysis and consequently are not supported well by the traditional planar network model (see chapter 3).

The OO approach also has some disadvantages. One problem is the loss of a standard template for interacting with data, particularly with respect to database queries. Although the relational data model can be cumbersome, it does support a standard object (the relation) and a well-known template for database queries (SQL). In OO this standardization is difficult since (by definition) the user defines all data types and objects. This does not imply that OO standards do not exist (see Khoshafian and Abnous 1995, pp. 437–458); however, these standards are by necessity general: users need the flexibility to design objects suitable for the application at hand. A lack of "tight" standardization may become less of an issue as object interoperability and object sharing becomes more developed.

A second problem is the performance of OO systems. In traditional programming languages and databases, code is linked to the proper memory addresses at compile time. Since data types and behaviors are builtin, the compiler knows which type of procedure to use at compile time. For example, the compiler will know whether a particular line with the arithmetic "add" operator will involve integers or real numbers; therefore, it can link this code to the correct procedure during compile time. In contrast, OO systems require *dynamic binding* (run-time binding) rather than *static binding* (compile time binding) since the compiler cannot know which type of object will be used in a procedure. There is no way for the compiler to know whether a `neighborhood` link or a `major arterial` will be present when the procedure `calculate flow cost` is invoked. Therefore, the compiler must wait until that moment in run-time before linking to the appropriate executable. This can slow the run-time performance of OO code or queries. However, this may become less of an issue as processor speeds continue to increase at exponential rates.

## Other data modeling and design issues

### Distributed databases and interoperability

Distributed databases and interoperability are two related concepts. A *distributed database* is a data collection that belongs logically to the same system but is distributed over different sites linked by a communications network. Advantages of distributed databases include (Elmasri and Navathe 1994):

i. *Distributed nature of some database applications.* Often, a database can be partitioned into entities relevant only to *local users* and entities relevant to *global users* or everyone interacting with the database. Data accessed by local users can be kept locally while data used by global users can be kept at a central site.

ii. *Data sharing with local control.* A closely related concept to the distributed nature of some database applications is controlled data sharing. This refers to allowing controlled access by local users to local data.

iii. *Reliability and availability.* "Reliability" refers to the probability that a system is up at a particular moment whereas "availability" refers to the probability that a system is continuously available over some time period. With a distributed database, a site can fail even although the rest of the system is still operating. If data are carefully replicated (at the physical level), then the reliability and availability of the entire database increases.

iv. *Improved performance.* Data distributed over several sites means that each site has a smaller database and receives a smaller number of transactions. This improves the system performance, especially as perceived by users.

*Interoperability* is a broader concept. While "distributed database" usually refers to dispersing an integrated and homogeneous database system (i.e., a database with the same conceptual and logical structure) among different nodes in a computer network, "interoperability" refers to integrating independent and heterogeneous database systems. Interoperability is much more complex since it spans the data modeling levels from semantics to data structures as well as encompasses hardware, software, and network protocol compatibility.

*Distributed geographic databases*

A distributed database involves the following design principles: (i) *partitioning*, (ii) *allocation*, (iii) *replication*, and (iv) *transparency*. *Partitioning* refers to separating the database into fragments suitable for distribution. We can partition any database in two ways, namely *horizontal* or *vertical*. A *horizontal fragment* is selected tuples or records of the database with the records selected based on one or more attribute values. A *vertical fragment* is selected attributes of the database. It is critical that each vertical fragment retains the primary key for the records so that the original database can be reconstructed (Elmasri and Navathe 1994). *Mixed partitioning* applies both methods simultaneously.

In digital geographic databases, horizontal fragmentation translates into partitioning according to geographic space. We divide the geographic database into fragments corresponding to different regions or geographic domains. Each fragment contains the entire cartographic structure and attributes for that region. Vertical fragmentation in digital geographic databases involves separating the database into thematic layers in a manner similar to the classic layer model of GIS database organization. Each thematic layer can serve as a fragment to be distributed; the fragment will contain the cartographic structure for that attribute as well as the recorded attributes. One fragment will also consist of the datum and registration for the geographic database. Figure 2-9 illustrates these strategies (after Cova and Goodchild 1994).

The database fragments must be *allocated* among the various component sites in an effective manner. The *allocation schema* depends on factors such as the degree of local versus global uses for each fragment and the desire to improve reliability and availability. *Replication* refers to storing a fragment at more than one site. The database can be *fully replicated* at each site, have a *nonredundant allocation* or involve *partial replication* of selected fragments. Full replication obviously maximizes reliability and availability but requires substantial overhead

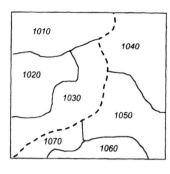

Parcels (stored at Site A)

| polygon id | owner | land use code |
|---|---|---|
| 1010 | Lowry | 100 |
| 1020 | Siebeneck | 200 |
| 1030 | LaBonty | 300 |

Parcels (stored at Site B)

| polygon id | owner | land use code |
|---|---|---|
| 1040 | Zajkowksi | 300 |
| 1050 | Granberg | 400 |
| 1060 | Hofmann | 100 |
| 1070 | Wu | 200 |

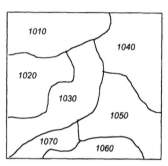

Parcels (stored at Site A)

| polygon id | owner |
|---|---|
| 1010 | Lowry |
| 1020 | Siebeneck |
| 1030 | LaBonty |
| 1040 | Zajkowski |
| 1050 | Granberg |
| 1060 | Hofmann |
| 1070 | Wu |

Parcels (stored at Site B)

| polygon id | land use code |
|---|---|
| 1010 | 100 |
| 1020 | 200 |
| 1030 | 300 |
| 1040 | 300 |
| 1050 | 400 |
| 1060 | 100 |
| 1070 | 200 |

**Figure 2-9:** Partitioning a digital geographic database for distribution (Cova and Goodchild 1994)

for transaction and concurrency control. Also, updates must occur at all sites if the database is to remain consistent (Elmasri and Navathe 1994).

An important distributed database design principle is *transparency* (Date 1995). Ideally, the user should be unaware that the database is distributed among multiple sites, that is, the distributed database should seem exactly like a nondistributed database. The problems of a distributed database should be internal or implementation-level problems not external or user-level problems. This is achieved if several subsidiary rules are followed; these include (Date 1995):

i. local autonomy of sites (i.e., all operations at a site are controlled by that site)
ii. avoiding a central "master" site (i.e., the entire system should not depend on a single site)

iii. continuous operation (i.e., planned shutdowns should not be required, although unplanned shutdowns can be difficult to avoid)
iv. location independence (i.e., users need not know where data are physically stored)
v. fragmentation independence (i.e., users need not know that a database entity is fragmented)
vi. replication independence (i.e., users need not know that a fragment is replicated)
vii. support for distributed query processing and transaction management
viii. independence from hardware, operating systems, communication networks, and the DBMS at each site (i.e., the system should support heterogeneous implementations)

These principles may be difficult to achieve in practice but are critical for user accessibility and friendliness.

*Interoperability*

As noted previously, a distributed database typically involves distribution of a homogeneous DBMS among component sites. By design, local data schemas are compatible with each other and with the global schema. In contrast, interoperability typically involves a bottom-up integration of existing systems that are not necessarily intended to be compatible (UCGIS 1998). The boundary between distributed databases and interoperability systems can be fuzzy in practice: for example, an intelligent transportation system can involve both a distributed homogeneous DBMS as well as interoperable heterogeneous DBMSs.

DBMS heterogeneity occurs in three realms (Bishr 1998). *Semantic heterogeneity* occurs when two or more databases describe a fact in the real world in different ways. In other words, the same real world thing has different meanings among several databases. For example, a "road segment" in the real world may mean a topological connection between two intersection nodes in a travel demand database but may simultaneously refer to an area with uniform pavement quality in a pavement management database. *Schematic heterogeneity* results from differences in conceptual data models. For example, an entity or object in one database may be a relationship or property in another database. Similarly, two object-oriented databases may have different object class hierarchies even though they describe the same real world facts. *Syntactic heterogeneity* results from different logical data models. For example, in GIS, population density may be represented in the vector schema as discrete polygons or in the raster schema as a lattice.

Interoperability can occur at varying levels of sophistication, with higher levels addressing the three dimensions of DBMS heterogeneity more effectively (Bishr 1998). The lowest sophistication level is *network protocol interoperability*. Using a system such as TELNET, users can download files from a remote machine if they know its operating system (OS). The next highest level is *hardware and operating system interoperability*: users can interact with a remote machine using standard rather than OS-specific commands. An example is the *File Transfer Protocol* (FTP) system. Although file transfer is easy, users must know the file format and have an appropriate converter (if available). The next highest level, *spatial data file interoperability*, allows the user to download files from a remote site in a standard format and automatically convert it to the desired format. A disadvantage is that the user must know the user interface, query language, and metadata at the remote site to interact effectively. *DBMS interoperability* resolves this problem by allowing users to access data at a remote DBMS using the user interface and query language of their local DBMS. However, users must have prior knowledge of the data model and data semantics at the remote site. Data models are encompassed in the next level, namely, *data model interoperability*. Users have access to a virtual global data model that abstracts all remote databases. Queries are structured based on the global data model but are mapped automatically into the data models of the remote sites. The highest level, *application semantic interoperability*, allows seamless integration among remote systems without requiring prior knowledge of the data assumptions and semantics.

Research on GIS interoperability tends to be focused on the highest levels (*data model* and *application semantic* interoperability). While short- and medium-term research on problems such as data transfer standards and metadata are crucial, the emerging frontier of interoperable GIS will require development of a simple and robust model of geographic information (UCGIS 1998). Searching for geographical information should be based on what the data represent in the real world rather than its presentation in the database (Bishr 1998). This requires clear understanding and agreement on geographic information semantics.

One strategy for geographic information interoperability is the OpenGIS (OGIS) project (Gardels 1996; Open GIS Consortium 1998). OGIS intends to replace the current GIS paradigm of specific GIS applications tightly coupled to internal data models and data structures. OGIS provides a detailed design framework so that developers can create software that allows users to access and process geographic data from varying sources across a generic computing interface based on open technology. This includes:

i. the *Open Geodata Model*, a common set of basic geographic information types that can be used to model the data needs of more-specific applications using object-oriented or conventional programming methods
ii. *OpenGIS Services*, the set of services required to access and process the geographic data types in the Open Geodata Model and provide capabilities to share geographic data within a community of users and translate data among different communities
iii. an *Information Communities Model* that uses the Open Geodata Model and OpenGIS Services to catalog and maintain data definitions within a community and share data among communities with different geographic data semantics

The OGIS approach facilitates exchange of information among diverse data models, GIS applications, non-GIS applications (e.g., statistical packages and document management) and users.

Figure 2-10 provides the generic OGIS architecture. Figure 2-10 suggests an architecture that provides interfaces among data stores, applications and

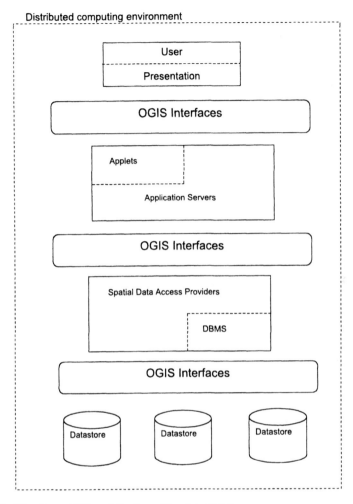

Figure 2-10: The Open GIS architecture (Gardels 1996)

users in a distributed computing environment. However, this is misleading: the OGIS architecture is object-oriented: each component is an object with a specified interface method for exchanging information with other components. OGIS specifies a set of operational software tools for transferring geographic data from various sources into an integrated object model that can be accessed directly from applications. Figure 2-10 highlights the structure of the information flows and the role of OGIS interfaces in facilitating communication among GIS components.

The core of the OGIS model is the Open Geodata Model (OGM). Each OGM data object consists of three sets of components (Gardels 1996). The first are *spatial components*; these are geometric entities such as points, lines, grids, and possibly spatiotemporal referencing. The spatial components can accommodate both the discrete entity (vector) and field (raster) views of spatial phenomenon; a geodata object can be defined in either representation. The

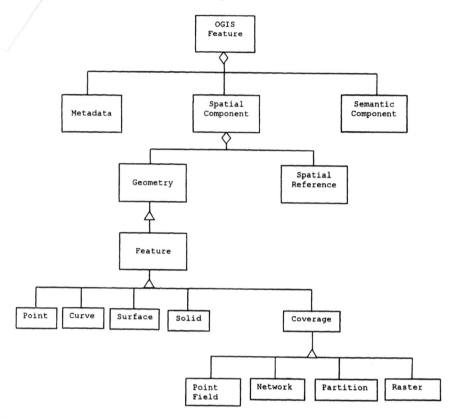

Figure 2-11: The Open GIS Geodata model

second are *semantic components* that define the meaning of an object's elements using a catalog or data dictionary. Finally, *metadata components* provide any additional information required for members of an information community to properly interpret the data object. Figure 2-11 illustrates a portion of the OGIS geodata model as an object class hierarchy. For more details, see Gardels (1996).

### Spatiotemporal data modeling

Most transportation phenomena have a temporal dimension. For example, land use and transportation infrastructure changes over time; we may wish to relate these to changes in travel patterns. A traffic accident on a highway often causes congestion and queuing in traffic flow. Recording traffic volumes during a certain time period (e.g., peak hour) as static data in a GIS database does not meet the needs of developing intelligent transportation system (ITS) applications. Furthermore, traditional travel demand models (e.g., spatial interaction models) treat travel costs as a static variable. Mikula et al. (1996) suggest that the retroactive effects of state variables should be incorporated in dynamic modeling of spatial interaction. These examples of dealing with the temporal dimensions of geographic phenomena are not unique to GIS-T applications. Goodchild et al. (1993, p. 278) argue that "analysis of social process requires a time-dependent, multidimensional, relative space, rather than the static, 2-D absolute view embedded in current GIS technology."

Geographic information consists of three components: *location*, *attribute*, and *time*. Sinton (1978) suggests that one of the three components must be *fixed* and another component is *controlled* in order to *measure* the third component. For example, the traffic volume data collected from an automatic counter is at a fixed location for a controlled period of time. On the other hand, pavement condition data recorded as mileposts along a highway from a field survey performed on a particular date has a "fixed" time component and "controlled" attribute values (e.g., good, average, or poor) in order to measure their locations along the highway. It is also possible to use the location as a control. The Landsat Thematic Mapper (TM) satellite imageries are composed of pixels at 30-meter resolution taken at a fixed time. In this case, attribute values are measured for each controlled spatial unit (i.e., 30 m × 30 m pixels) at a fixed time. In transportation studies, spatial units organized by political boundaries or traffic analysis zones can also be used as the control in the measurements of attribute values. Chrisman (1997) provides good discussion of this topic in relation to different measurement frameworks for GIS.

*Temporality concepts*

Langran (1992, p. 32) defines temporality as "a sequence of states punctuated by events that transform one state into the next." A state has duration, while events are treated as instantaneous points in time. For example, the 1990 census survey (i.e., an event) changed the census tract boundaries that were used during the 1980s (i.e., a state). An event therefore often produces a new object version and leads to a new map state. The temporal changes of object versions and map states present a major challenge to the design of temporal GIS databases.

Each time a new object version is created in the real world, it is associated with a *world time*. The GIS database, however, may or may not be updated at the instant when the change takes place. Consequently, it is also necessary to record the *database time* that reflects the time when the database update is completed. Inclusion of both world time and database time allows database users to identify the time lag between the time when an event took place and the time when the database was updated. This is especially useful to ensure no decision is made on the databases that do not reflect the changes in the real world.

Relational database management system (RDBMS) provides three options of *versioning* for the handling of temporal components (Langran 1992). *Relation-level versioning* creates a new table when one or more attributes in the table are changed. This option is simple to implement and corresponds to the snapshot approach of maintaining GIS databases. Data redundancy is the major shortcoming of this option since the entire table is duplicated even though only one attribute needs to be updated. *Tuple-level versioning* does not create new tables. Instead, it adds new tuples (i.e., records) into the existing table to reflect the updates such as additions, alterations, deletions, or restorations of individual records. This option requires less storage space and permits more frequent updates of the database. Nevertheless, its structure is more prone to data inconsistency and it is difficult to establish relational joins across time. *Attribute-level versioning* attaches time stamps to individual attributes in a table to record their temporal changes. It is more efficient in data storage since there is no duplication of tables or tuples. However, this option requires the use of variable-length data fields to record the time stamps. This requirement violates the basic relational database principles (Date 1995) and makes data queries more difficult.

*Spatiotemporal data models*

Due to the long life of transportation infrastructure and the dynamic nature of traffic flows, temporal data management has been identified as a key technology for GIS-T (Josephson et al. 1995). An ideal GIS data model should be able to handle all three components of geographic data (i.e., location, attribute, and time) effectively and efficiently. Also, for applications that are not real-time, the data model must recognize two types of time, namely, *world time* and *database time*, the former recording time in the real world and the latter recording when the change was made in the database.

It is common practice to treat time as the fixed component in GIS databases and reducing data handling to a *snapshot model*. Under the snapshot model, each map layer represents a given state of geographic data. This creates not only data redundancy concerns but also a lack of temporal connections between the features stored in different map layers. For example, temporal changes of roadway alignment become difficult to retrieve other than a display of the map layers representing that state at different times.

There are alternative data models handling spatiotemporal data in GIS (Zhao and Shen 1997) The *amendment model* stores only changes (i.e., amendments) to the base state data (Langran 1992; Peuquet and Duan 1995). The amendments can then be superimposed on the base state layer to show the changes over time. This approach maintains temporal structure with minimal storage requirements. Zhao et al. (1997) provide an example for maintaining public transit service data. They design a relational database within the ArcInfo GIS software that supports *versioning*. This creates a new version of an entity from the base state for each temporal change. The system records time intervals for both world and database time. Versioning is available at the granularity level of tables, records and attributes.

The *space-time composite model* carries the amendment model one step further. This creates a discrete object, with its own history, from the parent object when a change occurs in a GIS map layer (Langran and Chrisman 1988). The new objects reflecting the changes are stored in the original GIS map layer. As a result, topology between past and present features is maintained in the same database. Each record also contains the temporal data that indicates the time period for which the attributes associated with a particular feature is valid. An example of implementing the space-time composite model for airline network evolution can be found in Shaw and Donnelly (1994) and Donnelly (1993).

Yearsley and Worboys (1995) propose a *deductive model* of planar spatiotemporal objects. Their model integrates abstract spatial data types over the geometric layer to construct a higher-level topological data model (i.e., *spatiotemporal object layer*). In the geometric level, one object may belong to several higher-level spatiotemporal objects. Each geometric layer object is attached to *bitemporal elements* (i.e., database time and world time, see Snodgrass 1992) that represent the union of all times at which these higher level objects exist. Then, a deductive database system designed to be a fusion of object-oriented and deductive technologies is used to implement the operations of this multiple-level spatiotemporal data model.

Peuquet and Duan (1995) develop an Event-based Spatio-Temporal Data Model (ESTDM). This data model uses location in time as its primary organizational basis to record changes. The sequence of events through time of a spatio-temporal process is organized as a time line or temporal vector. This time line is structured as an event list that represents an ordered progression of changes from a known starting time instant to a later point in time. Each time point along the time line can associate with a particular set of locations and features in space-time that changed at the particular time instant. Since the ESTDM is designed to represent spatial changes relative to time, it can facilitate queries of temporal relationships and comparisons of different sequences of change through time.

The above-mentioned temporal GIS data models are still inadequate for handling dynamic transportation phenomena such as traffic flows and vehicle routing. In these cases, the dynamic processes are at least as important as the spatial distribution patterns. For example, the peak traffic volume tends to take place on different street segments at different time in a dynamic manner. Although it is feasible to visualize the dynamic process with the use of real-time data, analysis and modeling of such dynamic phenomena presents a challenge to the design of a robust temporal GIS data model. Raper and Livingstone (1993) propose an object-oriented approach that represents process as an object to model geomorphological systems. Yuan (1999) suggests that a three-domain model (spatial domain, temporal domain, and semantic domain) is needed to represent spatiotemporal behavior of geographic entities. Through the treatment of time as a temporal object instead of an attribute in GIS, both static and dynamic spatial changes can be modeled by dynamically linking to the objects from location-centered, entity-centered, and time-centered perspectives. Research on the handling of dynamic transportation data and processes is rather limited in the literature. However, this is one research topic that could benefit many GIS-T applications such as intelligent transportation system (ITS) implementations.

## Metadata

A critical feature of any DBMS is its ability to "self-describe," that is, maintain not only the database but also a description of the database. *Metadata* are "data about the data," that is, a complete definition or description of the database at a given time. Typical metadata consists of information on the structure of files, type and storage format of data items, and integrity constraints. Metadata are often maintained within a DBMS as part of the *system catalog* or as a stand-alone *data dictionary* (Elmasri and Navathe 1994).

Metadata is especially critical for digital geographic databases. Geographic databases are often very large and integrate data from disparate sources. Each source can have different data definitions, map scales, resolutions, and accuracy. Metadata can provide the navigational aids to allow users to comprehend a large spatial database, answering the question "What are the data and do they meet our needs?" (Ganter 1993; Worboys 1995). Metadata can also support data transfer, providing the information required to process and interpret data from external sources (FGDC 1994). Geographic databases also have limited shelf-life since geographic reality can change quickly. The *change traffic* (Marble 1991) introduced by regular update procedures can introduce cumulative error into the database. Changes in source data must also ripple through to derivative data to ensure that these are also up to date. Metadata can track the evolution of data for potential roll-back if update problems occur as well as keep track of the lineage of derived data (Ganter 1993).

Most research on metadata management focuses on *source metadata*; these are information that relates directly to source data. Less attention has focused on *derived metadata*, that is, information about new data generated through operations on source data (Ganter 1993). An example of source metadata is the Content Standards for Digital Geospatial Metadata developed by the U.S. Federal Geographic Data Committee (FGDC 1994). These metadata standards intend to provide a common set of terminology and definitions for documenting digital geospatial data. The standards correspond to fundamental metadata roles, including:

i. *availability*—data needed to determine the sets of data that exist for a geographic location
ii. *fitness for use*—data to determine if a set of data meets a specific need
iii. *access*—data needed to acquire an identified set of data
iv. *transfer*—data needed to process and use a set of data

Chapter 4 discusses the FGDC Content Standards in more detail. An example of derivative metadata is the *GEOLINUS* system (Essinger and Lanter 1992; NCGIA 1992). GEOLINUS is a metadata management addition to the ArcInfo GIS software. It provides a graphical flowchart representation of the source data and derived data coverages. For examples of other metadata management systems, see Medyckyj-Scott et al. (1991).

Metadata management also involves specifying the information appropriate for each user. Ganter (1993) identifies three levels of GIS user types and the metadata types relevant for each. The *GIS user* is typically concerned only with high-level metadata. This can include a basic description of the data, map scales, timeliness, quality, and status (available/not available). The *GIS administrator* is concerned with medium-level metadata that aids in maintaining database integrity and security. This can include the data owner, its original format and media, storage location, update status, revision dates, security level, and permissions. The *system administrator* is concerned with low-level metadata that relates to maintenance of a stable computer system. This information includes available analytical routines, variable names, field types, field units, field widths, windowing, color palettes, font locations, and output devices. An effective metadata management system will tailor the information provision to the specific user type to avoid providing irrelevant information and therefore degrading its effectiveness. The *General GIS Interface* (GGI) system described by Ganter (1993) provides customized metadata for different user types.

## Data warehousing

### Overview

The growth of sensing, communication, and information processing technologies is creating an explosion of data in many domains, including geographic and transportation data. The scope, coverage and volume of digital geographic data are growing rapidly due to new high-resolution satellite systems, national spatial data infrastructure initiatives and georeferenced point-of-sale, logistics, and behavioral data often collected by automated devices. The geographic data explosion is only in its initial stages. Many experts are predicting embedded microprocessors and the development of information architectures where many artifacts are simultaneously IP-addressed, georeferenced and can communicate status and other information over wireless networks (see the volume by Denning and Metcalfe 1997).

Transportation is moving from data-poor to a data-rich environment. ITS development and deployment is creating the ability to capture transportation data in real-time or near-real-time across scales from individual vehicles to system-wide flows (see chapter 9). The increasing resolution of satellite-based remote sensing platforms can capture information on individual vehicles, transportation facilities, natural and built environments at regional and greater scales. Global positioning system receivers, soon to be embedded into cellular phones, personal digital assistants and wireless web clients, can accurately record individual space-time trajectories at high resolutions.

The coming tsunami of information will strongly affect GIS-T practice. Since georeferencing will be common, particularly for transportation-related data, GIS are well positioned to serve as a central data integration, archiving and analysis platform. Most of the modeling and analysis techniques discussed in the second half of this book will benefit from the information explosion, particularly computational approaches. Also emerging are scalable *knowledge discovery from databases* (KDD) techniques that can handle large data volumes, do not require strict statistical assumptions and can uncover hidden and surprising patterns in very large databases. These *data mining* techniques that explore very large databases looking for deeply buried knowledge through pattern, association, cluster, and trend exploration. See Adriaans and Zantinge (1996) for an accessible introduction to KDD and Fayyad et al. (1996) for a collection of higher-level readings. Techniques are also emerging for *geographic knowledge discovery* (GKD) and *geographic data mining*. These exploratory techniques recognize and exploit the unique information in geographic and temporal data (see Miller and Han

2000, 2001). At this point, there are few transportation data-specific techniques or applications; however, see Roddick and Lees (2001), Marble et al. (1997), Smyth (2001), and Bernard et al. (1998) for relevant work on exploring spatiotemporal and interaction data.

As we emphasized earlier in this chapter, data require effective storage after recording and capture. The database design processes described earlier in this chapter are adequate to store "operational" data for daily or routine transactions. However, storing *historical data* (i.e., data over longer time periods) introduces other challenging database design issues, particularly given the volume of data available from real-time and other sensors. A huge volume and diverse spectrum of geographic data can easily overwhelm query and retrieval operations, limiting our ability to exploit these data.

A *data warehouse* (DW) is a database specially designed as an efficient repository for storing and querying from very large databases. Unlike the operational databases discussed earlier in this chapter, a DW is designed to be *nontransactional*. In other words, we do not expect the DW to be used for regular data retrieval and updating. This allows us to relax some of the strict design guidelines required for standard database management systems (DBMS). The normalization process discussed earlier in this chapter refines a relational data model so that a data item occurs logically in as few places as possible (ideally, one). This reduces data storage requirements and also minimizes the possibility of update anomalies (i.e., an item is updated in one place but not others, creating a database inconsistency). However, normalization is a bad idea when designing a DW. Since we need to retrieve large volumes of related items from the data warehouse, particularly in exploratory analyses such as mining and visualization, we need the DW to be as "flat" as possible. Minimal data storage is a secondary, sometimes even irrelevant, concern.

*Data warehouse design issues*

Bedard, Merrett, and Han (2001) provide a good review of design fundamentals for standard and geographic DWs. The next few paragraphs provide a summary. For a more general reference, see Jarke et al. (2000).

A DW is *enterprise oriented*: ideally, it is a single, homogeneous source for enterprise-level strategic decision making. Because of this objective, the DW must *integrate* data from often a wide range of heterogeneous operational DBMS and platforms. This can be challenging, particularly for geographic data. The DW must complement the existing operational DBMS with minimal disruption. Therefore, the DW usually consists of *read-only copies* of the operational database, downloaded at regular time intervals. The inability of the DW to modify the contents of the operational DBMS assures minimal disruption to the operational system and can even mitigate data ownership and management conflicts (since the DW consists of copies). The DW is *nonvolatile* in the sense that once inserted, the data cannot be updated. This means that new copies of the operational DBMS do not update or overwrite existing copies. Instead, current and historical copies of the operational data are maintained in a DW to support trend detection and similar temporal queries. Nonvolatility is relative: Data storage constraints may require purging (perhaps after summarizing) the oldest data in the DW.

Strategic decision-making requires rapid querying of summary data and "zooming-in" to detailed data when an interesting trend or pattern is identified. These techniques, often referred to collectively as *online analytical processing* (OLAP) tools, include *drill-down* (zoom-in), *drill-up* (generalize) and *drill-across* (show related data) retrievals. A DW often stores data at varying levels of detail to facilitate both fast retrieval for OLAP tools and detailed case exploration (including mining). This requires storing data in summary form at different levels of generalization. Logical data models that support OLAP include *data cubes* that store summary cross-tabulated data (equivalent to the SQL "group by" all possible cases query) with hierarchical links among these synthetic data items (see Harinarayan, Rajaraman, and Ullman 1996; Shekhar et al. 2001).

The typical DW architecture consists of a centralized and integrated DW that receives data copies from several heterogeneous operational databases and supports distributed clients with OLAP and data mining tools through a client-server relationship. There are several variations on this generic architecture, including several tiers of DW servers (with increasing levels of detailed data maintained in each DW tier) and *data marts* or smaller, limited-purpose DWs. DWs can also be *virtual* if its size is modest enough for "on-the-fly" integration from the operational databases (see Bedard, Merrett, and Han 2001).

*Geographic data warehouse design challenges*

Geographic data introduces some particular challenges, particularly with respect to the most difficult aspect of the DW design process, namely, integrating heterogeneous data into an efficient homogeneous database. Most of the geographic data interoperability issues discussed previously in this chapter are issues in this problem. Semantic differences and changes over time are particularly difficult with geographic data. Initiatives such as spatial data infrastructures and spatial data transfer standards (see chapter 4) are helpful only if the enterprise and its organizational subunits follow these standards in their operational databases. Differences in map scale, geometric detail and spatial accuracy can also be difficult to resolve in the GDW integration. Metadata can play a key role by tracking these parameters for restricting data integration to the appropriate levels of generalization with the GDW hierarchy.

As with most digital geographic databases, location referencing serves as the central integrating factor in GDW. As we will discuss in the next two chapters, location referencing for transportation data can consist of quantitative (e.g., coordinates), qualitative (e.g., place names) and hybrid (e.g., street addresses, and linear references) systems. Summarizing geographic data along generalization hierarchies requires methods for aggregating both the attributes and location references for geographic units. In chapter 4, we will discuss issues and methods for conflation, spatial aggregation and spatial interpolation tools of attribute data.

We can generalize location references in three ways (Bedard, Merrett, and Han 2001). *Nongeometric* generalizations consist only of qualitative spatial units such as municipalities, administrative units and route segments, nested according to inclusion relationships. *Geometric-to-nongeometric* generalizations retain quantitative referencing at lower levels but switches to qualitative spatial units at higher levels. Examples include coordinate or address-based georeferencing converting to traffic analysis zones or linear referencing converting to routes at higher levels of spatial generalization. *Fully geometric* hierarchies maintain quantitative location referencing at all levels. The GDW should also support derivation of alternative geographic generalization paths for the georeferenced data, allowing flexible integration of data across referencing systems. These strategies imply different data storage requirements since quantitative referencing systems generally require more storage space.

As might be guessed, summarizing and maintaining data in a geographic data cube or similar logical data model is critical for efficient querying and exploration using spatial OLAP tools. This is another GDW design challenge: Spatial summary measures can be computationally intensive since they involve topological and geometric computations such as merges and overlays. One possibility is to just maintain database pointers to the items required for computing the summary measure. A summary measure is calculated only when needed. This is adequate for modest tasks such as basic cartographic visualization but quickly bogs down for exploratory querying and mining. Another strategy is to selectively precompute and maintain geographic summary data, typically only for at certain levels of generalization and for related attributes ("themes" using the cartographic term). We can also save some computational effort by approximating the geographic summary measures at some levels using heuristics (such as minimum bounding rectangles rather than a full polygon merge). The system can compute more-accurate measures during "drill-down" operations (Bedard, Merrett, and Han 2001).

Han, Stefanovic, and Koperski (1998) discuss methods for designing spatial data cubes to support OLAP with geographic data. *Map cubes* are extensions of this logical data model that includes spatial data and supports visualization by creating albums of summary maps (see Shekhar et al. 2001).

## Conclusion

This chapter reviewed general issues concerning data modeling, that is, capturing information about a "miniworld" in a computational platform. Although these are general issues, we tried to show linkages to data modeling issues in GIS-T by providing particular transportation examples and highlighting transportation data issues. In the next chapter, we will discuss data models that are specific to GIS-T. We encourage the reader to keep the general data modeling issues in mind while reviewing these GIS-T data models. This will help maintain a critical perspective on these models and hopefully indicate the strengths but especially the weaknesses of these models.

# 3

# GIS-T Data Models

In this chapter, we discuss data models that are specifically oriented to GIS-T applications. We first discuss *graph theory*; this is the mathematical basis for representing networks, particularly in modeling. We will then discuss *network representation of a transportation system*, including the node-arc network model for representing a complex transportation system and logical data models for maintaining this representation in a computer database. We will then discuss *linear referencing systems* and the *dynamic segmentation* data model for recording and storing one-to-many relationships in a transportation network. Following this is a discussion of advanced *data models for intelligent transportation systems and related applications*, including data models that support navigation. Finally, we will review *distributed and interoperable GIS-T data models*.

We should note at the onset of this chapter that a large gulf exists between the rich features and attributes of transportation systems in the real world and the limited models used for their representation within a computer. First, many GIS software packages only recognize predefined, simple geometric entities (i.e., points, lines, and polygons). These are inconsistent with richer entities often found in transportation systems. Most GIS software are still inadequate at handling data such as origin-destination flows, complex paths, and temporal changes (Bernstein and Eberlein 1992; Goodchild 1998). Many existing GIS have limited and clumsy representations of topology. A GIS typically computes topology based on consistency rules (such as the node-arc-area model discussed in chapter 2) and the geometric representations of data, ignoring the data semantics. Therefore, the topology of underpass/overpass and intermodal transfers between a commuter rail line and highways can be misrepresented (also see Spear and Lakshmanan

1998). The separation of transportation features into different layers results in a lack of topological information between entities stored in different layers (Bernstein and Eberlein 1992; Shaw 1993).

As will be discussed in this chapter, the basic GIS data model has been extended to include complex features of transportation systems. This includes representations of intersection turning properties (turn tables), one-to-many relationships (dynamic segmentation) and even explicit relationship of traffic lanes. Improving support for object-oriented data modeling and design allows semantic information to be incorporated when computing topology among different system components and multiple geometric representations of the same transportation entity (e.g., interchanges).

## Mathematical foundations: Graph theory and network analysis

The underlying mathematical model for networks derives from *graph theory*. Graph theory examines the relationships or connections among members of a set. A graph consists of a set of *vertices* (loosely, "points") representing members of the set. A unique label, typically an integer, identifies each vertex. *Edges* (loosely, "line segments") represent the logical relationships or physical connections among vertices. We identify an edge using the labels of the vertices it connects. For example, $\{i, j\}$ represents an edge between vertex $i$ and vertex $j$.

Edges can be one-way or *directed*; this means that the corresponding relationship holds only in the direction specified. Otherwise, the edge is undirected and we assume that the relationship holds in both directions. Since graph theory focuses only on connectivity, two graphs are *isomorphic* if they have the same connectivity pattern regardless of their geometric appearance (see Worboys 1995, p. 135 for an example).

A *planar graph* is embedded within the Euclidean plane. This means that a vertex must exist whenever two edges cross; that is, one edge cannot "float above" or "tunnel under" the other. The planar graph also creates a subdivision of the plane; each subregion or *face* can be related to the elements of the graph. In contrast, a *nonplanar graph* does not require vertices at edge crossings and does not subdivide the plane; it exists independently of any mathematical plane.

A *network* is a graph that accommodates interaction or movement behavior explicitly. *Nodes* are locations where flow originates, terminates or relays while *arcs* are the conduits for flow between nodes. Arcs can represent physical conduits (e.g., a road segment) or logical relationship (e.g., airline service between two cities). Similar to graphs, each node in a network must have a unique (typically, integer) label and we identify an arc by the labels of the two nodes that it connects. Arcs are directed or undirected. If the arc is directed, the node ordering indicates the flow direction. *Planar networks* require a node at every arc crossing while *nonplanar networks* have no such requirement.

An important difference between a network and a graph is that a network can accommodate *weights* associated with each arc. In general, an arc weight is a function that represents the unit flow cost for that arc (i.e., the cost incurred by one unit of flow when traversing the arc). We require each arc weight to be nonnegative since negative costs can create complications when computing network properties such as shortest paths. Appropriate scaling of the weights can easily circumvent this restriction (Biggs 1985).

## Network representation of a transportation system

### The node-arc representation

Sheffi (1985, pp. 10–18) provides an excellent discussion of the basic issues involved in using the traditional planar network model (sometimes referred to as the *node-arc representation*) to represent a transportation system. In general, we deal exclusively with directed networks (that is, a network consisting of directed arcs) since transportation systems typically have important directional flow properties (e.g., one-way streets, differences in directional travel times depending on the time of day).

The node-arc representation traditionally partitions a transportation system into separate, modal-specific subnetworks. For example, an urban transportation system may consist of separate networks corresponding to the private transportation network (i.e., street network) and each public transportation network (e.g., bus system, subway system, commuter rail system). *Transfer arcs* link the separate subnetworks; these arcs represent modal transfers. Although the traditional node-arc representation is still widely used, emerging transportation network models treat multiple modes in a more-sophisticated manner that does not impose an artificial separation. These will be discussed in subsequent sections.

Within the private (street) transportation network, nodes generally correspond to street intersections while arcs correspond to street segments between intersections. Similarly, nodes correspond to interchanges and arcs correspond to highway segments when representing limited access highways. Two directed arcs oriented in opposite directions represent a two-way street or parallel limited access highway segments with opposite directional flow. A *generalized cost function* represents the unit flow cost for traversing the arc. Chapter 6 discusses these functions in greater detail.

Although nodes correspond to the street network's intersections, these can be represented at varying levels of resolution using the node-arc model. Figure 3-1 illustrates two methods for network representation of a street network. In figure 3-1a, the intersection is aggregated to a single node. Although parsimonious, this method is simplistic and does not capture a critical intersection property, namely, the varying *turn impedances* associated with different directions of travel through the intersection. For example, a left turn can require more time than a right turn or traveling straight through the intersection. In addition, turn restrictions may be present (e.g., "no left turn"). To capture these features, we can use the expanded representation in figure 3-1b. This method expands the intersection to four nodes with connecting arcs representing direction-specific travel. Although this method can capture the necessary turn properties of the intersection, a problem is the large increase in the number of nodes and arcs in the network. Note that each intersection in the street network can potentially contribute four

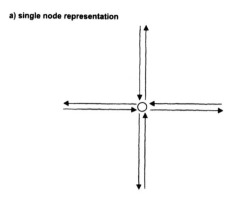

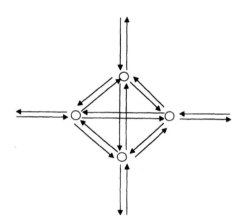

Figure 3-1: Network representation of a street intersection (after Sheffi 1985)

nodes and 12 additional arcs to the database: the total can be quite large given a detailed representation of the street system. This not only can rapidly expand the data storage requirements for the network but can also degrade the performance of network procedures. For example, the run-time for shortest path routines generally increases as a function of the number of nodes in the network.

The node-arc model can represent a public transit network using a simple linear network with nodes representing public transit stops. Figure 3-2a illustrates this strategy. Nodes represent transit stops while arcs represent in-transit or *linehaul* links. Figure 3-2b illustrates the transit line connected to an origin and a destination. In this case, we include special links that allow flow to enter the transit system from an origin (an *entrance arc*) and to leave the transit system at a (final) destination (an *egress arc*). Figure 3-2b also illustrates connecting two transit lines. Model transfer arcs connect the corresponding stops that allow transfers between the two transit lines (e.g., a subway station where travelers can transfer

### GIS-T Data Models 57

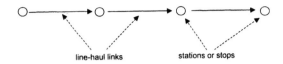

a) "line-haul" movement

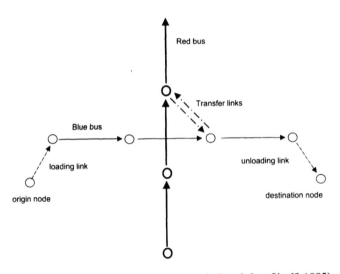

b) system entrance, egress and transfers

Figure 3-2: Network representation of a public transit line (after Sheffi 1985)

among two independent lines). Transfer arcs can also connect the street network to a public transit system (e.g., a "park-and-ride" station).

Origins and destinations are locations where trips originate or terminate. In reality, these locations are distributed almost continuously in a given study area. However, these locations are often spatially aggregated into *traffic analysis zones* (TAZ) for computational tractability. The traditional method for linking these zones into the transportation network is to collapse each zone to a single point, typically the zone's *centroid* or center of gravity. *Centroid connector arcs* link these centroids to the transportation network; figure 3-3 provides an example. These connectors or *dummy arcs* represent the transportation network within the given TAZ. If the actual origins and destinations are uniformly distributed within the TAZ, the centroid connectors can be weighted with the travel cost of time required from the TAZ center to each node. If these locations are not evenly distributed then the arc weights must be weighted accordingly (Sheffi 1985). Nevertheless, this type of aggregation can introduce error into a transportation

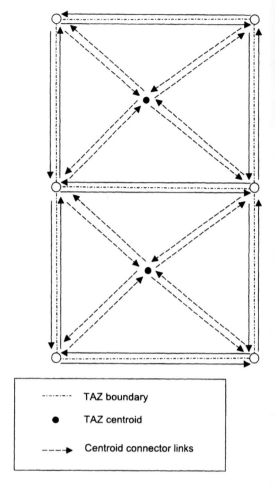

Figure 3-3: Connecting traffic analysis zones to a transportation network (after Sheffi 1985)

analysis. Chapter 4 will discuss general spatial aggregation issues and chapter 8 will discuss techniques specific to travel demand modeling.

To maximize database integrity, a GIS often requires planar embedding of the node-arc model; that is, the resulting formal model is a planar network. Planar embedding ensures that the topology (connectivity) of the resulting spatial data layer is correct; for example, all polygons are enclosed by arcs. (Refer to figure 2-4 and the related discussion.) However, the planar network requirement that all arc intersections correspond to network nodes does not match the real-world properties of a transportation network (Goodchild 1998; Spear and Lakshmanan 1998). For example, a limited access highway may pass over or under a surface street. Placing a node at that over- or underpass implies that automobile traffic can turn off or onto the highway. This can cause problems during network routing.

A partial resolution of the planar network problem is to relax enforcement of

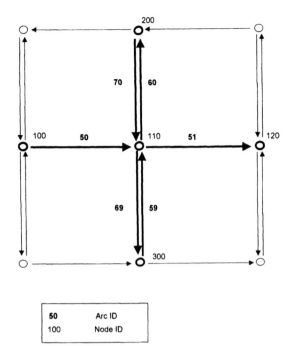

Figure 3-4: Example network for relational model example

planar topological consistency. Some GIS software allow varying levels of enforced topological consistency when building a spatial data layer (such as not "building" the polygons implied by the node-arc-area topology). However, this can lead to data integrity problems. Another strategy is to use the expanded intersection representation (figure 3-1b) and restrict turns at over- or underpasses, although in many respects this is a "work-around" rather than a true extension beyond the planar network. Because of this problem, several emerging GIS-T data models greatly expand on the traditional node-arc model; these are discussed below.

### Data models for the node-arc representation

The most common logical data model used to support the node-arc representation is the relational model. Figure 3-4 provides a simple example network and figure 3-5 provides the normalized relational structure for this network. Note that the relational schema in figure 3-5 is not the full NAA model from figure 2-4. For clarity, we will focus on the more limited planar network model in this section. However, the planar network model can be easily extended to the NAA model to achieve a topologically consistent spatial data model.

The relational structure to support the planar network model typically consists of the following relations (Goodchild 1998). First, an Arc relation

**Arc**

| arc id | from node | to node | (other attributes) |
|--------|-----------|---------|--------------------|
| 50 | 100 | 110 | |
| 51 | 110 | 120 | |
| 59 | 300 | 110 | |
| 60 | 110 | 200 | |
| 69 | 110 | 300 | |
| 70 | 200 | 110 | |

**Node**

| node id | (attributes) |
|---------|--------------|
| 100 | |
| 110 | |
| 120 | |
| 200 | |
| 300 | |

**Turn Table**

| from arc | to arc | impedance |
|----------|--------|-----------|
| 50 | 60 | -1 |
| 50 | 51 | 1 |
| 50 | 69 | 2 |
| 70 | 69 | 1 |
| 70 | 51 | 3 |
| 59 | 60 | 1 |
| 59 | 51 | 2 |

Figure 3-5: Relational data model representation of the network in Figure 3-4

maintains information on the network arcs and their relevant properties. Required fields for a directed network include an arc ID (the primary key) and the from node and to node for the arc. The from node and to node fields identify the direction of travel through the directed arcs. Other fields are optional, but commonly included fields are arc length, free flow travel time, base flow, estimated flow, and flow cost function. The "base" and "estimated" flows usually refer to observed flow and flow estimated from some modeling exercise. The flow cost function field often contains a pointer to some mathematical function rather than the function itself. The node relation typically contains a node ID field and any relevant attributes of the nodes.

As noted previously, we often want to expand our representation of a network intersection to include different travel costs associated with different directions of travel through the intersection. Our formal model for this representation is the expanded network model illustrated in figure 3-1b. An efficient way to represent this information within the relational data model is through *turn tables*. Turn Table contains a tuple corresponding to each direction of travel through an intersection. An additional field maintains the travel cost associated with that direction of travel (or perhaps a pointer to a flow cost function). A reserved character (such as a negative number) can indicate a turn restriction. Similar to the expanded intersection representation, the turn table strategy is effective but not efficient. Turn tables require adding twelve tuples to the database for each intersection in the street network. The total can be quite large for a detailed urban street network.

We often want to include information on address locations within the network. This is useful for *address matching* within the network, i.e., georeferencing entities (such as home addresses, businesses) based on their street address. To maintain address information, we can include a Reference Address relation in our data model. This relation can include the following fields. A from address field and a to address field provide the address range for the given arc. The side field indicates the side of the street to which the address range applies. A parity field indicates whether the address numbers on each side are always even or always odd. (When referring to sides of an arc, we use the *topological left* and the *topological right*. These are the left and right sides based on the direction implied by the from node, to node sequence.) The street name corresponding to the arc often must be partitioned into the following fields: (i) a street prefix (e.g., "North"), (ii) street name (e.g., "Oak"), (iii) street type (e.g., "Avenue"), (iv) street suffix (e.g., "East"). Some GIS software can address match polygons as well as points.

## Weaknesses of the node-arc representation

The popularity of the network model results from the strong foundation provided by graph theory and the ease of manipulating the discrete mathematical structures representing networks. Networks are easily stored and manipulated using traditional, procedural coding methods. Despite its popularity, the basic network model has substantial weaknesses.

One problem with the network model has been discussed previously, namely the representational difficulties imposed by the planar embedding requirement. Forcing nodes to exist at all arc intersections does not represent well the real-world properties of transportation networks that can contain features such as "underpasses" and "overpasses." We can resolve this problem by relaxing the topological consistency of the planar network or by adding turn tables. The former can allow violations of database integrity while the latter is an inefficient workaround.

Another problem with the traditional network model is the assumption that arcs are homogeneous, i.e., arc characteristics do not change between its end nodes (Goodchild 1998). This is not the case for many transportation applications. An obvious example is pavement management: Pavement quality can vary substantially within a given street segment. A less-obvious example is traffic flow modeling: The node-arc model implies that network flow levels and related properties (such as travel time) are uniform within the arc. Also, note that this imposes a fixed level of spatial resolution, i.e., we cannot exploit information on lane configurations or street geometry. This information is often useful for advanced applications such as ITS.

A third problem is difficulty in supporting one-to-many relationships among transportation entities. As Fletcher (1987) notes, the relationships between real/physical transportation entities (e.g., real world physical entities such as highways, intersections, and interchanges) and real/logical transportation entities (e.g., state or federal routes) are often one-to-many. The standard planar network model and its relational counterpart cannot easily accommodate these relationships.

An important type of one-to-many relationship occurs with *linear referencing systems* (LRS). State Departments of Transportation (DOTs) and metropolitan planning organizations (MPOs) often use LRS (e.g., mileposts) for locating incidents or infrastructure within the network. Since these referencing systems exist at the sub-arc and supra-arc level they cannot be supported effectively using the traditional representation.

### Linear referencing methods and linear referencing systems

In attempting to maintain information on transportation infrastructure, many government agencies (e.g., Departments of Transportation or DOTs) and planning organizations (e.g., metropolitan planning organizations or MPOs) have developed *linear referencing systems* (LRS) for transportation facilities. LRS support the storage and maintenance of information on *events* that occur within a transportation network; these can include phenomena such as pavement quality, accidents, functional classes, traffic flow and maintenance districts. As these examples suggest, events can be points or lines (and, in some cases, areas; see below) referenced within the transportation facility.

A LRS typically consists of the following components (Dueker and Butler 1997; Nyerges 1990; Sutton 1997; Vonderohe and Hepworth 1996): (i) a transportation *network*, (ii) a *location referencing method* (LRM), (iii) *datum*. The transportation network consists of the traditional node-arc topological network. The LRM determines an unknown location within the transportation network using a defined path and an offset distance along that path from some known location. This provides the basis for maintaining event data within the network. The datum is the set of objects with known (directly measured) georeferenced locations (see chapter 4). The datum ties the LRS to the real world and supports the integration of multiple-networks, multiple LRMs for a given network, multiple-event databases and cartographic display of the data.

#### Linear referencing methods

Nyerges (1990) identifies three major LRM strategies, namely, *road name and milepoint, control section* and *link and node* systems. Figure 3-6 illustrates these systems. *Road name and milepoint* is a system familiar to anyone who has driven on limited access highways in the United States and Europe. This system consists of a road naming convention (i.e., a standard procedure for assigning names to highways and streets) and a series of milepoint references (i.e., distance calculations along the transportation facility, typically measured in fractions of a mile or kilometer). Milepoint referencing requires a designated "point-of-origin" (e.g., a "mile 0") as a datum; this is often an end point of the route or where the route

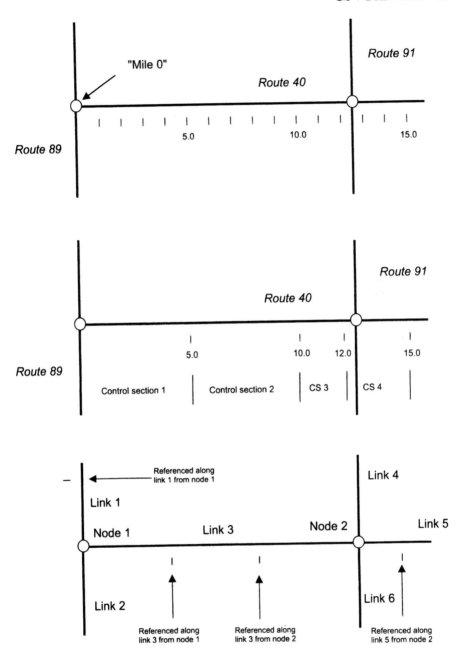

Figure 3-6: Major linear location referencing methods (after Nyerges 1990)

crosses a state, county, province, or national boundary. The LRM designates milepoints as distances from this origin through the transportation route. This requires accurate measurements from the route point-of-origin when referencing events.

Although accurate when originally measured, the milepoint referencing system can become increasingly inaccurate over time as road modifications and other changes in road geometry occur (e.g., road straightening). In other words, the reference milepoint may not correspond to the actual distance from the point of origin. This can create problems when attempting to maintain historical records of events. For example, automobile accidents that occurred exactly at the same location at two different times might have different milepoint location references if realignment occurred during the intervening period. This requires some type of translation function or identifier to ensure that these events are understood to be at the same real-world location (Bureau of Transportation Statistics 1998). Because of this property, perhaps a more appropriate name for this system is *road name and reference point*; this name does not imply road distance (Nyerges 1990).

An LRM closely related to the milepoint system is the *route reference post offset* system (Bureau of Transportation Statistics 1998). In this system, posted signs at known locations along the feature serve as multiple points of origin, i.e., each posted sign is a datum. This allows an event to be referenced as an offset from any posted reference point (not just the point-of-origin of the entire linear feature). Offsets can also be negative as well as positive. Reference posting is more flexible than the milepoint LRM; for example, it does not require a reference sign at a fixed metric such as every mile (although in practice this often is the case). A disadvantage is that the reference signs must be maintained carefully, particularly with respect to their location. For example, workers must be very careful to replace a missing or damaged sign at exactly the same location as the previous sign.

*Control section* LRM designates transportation facility segments such that the data of interest are homogeneous over each segment. Control sections are often used in transportation facility performance monitoring and maintenance funding; for example, monitoring the pavement quality for a control section can help determine the pavement maintenance funds allocated to the facility. Control sections typically use milepoints as their referencing systems; that is, control sections begin and end at specified milepoints (Nyerges 1990).

The *link and node* LRM directly represents the topology (connections) in the transportation system. Chains of segments represent the links of the transportation system while nodes represent intersections and terminal locations (e.g., dead-ends) in the network. Referencing nodes in some geographic coordinate system provide information on both the topology and geometry of each link in the transportation system (Nyerges 1990). This method does not require complete resurveying and reassignment after realignment of the linear facility: The distance between two nodes may change but this may not affect the remaining reference system. A disadvantage of this approach is the identification and communication of event locations. An individual must have a node map available to identify locations in the field. This means that event locations cannot be communicated to the public unless node maps are widely available (and used). Although not a problem for traditional LRM usage, this creates substantial problem for emerging applications such as intelligent transportation systems (ITS) (Bureau of Transportation Statistics 1998).

### Fixed-length and variable-length (dynamic) segmentation

Maintaining data on spatial objects or *events* referenced using an LRM requires developing a *segmentation* scheme for the network data model and linking the linearly referenced data to the network model using the segmentation scheme. Two basic types of segmentation schemes are available. These differ with respect to whether the location or attribute dimension is being controlled or measured (Nyerges 1990).

*Fixed-length segmentation* controls location (i.e., holds the spatial units constant) and measures the attribute of interest for each segment. Fixed-length segmentation subdivides each network arc into segments of uniform length and records attribute value for each segment. Each attribute can have its own segment length, although the segment length for that attribute is constant across all locations in the network. An advantage of this approach is its simplicity: since we subdivide the arcs into fixed intervals, there is a one-to-one correspondence between a segment and an arc (each segment is linked to only one arc). A disadvantage is that it imposes a fixed level of spatial resolution on the linear data. For example, we cannot determine the spatial distribution of an attribute at a higher level of resolution than designated by the fixed segment length. Also, this approach implies redundant data if an attribute occupies a large contiguous linear region within an arc (i.e., several contiguous segments may record the same attribute value). Because of this inflexibility, fixed-length segmentation only can support data referenced using the control section LRM.

*Variable-length or dynamic segmentation* controls the attribute (i.e., holds the attribute value constant) and measures the locations where this attribute has the specified value. In this case, we prespecify the condition of interest and allow each segment to vary in length to encompass all contiguous locations that exhibit that value. A segment can transcend arcs and therefore have one-to-many relationships with arcs. For example, a segment can begin in any location in an arc, traverse several arcs, and then end in another arc. This can maintain data referenced using all three major LRMs. Dynamic segmentation can also easily maintain *route structures* such as bus routes since these are similar in structure to the route milepost LRS. Because of this flexibility, the dynamic segmentation scheme is more popular than fixed-length segmentation (Dueker and Vrana 1992; Nyerges 1990).

Several commercial GIS packages provide a dynamic segmentation capabilities, typically maintained at the logical level using the relational data model. Differences exist with respect to the handling of spatial objects and topological data (Dueker and Vrana 1992). In general, users will start with a base network that consists of all nodes and arcs of interest. On top of this base network GIS layer, users can define different *route systems* that represent, for example, state highway routes, transit bus routes, school bus routes, and so forth. Each *route* in a route system represents a collection of whole and/or partial links in the network. A route can be continuous, have gaps and/or branches, or repeat itself on certain links (see figure 3-7).

Since routes are assembled from network arcs or partial arcs, it is important to keep track of the relationships between each route and its associated network arcs. For example, ArcInfo creates a *route attribute table* (RAT) and a *section*

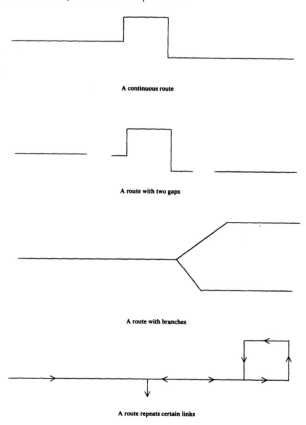

Figure 3-7: Different types of routes

*table* (SEC) to store the relationships between individual routes and network arcs (see figure 3-8). The RAT has one record for each route and stores an identifier, a "begin milepost" and an "end milepost" for each route. The SEC keeps track of the network arcs that are associated with each route. For example, the first record in the SEC in figure 3-8 indicates that the Arc 292 is 6 miles long (defined by the F-MEAS and the T-MEAS data items) and the entire arc (defined by the F-POS and the T-POS data items) is part of the Route 2901 (where the Route# in RAT and the RouteLink# in SEC serve as the common data field of matching records in these two tables). If a route consists of a partial network arc, the partial arc will be defined by its respective F-POS (i.e., from position) value and its T-POS (i.e., to position) value. For example, a record with F-POS = 50 and T-POS = 100 suggests that only the second half of the network arc is part of the route.

Once a route system is defined under the dynamic segmentation data model, it can be used to associate the locations of event data with the base network GIS map layer. An *event* is an attribute that is associated with a portion of a route or

### a. Route attribute table (RAT)

| Record # | Route# | Route-ID | ROADWAY | BEGIN_POST | END_POST |
|---|---|---|---|---|---|
| 1 | 1 | 2901 | 87000001 | 0.000 | 6.000 |
| 2 | 2 | 2902 | 87000002 | 0.000 | 5.200 |
| 3 | 3 | 2903 | 87000003 | 0.000 | 5.400 |
| 4 | 4 | 2904 | 87000004 | 0.000 | 4.200 |
| 5 | 5 | 2905 | 87000005 | 0.000 | 7.400 |
| 6 | 6 | 2906 | 87000006 | 0.000 | 4.100 |
| 7 | 7 | 2907 | 87000007 | 0.000 | 2.900 |
| 8 | 8 | 2908 | 87000008 | 0.000 | 1.300 |
| 9 | 9 | 2909 | 87000009 | 0.000 | 2.400 |
| 10 | 10 | 2910 | 87000010 | 0.000 | 3.600 |
| 11 | 11 | 2911 | 87000011 | 0.000 | 2.500 |
| 12 | 12 | 2912 | 87000012 | 0.000 | 4.984 |
| 13 | 13 | 2913 | 87000013 | 0.000 | 2.400 |
| 14 | 14 | 2914 | 87000014 | 0.000 | 5.800 |
| 15 | 15 | 2915 | 87000015 | 0.000 | 1.013 |
| . | . | . | . | . | . |

### b. Section table (SEC)

| Record # | Route Link# | ArcLink # | F-MEAS | T-MEAS | F-POS | T-POS | Section # | Section -ID |
|---|---|---|---|---|---|---|---|---|
| 1 | 1 | 292 | 0.000 | 6.000 | 0.000 | 100.000 | 1 | 1750 |
| 2 | 2 | 293 | 0.000 | 5.200 | 0.000 | 100.000 | 2 | 1751 |
| . | . | . | . | . | . | . | . | . |

**Figure 3-8:** Dynamic segmentation data model in ArcInfo GIS software

a single location on a route. Event data are commonly stored in separate data tables known as event tables. There exist different types of events. *Linear events* represent the attribute values associated with discontinuous linear segments along a route. For example, curb-side parking is available along certain segments of a street only. In this case, the linear event of curb-side parking must be referenced by both the from milepost and the to milepost of each street segment with available curb-side parking. Figure 3-9a indicates that curb-side parking is available along the segments of 2.53–2.75 miles, 3.64–4.27 miles, and 6.82–7.35 miles on the Route 441. *Continuous events* are associated with the entire route; therefore, they only need to be referenced at the locations where the event value changes. Posted speed limits and traffic lane width along a highway route are two examples. Route 441 in figure 3-9b, for example, has a posted speed limit of 35 mph along the 0 mile to 2.5 miles segment, 45 mph between the 2.5-mile and the 6.7-mile segment, and 35 mph again for the segment of 6.7 mile to 12.6 mile. The third type of event data is known as *point events*. Point events occur at a specific point location on a route. Examples include accident locations, bus stops, and intersection locations. Since a point event is defined by a single locational measure along a route, its data table structure is similar to that of continuous events (figure 3-9c). However, the milepost data in a point event table are interpreted for points rather than continuous line segment. For example, the bus stops shown in figure 3-9c are located at 0.8 mile, 1.4 miles, 2.1 miles, 3.3 miles, and 4.5 miles along the bus route 10.

(a) Curb-side Parking:

| Route ID | From Milepost | To Milepost | Parking Fee |
|---|---|---|---|
| 441 | 2.53 | 2.75 | $0.75/hr |
| 441 | 3.64 | 4.27 | $0.50/hr |
| 441 | 6.82 | 7.35 | $0.75/hr |
| 68 | 1.25 | 1.67 | $1.00/hr |
| . | . | . | . |

(b) Posted speed limit:

| Route ID | To Milepost | Posted Speed Limit |
|---|---|---|
| 441 | 2.5 | 35 mph |
| 441 | 6.7 | 45 mph |
| 441 | 12.6 | 35 mph |
| 68 | 4.3 | 30 mph |
| . | . | . |

(c) Bus stops:

| Bus Route ID | Milepost | Shelter |
|---|---|---|
| 10 | 0.8 | Y |
| 10 | 1.4 | Y |
| 10 | 2.1 | N |
| 10 | 3.3 | N |
| 10 | 4.5 | Y |
| . | . | . |

Figure 3-9: Event tables in a dynamic segmentation data model

## Using dynamic segmentation for multimodal routing: The vnet design

As mentioned earlier in this chapter, the traditional method for maintaining multimodal networks is to develop separate subnetworks for each mode and connect these networks using transfer arcs. While this is straightforward, maintaining these data using a relational database scheme results in a non-normalized design that can lead to database update anomalies. The problem is that functional dependencies among the different modal networks are not maintained. For example, bus routes usually depend on the street network. Separating these into separate subnetworks loses information on these dependencies. Consequently, if a street segment is closed, there must be corresponding updates to the bus routes that use that segment. If one or more of these updates do not occur, the database will be in an anomalous state.

A *virtual network* (vnet) database design based on route systems can capture the functional dependencies in multimodal transportation network (Miller and Storm 1996; Miller, Storm, and Bowen 1995). This strategy separates the topological network from the network attributes by maintaining these attributes using routes and point events. For example, in the case of automobile travel, a route represents each direction of travel through a network arc and point events represent the street intersections. In the case of public transit, routes correspond to an entire bus (for example) route while point events are the stops along that route. All routes are based on the same underlying topological networks. This conceptually separates the topology of the network from attributes. Database integrity is more likely since any changes in the underlying network must propagate to the travel modes using the network.

The vnet design maintains the multimodal network in a relational database using point events and routes. Two different vnet representations are possible. The *2-vnet* is a parsimonious design appropriate for situations where data on direction-specific travel through an intersection is not required. In this case, the key fields for each vnet arc are:

    2-vnet (From Route Class, From Event ID, To Route Class, To
    Event ID)

where a "Route Class" is a type of transportation modal service (e.g., automobile, a particular bus route) and "Event ID" is the identifier for a point event on that "Route Class." Figure 3-10 provides an example network: figure 3-10a shows a base network intersection with routes defined for each direction of travel for automobiles while figure 3-10b shows the same intersection with two bus routes. Figure 3-11 provides the resulting 2-vnet relation. If we wish to track movement direction through intersections, we can add route identifiers to each route within a "Route Class" and specify a *3-vnet* relation as:

    3-vnet (From Route Class, From Route ID, From Event ID, To
    Route Class, To Route ID, To Event ID)

See Miller and Storm (1996) for more detail.

Although effective at maintaining the functional dependencies of modal services on the network, the vnet strategy has two weaknesses. First, it implies a large database since the network, section tables and routes must be maintained for each movement direction within each arc and for each bus route through each arc. Second, it requires a very-efficient computational implementation of the dynamic segmentation data model if it is to scale upward and handle large transportation networks. Continuing improvements in geodatabase servers for large datasets (e.g., ESRI's SDE) may make this feasible.

### Enterprise LRS data models

The flexibility of linear referencing systems and the dynamic segmentation data models supports their application in a wide variety of transportation domains, including infrastructure management, public transit, freight, intelligent transportation systems, waterway navigation, hydrological analysis, utilities management, and seismological sensing (Vonderohe et al. 1995). Although flexibility is desirable, less desirable are independent, application-specific LRS and data models among different departments within a DOT or MPO, different political jurisdictions, or different government levels. Ideal is a common LRS and data model that can support all relevant applications. This can support data sharing among agencies as well as *interoperability* among GIS-T applications. In short, a desirable goal is the development of an *enterprise data model* for LRS, that is, a data model with common features that support applications across an entire enterprise (e.g., agency, corporation, and MPO) or across several enterprises.

This section reviews two examples of enterprise LRS data models, namely, the

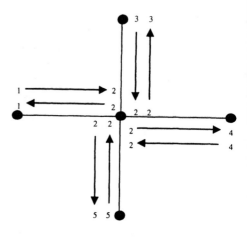

a) street network (automobile travel)

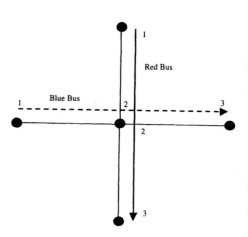

Figure 3-10: Example street intersection for vet example

b) Bus routes

National Cooperative Highway Research Program (NCHRP) LRS data model and the Dueker/Butler enterprise GIS-T data model. The NCHRP model results from a cooperative effort among academics, the federal government and practitioners in the United States. The model supports multiple transportation networks, LRS, event data, and cartographic representations through a single linear datum. The NCHRP model provides a generic data model than can be used as a core for more-specialized applications and implementations (Bureau of Transportation Statistics 1998). The Dueker/Butler model is more inclusive (and consequently more complex): in addition to multiple networks, LRS, events and cartographic representation, the model supports other, integrated spatial data (in

GIS-T Data Models    71

| | 2-vnet | | | | |
|---|---|---|---|---|---|
| | From Route Class | From Event ID | To Route Class | To Event ID | (Non-key attributes) |
| Arcs for "street network" (automobile travel) | Road | 1 | Road | 2 | ... |
| | Road | 2 | Road | 1 | ... |
| | Road | 2 | Road | 3 | ... |
| | Road | 2 | Road | 5 | ... |
| | Road | 2 | Road | 4 | ... |
| | Road | 3 | Road | 2 | ... |
| | Road | 4 | Road | 2 | ... |
| | Road | 5 | Road | 2 | ... |
| Arcs for bus travel | Blue Bus | 1 | Blue Bus | 2 | ... |
| | Blue Bus | 2 | Blue Bus | 3 | ... |
| | Red Bus | 1 | Red Bus | 2 | ... |
| | Red Bus | 2 | Red Bus | 3 | ... |
| Street network to bus transfer arcs | Road | 1 | Blue Bus | 1 | ... |
| | Road | 2 | Blue Bus | 2 | ... |
| | Road | 4 | Blue Bus | 3 | ... |
| | Road | 3 | Red Bus | 1 | ... |
| | Road | 2 | Red Bus | 2 | ... |
| | Road | 5 | Red Bus | 3 | ... |
| Bus to street network transfer arcs | Blue Bus | 1 | Road | 1 | ... |
| | Blue Bus | 2 | Road | 2 | ... |
| | Blue Bus | 3 | Road | 4 | ... |
| | Red Bus | 1 | Road | 3 | ... |
| | Red Bus | 2 | Road | 2 | ... |
| | Red Bus | 3 | Road | 5 | ... |
| Bus to bus transfer arcs | Red Bus | 2 | Blue Bus | 2 | ... |
| | Blue Bus | 2 | Red Bus | 2 | ... |

Figure 3-11: 2-vnet relation based on Figure 3-10 (after Miller and Storm 1996)

particular, areal transportation facilities and events). Both data models support higher-level GIS operations include network-based operations such as routing. Efforts are also underway to extend this data model to support referencing and representation in three-dimensional space and across time (see Adams et al. 1999).

*NCHRP LRS data model*

The NCHRP sponsored the development of the NCHRP 20–27 LRS data model in response to multiple independent efforts at developing LRM and LRS data models in the United States (NCHRP 1997). Vonderohe et al. (1995) present a draft version of the data model while Vonderohe and Hepworth (1996) provide a review that includes some minor revisions and clarifications relative to the initial report. NCHRP (1997) provides a concise summary of the activities that led to the data model as well as the data model itself.

The NCHRP data model supports the following fundamental operations (NCHRP 1997):

i. *locate*: Establish the location of an unknown point in the field by reference to other objects in the real world
ii. *position*: Translation of a real-world location into a database location

iii. *place*: Translation of a database location into a real-world location (i.e., the inverse of "position")

iv. *transform*: Conversion between various LRMs represented by database locations, between various cartographic representations, and between LRMs and cartographic representations

By supporting these fundamental operations, the data model also supports higher level operations. These include GIS operations such as overlay, connectivity and proximity as well as network analysis operations required in transportation analysis (pathfinding, routing, facility location and allocation of network resources).

Figure 3-12 provides a conceptual overview of the NCHRP data model. The data model separates the data elements into three layers. These layers are: (i) *linear referencing system* (LRS), (ii) *business data*, (iii) *cartographic representation*. The LRS consists of three sublayers that comprise its standard components, namely *location referencing methods* (LRM), a *network*, and *datum*. "Business data" refers to the event data referenced through the LRS.

Figure 3-13 provides the NCHRP data model using the OMT language (refer to figure 2-8). The central idea of the NCHRP data model is that a linear datum supports multiple cartographic representations at any scale and multiple networks for various application domains. The datum provides the fundamental referencing space for transformations among the different LRM, networks and cartographic representations. The datum also provides the link to the real world through its georeferencing attributes and empirical measures (NCHRP 1997).

The key datum objects are the Anchor Points and Anchor Sections. An Anchor Point serves the same function as a geodetic control point in 2-D

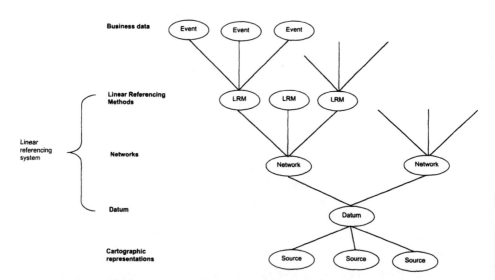

Figure 3-12: Conceptual overview of the NCHRP data model (Vonderohe and Hepworth 1996)

GIS-T Data Models   73

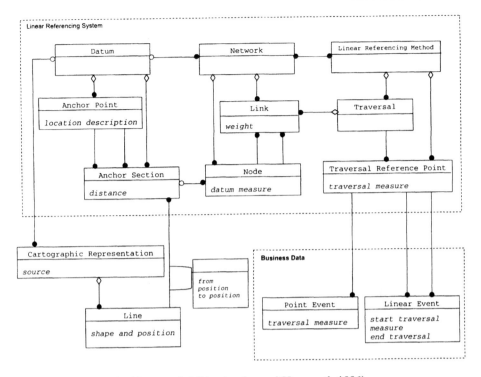

Figure 3-13: NCHRP object model (Vonderohe and Hepworth 1996)

or 3-D georeferencing systems: It provides known locations to tie the remaining data to the real world. The Anchor Section extends this concept to linear transportation data by defining linear features with highly accurate measured distances between a "from" Anchor Point and a "to" Anchor Point. Transportation networks are tied to the real world through these datum objects. Each network Node is georeferenced by determining its "distance" from an Anchor Point along the appropriate Anchor Section.

The Traversal object provides the basis for locating events such as a Point Event or a Line Event within a network using the LRM. A Traversal roughly corresponds to a network path, although the ordered Links that comprise a Traversal need not be connected. A real-world phenomenon such as an accident can be placed in the database as a Point Event by measuring an offset distance (a Traversal Measure) from some Traversal Reference Point along a Traversal; this may be positive or negative. The location of the Traversal Reference Point is determined using a Traversal Measure from the initial Node in the Traversal. Since the location of the Node is known with respect to some Anchor Section and Anchor Point, we have georeferenced the Point Event.

Vonderohe and Hepworth (1996) also specify accuracy standards for the NCHRP LRS data model. LRS input parameters (lengths of anchor sections and offsets of traversal reference points and nodes along anchor sections) typically have some level of random error due to measurement inaccuracies. This can affect the absolute and relative accuracy of the LRS. The former relates to uncertainty in the location of a point or linear event with respect to datum objects while the latter refers to the uncertainty in the location of a point event with respect to another point event or uncertainty in the length of a linear event. Error propagates from the LRS parameters based on their variances (i.e., measurement accuracy) and the covariances (i.e., dependency among the measurement accuracy of two system parameters). Using a linear law of error propagation, Vonderohe and Hepworth (1996, 1998a, 1998b) formulate a system of *observation equations* that specify relationships between errors in measurements and system parameters in the LRS. This allows computation of standard errors for event linear referencing and (working in the opposite direction) requirements for measurement accuracy to reference events with a given accuracy level.

### *Dueker/Butler enterprise data model*

Dueker and Butler (1997) develop an enterprise GIS-T data model that extends the enterprise data concept beyond LRS to include other transportation features. Similar to the NCHRP LRS data model, the Dueker/Butler model is based on independence among geographic datum, events and geometry for cartographic display and link-node topology. The model can also accommodate areal transportation features (e.g., airports, railyards) as well as areal events (e.g., "park-and-ride" lots).

Figure 3-14 provides the conceptual data model (in E-R notation; see figure 2-1) for the Dueker/Butler enterprise GIS-T data model. The core of the model is the entities Transportation Feature, Jurisdiction and Event Point. A Transportation Feature is some identifiable element of the transportation system; this can be a point, line or area. Jurisdiction refers to a political or similar entity for designating transportation features and their labels (e.g., ZIP Codes for streets, airport designators such as LAX). An Event Point is the location on the transportation feature where some portion of a point, line or area event occurs. An offset distance from the beginning of the Transportation Feature designates the Event Point. A Transportation Feature can have any number of Event Points but an Event Point is associated with exactly one Transportation Feature.

Point Event, Linear Event, and Area Event correspond to point, line and areal phenomena occurring within a Transportation Feature. A Point Event is a physical component or attribute that is found at a single location. Point Events may occur independently or on linear or areal Transportation Features. Point Events are related to a Transportation Feature based on its location on that feature, i.e., through an Event Point. An Event Point may have any number of Point Events but a given Point Event is associated with exactly one Event Point. A Linear Event is a physical component or attribute that is found along a segment of a linear Transportation Feature. Linear Events are defined by exactly one *begins* Event Point and exactly one *ends* Event Point. However, an Event

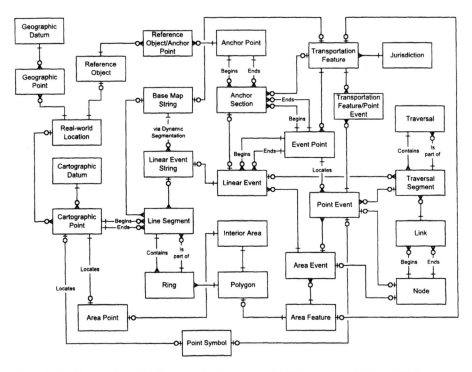

Figure 3-14: Enterprise GIS-T conceptual data model (Dueker and Butler 1997)

Point may be the *begins* or *ends* Event Point for any number of Linear Events. An Area Event is a transportation system component or a non-transportation entity that affects a Transportation Feature. An Area Event is expressed as an attribute of a Transportation Feature using the Linear Event(s) and Point Event(s) defined by its intersection with a Transportation Feature. For example, an Area Event such as a city can be expressed by creating a Linear Event for the portion of the Transportation Feature located within it. Conversely, a "park-and-ride" Area Event can be expressed as a Point Event where its driveway intersects with a linear Transportation Feature.

Figure 3-14 also specifies a special type of point event entity, namely, a Transportation Feature/Point Event. This special type of point event represents Transportation Feature intersections or multi-modal junctions. If we define these as Point Event entities, each Transportation Feature involved in the relationship will have an independent representation of the intersection or junction. Adding a Transportation Feature/Point Event entity still requires each Transportation Feature to maintain its own information about the intersection or junction. However, this entity allows the creation of a single Transportation Feature/Point Event relation in the logical data model. This relation can maintain information about intersections and junctions across all Transportation Features, meaning that their attributes need not be duplicated for each `Transportation Feature`. The `Point Event` associated with each `Transportation Feature` now only maintains the location of the intersection, its status as an intersection and a unique identifier.

The Dueker/Butler model uses a georeferencing strategy similar to the NCHRP data model. However, the Dueker/Butler uses an expanded view of real-world objects and directly georeferenced objects. Datum now consist of `Real World Location`, `Geographic Point`, `Geographic Datum`, and `Reference Object` in addition to the `Anchor Point` and `Anchor Section` entities in the NCHRP model.

The entity `Real World Location` describes the physical object being represented in the database. This allows the inclusion of metadata within the database, particularly with respect to the semantics (meaning) of database entities with respect to the features in the real world. The entity `Geographic Point` maintains the real-world location (latitude/longitude/elevation) of point locations expressed in some `Geographic Datum` (e.g., North American Datum 1983). More than one `Geographic Datum` can be accommodated. A `Reference Object` may be associated with a `Geographic Point`; this is a reference object for at least one (possibly more) `Anchor Point`. Since each `Anchor Point` is georeferenced through a `Reference Object`, it only maintains a unique, real-world identifier (e.g., the name of the "real world" geographic object such as a street intersection). These expanded datum features improve the flexibility of the georeferencing system and allow easier integration of the transportation data model with other georeferenced data sets.

There are other differences between the NCHRP and Dueker/Butler data models. Dueker/Butler ties event referencing directly into the datum rather than linking it to the datum through the topological structure of the network. This creates a cleaner separation between the datum and topological entities of the data model. Cartography in the Dueker/Butler model is more appropriately related to the transportation features rather than the datum. Dueker/Butler also does not require transportation features to have a topology; this can support data on a wider variety of nontopological transportation features (Dueker and Butler 1999).

*Implementation issues*

Since enterprise LRS data models are typically generic, there are often difficulties in implementing these models in an operational database system. Some of these issues are case-specific and cannot be addressed in a systematic manner: for a review of several implementation case studies, see Bureau of Transportation Statistics (1998). This section reviews selected, systematic implementation issues.

Several linear referencing pathologies can occur due to route definition problems when implementing a LRS. These pathologies include *discontinuous routes*, *dog-leg routes*, *split roads*, *cul-de-sacs*, and *ramps* (Sutton and Bespalko 1995; Vonderohe and Hepworth 1996). A discontinuous route (traversal) occurs when designated or logical routes stop and start, creating gaps. This requires a judgement call as to whether to reference the route as if the missing portion was in place or start the offset measure over again where the route restarts. The latter case is the most common strategy used in practice.

Dog-leg routes (traversals) occur when designated or logical routes share common sections of a physical transportation facility. A decision must be made with respect to the assignment of events to individual routes along the shared sections unless there is some administrative reason to assign the event to only one of the traversals. The NCHRP and the Dueker/Butler data models reconcile this problem by allowing transformation or sharing of events among traversals.

The split road problem occurs when divided highways have two roadways of unequal length. Vonderohe and Hepworth (1996) point out that this is not a dilemma with the NCHRP data model (and, we could add, the Dueker/Butler data model). Each of the two roadways will have independent datum objects, links and traversal designations.

Cul-de-sacs (i.e., a street closed at one end with a circular feature; these are often found in residential areas) can create problems since offset measurements can be arbitrary or nonunique. Reconciling this problem simply requires a linear referencing convention for coding these features (e.g., clockwise or counter-clockwise). Ramps are often transitions among routes and therefore must be dealt with in a special manner. Some transportation agencies (such as Colorado DOT and Washington DOT in the United States) have developed ramp coding strategies that associate these with particular routes. These features can be accommodated by representation at the datum level as an anchor section (Vonderohe and Hepworth 1996).

## Transportation data models for ITS and related applications

*Intelligent transportation systems* (ITS) are an attempt to improve the safety, efficiency, and capacity of surface transportation systems through the use of information and telecommunications technologies. ITS reflects a shift in thinking about transportation systems. Rather than physically increasing the capacity of transportation infrastructure, ITS intends to use existing capacity more effectively by collecting detailed spatial and temporal data about the transportation system and using this information in transportation system management. This can include techniques such as electronic toll collection, advanced traffic control, information provision to travelers, navigation, and vehicle guidance (Branscomb and Keller 1996; Nwagboso 1997).

Chapter 9 of this text reviews ITS in more detail. The current goal is to discuss the data modeling requirements of ITS. ITS introduces transportation database requirements beyond the traditional requirements of maintaining arc/node topology, 2-D georeferencing and linear referencing of events within transportation features. A fully developed ITS requires a high-integrity, real-time information system that will receive inputs from sensors embedded within transportation facilities and from vehicles equipped with Global Positioning System (GPS) receivers. Information will be updated continuously in a system of databases that maintains a dynamic model of the integrated, multi-modal transportation system.

This data will be used to provide route information and navigation to travelers as well as update traffic control devices such as timed traffic lights and variable message signage in real-time (Branscomb and Keller 1996).

ITS data requirements can also go well beyond maintaining the performance of transportation system components. Ancillary information that enhances information provision and route guidance for travelers includes (Golledge 1998):

i. disaggregate data on travelers' characteristics (socioeconomic attributes, attitudes/perceptions with respect to decision making and information provision, criteria for route selections)
ii. data on information system characteristics (ITS availability, type of information provided, access costs, reliability)
iii. trip/transportation system characteristics (usual trip times, network performance, travel times on each link, accident frequency)

These data must be integrated with the system performance data and accessed often in real-time.

This brief description of ITS implies the following functional requirements for ITS databases. First, ITS requires *navigable* data models, that is, data models that can locate a vehicle within the map reference frame and provide navigation functions as well as other information about current and anticipated system performance. Second, the diverse ITS databases must be integrated in a manner that is seamless to travelers. Seamless integration must occur among the diverse data within an ITS jurisdiction as well as across jurisdictions. Finally, the data must be *interoperable*, that is, easily exchanged and accessed among heterogeneous system components and among different ITS.

Navigable data models

As the name implies, a navigable data model is a digital geographic database of a transportation system that can support vehicle guidance operations at a high level. For ITS, this typically involves the following functions (Dane and Rizos 1998). First, the data model must be able to translate a given latitude and longitude into a street address and vice versa. Travelers use address systems for location referencing while the ITS typically tracks a vehicle location using its GPS receiver. Unambiguous translation among locations referenced using both systems must occur if the system is to provide meaningful navigational directions to the traveler. We will discuss these issues in more detail in chapter 9.

Second, the data model must support *map matching*, that is, the ability to "snap" a vehicle's position to the nearest location on a network segment when its reported or estimated location is outside the network. This can occur because of differences in accuracy between the GPS and the digital network database. It can also occur if the ITS uses a dead-reckoning system for estimating a vehicle's location between GPS "fixes."

Third, the data model must support *best route calculation*. This refers to assisting the traveler in selecting an optimal route based on stated criteria (e.g., travel time, cost, navigation simplicity). This requires a high level of spatial data accuracy including traffic regulations such as turn restrictions. The fourth function is closely related: The data model must support *route guidance*. This refers to navigation instructions as the route is executed. This is a challenging task in real time since the ITS must process the traveler's current position, perform address and map matching and then provide navigational cues before the landmark or decision juncture is reached. If a traveler misses a turn, the system should recover gracefully and calculate a new best route and provide directions along this new route.

The traditional node-arc model can support these capabilities, albeit nominally. Much of the high-resolution positional information provided by in-vehicle GPS units and in-facility sensors is lost when referenced within the traditional network structure. While enhancing the network with dynamic segmentation can improve its ability to support positional information along the length of each arc, this cannot fully capture complex road geometry such as ramps and nonplanar features such as underpasses and overpasses. Also required to support advanced traffic control and navigational devices is positional information resolved to the individual lanes within each segment. Emerging data models for ITS attempt to represent these complex spatial features to support more fully the high-resolution positional information and the detailed route guidance and traffic control features available through ITS.

*Lane-based network data models*

A straightforward way to enhance the traditional node-arc model for ITS applications is to add information on lanes within each arc. Lane information can be used in ITS to provide turn directions, instructions for avoidance of obstructions or restrictions in particular lanes, monitoring lane-specific flow for analysis and modeling and representing beginning and ending points for lanes for enhanced route guidance. This requires a data model that can represent the appearance/disappearance of lanes, connectivity among parallel lanes and at turns, and movement obstructions and restrictions including forward, lateral and turn barriers or restrictions (Gottsegen, Goodchild, and Church 1994).

Several strategies are available for enhancing the node-arc model with lane information (Fohl et al. 1996; Gottsegen, Goodchild, and Church 1994). One possibility is to work directly with the arc-node topology and add a node to "split" the arcs when the number of lanes change. However, as noted previously in this chapter, this approach to capturing one-to-many relationships in the traditional node-arc model results in a rapid proliferation in the number of arcs and nodes, increasing the data storage requirements. Also, updating the database to reflect lane changes (e.g., temporary lane restrictions during construction) would be cumbersome and require rebuilding network topology.

A second strategy is to use dynamic segmentation. In this case, a change in the number of lanes is recorded using an offset distance or a percentage coverage from the beginning of an arc. These route systems cannot traverse more than one arc since this would alter network topology. Two options exist within this general strategy. The first is to store a route corresponding to each lane. Each route would have an independent beginning and ending offset within the arc. The second option is to store the number of lanes as the phenomenon that occurs along the arc. In other words, we would subdivide the arc into portions that have two lanes, one lane, and so forth, with the number stored as an attribute of the route. While the second method is more parsimonious with respect to data storage, a disadvantage is that it does not retain information on which lanes continue or terminate after a location where the number changes.

Another property that must be captured by the data model is connectivity among lanes. Two types of connectivity must be captured, namely, connectivity among lanes at turns and connectivity between parallel lanes within an arc (for representing lane-changing behavior). With respect to connectivity at turns, one possibility is to record, as an attribute of each lane, the possible movement from each lane at a junction to the intersecting arc as a whole. A second approach is to expand the turntable structure used to represent intersections in the traditional node-arc model. In this case, each record of the turntable represents a lane and possible turns on to the intersecting arc. This requires an even larger increase in data storage than the standard turntable structure. Also, since it does not record the lane it cannot handle special-use lanes. We can improve this by an even greater expansion of the turntable; that is, store a record corresponding to a movement from each lane in an arc to each lane in the intersecting arc. Although it captures the full topology, it results in an explosion in the database storage requirements.

Movement restrictions between adjacent parallel lanes within an arc can be captured either in the geometry or as an attribute of the arcs. In the former case we would treat each lane with restricted movements as a separate arc. Portions with free travel among lanes can be modeled as a single arc. We could also maintain an attribute associated with each lane record indicating whether movements from that lane are possible. This latter method is better at handling changes in lane movement restrictions but is less effective for routing algorithms.

Figure 3-15 describes the relations for a prototype lane-based navigable data model (Fohl et al. 1996). Lane maintains information on the lanes within an arc. Each lane is an independent route structure in a dynamic segmentation-type data schema. The lane is the basic element of the data schema; this is a subtle but important difference with standard dynamic segmentation schemas in which a lane would be a linear event referenced within a route structure. Therefore, lane ID is the key field for the relation while street ID is a foreign key linking Lane to a street relation maintaining spatial information on the polylines corresponding to each street. Point Turn Table reflects fully expanded turntable strategy discussed above; that is, a record corresponds to a turn from each lane in the origin street to each lane in the destination street. Since lanes are the basic data element, each lane ID is unique and the turn table needs not maintain street identifier data. The turn ID field is the key field for this relation, although this is not required since a key can be derived from the combination of the lane ID, to lane fields. The position and to position fields maintain the offset distance along the street where the turn is located. Finally, Linear Turn Table maintains movement possibilities along lanes. However, rather than a single turn location as in the Point Turn Table, turns now have a start position and an end position where the turns are allowed. The to offset field maintains the location on the to lane corresponding to the start position of the origin lane,

## GIS-T Data Models

| Lane | | | |
|---|---|---|---|
| lane id | street id | from<br>(start position of lane) | to lane<br>(end position of lane) |
| | | | |

| Point Turn Table | | | | | |
|---|---|---|---|---|---|
| lane id | turn id | position | to lane | to position | impedance |
| | | (start position in origin lane) | (turn destination lane) | (end position in destination lane) | |

| Linear Turn Table | | | | | | | |
|---|---|---|---|---|---|---|---|
| lane id | turn id | start | end | to lane | to offset | distance | impedance |
| | | (location within lane at beginning of turn) | (location within lane at end of turn) | (destination lane of turn) | (location on "to lane" corresponding to "start" position of turn) | (linear distance required to complete turn) | |

Figure 3-15: Relations for a lane-based navigable data model (Fohl et al. 1996)

that is, the locations along the `to lane` that could theoretically receive flow from the origin lane. The `distance` field maintains the actual distance required for the lane change.

### Three-dimensional data models

A more radical approach to navigable data models for ITS is to abandon the node-arc model entirely. Note that the classical node-arc model breaks down when applied to the following domains: (i) true distance measurement across sloping or hilly terrain, (ii) representation of 3-D structures such as overpasses and on-ramps, and (iii) assigning multiple routes over a single arc.

These are fundamental flaws of a 2-D geospatial representation. Rather than address these fundamental flaws, the GIS-T community attempted to fix the 2-D model by adding additional layers of new topologies, particularly routes and LRS. However, the network pathologies reviewed by Sutton and Bespalko (1995) suggest that these problems are not completely resolved and indeed will be even more problematic when attempting to support ITS applications. Bespalko et al. (1998) argue that we should abandon the 2-D paradigm resulting from the origins of GIS in automated cartography. A shift from 2-D to 3-D GIS can eliminate the pathologies that plague implementation of LRS data models.

Bespalko, Ganter, and Van Meter (1996) discuss the characteristics of a 3-D, object-oriented GIS-T data model suitable for ITS. This model would represent navigable areas as thin, articulated surfaces that are coaxial and branching. For example, several side-by-side ribbons would represent multiple lanes that at some point could be joined by additional ribbons or could branch off. Ribbons could run above or below other ribbons to represent 3-D structures such as on- and off-ramps. This improves the positional accuracy allowed by the node-arc-route structure of a dynamic segmentation-enhanced network.

By capturing the third spatial dimension, a GIS can support effectively advanced ITS applications such as route guidance, vision enhancement and automated piloting. Guidance applications provide suggested routes to travelers; this requires clear communication of travel directions at the appropriate decision junctures en-route. A 3-D object-oriented data model can distinguish between overpasses, underpasses and intersections, thereby providing guidance through complex intersections or ramp structures. Vision enhancement technology estimates vehicle speed, direction and proximity information using a heads-up display. Incorporating 3-D information allows more accurate predictions of potential vehicle conflicts in real-time. Long-term advanced ITS planning calls for automated piloting of vehicles in dedicated highway lanes. Automatic piloting requires accurate information on in-road sensors and the ability to react to changing conditions (including terrain) in real-time (Bespalko, Ganter, and Van Meter 1996).

### Distributed and interoperable transportation databases

In chapter 2, we discussed general issues involving distributed databases and interoperability for GIS-T. Distributed and interoperable databases are critical for ITS since data and system components are likely to be distributed over many sites both within a jurisdiction and among different jurisdictions. Also, support for *integrated highway information systems* (IHIS) requires a high degree of data sharing among agencies and interoperability among disparate and heterogeneous DBMS.

ITS deployments are shifting increasingly from centralized to distributed and interoperable configurations. This reflects the widening scope of ITS in terms of geographic coverage and functionality. In the past, ITS configurations have mirrored the "integrate and centralize" principle in nondistributed database systems. Transportation operations centers (TOCs) served as collectors and disseminators of information in computerized traffic control systems. The TOC maintains a high degree of control over the database. Users include transportation agencies, local governments, transit authorities, private businesses, researchers, and citizens. Users interacted with the database for diverse purposes such as data viewing (e.g., drivers using in-vehicle navigation systems), transferring data to local sites for manipulation and integration with local data, integrating local data into the TOC database for availability to others and modifying the TOC database. This requires users to relinquish rights and control over data and receive access permissions from the TOC database administrator. This impedes participation and removes quality control over local data from the users who are most familiar. Also, large centralized databases can not meet the performance levels necessary for handling dynamic transportation data particularly for ITS operating at a regional or national level (Cova and Goodchild 1994).

A distributed database facilitates data sharing among users; this is essential in an ITS environment where dynamic data from diverse sources must be integrated in real-time. Local autonomy facilitates more effective quality control for local data and increases the likelihood of continuous operation and data availability and reliability. The smaller databases and smaller number of transaction requests at distributed sites facilitates real-time data acquisition and visualization. However, a distributed database potentially introduces problems with respect to data accuracy and integrity. This is a concern since ITS demand very high data accuracy and integrity, particularly for navigational functions (Cova and Goodchild 1994).

Given the diverse and multifaceted nature of ITS applications, it is likely that data sharing will involve multiple users with diverse semantic, schematic and syntactic data models and implementations. ITS data sharing can range from simple communication of location references that must be reconciled within a local database to exchanging and integration of entire databases. The players involved in this exchange can include a multitude of public and private sector organizations over a wide geographic area, even up to national scales. A common language for ITS can be difficult to achieve due to the diversity of data semantics even within a single organization. Schema interoperability is even more problematic since proprietary interests arise with the use of private databases (and, in some cases, public databases). Some schema interoperability is possible through data transfer standards (such as SDTS) and architectures such as OGIS. However, data transfer standards and interoperable architectures do not address many of the institutional and organizational barriers to data sharing that can plague full ITS deployment (Ganter, Goodwin, and Xiong 1995; Goodwin 1994). Chapter 9 will discuss some of these issues in more detail.

*Integrated highway information systems* (IHIS) also require interoperability of geographic information. An IHIS is a system of linked transportation databases to support analysis, reporting and decision making functions needed to manage transportation infrastructure (Alfelor 1995; NCHRP 1987). As the name implies, much of the discussion surrounding IHIS has focused on road and highway infrastructure. However, the movement towards greater intermodal connectivity and coordination in the United States, Europe and elsewhere means that these systems will have wider scope as they develop (similar to the change from "Intelligent Vehicle Highway Systems" to "Intelligent Transportation Systems" in the United States).

IHIS attempt to encompass "cradle-to-grave" information management for transportation projects, that is, from the design phase through construction to operation and maintenance. Traditional transportation agency organizational structure has delegated these activities to more-or-less independent divisions or departments. Separate *management information systems* (MIS) and GIS have evolved within each division to store data relevant to its domain and provide decision support over a range of needs from short-term operational goals to long-term strategic planning. This has resulted in heterogeneity among information systems, duplication of data and suboptimal strategic decision-making (Alfelor 1995).

An IHIS consists of a collection of databases organized according to their role in transportation planning. The functional levels of information required by a transportation agency includes (Alfelor 1995) (i) *sectoral* (transportation system viewed as a whole), (ii) *network* (planning, programming, and budgeting of network links and other components), (iii) *project* (technical and functional design), and (iv) *operational* (construction, maintenance, traffic, and safety).

*Research and development* is another functional level often considered, but this can cut across the levels listed above as an auxiliary activity within each of the levels. The detail of the spatial data required for supporting these functions increases from sectoral to operational; this translates into differences in scale and resolution.

Integrating databases into a comprehensive IHIS involves issues of semantic, schematic and syntactic heterogeneity discussed previously. This is particularly the case during the initial phasing-in of an IHIS when existing policy and practices, data and DBMS within an agency must be reconciled. If these problems are surmounted, IHIS implementation may reduce to the relatively tractable problems associated with distributed DBMS since departments and divisions within an agency could be brought into a single integrated system, at least in theory. However, similar to ITS, the semantic heterogeneity problems are difficult to overcome. The biggest obstacle to achieving a comprehensive IHIS is people, not technology (NCHRP 1987).

## Conclusion

This chapter provided a review of data models specifically oriented to GIS-T applications. We reviewed the mathematical foundations of network representations and discussed the node-arc model and its implementation using the relational data model. We discussed linear referencing systems and the dynamic segmentation data model, a major extension of the node-arc model that allows one-to-many relationships to be maintained. We also discussed navigable data models and distributed/interoperability issues in ITS.

While careful database design is important for GIS-T, it is also critical to find and integrate data for populating the database. In the next chapter, we will discuss transportation data sources and data integration.

# 4

# Transportation Data Sources and Integration

After designing a GIS database, the analyst must find data sources to populate the database. This chapter discusses issues regarding populating a GIS for transportation-related applications.

Using geographic data effectively requires a solid foundation in basic mapping concepts. Although a GIS transcends the traditional paper map, many of the principles involved in collecting and representing digital geographic data are directly related to centuries-old concepts and techniques of measuring the surface of the globe and projecting this information to the 2-D Cartesian plane.

After reviewing basic mapping concepts, we will discuss primary and secondary sources for GIS-T data. We will discuss methods for collecting transportation data and briefly mention some major secondary data sources and products in the United States and Europe.

When data are obtained from different sources, they are likely to be in different formats and have varying quality. Although several standards have been established to facilitate spatial data exchanges, data integration remains as a challenge in GIS-T. This chapter discusses some of the most frequently encountered GIS-T data integration issues. We will discuss the closely related issue of spatial data aggregation and the modifiable areal unit problem (MAUP). Data integration often requires aggregation; as we will see, aggregating spatial data can create confounding effects on the results of any analysis that uses these data. This is long recognized and serious problem that is beginning to yield due to GIS' ability to handle and reconfigure spatial data.

## Basic mapping concepts

Although a GIS transcends the traditional paper map, many of the principles that dictate mapping locations on the Earth's surface also apply to a GIS. Without a good understanding of the basic map concepts, it is very easy to misuse geographic data. For example, if we use GIS software to measure the length of a highway segment using data derived from source maps at 1:100,000 scale, the answer will be a rough approximation of the same answer derived from a source mapped at a larger scale. Most GIS software will not communicate this error or even its possibility to the user. We could also use a GIS data set to measure spatial proximity to map features in another GIS data set that was created from maps with a different geodetic datum and at a different scale. Again, most GIS software will generate this output without comment even though it may have error that is greater than the accuracy requirement of an application. It is the responsibility of the analyst to understand these limitations; this requires a basic understanding of mapping.

This section will briefly review four important map concepts, namely, *geodetic datum*, *map projections*, *coordinate system*, and *geographic scale*. For a complete review of basic map concepts, we refer the reader to a good introductory cartography book (e.g., Campbell 1997; Dent 1998).

### Geodetic datum

The starting point of any mapping is a *geodetic datum*: this is a model of the Earth's shape. The geodetic datum is the foundation for any locational coordinate system (Dana 2000a). Although we often describe the shape of the Earth as a sphere, this is not a precise description of the true shape of the Earth. The Earth is more like a slightly flatten sphere that bulges at the equator due to its rotation. To represent this slightly flattened Earth's surface, a mathematical equation based on the semimajor axis and the semiminor axis of an ellipse can be developed to derive a simple geometric surface. This mathematical representation of the earth's surface is known as an *ellipsoid* (French 1996).

In many GIS applications, the topographic surface of the Earth is an important variable. Topography is measured against some predefined reference levels in a survey. In plane and geodetic surveys, measurements are based on a plane perpendicular to the gravity surface of the Earth (Dana 2000a). Due to local variations, the gravity surface of the Earth is irregular. As a result, it is different from the simple geometric surface represented by the ellipsoid. This irregular gravity surface of the Earth is known as *geoid* (figure 4-1). The geoid represents a surface of which the gravity potential is equal everywhere and the direction of gravity is always perpendicular (Defense Mapping Agency 1984). The geoid model is significant because many geodetic measurement instruments consist of a leveling device that is perpendicular to the direction of gravity.

Table 4-1 lists some frequently encountered datums for different parts of the

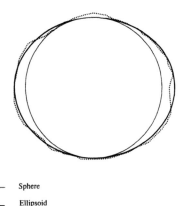

— Sphere
— Ellipsoid
······ Geoid

Figure 4-1: Sphere, ellipsoid, and geoid

world. For example, many maps and GIS data sets of the United States are referenced to the North American Datum 1927 (NAD 27) or the North American Datum 1983 (NAD 83). Offsets exist between the coordinates of the same location on the Earth's surface that are referenced to different datums. In other words, performing GIS functions on two data sets of different datums will likely result in positional errors. The amount of error varies among the locations on the Earth's surface and between the datums. In a 3-D space, potential positional error could be as large as 1 km (approximately 3,000 ft). Therefore, datum conversion

Table 4-1: Frequently Encountered Datums in Selected Parts of the World

| Datum | Ellipsoid | Region of Use |
| --- | --- | --- |
| Australian Geodetic 1966 | Australian National | Australia, Tasmania |
| Australian Geodetic 1984 | Australian National | Australia, Tasmania |
| European 1950 | International 1924 | Europe (excluding U.K.) |
| European 1979 | International 1924 | Europe (excluding U.K.) |
| Indian | Everest (India 1956) | India, Nepal |
| Korean Geodetic System | GRS 80 | South Korea |
| North American Datum 1927 (NAD 27) | Clarke 1866 | Canada, U.S.A., Mexico, Central America |
| North American Datum 1983 (NAD 83) | Geodetic Reference System 1980 (GRS 80) | Canada, U.S.A., Mexico, Central America |
| Ordnance Survey Great Britain 1936 | Airy 1830 | U.K. |
| S-42 (Pulkovo 1942) | Krassovsky 1940 | East Europe |
| South American 1969 | South American 1969 | South America |
| Tokyo | Bessel 1841 | Japan, South Korea |
| World Geodetic System 1972 (WGS 72) | WGS 72 | Global definition |
| World Geodetic System 1984 | WGS 84 | Global definition |

*Source*: After Dana 2000a.

is a necessary step when multiple GIS data sets of different datums are used together in a GIS application. For additional information on geodetic datums, consult an introductory geodesy book (e.g., Wolfgang 1991; Smith 1997).

World Geodetic System 1984 (WGS 84) is the most recent ellipsoid. WGS 84 has several valuable properties. First, it supports elevation as well as latitude and longitude measurements. Second, it is geocentered and therefore can support GPS. Third, it is a world standard and therefore supports integration of international data. Fourth, it is compatible with the common North American mapping datum, NAD 83.

### Map projections

A map projection is a mathematical transformation of locations on a curved surface to locations on a flat plane. During a transformation from a curved surface to a flat plane, it is not possible to maintain all of the geometric properties. As a result, all map projections include some geometric distortions in shapes, areas, distances, angles, or directions. When maps are used as the data sources in GIS, it is critical to examine the distortions embedded in the maps.

We can classify map projections based on their properties of geometric distortions. *Conformal map projections* preserve the angles in the original features. For small areas, this group of projections also preserves the shapes of original features. In general, however, conformal map projections do not preserve area. The commonly used and well-known Mercator projection is an example of a conformal map projection. *Equal-area map projections* preserve the areas of original features; therefore, all features show correct relative sizes on a map. However, angles and shapes are distorted. The Lambert equal area projection is a frequently used equal-area map projection. *Equidistant map projections* attempt to preserve distances between different locations. However, it is not possible to preserve all distances between any two points. An equidistant map projection only preserves the distance measures from one or two selected points. One example is the Conic equidistant projection.

Due to the different types of geometric distortions embedded in various map projections, it is a common practice of transforming all GIS layers into the same map projection before any GIS analysis and display functions are applied to the map layers. Selection of a map projection is dependent on locations of interest on the Earth's surface and project requirements. For projects covering a small geographic area (e.g., a city), distortions in a map projection are relatively small and might not be a concern for a project that does not require a high level of positional accuracy. For projects at regional, state, national, or global scales, distortions embedded in a map projection should be evaluated to ensure that they meet the requirements of a study. In addition, different map projections introduce various levels of distortions to locations on the Earth's surface. Each map projection selects its specific meridian(s) and/or parallel(s) as the central reference locations in the development of a 2-D map from the curved Earth's surface. As a consequence, distortions generally increase as we move away from these reference meridian(s) and/or parallel(s).

## Coordinate systems

Coordinate systems reference locations in continuous space. There are two major types of coordinate systems: *geographic* and *plane* coordinate systems. The geographic coordinate system uses latitudes and longitudes to identify locations on the Earth's surface. This is an implicit 3-D grid restricted to the surface. A plane coordinate system, on the other hand, is a 2-D grid developed on a flat plane. It first establishes an origin point, followed by setting up two axes that pass through the origin point and perpendicular to each other. This type of planar coordinate system is commonly known as a Cartesian coordinate system. Locations in a Cartesian coordinate system are identified by their offsets from the origin along the two axis directions. Another type of plane coordinate system is known as the polar coordinate system, which identifies a location by its distance offset from the origin and an angle measured from a fixed point. Figure 4-2 illustrates both systems.

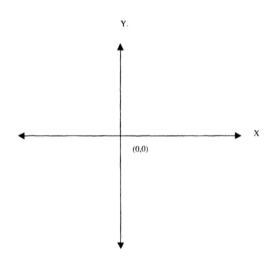

(a) Cartesian coordinate system

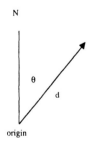

(b) Polar coordinate system

**Figure 4-2:** Cartesian and polar coordinate systems

Most GIS data are based on either geographic or Cartesian coordinates, although some engineering and land survey data use polar coordinates. Several Cartesian coordinate systems deserve a brief review here. Many national and international mapping agencies use the Universal Transverse Mercator (UTM) coordinate system. This provides georeferencing for the entire globe at a high level of precision. This coordinate system is based on the transverse Mercator map projection. It divides the globe into sixty north-south zones (i.e., 6 degrees of longitude per zone). These zones are numbered 1–60 eastward starting from the 180° W longitude. Due to the significant distortions near the north and the south poles, UTM coordinate system covers the area between latitudes of 80° south and 84° north only. This range of latitudes is divided into 20 zones that are labeled northward from C through X, with letters O and I omitted to avoid potential confusions with numbers 0 and 1. Each of these zones covers 8° of latitude except for the zone X which covers 12° latitude. By combining the two sets of zone labels, each UTM zone can be uniquely identified. For instance, UTM zone 3N covers the Earth's surface between longitudes 168° W to 162° W and latitudes 0° N to 8° N. See figure 4-3 for an illustration.

In addition to the zone classification, the UTM coordinate system uses false origin points to define coordinates in each zone. In the Northern Hemisphere, a false origin is established at the intersection point of 500,000 m west of the central meridian in each UTM zone and the equator. In the Southern Hemisphere, the false origin of each UTM zone is at the intersection of 500,000 m west of the central meridian of the zone and 10,000,000 m south of the equator. Locations in each UTM zone are measured as easting and northing from the false origin of the zone. Since locations in each UTM zone are referenced to its own false origin point, UTM coordinates need to be adjusted when dealing with features in different UTM zones.

The U.S. Military Grid Reference System is an extension of the UTM coordinate system. It further subdivides each UTM zone (e.g., UTM zone 18S) into a grid of $100 \times 100 \text{ km}^2$ cells. A system of two-letter designations is used to identify each cell in a UTM zone. Locations in each cell are measured in meters as easting and northing from the southwest corner of the cell. By concatenating a UTM zone number (e.g., 18S) with a two-letter military grid designation (e.g., UT) followed by easting and northing without punctuation (e.g., an easting of 92106 and a northing of 20137 is written as 9210620137) will result in 18SUT9210620137. This code identifies a specific location on the Earth's surface. The precision of locational referencing of the military grid coding system depends on the number of digits used in measuring easting and northing. The U.S. military grid system allows a maximum of ten digits in the codes (i.e., five digits each for easting and northing). This translates into a 1-m precision level.

Another commonly used coordinate system in the United States is the State Plane Coordinate System (SPCS). The SPCS is designed to simplify local survey computations by establishing a SPCS for each state. Depending on the size and shape of each state, two map projections (Transverse Mercator Projection and Lambert Conformal Conic Projection) are used to establish the SPCS of the 50 states and the territories of the U.S. The Transverse Mercator Projection is

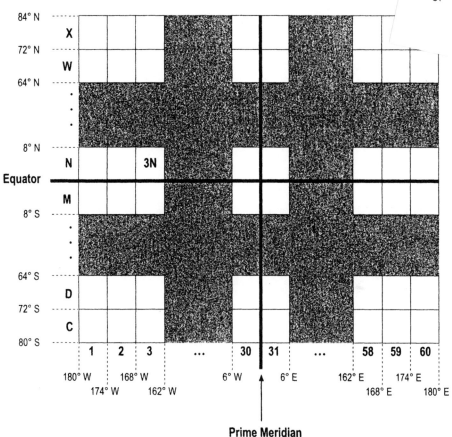

Figure 4-3: Universal Transverse Mercator (UTM) coordinate system

used for states, or parts of a state, that have shapes elongated in the north-south direction, while the Lambert Conformal Conic projection is for states, or parts of a state, with shapes extended in the east-west direction. Alaska is an exception that also employs the Hotine Oblique Mercator Projection. (See Snyder 1987 for a state-by-state listing of SPCS zones and map projections.) Each SPCS has its own false origin that is located outside the zone such that only positive coordinates are used. The typical false-origin location is at 2,000,000 feet west of the central meridian under the Lambert Conformal Conic Projection and at 500,000 feet west of the central meridian in the case of the Transverse Mercator Projection (Campbell 1997). The false origin is located south of the southern edge of each zone under both projections. A full location designation in SPCS is written as, for example, 737,408 feet east, 699,854 feet north, Florida, East Zone.

The British Ordnance Survey uses the British National Grid System. The National Grid is based on a hierarchical structure. At the highest level, it divides the United Kingdom into a set of 500 km × 500 km grid squares, with each grid square assigned with a letter. Each 500-km square then is subdivided into 100 km × 100 km grid squares. The twenty-five 100-km grid squares are assigned with letters from A to Z (with the letter I omitted) in a row-by-row pattern starting from the 100-km grid square at the northwest corner (see figure 4-4a). Each 100 km × 100 km grid square therefore is referenced by a unique two-letter designation. Within each 100-km grid square, it is further subdivided into one hundred

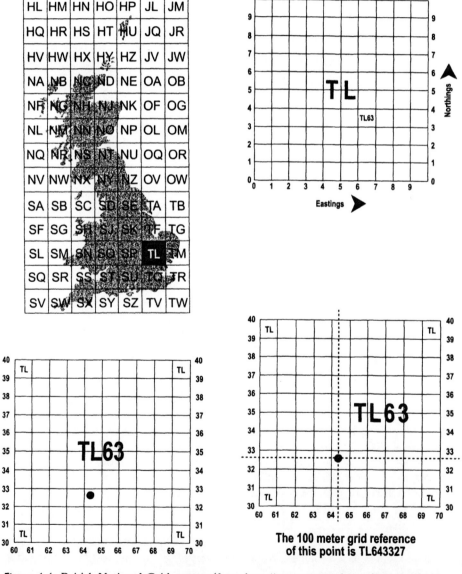

Figure 4-4: British National Grid system (from http://www.o-s.co.uk/products/natgrid/)

10 km × 10 km grid squares that are referenced by two-digit designations of 00–99 (see figure 4-4b). The same subdivision process is repeated to create 1 km × 1 km grid squares with a four-digit designations (see figure 4-4c) and 100 m × 100 m grid squares with six-digit designations (see figure 4-4d). For example, the designation of TL643327 in figure 4-4d indicates that the point is located at 4,300 m east and 2,700 m north of the southwest corner of the TL63 100 km × 100 km grid square.

### Geographic scale

The ability to display a digital map at different scales, using the zoom in and zoom out functions available in a GIS software, often lead novice users to think that map scale is an unimportant subject in GIS. In fact, the resolution and positional accuracy of a GIS database is directly related to the scale of the source map used to create the database. *Map scale* is the ratio of map distance to ground distance. For example, on a 1:24,000 scale map, one inch on the map is equivalent to 24,000 inches (or 2,000 ft) on the ground. At this scale, it is impossible to graphically represent the details on the ground such as the detailed alignment of a highway. On the other hand, a map of 1:2,400 scale can show more details of a highway alignment since 1 inch on this map now represents only 2,400 inches (or 200 ft) on the Earth's surface.

Transportation applications can cover a wide range of geographic scales. These vary from a global scale from global transportation logistics analysis to an engineering study of one highway interchange. When we design and populate GIS databases to meet the needs of various transportation applications, considerations must be given to the positional accuracy level appropriate to the application at hand. For applications that require a higher level of positional accuracy and ground details, larger scale maps (e.g., 1:10,000) are needed as the source maps for the creation of GIS databases. Applications that require less ground details and a lower positional accuracy level can use smaller scale maps (e.g., 1:100,000) as the source maps during the database creation process. Given the same physical dimension, a smaller scale map covers a larger ground area and a larger scale map covers a smaller ground area. Therefore, a transportation study at the national or state level normally requires smaller scale maps than a transportation study at the local or site level. Vonderohe et al. (1993) summarize the relationships between different GIS-T applications and their associated map scales and accuracy requirements; see table 4-2.

### Transportation data capture and data products

As will become very clear in the remaining chapters of this book, transportation analysis and planning are data-intensive endeavors. Data requirements can range from the transportation system itself, the flows within the system, the characteristics of geographic locations that generate these flows and the physical environment encompassing the transportation system. Given the wide thematic range, map scales, and accuracy requirements for transportation analysis, it is

Table 4-2: GIS-T Applications and Their Associated Data Requirements

| Geographic Extent | Activity | Map Scale | Data Precision (feet) | Generalization Level |
|---|---|---|---|---|
| State | Statewide planning | 1:500,000 | 830 | High |
| Multidistrict | Corridor selection | 1:500,000; 1:100,000 | 830; 170 | |
| District | District planning | 1:100,000 | 170 | |
| Metropolitan area | Facilities management | 1:100,000; 1:24,000– 1:12,000 | 170; 40 to 30 | |
| Project | Corridor analysis | 1:24,000– 1:12,000 | 40 to 30 | |
| | Engineering design Construction | 1:1,200– 1:120 | 3 to 0.33 | Low |

*Source*: After Vonderohe et al. 1993.

often the case that these data will be a mixture of detailed primary data tailored for the specific project and secondary data acquired from public and private sources.

In this section, we will discuss methods for collecting primary transportation data and some sources of secondary transportation data. We focus discussion on geographic data capture and sources. We refer to reader to other transportation (e.g., Ortúzar and Willumsen 1994) or behavioral geography (e.g., Golledge and Stimson 1997) texts for discussions of broader transportation data collection issues, such as sampling socio-economic characteristics from a heterogeneous population or collecting travel diary data (although we will touch on the latter issue in chapter 8).

### Geographic data capture systems

*Overview*

In a U.S. National Cooperative Highway Research Program (NCHRP) report, Karimi, Hummer, and Khattak (2000) review existing and emerging technologies for capturing roadway inventory data. Although this is a limited domain, the guidelines can be generalized to other GIS-T applications. We will summarize their report below and provide some additional details on three key technologies. Their report discusses potential applications, costs, accuracy issues, technology status, and (U.S.) state DOT practices in considerable detail.

*Geographic data capture systems* (GDCS) include some combination of three major components. Since transportation systems are dispersed geographically, a key component is a *transport technology* to move the other system components to a location where data can be captured. There are four types of transport technologies used in transportation data collection modes, namely, *person-based technologies, vehicle-based technologies, airborne technologies* and *satellite-based technologies*. Person-based technologies include backpacks and (increasingly) handheld devices. Since it relies on human motive power, it can be slow and time consuming (although possibly enhanced using bicycles, golfcarts or similar personal transport devices). Vehicle-based and airborne technologies involve driving or flying the sensors to the locations being inventoried. Satellite-based methods include remote sensing imagery; the increasing resolution of these devices is leading to their greater use in collecting data across a wide range of transportation applications.

A second component in a GDCS is a *georeferencing technology* for recording the location of a geography entity relative to known locations with known georeferences. Appropriate georeferencing technologies for transportation data capture include *distance measuring instruments* (DMI), *inertial navigation systems* (INS), *rangefinders*, and the *global positioning system* (GPS). A DMI is a mechanical device attached to a vehicle that measures distance (but not direction) from a reference location based on the number of wheel rotations. While inexpensive, DMIs provide only relative locational references and require frequent calibration. INS use accelerometers and gyroscopes to capture pitch, roll and heading information to provide relative location based on distance and direction from a known point. An INS provides much higher positional accuracy but is more expensive. Rangefinders use the Doppler effect of a radar or laser beam to calculate distance from the device to a target. GPS uses time differences from several broadcast satellites to the receiver to estimate its location.

A third GDCS component is a *geographic description technology* for recording the attributes associated with the referenced location or entity. This can include *keyboards* or *voice recognition systems* for person-based data entry in the field. Digital image capture devices include still and video *digital cameras*. These can be personal, attached to a vehicle or an airborne platform, or mounted in fixed locations such as major intersections or modal transfer locations. Other image capturing devices include passive technologies that record electromagnetic energy reflected from the sun or thermal infrared energy emitting from the target, or active technologies that transmit microwave or laser energy to the target. We often refer to the scientific and engineering activities that include latter technologies as *remote sensing*. We can also include *flow sampling and monitoring devices*, such as traffic counters. Through communication and other information technologies, digital cameras and flow monitoring devices are being increasingly integrated with georeferencing technologies into intelligent transportation systems (see chapter 9).

In the next subsections, we will discuss three key GDCS technologies that are highly relevant to transportation analysis, presently and in the future. These technologies are the *GPS, remote sensing*, and *flow sampling and monitoring devices*.

*The global positioning system*

GPS, funded and controlled by the U.S. Department of Defense (USDOD), provides positioning and navigation service to virtually every location on the globe. GPS operates on the principle of trilateration. That is, an unknown location on the Earth's surface "is determined by measuring the length of the sides of a triangle between the unknown point and two or more known points (i.e., the satellites)" (French 1996, p. 33). The distance between a satellite and an unknown point on the Earth's surface is calculated by multiplying the speed of light and the time lag between a signal transmitted from the satellite and the signal received at the unknown location. In theory, communication between three separate satellites and the unknown location will be sufficient to derive the $(x, y, z)$ coordinates of the unknown point. However, due to the less-than-perfect readings from the clock in a GPS receiver, we require signals from a fourth satellite to derive an unambiguous $(x, y, z)$ reading.

GPS reached its initial operational capability in December 1993 and its full operational capability in July 1995. It consists of three segments: space segment, control segment, and user segment. The *space segment* now consists of 24 satellites orbiting the Earth, with 21 navigational satellites and three active spare satellites. They operate in six orbits, which are located at 20,200 km (10,900 nautical miles) above the Earth's surface, and repeat the same orbit with a 12-hour period. The *control segment* includes five monitor stations and three ground antennas. The monitor stations receive signals from the GPS satellites and upload data to the satellites for necessary corrections. The *user segment* consists of the GPS receivers and the user community. In 1984, U.S. President Reagan guaranteed free civilian access to the GPS signal.

In the past, GPS provided two types of services (Dana 2000b). The *Precise Positioning Service* (PPS), which offers a 22-m predictable horizontal accuracy and a 27.7-m predictable vertical accuracy, was available primarily to U.S. and allied military users. Civil users worldwide have access to the *Standard Positioning Service* (SPS), which provides a predictable horizontal accuracy of 100 m and a predictable vertical accuracy of 156 meters. The PPS is based on the P-Code (*Precise Code*) transmitted by the GPS satellites, while the SPS is based on the C/A Code (*Coarse Acquisition Code*). The difference in the accuracy levels between the PPS and the SPS is due to the intentional degradation of the SPS signals through the *selective availability* (SA) controlled by the USDOD. On May 1, 2000, the USDOD eliminated the SA program, allowing civil users full access to the PPS (with possible "grey-outs" or SA in some sensitive areas).

GPS has many applications in transportation. In addition to its use as a navigational tool, GPS can be used in an automated vehicle location system for hazardous material transport, fleet management, emergency vehicle dispatch, as well as intelligent transportation systems. GPS readings of longitude and latitude coordinates also provide an efficient method for collecting locations to create various point, line, and polygon GIS map layers. The GPSVan technology, which uses GPS to record locations and video logs along a highway, is a good example of using the GPS to help collect transportation infrastructure data in a moving vehicle (Novak and Nimz 1997). By mounting GPS receivers on test vehicles that record positions at frequent intervals, more accurate network travel time information can be collected (Guo and Poling 1995).

A major weakness of GPS for transportation data collection is the requirement for line-of-sight communication between the GPS receiver and a sufficient number of satellites. Tall buildings in urban area often block communication and make georeferencing via GPS impossible. In these cases, GPS must be combined with other georeferencing technologies.

*Remote sensing of transportation data*

*Remote sensing* (RS) refers to the scientific, engineering and application issues associated with capturing data about geographic features from distant locations. These "distant locations" are usually orbital (satellite) or suborbital (airborne) platforms, although some also include more proximal platforms such as data capture and georeferencing vehicle such as the GPSVan and mobile image platforms for recording pavement conditions.

RS is a large and sophisticated domain. While we will provide an overview, we cannot give remote sensing more than a cursory treatment. We refer the reader to good introductory texts such as Jensen (1996), Lillesand and Kiefer (2000), or Richards and Jia (1999). We will focus more attention on the specific problem of extracting transportation network data from this imagery and identify some potential transportation applications of new high-resolution RS systems. (Later in this chapter, we also discuss digital orthophotos, a RS product available from the U.S. Geological Survey).

GIS and RS are highly complementary. RS is a rich source of data for GIS. In addition, GIS can enhance the RS image correction and classification processes by supporting image processing, visualization and integrating ancillary information. Integrated RS and GIS can support analysis and decision making, traditionally for environmental applications but increasingly for transportation applications given new developments in the spatial, temporal and spectral resolution. We will briefly mention some of these issues below; for a more detailed discussion, see Wilkinson (1996).

*Choosing among imaging sensor technologies and platforms* There are two basic types of imaging sensor technologies, namely, active versus passive sensors. The two traditional RS platforms are orbital (satellite based) and airborne (aircraft based). Within this basic typology, there is a considerable amount of variation in RS products due to factors such as spatial, temporal, and spectral resolutions, accuracy, and scene size or coverage. We will briefly review the issues in choosing among RS sensors and platforms. We draw this discussion from Jensen (1996), Jensen and Cowen (1999), and workshop materials from Turner and Hansen (2000). Since these issues are considerably more involved than the cursory treatment here; we refer readers to the original sources. Also, since this is a rapidly changing field we also encourage readers to explore WWW sources for new material and products.

*Passive* RS devices are sensors that record electromagnetic energy reflected from the sun or thermal infrared energy emitting from the target. Passive systems are typically used to extract land cover and geological features in a study area. Passive systems include aerial and satellite-based photography as well as sensors that sample other (nonvisible) bands from the electromagnetic spectrum (see below). *Active* imaging sensors transmit energy to the target and record its reflectance. Active systems includes *radio detection and ranging* (RADAR) systems that use radio waves and *light detection and ranging* (LIDAR) systems that use laser energy. These systems are effective at extracting terrain and topographic information. RADAR and LIDAR can operate at night. RADAR can penetrate clouds and vegetation cover; LIDAR can penetrate some clouds.

Choosing among RS imaging sensors and platforms involves the following considerations. One issue is the *spatial resolution* or the area on the ground represented by one picture element or *pixel*. A pixel contains average reflectance values of spectral bands for a small area on the ground. Traditionally, passive sensors have limited spatial resolution, generally not higher than 10 m (e.g., SPOT, Landsat Thematic Mapper). However, new high-resolution passive sensing systems (e.g., IKONOS) can achieve spatial resolution down to 1 m, greatly increasing their potential for transportation analysis and other socioeconomic applications (see below). The platforms mentioned above are orbital; aerial photography can achieve spatial resolution to $0.25 \, m^2$ but for more limited spatial coverage. Many RADAR-based sensor systems have resolutions ranging from 8 to 50 m. *Synthetic aperture radar* (SAR), a system that uses the geometry of sensor movement on an aircraft to create a large "virtual" RADAR antenna, can presently achieve spatial resolutions down to the sub-meter level (see Lillesand and Kiefer 2000). LIDAR-based systems promise much greater than sub-meter accuracy.

A second consideration is the *spectral resolution* or the number and type of bands of the electromagnetic (EM) spectrum captured by the sensor. The major issue here is the type of information desired by the analyst. Different bands highlight different types of land cover, infrastructure, atmospheric conditions and geological features. More bands provide more information. The portions of the EM spectrum captured by passive systems includes *visible color* (V) (wavelength of 0.4–0.7 μm), *near-infrared* (NIR) (0.7–1.1 μm), *middle-infrared* (MIR) (1.5–2.5 μm) and *thermal infrared* (TIR) (3–12 μm). *Panchromatic* (P) sensors capture one broad band in the visible color range (0.5–0.7 μm) at a high level of spatial resolution. *Multispectral* sensor systems that capture three to seven bands across these ranges. New *hyperspectral* sensor systems such as the *Airborne Visible InfraRed Imaging Spectrometer* (AVIRIS) capture over 200 very narrow bands within the visible and infrared spectrum. This creates a very detailed spectral signature for each pixel, allowing discrimination to the subpixel level (i.e., the groundcover "mix" within a pixel). RADAR systems typically capture three different bands depending on the desired spatial resolution and depth of penetration through the vegetation cover. LIDAR is a single band.

A third consideration is the *temporal resolution* or revisit rate of the RS platform for the given study area. Airborne platforms theoretically have frequent revisit rates, subject to the difficulty and expense of scheduling flights. Satellite-based sensors traditionally have fixed orbits with visit rates ranging from 1 day to several weeks. As the name implies, geosynchronous satellites can constantly monitor a given study area. It is possible and will become more financially feasible in the near future to dedicate a geosyncronous RS platform to a particular area (e.g., the New York metropolitan area).

A fourth consideration is *scene size* or "footprint" of the coverage. Swath widths are variable depending on the platform and sensor configuration. Widths can range from several meters for airborne aerial photography/LIDAR to tens and hundreds of kilometers for satellite-based systems.

Jensen and Cowen (1999) and Karimi, Hummer, and Khattak (2000) provide good overviews of RS technologies and platforms for capturing socioeconomic and transportation data. Table 4-3 provides minimum resolution requirements for capturing human-made features; this is selected from table 1 in Jensen and Cowen (1999). The spectral abbreviations are as introduced above, with R referring to RADAR (the authors do not discuss AVIRIS or LIDAR). See Jensen and Cowen (1999) for additional detail and explanation of the spectral requirements.

*Land-cover/land-use* classes refer to the U.S. Geological Survey standard classifications; these range from generalized land-use/land-cover (Level I) to building and property line information (Level IV); see the USGS website for more information. *Digital elevation models* (DEMs) refer to elevation data sampled at regular locations (e.g., a "lattice" or *regular square grid* (RSG) in the 2-D plane). *Building and property* data include building footprints, property lines and building heights (using stereoscopic imagery). This can be combined with ancillary census or other socio-economic data to infer detailed land-use patterns. *Socioeconomic attributes* include population estimates based on building counts, urbanized land and land-cover/land-use. With sufficient spatial resolution, it is also possible to measure quality-of-life indicators based on housing characteristics and situational attributes (e.g., access to schools and open space). Stable and

Table 4-3: Minimum Resolution Requirements for Extracting Socioeconomic and Transportation Data from Remotely Sensed Imagery

| Human Features | Temporal | Spatial (m) | Spectral |
|---|---|---|---|
| Land cover/land use USGS level I | 5–10 y | 20–100 | V, NIR, MIR, R |
| USGS level II | 5–10 y | 5–20 | V, NIR, MIR, R |
| USGS level III | 3–5 y | 1–5 | P, V, NIR, MIR |
| USGS level IV | 1–3 y | 0.25–1 | P |
| Elevation Large-scale DEM | 5–10 y | 0.25–0.5 | P, V |
| Large-scale slope map | 5–10 y | 0.25–0.5 | P, V |
| Building and property | 1–5 y | 0.25–0.5 | P, V |
| Socio-economic attributes | | | |
| Local population estimates | 5–7 y | 0.25–5 | P, V NIR |
| Regional/national population estimates | 5–15 y | 5–20 | P, V, NIR |
| Sensitive environmental areas Stable | 1–2 y | 1–10 | V, NIR, MIR |
| Dynamic | 1–6 mo | 0.25–2 | V, NIR, MIR, TIR |
| Emergency response Pre-emergency | 1–5 y | 1–5 | P, V NIR |
| Post-emergency | 12 hr–2 d | 0.25–2 | P, V, NIR, R |
| Damaged transportation, housing, utilities, and services | 1–2 d | 0.25–1 | P, V, NIR |
| Transportation infrastructure | | | |
| Road centerline | 1–5 y | 1–30 | P, V, NIR |
| Road width | 1–2 y | 0.25–0.5 | P, V |
| Traffic counts | 5–10 min | 0.25–5 | P, V |
| Parking studies | 10–60 min | 0.25–0.5 | P, V |

*Source*: After Jensen and Cowen 1999.

dynamic (changing) *sensitive environmental areas* includes wetlands, endangered species habitats and parks. *Emergency response* attributes include a predisaster imagery database for emergency planning and rapid postdisaster imagery to assess damage to transportation, utilities and other service lifelines required for emergency response and disaster recovery.

Jensen and Cowen (1999) specify the minimum resolution requirements for deriving transportation features such as road centerline, road width, traffic counts and parking studies. Road centerline data are a possible source for a road network database (or updating secondary network data sources). Road width data allow extraction of geometric data for some of the advanced navigable databases discussed in the previous chapter. Extracting individual vehicles for traffic and parking studies requires high spatial and temporal resolutions that are only now being achieved by RS systems. Karimi, Hummer, and Khattak (2000) also provide a detailed list of roadway geoemtric features obtainable by high spatial resolution RS (spatial resolution greater than 1 m).

*Extracting transportation information from remotely sensed data*   General steps for extracting thematic and feature information from RS imagery include assessing image quality, image correction/registration and image interpretation (Jensen 1996; Richards and Jia 1999). *Assessing image quality* includes using simple statistics and visual inspection to see if the general image properties are inline with expectations. *Image correction/registration* includes radiometric and geometric correction. Radiometric correction attempts to compensate for differences between the recorded spectral signature and the true spectral signature on the ground due to atmospheric scattering and absorption, attenuation caused by terrain, and noise in the scanner. Geometric correction attempts to compensate for distortions caused by the motion of the sensor platform, the rotation of the Earth, changes in altitude (e.g., due to platform motion or changes in topography) and the perspective of the sensor. *Image registration* refers to referencing the imagery to a geographic coordinate system so that the pixels can be addressed as geographic locations. This can be done with respect to a map or another registered image.

*Interpreting* digital imagery includes classification of pixels and feature extraction based on the resulting pixel classes. *Classification* translates each pixel into a group or class based on its spectral signature. There are three general strategies: (i) *hard classification*, (ii) *soft classification*, (iii) *hybrid approaches* using ancillary information (Jensen 1996). Hard classification methods are usually statistical. Hard classification can also be supervised or unsupervised. *Supervised* methods require the analyst to calibrate the imagery spectral signatures to known geographic features in the scene. *Unsupervised* methods do not presume this type of knowledge; these methods generally examine relative differences among each pixel's signature. Fuzzy set theory provides the basis for soft classification; this allows a pixel to have a continuous membership function describing the likelihood (but *not* the probability) of a pixel being in each class (see chapter 10 for a brief discussion of fuzzy set theory). A GIS enhances the ability to integrate ancillary information (e.g., a soils map, census data) into the classification process using hybrid methods (see Mesev, Longley, and Batty 1996; Wilkinson 1996).

Extracting transportation network information is a special case of the more general problem of extracting linear features from RS imagery (Wang, Treitz, and Howarth 1992). Linear feature extraction methods include *filtering techniques* that use numeric masks (such as a raster data layer or a small raster window) to convolute the image and detect "edges" or abrupt changes in spectral signatures (e.g., Nevatia and Babu 1980). This can be enhanced using contextual information such as the size, direction and connectivity relationships among linear features (see Guerney 1980). *Mathematical methods* include calculating the gradient or rate of steepest descent in brightness from a pixel. Since roads are linear, they exhibit strong and abrupt changes in brightness with direction (see Wang, Treitz, and Howarth 1992). *Knowledge-based systems* use codified knowledge such as rules derived from experts to automate the network extraction process (see Swann et al. 1988; also see chapter 10 for a discussion of knowledge-based GIS).

*Extracting geographic measurements from imagery* Photogrammetry is the science and technology of obtaining spatial measurements and other geometrically reliable products from aerial and satellite-based photography (Lillesand and Kiefer 2000). The high spatial resolution of aerial photographs makes this a useful source for data to support transportation design and engineering. Air photo imagery is also increasingly used as image backdrops for other GIS data and to create 2-D and 3-D animated "fly-throughs" for geographic visualization. Although other RS systems are catching up with respect to spatial resolution, the low cost and flexibility of aerial photography means that these systems will still be competitive for transportation project data acquisition for the foreseeable future.

Aerial photography generally captures the panchromatic, color and infrared portions of the electromagnetic spectrum. We can also classify aerial photography platforms based on their geometric relationship with the ground. *Oblique* photographs have an intentional inclination of the camera axis with respect to the ground. High oblique photography captures the horizon while low oblique photography does not. *Vertical* photographs are made with the camera axis as orthogonal as possible with respect to the ground (Lillesand and Kiefer 2000).

Before vertical aerial photography can be used for engineering calculations, we must compensate for distortions caused by the geometry of the photography platform, terrain and objects within the terrain. Although ideally orthogonal, minor tilt (on the order of 1–3 degrees) can occur due to unintended aircraft movement. Also, unless the plane is flying over uniformly flat terrain, variations in terrain elevation create variations in scale (the ratio of distance on the photograph to distance on the ground). Apparent scale varies proportionally to the height of terrain on the ground. *Relief displacement* is the apparent "leaning" of vertical features (e.g., trees and building) away from the image center with distance from that center. While this is a distortion, it can also be exploited to measure the height of those objects. All of this requires accurate registration to ground control points with known locations so that the exact position and orientation of the camera at the time of exposure can be determined (Lillesand and Kiefer 2000).

Vertical aerial photographs can be the basis for accurate, high-resolution maps. *Stereopairs* are paired photographs that cover the same geographic extent at slightly different angles. This creates a phenomenon known as *image parallax* or the apparent change in relative position caused by a change in observer angle. Since the change in observer angle is known in stereopairs, this phenomenon can be exploited to accurately measure object heights, terrain, and plot maps. Vertical photographs can also be combined with ancillary terrain information available to derive photographic images with geometric accuracy similar to planimetric maps. These images, known as *orthophotos*, have constant scale and no tilt or relief distortion. These products have traditionally been derived from analog methods such as stereoplotters. However, GIS and software for processing digital imagery have made this process easier and less error-prone (Lillesand and Kiefer 2000).

*Flow and interaction sampling*

Capturing demand data for transportation systems involves sampling and monitoring flow and interaction data. These flow data are usually captured using traffic counters or embedded loop detectors at selected network locations. Flow sampling in geographic space often occurs by designating screen-lines, cordons, or other interception points at *Traffic Analysis Zones* (TAZ) boundaries in the study area. New technologies such as digital video cameras with automated image processing and self-reporting from "smart vehicles" at toll booths or similar stations are increasing the volume and spectrum of flow data available to the transportation analyst. Since flow and interaction often exhibits strong geographic patterns, this requires choosing the locations for sampling or monitoring so that the sampled flow is representative of the population. A related problem is estimating flow and interaction patterns given observations at fixed geographic locations.

Determining good locations for flow sampling or estimating flows based on geographic patterns are related to the general spatial analysis problems of *boundary effects* and *spatial sampling*. Miller (1999b) reviews these related problems. We will summarize this discussion below; for more detail and references, see the original source.

Boundary problems in spatial analysis result when we impose artificial and arbitrary boundaries on an unbounded geographic process. An example is delimiting a study area and the composite traffic analysis zoning system for travel demand forecasting. (This is also related to another spatial analysis problem known as the modifiable areal unit problem; we discuss this later in the chapter.) Arbitrary boundaries create edge effects by ignoring interdependencies and interactions outside the study area (Fotheringham and Rogerson 1993; Griffith 1983, 1985; Martin 1987). Boundaries also impose arbitrary shape effects that affect the measurement and analysis of flow patterns and other spatial properties such as clustering (Ferguson and Kanaroglou 1998; Griffith 1982). A possible solution to this problem is to reestimate the geographic process under random realizations of the boundaries to generate an error distribution for the measure (see Haslett, Wills, and Unwin 1990; Openshaw, Charlton, and Wymer 1987).

With cordon-based sampling, boundary effects occur since flow is only recorded when it crosses a boundary. As the size of a zone increases, the likelihood that shorter trips will be missed and underrepresented in the sample also increases. Methods are available for adjusting flow property estimates (such as mean travel distance) based on zonal geometry and assumptions regarding the pattern of travel demand within each zone. (see Ord and Cliff 1976; Rogerson 1990). Kirby (1997) develops methods for the special case of a regular grid of interception boundaries; this analytical approach can be modified for irregular zonal boundaries using GIS tools and ancillary geographic demand data (e.g., a high-resolution RS image showing housing density).

Locating flow monitors within a network or estimating demands based on flows sampled at given network locations is a specialized type of spatial sampling problem. Flow detectors provide an inexpensive mode for estimating demand data such as origin-destination matrices. If a study area consists of $m$ TAZs (where $m$ is some integer), the required O-D matrix will have either $m^2$ or $(m^2 - m)$ cells depending on whether intra-zonal flows are included. Theoretically, at least $m^2$ or $(m^2 - m)$ consistent and independent samples are required to uniquely determine the O-D matrix. In practice, the number of traffic counters is lower and the samples are not consistent or independent. Inconsistent samples occur due to measurement error in the recording device and nonsimultaneous sampling. Flow sample dependencies can occur when a detector is "downstream" along some traveled path from another detector.

Methods for estimating O-D flows based on limited samples include statistical estimation, Bayesian updating and entropy maximization (see Cascetta and Nguyen 1988; Ortúzar and Willumsen 1994 for reviews of these methods). Inconsistent samples can be accommodated by treating the flow on each arc as a random variable in the estimation process (see Cascetta and Nguyen 1988). Bell (1983) discusses a method for detecting dependencies among flow samples.

Instead of detecting flow sample dependencies, a better strategy is to locate the detectors so that flow sample dependencies are minimized. Unfortunately, there are no clear guidelines or methods for this task. In chapter 6, we will discuss some *flow intercepting facility location methods* that attempt to site facilities to capture the maximum amount of flow in the network (e.g., for locating vehicle inspection stations). These methods can be adapted for the sampling problem as well. Black (1992) develops a network autocorrelation statistic that can be adapted for this purpose to identify independent links based on a previous flow matrix. We will discuss this method briefly in chapter 10.

### Transportation data products

Transportation data products refer to the data sets collected by public sector or private sector entities and made available to the GIS-T user community free of charge or with a fee. Since data collection and database creation often constitute the most time-consuming and labor-intensive parts of a GIS project, it is normally more efficient to use secondary data sources whenever they are available and meet the accuracy requirements.

This section illustrates the major GIS-T data products that are available from public and private sector sources. Given space limitations, we cannot cover all data products for all regions of the world. Instead, we will focus on major and relatively stable (persistent over time) data products available in the United States and Europe. Hopefully, this will still be helpful to readers in other regions of the world by suggesting the types of products that can be available.

A challenge of reporting the available data sources is the nature of constant changes in their status. We anticipate that some of the information reported in this section may be out of date by printing. Fortunately, most of the data sources listed below have a Web site that readers can retrieve the latest information on data availability. Since WWW universal resource locators (URLs) are volatile, we do not provide these addresses here. However, the WWW sites can be found easily using a good search engine and the keywords gleaned from the discussion below.

### Public sector data products

*Topologically Integrated Geographic Encoding and Referencing (TIGER) files*
To support its mapping needs for the 1990 decennial census and other programs, the United States Census Bureau developed a digital database, known as the *Topologically Integrated Geographic Encoding and Referencing* (TIGER) system, during the 1980s. The TIGER digital database replaced the *Dual Independent Map Encoding* (DIME) files used in the 1980 decennial census by the Census Bureau. The TIGER system supports the following operations at the Census Bureau:

- creation and maintenance of the digital geographic database that includes complete coverage of the United States and its territories
- production of maps from the TIGER database for all Census Bureau enumeration and publication programs
- assignment of individual addresses to geographic entities and census blocks based on polygons formed by features such as roads and streams

The TIGER database employs a topological structure that defines the location and relationships of streets, rivers, railroads, and other features to each other and to various geographic entities used in the census enumeration.

The U.S. Census Bureau releases periodic extracts of the TIGER database to the public. These extracts are known as the TIGER/Line files. Each new release incorporates data updates and other changes made to the previous version. Coordinates used in the TIGER/Line files are longitudes and latitudes. For the 1994 and earlier versions, the North America Datum 1927 (NAD27) was used. It was changed to the North America Datum 1983 (NAD83) starting from the 1995 version of TIGER/Line files.

The TIGER/Line files use a collection of spatial objects represented as points, lines, and polygons. One major advantage of using the TIGER/Line files in a GIS environment is that the spatial objects are identified with *Federal Information Processing Standards* (FIPS) codes and Census Bureau codes of legal and statistical entities. Therefore, it is easy to link the TIGER/Line files with census data, such as the Summary Tape Files (STFs) and the Public Law (PL) 94–171 data files. Transportation features included in the TIGER/Line files can be identified through the *Census Feature Class Codes* (CFCC).

Since TIGER/Line files include transportation features and census enumeration units, they are frequently used in GIS-T projects. TIGER/Line files can be used to meet three major needs in a GIS-T project. First, they can be imported into GIS software to create various GIS map layers of transportation facilities and census enumeration boundaries. These GIS layers created from TIGER/Line files can serve as initial GIS databases for quick analysis and, if needed, additional enhancements. Second, since TIGER/Line files consist of street address range data, they are also a readily available source for performing address-matching. Third, once census enumeration boundary layers are created from the TIGER/Line files, they can be linked to census data sets through the unique identification codes for different levels of census enumeration units. The linkages to census data sets offer a wide range of demographic and socioeconomic data items that could be useful in various GIS-T projects.

*National Transportation Atlas* The Bureau of Transportation Statistics (BTS) is an operating administration of the U.S. Department of Transportation (USDOT). Due to the mandates of the Intermodal Surface Transportation Efficiency Act (ISTEA) of 1991, BTS was established in December 1992 to collect, compile, analyze, and make accessible information on the nation's transportation systems. Various kinds of transportation data are available free at the Bureau of Transportation Statistics. One data set, the National Transportation Atlas (NTA) databases, is discussed here. Readers are encouraged to visit the BTS web site for a complete list of available data.

The *National Transportation Atlas* (NTA) is a collection of geo-spatial databases developed by USDOT and other federal agencies that depict nationally significant transportation facilities, networks, and geographic entities. Transportation facilities are normally represented as point features that depict the location of a transportation terminal, intermodal transfer facility, or key transportation structure such as a bridge, tunnel, or navigation aid. Transportation networks are topologically connected line databases of the locations and centerline alignments of nationally significant roads, railroads, waterways, airways, and transit routes. Attributes associated with transportation networks include route names or numbers, capacity measures, various network classifications, and traffic volumes. Network files available at the BTS web site include the National Highway Planning Network, National Railway Network, National Waterway Network, and National Transit Network. All U.S. transportation networks in the NTA are constructed at a geographic accuracy level equivalent to a 1:100,000 scale map.

*Digital line graphs* The U.S. Geological Survey (USGS) is a lead U.S. federal agency for the collection and distribution of digital geospatial data sets. Three major types of digital geospatial data sets are of interests to transportation applications: digital line graphs (DLGs), digital elevation models (DEMs), and land use and land cover data sets. Both the horizontal and the vertical positional accuracy levels of USGS digital data sets are compiled to meet the *National Map Accuracy Standards* (NMAS). NMAS horizontal accuracy requires that at least 90 percent of well defined points tested be within 1/50 inch of the true position for maps at 1:20,000 scale or smaller, and be within 1/30 inch of the true position for maps larger than 1:20,000 scale. NMAS vertical accuracy requires that at least 90% of well-defined points tested be within one half contour interval of the correct value. In general, the scale and accuracy levels associated with the USGS digital geospatial data sets are more appropriate for transportation planning studies than for transportation engineering studies.

*Digital Line Graph* (DLG) files are digital vector representations of topographic and planimetric map features that are derived from USGS maps or aerial photographs. All DLG data distributed by USGS are DLG-Level 3 (DLG-3), which means the data contain a full range of attribute codes, have full topological structuring, and have passed certain quality-control checks.

DLG files are available in *optional* and *SDTS* (Spatial Data Transfer Standards, discussed in the data standards section of this chapter) formats. The optional format consists of topological linkages that are explicitly encoded for node, line, and area elements. The SDTS format is designed to transfer spatial data with complete content transfer (no loss of information). It also includes data quality reports that provide metadata and documentation of data processing.

*Digital elevation model (DEM)*  Digital Elevation Model (DEM) files distributed by the USGS consist of elevation data at regularly spaced points. They are available at 7.5-minute, 15-minute, 30-minute, and 1-degree corresponding to the various scales of USGS topographic quadrangle maps. The 15-minute and 30-minute DEMS are referenced to geographic coordinates (latitude/longitude) while the 7.5 minute DEM is referenced to the UTM coordinate system. The U.S. Defense Mapping Agency produces the 1-degree DEM data. These are referenced to the geographic coordinate system of the World Geodetic System 1972 Datum. A few units are also available in the World Geodetic System 1984 Datum. DEM files can be used in the generation of graphics such as isometric projections displaying slope, aspect, and terrain surfaces. They may also be used in analysis such as cut-and-fill, environmental impact studies and other terrain modeling.

The USGS term DEM is sometimes co-opted to refer to all elevation models with data sampled as a *regular square grid* (RSG) in the 2-D plane. Some in the GISci literature (particularly computer scientists) use the more appropriate generic term of RSG to refer to this data model. Another possibility is to sample elevation data at irregular locations, for example, to record more information at locations with higher relief using a *triangulated irregular network* (TIN) (see Boots 1999; De Floriani and Magillo 1999). We will use RSG and TIN when referring to surface data sampled at regular and irregular elevation data models, respectively. The surface will generally (but not always) be a terrain. *Digital terrain model* (DTM) will be the generic term referring to both approaches to modeling terrain. DEM is reserved for the USGS product.

*Land use and land cover*  The USGS *Land Use and Land Cover* (LULC) digital data sets describe the vegetation, water, natural surface, and cultural features on the land surface. Manual interpretation of aerial photographs ac[quired from] high-altitude platforms serves as the primary source of compiling [these] maps, supplemented by earlier land use maps and field surveys as [second]ary data sources. These maps were then digitized to create the digital databases.

The LULC mapping program is designed so that standard USGS 1:250,000 topographic maps can be used for compilation and organization of the land use and land cover data. In some areas, such as Hawaii, 1:100,000 scale topographic maps are also used. All LULC digital data conform to the UTM projection. The minimum area representing the human-made features of the LULC polygons are 10 acres (approximately 4 hectares) that have a minimum width of 600 feet (approximately 200 meters). Nonurban and non-human-made features may be mapped with polygons with a minimum area of 40 acres (approximately 16 hectares) that have a minimum width of 1,320 feet (approximately 400 meters).

The LULC digital data are released in two different formats. The first format was developed as part of the *Geographic Information Retrieval and Analysis System* (GIRAS). Some commercial GIS packages provide a conversion program to read the GIRAS files. The other format is the *Composite Theme Grid* (CTG) format. This format is grid cells instead of polygons. The center point of each cell is 200 m apart from other center points in adjacent cells.

*Digital orthophotos*    A digital orthophoto is a geo-referenced image derived from a digitized perspective aerial photograph, or other remotely sensed image data, by differential rectification so that image displacements caused by camera tilt and terrain relief are removed. USGS and several other U.S. federal agencies are partners of the National Digital Orthophoto Program (NDOP). This program is oriented primarily toward the production of 1-m ground resolution gray-scale and color-infrared *digital orthophoto quarter-quadrangles* (DOQQ, 3.75-minute × 3.75-minute in extent) from 1:40,000-scale National Aerial Photography Program (NAPP) or NAPP-like photography. However, 1- or 2-m ground resolution digital orthophoto quadrangles (DOQ, 7.5-minute × 7.5-minute in extent) are also available at the USGS. It should be noted that the terms of 1-m or 2-m ground resolution refer to the distance on the ground represented by each pixel in the *x* and *y* directions. They do not imply the minimum distance that can be detected between two adjacent features.

Orthophotos combine the image characteristics of a photograph with the geometric qualities of a map. They are produced by scanning aerial photographs and then processed with additional information to correct the image displacements. Types of input data required to complete the process include: (i) the unrectified image, (ii) a digital terrain model of the same geographic coverage, (iii) the coordinates of ground control points, (iv) calibration information about the sensor collector device. These inputs are used to register the image file to the ground locations and to remove the image displacements. Therefore, digital orthophotos can be used as a background display layer in GIS or serve as the basis for the creation or updating of vector GIS layers.

Both digital orthophoto quarter-quadrangles (3.75-minute) and quadrangles (7.5-minute) are based on the UTM projection, with coordinates in meters.

owever, quarter-quadrangle orthophotos are registered to NAD83 and quadrangle orthophotos are cast on either NAD27 or NAD83. Digital orthophoto quadrangles and quarter-quadrangles meet the horizontal NMAS at 1:24,000 and 1:12,000 scale, respectively. The vertical accuracy of the source DEM must be equivalent or better than a level 1 DEM, with a root-mean-square-error (RMSE) of no greater than 7 m.

*U.S. National Spatial Data Infrastructure (NSDI)*   Executive Order 12906, signed by U.S. President Clinton in 1994, calls for the establishment of the *National Spatial Data Infrastructure* (NSDI) in the United States. The NSDI is defined as the technology, policies, standards, and human resources necessary to acquire, process, store, distribute, and improve utilization of geospatial data (U.S. Government 1994). The *Federal Geographic Data Committee* (FGDC) is charged to take the lead role of implementing the NSDI by involving state and local governments, the academia, and the private sector in the United States. The Executive Order also requires the FGDC establish an electronic *National Geospatial Data Clearinghouse* for the NSDI. The Geospatial Data Clearinghouse now consists of a collection of over 100 spatial data servers and is available on the Internet for searching for available digital geographic data. The search is based on the metadata (i.e., "data about data" discussed in chapter 2 and later in this chapter) of the digital geographic data files and returns information on how to acquire or download the files.

*U.K. Ordnance Survey*   As the Britain's National Mapping Agency, Ordnance Survey provides mapping products and services that range from paper maps, digital cartographic data to GIS data. Table 4-4 summarizes the major Ordnance Survey products.

OSCAR consists of four data products that comprise a vector network of arcs and nodes for the entire United Kingdom. Several grid references are used to position each arc such that the true shape of the road is represented. In addition, every arc and node has a unique reference number that links to the relevant information such as road name, road classification, road length, junction identifier, among others. The scales of source data are at 1:1,250 for urban areas, 1:2,500 for rural areas, and 1:10,000 for mountain, moorland, and some coastal areas.

The *Asset-Manager* is the most detailed of the four road products, providing a comprehensive representation of all public and selected private roads. Its level of detail makes the product suitable for inventory and maintenance of the road network and management of national street gazetteers. The *Traffic-Manager* includes all public and selected private roads, but the representation of road layouts has been filtered and road junctions are structured to mirror the way vehicles pass through them. Therefore, this product is better for vehicle navigation and guidance applications. The *Route-Manager* is designed for regional routing and networking requirements. It includes all of the public roads, with the exception of short cul-de-sacs. Complex road junctions are simplified as single nodes and multi-carriageways are represented as single links. Market area analysis and logistics applications are good candidates for using this product. The *Network-*

Table 4-4: U.K. Ordnance Survey Digital Data Products

| Data Product | Characteristics |
| --- | --- |
| Land-line | Multiple features including road centerlines, road names, and numbers; 1:1,250 for urban areas, 1:2,500 for rural areas; and 1:10,000 for mountain, moorland, and some coastal areas |
| Meridian | Regionally focused digital map data (including roads and railways) that bridges the gap between large and small scale mapping |
| Strategi | 1:250,000 scale vector digital map data, including roads, railway, airports and ferries |
| BaseData.GB | 1:625,000 scale vector digital map data suitable for network analysis only at national scales |
| Boundary-Line | Electoral and administrative boundaries at 1:10,000 scale; can be linked to census data |
| ED-LINE | U.K. census boundary data sets at 1:10,000 scale |
| ADDRESS-POINT | National grid coordinates and unique reference codes for every postal address in England, Scotland, and Wales; location reference accuracy is 0.1 meter |
| PANRORAMA and PROFILE | Digital elevation data at 1:50,000 and 1:10,000 scales, respectively |
| OSCAR | A family of road network products for the entire U.K. |

*Source*: U.K. Ordinance Survey website.

*Manager* has simplified road geometry relative to the other three road products. The reduced data volume makes it an ideal product for network analysis and route planning in a national-scale study.

*Multi-purpose European Ground-Related Information Network (MEGRIN)*
*Multi-purpose European Ground-Related Information Network* (MEGRIN) is an organization of 19 European National Mapping Agencies (NMAs) that aims to meet the needs of the pan-European digital map data market.

There are two major types of activities performed by MEGRIN. The first activity is to provide a single point of contact for metadata information about the digital map data available in European countries, which is known as the Geographic Data Description Directory. The second activity is to create high quality pan-European data sets based on NMAs data, which include the Seamless Administrative Boundaries of Europe (SABE) and Pathfinder towards the European Topographic Information Template (PETIT) projects.

MEGRIN's *Geographic Data Description Directory* (GDDD) metadata service provides information about 250 digital data sets available from the National Mapping Agencies (NMAs) of 22 countries of Europe. The aim of the project is to provide users with an up-to-date information on availability of digital geographical data in Europe (Bitenc 1994). Web pages are derived from the database directly, and the information is made available free of charge to all Internet users. The GDDD is a kind of an Internet search engine gateway dedicated to European Geographic Information.

Users of the GDDD can access the data set descriptions by selecting a list of suppliers, a list of standard product types, or a country from map display on the web page. Information in the GDDD falls into the following categories:

- *overview*—short abstract, including contact address (organization, web site address and person)
- *commercial information*—contains some commercial details of coverage, copyright, format, price, and other conditions related to the use of the data sets
- *technical information*—describes the technical specifications of data sources, features and content, updates, data accuracy, and other data quality parameters
- *descriptions of the NMAs*

The GDDD enables international users to locate data sets of interest, and contact the appropriate NMA.

*SABE* The *Seamless Administrative Boundaries of Europe* (SABE) data set provides detailed administrative boundary data for 26 European nations from the County level to Ward/Commune/Gemeinde/Termino Municipal (NUTS 5) level, with a total of over 100,000 polygons. It also contains unique identifiers from National Mapping Agencies for links to European Union census data. The standard release format of SABE is in Environmental Systems Research Institute's ArcInfo format, while other formats are available as customized products.

*PETIT* *Pathfinder towards the European Topographic Information Template* (PETIT) is a project in the European Commission's INFO2000 Programme. The project is as a first stage in the development of a consistent pan-European topological dataset that would be easily accessible to users with a single license agreement. The objective of PETIT project is to assess the market requirement for pan-European topographic data and the feasibility of meeting these requirements.

The PETIT project places an emphasis on the understanding of the user requirements for pan-European topographic data. It creates a prototype dataset to test the suitability of its content and specification for the proposed pan-European topological dataset. This prototype topographic dataset is organized into 9 thematic layers (including transportation and utilities layers) and covers the parts of several European countries. A web demonstrator based on a subset of the PETIT prototype dataset is available for potential users to contribute to the definition of user requirements.

*Other sources*

There are many other professional associations, organizations, and private sector companies that provide GIS data. This section briefly describes a select set of these agencies. We should note that these sites are not specific to GIS-T data; however, they provide good starting points for searching GIS data available in different parts of the world.

*Geographical Information Systems International Group (GISIG)* *Geographical Information Systems International Group* (GISIG) is a nonprofit association of over 100 organizations from European countries. Its web site offers a Digital Data Inventory that can be searched by country or by other criteria.

*Etak, Inc.* EtakMap Premium digital maps are ready-to-use databases of geographic information covering the United States. They are offered in standard and enhanced versions. The standard version consists of streets, street address, and ZIP+4 data. The enhanced version also includes road classification and enhanced graphic features such as shape points, water areas, political and statistical geography, and major landmark point features. The positional accuracy level is based on 1:24,000 USGS quadrangle maps in urban areas (absolute accuracy of 40 feet) and 1:100,000 USGS quadrangle maps in rural area (absolute accuracy of 160 feet). EtakMap databases are available for other countries.

Etak also offers the EtakGuide; this is a digital travel guide for car navigation systems. It combines road information with travel guide information of parks, hotels, airports, cultural facilities, restaurants, and so forth. It allows the user in a car to select a destination and receive a navigational guidance. EtakGuide will show the car's progress in real-time as it navigates to the destination.

*Geographic Data Technology, Inc.* Geographic Data Technology (GDT) offers a family of Dynamap digital maps. Besides the main products of address-based street network databases for display as a basemap layer or for geocoding, GDT also offers a set of transportation related products such as Dynamap/Highways and Major Roads, Dynamap/Traffic Counts, Dynamic/Routing, and Dynamap/Transportation. These products are enhanced versions of the street network database that include additional features (e.g., turn restrictions, ramp structures, traffic count locations, etc.) to facilitate GIS applications such as mapping, geocoding, traffic analysis, and routing.

*Navigation Technologies, Inc.* Navigation Technologies sells the NavTech Database, which is widely used in car navigation and route guidance products. The database provides information on primary and alternate street names, address ranges, driving rules (including one-way and turn restrictions), and points of interest in more than 40 categories, such as restaurants, hotels, gas stations, among others. The NavTech U.S. database offers detailed, street-level coverage for most urban areas in the U.S. and inter-city coverage for the contiguous 48 states. The NavTech Europe database covers some European countries.

## Transportation data integration

Most GIS projects involve data from multiple sources with different formats and varying data quality. The *data integration* process attempts to bring diverse data sets into conformity and is therefore a critical step to ensuring a successful GIS project. Although GIS-T shares common data integration challenges with other GIS projects, it also presents some unique challenges with respect to integrating transportation data (Dueker and Butler 1999).

This section discusses data integration issues that are relevant to GIS-T. We will first discuss *data standards* that attempt to achieve consistency among different databases. We will then discuss methods for data integration in GIS-T; these include *areal interpolation*, *address matching*, and *network conflation*.

## Data standards

### U.S. data standards

Many transportation agencies recognize the value of standards to ensure consistency within an organization and over time as personnel and management change. For example, the Florida Department of Transportation proposed a set of GIS standards that encompasses a wide range of GIS-T activities. The proposed standards include street network data, cartographic outputs, linear referencing systems, metadata, data collection, data archiving, data conversion, application frameworks, and graphic user interfaces (GUIs) (FDOT 1996).

In recent years, the United States federal government has developed spatial data standards that are required for all federal spatial data activities and entities receiving federal support (particularly U.S. state governments). These standards are critical for any transportation agency that uses data from U.S. governmental agencies and can also serve as a basis for data sharing among nonfederal agencies.

*Spatial Data Transfer Standard (SDTS)* The purpose of the U.S. SDTS is to promote and facilitate the transfer of digital spatial data between dissimilar computer systems, while preserving meaning and minimizing information loss. Its philosophy is based on the concept of self-contained transfer that includes spatial data, attribute, georeferencing, data quality report, data dictionary, and other supporting metadata in the transfer process. Potential benefits of the SDTS are increased access to and sharing of spatial data, reduced information loss in data exchange, and increased quality and integrity of spatial data.

The SDTS is organized into the *base specifications* and *multiple profiles* (see the special issue on SDTS, *Cartography and Geographic Information Systems* 1992, vol. 19, no. 5 for an overview and additional information). The base specifications currently cover logical specifications, spatial features, and using the general-purpose International Organization for Standardization (ISO) 8211 data exchange standard in an SDTS transfer. An *SDTS profile* is a well-defined subset of SDTS created for translating a specific type or model of spatial data. For example, the *Topological Vector Profile* (TVP) applies to geographic vector data with planar graph topology that is commonly used in most vector GIS software. Additional profiles have been under development and are expected to become part of the SDTS in the future. Among them, the *Transportation Network Profile* (TNP) is of the most interest to the GIS-T user community. TNP is designed for use with vector geographic data with network topology. The data are represented as vector objects that comprise a network or planar graph.

Several leading commercial GIS software products have included conversion programs between the SDTS and their native formats. There also exists the SDTS++, which is a C++ toolkit, for developers to write applications that can read and write SDTS data sets. As more federal agencies make their digital spatial data available in the SDTS format, it is likely that the U.S. GIS community will use the SDTS as a common data exchange format.

*Content Standards for Digital Geospatial Metadata* Many agencies discover that over time or after personnel changes, they no longer know the content or quality of legacy data sets. It is also common that many agencies that acquired GIS data sets from other sources with minimal documentation about the data. In both cases, the use of such data sets in an analysis or for making a decision becomes questionable. To maintain an agency's investment in its geospatial data and to encourage data sharing, U.S. President Clinton signed Executive Order 12906, "Coordinating Geographic Data Acquisition and Access: The National Spatial Data Infrastructure (NSDI)," on April 11, 1994. This executive order instructs U.S. federal agencies to use a standard to document new geospatial data beginning in 1995, and to provide the metadata to the public through the National Geospatial Data Clearinghouse.

On June 8, 1994, the U.S. *Federal Geographic Data Committee* (FGDC) approved the "Content Standards for Digital Geospatial Metadata," originally initiated in June 1992. Metadata describe the history, content, quality, condition, and other characteristics of data. The purpose is to provide a common set of terminology and definitions for metadata documentation. Specifically, the following metadata are defined in the *Metadata Content Standards* (MCS):

- *Identification Information*—describes the basic information about the data set, such as the title, the geographic area covered, and rules for acquiring or using the data.
- *Data Quality Information*—provides an assessment of the quality of the data set, such as positional and attribute accuracy, data completeness, data consistency, data sources, and methods used to produce the data. This data quality information follows the specifications in the SDTS.
- *Spatial Data Organization Information*—describes the mechanism used to represent spatial information in the data set, such as the vector or raster representation, street addresses or administration unit codes, and the number of spatial objects in the data set.
- *Spatial Reference Information*—describes the reference frame and encoding methods, such as the name of and parameters for map projections or coordinate systems, horizontal and vertical datums, and the coordinate system resolution.
- *Entity and Attribute Information*—provides information about the content of the data set, such as the names and definitions of features, attributes, and attribute values.
- *Distribution Information*—provides information about data distribution, such as contact for obtaining the data set, distribution media, and cost of acquiring the data set.
- *Metadata Reference Information*—provides information on the currentness of the metadata information and the responsibility party.

Although the "Content Standards for Digital Geospatial Metadata" provides a detailed list of data element definitions, it does not specify how this information should be organized in a computer system and the means by which the information is transmitted or communicated to the user. These decisions are left to each organization. A copy of the MCS can be obtained from the FGDC website. Some GIS software (in particular, ArcInfo) have metadata maintenance capabilities that support these content standards.

*European data standards*

Several Pan-European developments reflect the GIS coordination efforts among the European nations (Rhind 1998). The EUREF system represents an effort by many national mapping agencies in Europe to create a common geographic referencing system across Europe. The GISDATA initiative funded by the European Science Foundation is modeled after the U.S. *National Center for Geographic Information and Analysis* (NCGIA). The EUROGI (*European Umbrella Organization for Geographical Information*), which consists of national GIS bodies and professional societies, is an umbrella group that intends to become the European focus of geographic information issues. The *European Committee for Standardisation* "has initiated a working programme to define European standards in relation to all aspects of geographic data such as data quality, metadata, and data transfer" (Rhind 1998, p. 304). The scope of this program is parallel to those key subjects defined in the U.S. SDTS and MCS.

*European Umbrella Organization for Geographical Information (EUROGI)*
EUROGI was established in 1993 as a result of a study commissioned by the European Commission's Directorate General (DG) XIII-E to develop a unified European approach to the use of geographic technologies. The objectives of EUROGI are:

- Define a European geographical information policy and facilitate a European geographic information infrastructure. Under this objective, it includes actions such as *GI2000-European Geographical Information Policy Framework*, *Global Spatial Data Infrastructure* initiative, and *Geographical Information Infrastructures* at regional, national, and local levels.
- Improve communication between members of affiliated associations.
- Ease data exchange at the national and European level through the development of standards and limit the impact of the constraints of the various legal issues affecting geographical information.

EUROGI has developed statements and strategies on European geographical information standardization and interoperability issues.

### Data integration methods

*Areal interpolation*

A common situation facing the transportation analyst is having spatial data collected using different zoning systems such as census units, ZIP Codes, traffic analysis zones, municipalities, and so forth. Integrating these data requires basis transfer or *areal interpolation* of attributes between different spatial zoning systems. Goodchild, Anselin, and Deichman (1993) and Goodchild and Lam

(1980) provide general discussions of the areal interpolation problem and solution methods.

A straightforward approach to areal interpolation is the *areal weighting method* (Flowerdew, Green, and Kehris 1991; Goodchild and Lam 1980). The method estimates an attribute value for a "target" zone based on the degree of overlap with "source" zones. If the attribute is *extensive* (e.g., count data such as total population), the value for the target zone is:

$$\hat{y}_t = \sum_{s \in S(t)} \frac{a_{st} y_s}{a_s} \tag{4-1}$$

where $\hat{y}_t$ is the estimated value for the target zone, $y_s$ is the measured attribute value for the source zone, $a_s$ is the area of the source zone and $a_{st}$ is the area of the overlap between the source zone and the target zone and $S(t)$ is the set of source zones that intersect with the target zone. Equation (4-1) computes the attribute value for the target zone as the weighted sum of the values for the target zones, with the weights provided by the ratios. If the attribute is *intensive* (e.g., rates or proportions such as trip generation rates per household), then the value for the target zone is:

$$\hat{y}_t = \frac{\sum_{s \in S(t)} a_{st} y_s}{a_t} \tag{4-2}$$

where $a_t$ is the area of the target polygon. Equation (4-2) estimates the rate or proportion for the target zone as the weighted average of the values for the target zones.

Equations (4-1) and (4-2) assume a uniform spatial distribution of the attribute within each source zone. This is rarely if ever the case in reality: Variables such as population, population density, trip generation rates, and so forth, vary within spatial zones although these data may not be available directly. Sometimes available is ancillary information on the attribute distribution within the source zones. For example, we may have a remotely sensed image that shows the housing density within each source zone; this can allow us to infer the uneven distribution of population.

There are several methods for exploiting ancillary information to improve the areal interpolation estimates. Flowerdew, Green, and Kehris (1991) generalize the areal weighting method using the *expectation/maximization* (EM) algorithm, a statistical technique for inferring missing data. This consists of two iterated steps. The first step calculates the expected distribution of the target data given the source data and a statistical model (e.g., Poisson regression) describing the likely distribution of the target data based on the ancillary data. The second step reestimates the parameters of the statistical model based on the now "complete" dataset using maximum likelihood techniques. The "E" and "M" steps are iterated until convergence.

Goodchild, Anselin, and Deichman (1993) develop a method that uses ancillary information to derive *control zones* that are used in an intermediate step of the areal weighting method. Control zones are areas of uniform attribute density of the source data based on the ancillary information. The method uses areal weighting to first estimate the attributes for the control zones from the source zones and then uses the control zones to estimate the attributes for the target zones.

Gan (1994) reviews the areal interpolation problem for the specific case of transferring data from a census system to traffic analysis zones (TAZs). Since most activities (houses, workplaces) are located proximal to the transportation network, Gan (1994) notes that network data can be used as ancillary data to enhance the basis transfer process for this case. A good weighting factor is the length of the local network within the TAZs.

*Network conflation*

A common situation encountered in GIS-T applications is several data sets representing the same transportation network. For example, a local government may use the census data files to perform address-matching tasks on the street network. A version of the street network GIS layer may also be digitized from larger scale maps for facility and infrastructure management applications that require a better geometric representation of the street network. It is desirable for the agency to consolidate these two versions of the street network GIS databases into one database that incorporates the street address data files into the street network layer with more-precise coordinates.

*Conflation* is a process that matches two GIS network layers and transfers the link attributes from one layer to their corresponding links in another layer. A decision must be made as to which layer is the better geometric representation before starting a conflation process. The geometrically accurate layer (called *geometric layer*) will receive attribute data from another layer (called *attribute layer*). Network conflation process normally involves two major steps:

- **Step 1:** Match links in the geometric layer with those in the attribute layer.
- **Step 2:** Transfer link attribute data from the attribute layer to the corresponding links in the geometric layer.

Matching arc between two network coverages can be accomplished by examining the arc endpoints. Two arcs are considered to be matching if the arc endpoints in the geometric coverage and the attribute coverage coincide with each other or are located within a predetermined tolerance distance. Arc attributes in the attribute layer can be copied to the corresponding arc in the geometric layer.

Network conflation can become more time consuming when matching arcs cannot be easily identified between the two layers. For example, an arc in the attribute layer may cover two or more arcs or partial arcs in the geometric layer, or vice versa. In this case, links in one of the two layers must be split to create matching arcs.

Another method for network conflation is to adjust the arc coordinates in the attribute layer such that they match with the more accurate coordinates of corresponding arcs in the geometric layer. This approach is accomplished by a rubbersheeting process that adjusts the arc coordinates in one GIS layer to match the corresponding arcs in another layer. Rubbersheeting does not transfer attribute data between the layers. It therefore is appropriate for situations that one of the two layers contains all of the attribute data needed for the GIS applications.

*Address matching*

Some of the data available for a transportation analysis may be referenced by street addresses. For example, the locations of major employment and activity centers that serve as important data in travel demand analysis are often referenced by their street addresses. Another example is a customer address database for pickup and/or delivery of goods.

*Address matching* is a method that converts street addresses to locations in geographic space. GIS can interpolate the location of a given street address on an address-attributed street network (see chapter 3). The point locations derived through an address matching process can then be overlaid with other GIS layers (e.g., traffic analysis zones layer) to perform additional spatial analyses (see chapter 7).

Figure 4-5 shows an example of trip diary database with each trip origin identified by a street address (Started_fr), along with other data items associated with the trip, such as household ID (Hh), trip date (Date), individual ID (Indv), trip ID (Trip_no), and trip starting time (S_time). (Note: the street address data are synthetic to protect privacy; the remaining data are real). Figure 4-6 shows the resulting georeferenced data obtained through address matching.

To find the location of each trip origin on the address-attributed GIS street network map layer, the street address (i.e., the Started-fr data field) of each record in the travel diary database is compared against the street address range data included in the street network database. The address matching process matches each of the address data components (i.e., address number, street prefix direction, street name, street type, and so forth.) between the trip diary database and the GIS address database. If it finds a match, a point location will be automatically interpolated along the matched street segment according to the street address range in the GIS street network database and the specific street number associated with a particular trip origin. Figure 4-7 illustrates this process.

Most GIS packages provide users with both batch and interactive address matching operations. The batch operation does not require user interactions

| Hh | Date | Indv | Trip_no | S_time | S_mi | Started_fr | S_city | S_cnty | Purpose | E_time | E_mi |
|---|---|---|---|---|---|---|---|---|---|---|---|
| 1040 | 319 | 1 | 1 | 1230 | 285 | 290 Point East Dr | NMB | DC | 3 | 1240 | 288 |
| 1040 | 319 | 1 | 2 | 1715 | 288 | 1075 NW 12 Av | MIA | DC | 5 | 1802 | 302 |
| 1147 | 319 | 2 | 1 | 1130 | 769 | 49301 Alton Rd | MB | DC | 8 | 1155 | 784 |
| 1147 | 319 | 2 | 2 | 1245 | 784 | 28795 Biscayne Blvd | NMB | DC | 5 | 1316 | 809 |
| 1172 | 320 | 1 | 1 | 0727 | 346 | 18974 117 Av | MIA | DC | 3 | 0734 | 348 |
| 1172 | 320 | 1 | 2 | 0826 | 348 | 18432 117 Av | MIA | DC | 1 | 0847 | 357 |
| 1172 | 320 | 1 | 3 | 0904 | 357 | 7750 115 Av | MIA | DC | 2 | 0908 | 358 |
| 1172 | 320 | 1 | 4 | 0917 | 358 | 6536 115 Av | MIA | DC | 2 | 0955 | 376 |
| 1172 | 320 | 1 | 5 | 1100 | 376 | 12355 NW 136 St | MIA | DC | 2 | 1122 | 388 |
| 1172 | 320 | 1 | 6 | 1150 | 388 | 18432 117 Av | MIA | DC | 8 | 1205 | 392 |
| 1172 | 320 | 1 | 7 | 1255 | 392 | 11035 NW 170 St | MIA | DC | 5 | 1324 | 403 |

**Figure 4-5**: Example of travel diary data with each trip origin identified by a street address

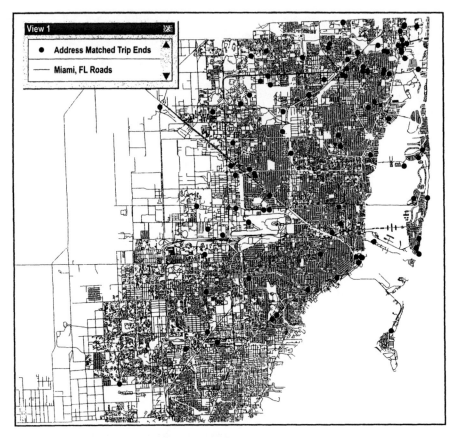

Figure 4-6: Georeferenced travel diary data based on address matching

and is suitable for matching a large number of address records. The unmatched address records are written into a separate file. Users then can use the interactive address matching option to correct the errors in the original data table and match a previously unmatched record to a particular street segment.

Although address matching is a powerful method of integrating address-based data with other types of spatial data, there are some limitations. First, the address matching process creates a separate point layer that is topologically independent from other GIS layers. In other words, the spatial relationships between the derived points and other geographic features must be generated through other GIS functions. Since many transportation analyses require network-oriented operations, it is desirable to have these point locations within the network topology for subsequent analyses. Most commercial GIS packages now employ a spatial search algorithm to automatically identify the nearest network node to a given point location before network analysis. The point locations are forced to match their closest network node locations. This may not be a desirable feature

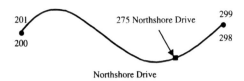

Figure 4-7: Interpolating the location of an address

when we deal with disaggregate data (e.g., travel diary) and we wish to preserve individual point locations.

Another problem with address matching is that street address numbering systems are not consistent everywhere. For example, the address numbers in Japan in some cases are assigned by their dates of construction; therefore, the interpolation method employed in address matching process will not work. Also, some address referencing and reporting can involve multiple names for the same street (e.g., "Main Street" and "U.S. Highway 89"). This will require an alias file to handle the streets with multiple names. Finally, survey data may be riddled with incomplete and vague addresses, including closest street intersections rather than the actual addresses.

*Integrating digital imagery*

Digital imagery, such as satellite imageries, digital orthophotos, scanned aerial photos, digital video files, scanned drawings and text documents are increasingly being used in transportation analysis and planning. A frequent use of these image files in GIS is in geographic visualization or as a background when editing existing data. Commercial GIS packages can display digital image files stored in a variety of formats (e.g., TIF, BMP, BIL, BIP, and JPEG). Some GIS software also allow users to link image files with specific geographic data elements as an attribute item in the GIS database. This design enables the user to click on a particular feature (e.g., railroad crossing) on the display to open a pop-up window of the associated image (e.g., a digital photo of the railroad crossing).

Aided with digital camera/video and GPS technologies, field crews can conveniently capture views at different locations for integration into GIS databases. This function is especially useful for applications such as safety evaluations at railway/highway crossings, traffic signage inventory and maintenance, pavement management, among others. With the scanned drawings and text documents linked to the GIS features, it also can be used to quickly review the drawings (e.g., engineering design, straight line diagram) or relevant documents (e.g., regulations, property deeds, and accident reports) for applications such as land acquisitions, right-of-way planning (see chapter 8), or transportation facility design and engineering.

Many commercial GIS software packages now provide functions for converting data between raster and vector representations. This allows imagery to be processed using vector GIS tools. Despite the ease of conversion, the integration and subsequent use of the data must occur with caution. First, imagery and raster data contains explicit information on its resolution (the pixel size); this is lost when converted to the vector format. Vectorization algorithms use a "thinning" process to convert pixelated lines into arcs. Even if the resulting vector data appears to have adequate positional accuracy, substantial topological errors may occur from small positional distortions. For example, very small shifts in the locations of two arcs may result in their intersection occurring on the wrong side of another arc. Extensive manual editing is often required (see Burrough and McDonnell 1998). Converting from vector to raster formats also creates similar errors, generally related to the difficulty representing detailed features present in the vector data using the visibly finite resolution of the raster model. See Carver and Brunsdon (1994) and Pavlidis (1982) for discussions of raster to vector conversion.

## Spatial data quality

### Data quality issues

Spatial data quality has been a concern for as long as maps have been used for solving geographic problems. The analog and manual methods required for spatial analysis with traditional maps creates a visible consciousness about data inaccuracies. For example, it is obvious when two paper maps do not register well against each other, forcing the analyst to make necessary adjustments to accommodate potential inaccuracies. However, once the maps are stored in a computer, there is a tendency to accept them without questioning their quality.

A major source of confusion about spatial data quality concerns the difference between precision and accuracy. *Precision* is related to the level of detail in data encoding. For example, milepost data recorded to 1/10 mile versus 1/100 mile represent different precision levels. *Accuracy*, on the other hand, refers to the closeness to ground truth. A traffic accident site recorded at 12.578 milepost along Interstate Highway 95 may represent a high precision level, but it is not accurate if the accident actually took place at 12.578 milepost on Interstate Highway 595. Computers can store data at a high precision level, but the accuracy of GIS data depends on how the data are collected, entered, manipulated, and presented. The edited volume by Goodchild and Gopal (1989) offers a comprehensive and in-depth discussion of data accuracy issues in spatial databases.

Data quality has a direct effect on the fitness of a data set for different applications. For example, an outdated land use map layer will lead to inaccurate estimates of travel demand patterns. Also, a highway base map layer with positional accuracy only to +/−30 feet can be adequate for some applications (such as travel demand modeling) but inadequate for others (such as transportation engineering design). Users of GIS data must be provided with essential data quality information about the data set such that they can make a judgement on the fitness of the data set for a particular application. The *Data Quality Specification* of the

SDTS defines five mandatory data quality modules: lineage, positional accuracy, attribute accuracy, logical consistency, and completeness (U.S. National Institute of Standards and Technology 1992).

*Lineage* describes the history of a data set, including its source materials and all processing steps used to produce the data set. It provides information such as dates of source materials, types of source materials (e.g., paper maps, aerial photos, satellite imageries, field surveys, and transaction records), methods of data entry (e.g., digitizing, scanning, and keyboard entry), and data manipulations (e.g., map projection, raster-vector data conversion, conflation, data classification, etc.). Lineage is an important aspect of metadata. In chapter 2, we mentioned the *GEOLINUS* system addition to the ArcInfo GIS software. This is an early attempt to maintain lineage metadata through a graphical flowchart representation of the source data and derived data coverages (see Essinger and Lanter 1992; NCGIA 1992).

*Positional accuracy* compares locations in the database to locations determined with higher accuracy (Goodchild and Hunter 1997). Comparison methods for point objects include the root mean square error and percentiles of the distance distribution (e.g., requiring 90% of points on a map to be within a specified error tolerance). Positional accuracy for linear and areal features is more complex. A common method is the *epsilon band*; this is a type of buffer or zone of uncertainty around a linear feature (see Veregin 1999). However, this method does not allow statistical assessments and is sensitive to outliers (i.e., the maximum error in the linear feature). Goodchild and Hunter (1997) describe a robust statistical method for linear features based on comparing sampled point locations. Shi and Liu (2000) develop a stochastic model of positional error in line segments. Leung and Yan (1998) also formulate a unified stochastic model for positional error in points, line segments and polygons. Methods for assessing error and uncertainty in digital terrain models include Hodgson and Gaile (1996), Hunter and Goodchild (1997), and Lopez (1997).

*Attribute accuracy* indicates the fidelity of nonspatial data. It is again based on a test of ground truth data (e.g., highway functional classification) collected at selected sample locations against the attribute data stored in the digital database. A misclassification matrix (figure 4-8) can then be constructed and statistical indices are computed to reflect the attribute accuracy level.

*Logical consistency* describes the fidelity of relationships among the map features encoded in a GIS database. In a fully topological spatial data model (e.g., the NAA model; see chapter 2), logical consistency is obtained if map features have correct topological relationships with each other. For example, all lines intersect at locations where intended, no lines are entered twice, all polygons are closed, each polygon has a unique label point, all inner rings are completely embedded in enclosing polygons, and so on.

*Completeness* concerns the exhaustiveness of both spatial and attribute components of a GIS database. For example, the minimum area and minimum width used in creating a map layer must be reported to show its spatial completeness. Tests for taxonomic completeness of attribute data indicate, for example, whether a classification system is exhaustive and covers all possible outcomes.

|  | Highway Functional Classification | Database | | | | | |
|---|---|---|---|---|---|---|---|
|  |  | Major Arterial | Minor Arterial | Major Collector | Minor Collector | Local | Sub-total |
| Ground Truth | Major Arterial | 25 | 2 | 1 | 0 | 0 | 28 |
|  | Minor Arterial | 3 | 17 | 2 | 0 | 0 | 22 |
|  | Major Collector | 1 | 3 | 42 | 2 | 0 | 48 |
|  | Minor Collector | 0 | 2 | 4 | 38 | 6 | 50 |
|  | Local | 0 | 0 | 1 | 5 | 26 | 32 |
|  | Sub-total | 29 | 24 | 50 | 45 | 32 | 180 |

Figure 4-8: Example of misclassification matrix

### Spatial data error

There are many types of potential data errors in GIS. Some of them are related to the nature of representing the real world in an abstract form (i.e., map, drawing, and database). These errors are difficult to remove completely.

For example, a simplification and generalization process is often employed to derive a manageable database that represents the real world entities. Theoretically, there exist infinite number of points along the Interstate Highway 95 between the south end in Miami, Florida and the north end at the U.S and Canada boarder in Maine. It is infeasible to record the coordinates of every single location along this highway in a GIS database. Even though we use a GPS device to collect the coordinates of every 1/100 mile along the I-95 to create the database, it still is a generalized representation of the I-95 in the real world.

Similar situations also exist with attribute data in GIS databases. Pavement condition data are often coded as categories (e.g., excellent, good, average, and poor). It is not uncommon to encounter situations that involve a judgement call between two consecutive categories. This is similar to fuzzy boundaries that are frequently associated with environmental data coding such as the transition from one vegetation cover zone to another vegetation cover zone. Both the generalization and the fuzzy boundary examples indicate the possible impurities in a GIS database. Strictly speaking, they can be considered as errors in the databases. We often have to accept these data errors since there is no easy way to resolve them.

A GIS database could also include other data errors that can be identified and corrected. For example, an arterial road is misclassified as a collector road, or a dead-end street is digitized as a through street. Data validation procedures are needed to identify and correct these mistakes.

### *Sources of spatial data error*

Aronoff (1989, p. 142) lists common sources of data errors that are encountered in six different stages of a GIS project (data collection, data input, data storage, data manipulation, data output, and use of results). Burrough (1986) also classifies data error sources as *obvious errors*, *measurement errors*, and *processing errors*. The critical point is that GIS data errors could be introduced from many sources and at every stage of a project. Whether field data, paper maps, or remotely sensed data are used as the sources to create a GIS database, errors could be embedded in the original data sources due to measurement errors, omissions, outdated data, incorrect interpretation, etc.

During the process of converting analog data to digital data, additional errors could be introduced through digitizing, edge-matching, raster-vector conversions, fuzzy boundary determination, data entry mistakes, and insufficient numerical precision of data storage. When map overlay operations are performed, errors in the original map layers can be propagated into the output map layer. Map projection transformations can cause distortions in the measurement of area, distance, and direction. The final output of a GIS analysis results could be misleading due to inappropriate data classification and cartographic presentation methods used. GIS output can be misused if the user ignores the various data error sources introduced at the different stages of a GIS project.

*Linear referencing errors*

In addition to the above common spatial data errors, there also exist some data errors that are more specific to the GIS-T applications. As discussed in chapter 3, transportation agencies deal with a wide range of attributes referenced along the linear transportation routes. Dynamic segmentation is a method designed for the handling of spatial data based on a linear referencing system. The basic design of dynamic segmentation model is that each route is defined individually (e.g., a state highway), and a collection of routes form a route system (e.g., a state highway system). Locations of different attributes associated with a particular route are referenced by the linear measures along that route and are independent of other routes in the route system. This design offers a flexible way of defining independent routes in a route system, but it also causes a potential data quality problem where two routes intersect with each other or share a common route segment.

Figure 4-9 shows a sample route system that consists of three separate routes. Route 1 and Route 2 cross each other at Intersection B, while Route 1 and Route 3 share a common segment between Intersection A and Intersection C. If Intersection B is referenced in both Route 1 and Route 2 in terms of its linear mileposts along the two routes, it is likely that Intersection B will show up as two separate, although often closely spaced, points on the display. This situation arises due to the independent linear interpolations performed on the two routes based on the digitized route lengths and the real world milepost data. Even small errors either in the digital route alignments or in the real world milepost measures could cause offsets from the exact intersection location in the digital database. Similar problems can occur when two routes share a common route segment when referencing route characteristics.

The data error described above could be fixed with a tolerance level built into the dynamic segmentation model that automatically resolves offsets smaller than a user-specified tolerance level. The theory to support these types of operations is emerging. For example, as mentioned in chapter 3, Vonderohe and Hepworth (1996, 1998a, 1998b) determine error propagation and tolerances in the NCHRP LRS data model. An automatic fix of such mismatches means that the GIS

124    Geographic Information Systems for Transportation

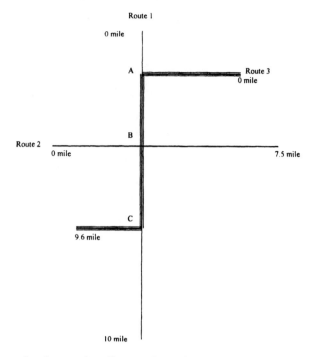

Figure 4-9: Example of errors in a linear referencing system

software must determine appropriate locations based on a predetermined rule. Similar to the fuzzy tolerance level used in creating a topologically consistent GIS map layer, using a single tolerance level may not be appropriate in all situations. With the absence of automated error-fixing algorithm in a GIS package, users will have to calibrate the individual route measures using selected locations along each route as the control points.

Handling spatial data errors

No GIS databases are perfect representations of the reality. While many mistakes made in the GIS database creation process (e.g., a missing line segment) can be corrected, certain errors in GIS databases are due to natural variations (e.g., fuzzy boundaries between soil zones) and cannot be completely eliminated. In addition, errors could be introduced through GIS data transformations (e.g., vector-to-raster data conversion) and GIS data analysis (e.g., overlay of two GIS layers created from maps at 1:2,000 and 1:24,000, respectively). These effects are complex and not well understood (see Veregin 1995). An appropriate approach to handling data errors is to try to eliminate the mistakes and to manage other errors.

*Quality assurance and quality control* (QA/QC) procedures can help identify and assess different types of data errors; therefore, QA/QC procedures need to be in place to evaluate the products at every stage of a GIS project. There are various methods of assessing data quality. For example, check plots could be produced and overlaid with original maps for the identification of digitizing errors. On the other hand, quality of a GIS database could be assessed by comparisons to sources of higher accuracy (Chrisman 1997). Higher accuracy data may require extensive fieldwork and significant resources to collect. A reasonable balance between higher data quality levels and resource requirements therefore must be reached. Choice of a reasonable balance is often an outcome of careful reviews of an organization's mandates and application needs.

Standards and documentation are other useful approaches of handling data errors. For example, the five data quality components (i.e., lineage, positional accuracy, attribute accuracy, logical consistency, and completeness) included in the MCS can be used as guidelines to minimize data errors in the QA/QC process as well as to document data quality. Standards also could be extended to other GIS data in order to help control data quality and minimize data errors. For example, the Florida Department of Transportation (1996) developed a set of proposed standards for data collection, street network files, linear referencing system, GIS layers, data conversion, database and archiving, metadata, cartographic outputs, and application framework and GUIs.

Documentation on how a GIS database was created and processed with respect to the set of standards can guide the appropriate use of the database relative to its data quality level. Data errors cannot be completed eliminated, but they can be controlled through the implementation of standards and they can be documented to ensure appropriate uses of the data.

## Spatial and network aggregation

### Spatial aggregation and the modifiable areal unit problem

It is a common practice in transportation analysis to collect data by geographic zones, such as census blocks, census block groups, census tracts, or traffic analysis zones. These data may also be further aggregated to a spatial zoning system that is chosen for a particular application. These spatial units are often arbitrary; in other words, they reflect the needs of the analyst and not any empirical geographic process.

A transportation application where spatial aggregation is common practice is travel demand modeling. While many transportation textbooks provide good guidelines for designing these zoning systems (e.g., Ortúzar and Willumsen 1994), the transportation analyst usually defines these systems based on intuition and other difficult to codify rules. This can affect the results from the travel demand analysis since each boundary change affects the proportion and pattern of interzonal trips versus intrazonal trips, with the latter trips not captured in the modeling process (Ding 1994).

The *modifiable areal unit problem* (MAUP) occurs when the spatial zoning system used to collect and/or analyze geographic data is "modifiable" or arbitrary. Since the spatial units are arbitrary, the results from analysis based on these units may be arbitrary, that is, an artifact of the spatial units rather than reflecting the true underlying geographic process. MAUP effects can be divided into two components. *Scale effects* result from the spatial aggregation of the data. *Zoning effects* relate to changes in the spatial partitioning at a given level of spatial aggregation (Openshaw and Taylor 1979; Wong and Amrhein 1996).

There is a large body of evidence on the MAUP (see Miller 1999b). A classic paper by Robinson (1950) demonstrates that the correlation measures can increase with increasing levels of data aggregation. A simulation analysis by Openshaw and Taylor (1979) confirms the MAUP. Fotheringham and Wong (1991) show the parameter estimates in a regression model became more unpredictable and less reliable with scale changes and zone definitions. Green and Flowerdew (1996) report similar effects related to the degree of spatial autocorrelation in the data. Putnam and Chung (1989) find aggregation and zoning system effects in spatial interaction and similar travel demand models.

In the past, most analysts ignored MAUP effects since the data and tools to deal with these effects were not available. However, the digital cartographic data and GIS tools are now available for assessing the effects of zoning systems and developing optimal zoning systems for particular applications. In chapter 8, we will discuss some of these methods for traffic analysis zone (TAZ) design in transportation planning and analysis. At present, we will discuss some of the general issues and methods. This discussion is based on Miller (1999b); we refer to reader to that paper for more detail and a more comprehensive bibliography.

One approach is to assess MAUP effects within a particular analysis. A straightforward, GIS-friendly strategy is to perform a sensitivity analysis by changing the zoning system and rerunning the analysis. For example, Chou (1991) recomputes a spatial autocorrelation method at different levels of map resolution to assess the aggregation effects. Haslett, Wills, and Unwin (1990) and Openshaw, Charlton, and Wymer (1987) discuss computational methods and software tools for performing spatial analysis under repeated random realizations of spatial units. The distributions generated from these simulations can be summarized and analyzed to examine the stability and robustness of results.

Some techniques exist for assessing MAUP effects within particular models. Moellering and Tobler (1972) develop a scale-dependent analysis of variance technique. Batty (1976) develops a general information statistic to measure aggregation effects in spatial analysis. Batty and Sikdar (1982a, 1982b, 1982c, 1982d) use these statistics to assess aggregation effects in spatial interaction models. More research on this topic is required, particularly with respect to transportation modeling.

Another approach is designing optimal zoning systems for particular applications; this attempts to reduce the "M" in the MAUP problem by making the spatial units less arbitrary. Openshaw (1978) develops a general statement of the problem of classifying $n$ areas into $m$ regions (where $n > m$) to maximize a function of the zoning system, subject to the $m$ regions being internally connected. Effective heuristic solution techniques are available for several special cases of this problem, including spatial interaction modeling (Openshaw 1977b), linear regression (Openshaw 1977a, 1978) and political districting (Horn 1995; Openshaw and Schmidt 1996). Lolonis and Armstrong (1993) discuss the use of location theoretic techniques as decision support tools for configuring administrative regions.

A strategy for handling MAUP effects suitable for transportation applications is to use more accurate interzonal distance or travel cost measures. In many transportation applications, much of the MAUP effects can be traced to using poor interzonal distance measures (see Beardwood and Kirby 1975; Webber 1980). The centroid-to-centroid distance is often a poor summary of the distribution of distances between two zones (see Francis et al. 1999 for a comprehensive discussion). Other possible summaries of this distribution is the *expected distance*, the *average distance* (this equals the expected distance only for the special case where interaction is equally likely between all locations in both zones), *minimum distance* and *maximum distance* (see Okabe and Miller 1996).

Okabe and Miller (1996) develop algorithms for computing the exact minimum, maximum and average distances between all possible pairings of points, lines, and polygons when stored in a vector GIS format. The minimum and maximum distance algorithms are tractable while the average distance algorithms are restricted to moderately sized databases. Still required is research on good average distance approximations as well as methods for expected distances that can handle a wide range of spatial object types.

If a network connects the zones in a transportation analysis situation, MAUP effects occur due to the choice of zonal representation points and the connection of these points to the network (see chapter 3). Forcing flow to load from a 2-D region onto a finite set of possibly arbitrary network locations introduces travel cost error measurement particularly when modeling flow in congested networks (see chapter 6). An extreme (and somewhat impractical, at least at present) solution is to eliminate the zonal system entirely and model travel as occurring across continuous space (see Angel and Hyman 1976 for a groundbreaking study). Although this is difficult analytically, computational approximations such as modeling shortest paths through lattices and polygons are very tractable (see chapter 5). Also, Daganzo (1980a, 1980b) develops a method for combining a continuous representation of spatial demand with a finite transportation network in congested network flow modeling. This method eliminates the need for aggregated origin and destination zones and thus can eliminate many MAUP effects in travel demand modeling.

## Network aggregation and generalization

The transportation networks used in many transportation analyses are selective representations that aggregate some arcs. Network data aggregation shares certain similarities with polygon data aggregation, but also introduces additional challenges.

Figure 4-10 illustrates a common network aggregation situation. The widely used node-arc data model in commercial GIS packages stores network attributes with fixed line segments (arcs) in a network. When two adjacent network links (e.g., AB and BC) share a common attribute value, the pseudo node (i.e., B) between them can be removed to form a longer network link. This is similar to the removal of the common boundary between two adjacent polygons with the

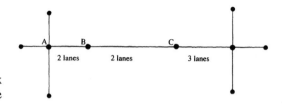

Figure 4-10: Network aggregation example

same attribute data value. Alternatively, some network data aggregation problems can be more effectively and efficiently handled by the dynamic segmentation data model, which does not require a predetermined segmentation of network links.

Another situation encountered with network data aggregation is when multiple links in a more detailed network correspond to a single link in a more-generalized network. Network conflation can be used to transfer the attribute data values of a network link in one layer to multiple matching links in another layer. However, transfer of different attribute values of multiple links in one layer to a single link in another layer presents a problem of combining the multiple data values into a single representative value. This problem is similar to the areal interpolation problem discussed above, with the network aggregation case involving a basis transfer among linear features rather than areal features. Techniques such as the areal weighting method can be adapted for this problem.

Sutton (1997) summarizes the three common approaches to performing network aggregation and conflation. *Hard coding*, or known as hard wiring or cold linkage, requires a correspondence table that relates the different segments in the two network layers. As a result, segments in the model network layer can be mapped to the corresponding segments in the street centerline network layer. This approach is straightforward and intuitive to follow. However, the coding process could be tedious and time consuming for a large and complex network.

*Soft coding*, also known as warm linkage, attempts to code the attribute network segments as a subset of the geometrically accurate network layer in a GIS. Although this approach appears to be a feasible alternative, it could encounter some major problems. First of all, the attribute network may not be topologically consistent with the underlying geometric network. For example, the attribute network may not take into account the complex connectivity patterns at highway interchanges that are present in the geometric layer. Second, data such as turn impedances must be coded in the GIS. Third, different attribute networks (e.g., highway versus transit) are likely to have different network topologies (e.g., HOV lanes and express bus lanes) on the same geometric layer. In addition, the coding effort is not a trivial task.

The *hot linkage* approach maintains separate attribute network and geometric network layers, but automates a seamless exchange of data between them. In other words, the modeling procedures will continue to function on the attribute network outside of a GIS. Linear referencing and dynamic segmentation capabilities of GIS are used to manage the correspondence between the attribute network and the geometric network. Results then can be brought into the geometric network for data query, map display, and other GIS analysis functions. The GIS can also be used for network editing when there are changes in either network. This approach is the most flexible solution among the three alternative approaches.

## Conclusion

In this chapter, we discussed data sources and issues regarding data integration in GIS-T projects. This discussion, combined with the previous two chapters on data modeling and specific GIS-T data models, provides a good foundation for designing and populating a GIS-T database. The question that follows is what to do with these data. The next two chapters will discuss some common network algorithms that are at the foundation of many GIS-T applications. This will include shortest path algorithms and network routing (chapter 5) and flow modeling and facility location (chapter 6).

# 5

# Shortest Paths and Routing

At the core of any GIS-T software are procedures or *algorithms* for conducting analyses and solving routing and location problems within a network. This is the first of two chapters that discuss network algorithms for transportation analysis. This chapter focuses on methods for calculating shortest paths and solving vehicle and fleet routing problems in a network. The next chapter will complete the survey of network algorithms by discussing methods for solving network flow problems and finding optimal or near-optimal facility locations.

We begin this chapter by discussing some crucial properties of networks and algorithms. This includes methods for mathematically representing graphs and networks and precise definitions of network connectivity properties such as "path" and "tree." We will then discuss fundamental properties of algorithms, including the difference between "exact" and "heuristic" algorithms and methods for predicting an algorithm's run time for a given problem size. Even if one is not involved in constructing new algorithms, knowing these basic properties is important for evaluating and choosing "off the shelf" algorithms (and, by extension, software).

Next we discuss the shortest path algorithm, perhaps the central algorithm in network analysis. We first discuss a generic shortest path algorithm and different methods for implementation on computational platforms. As we will see, different implementations of the same algorithm can vary widely in performance, particularly when applied to real-world transportation networks. Again, knowing this information is important for choosing GIS and other software systems that include shortest path routines. Also, another reason for reviewing shortest path implementation issues in detail is that it provides an excellent forum for discussing basic physical processing structures used in computers. Everyone working with computers should be exposed to this material (at least once) since they are at the heart of computer programming.

We will next discuss methods for computing shortest paths very quickly, i.e., in real time or near-real time. This will include a review of parallel processing implementations of the shortest path algorithm. Parallel processing implementations will become increasingly important as these platforms become more prevalent and filter down to the desktop. We will also review the $A^*$ algorithm, a fast but inexact method for computing shortest paths between specific origin-destination pairs. We will end our discussion of shortest path problems by reviewing methods for calculating shortest paths through surfaces such as terrain. This problem can arise when modeling off-road travel or in corridor analysis.

We then discuss two closely related vehicle routing problems, namely, the *traveling salesman problem* (TSP) and the *vehicle routing problem* (VRP). The TSP attempts to find the least cost route for a vehicle through a set of locations that returns to the starting location. The VRP attempts to determine the simultaneous routings of multiple vehicles through a set of locations where each vehicle starts and stops at a central location (the "depot"). These problems are very important due to their direct application in transportation as well for a surprising range of non-transportation problems. They are also fascinating in their own right since they are easy to state but notoriously difficult to solve.

## Fundamental network properties

The network algorithms discussed in this and the next chapter rely on a mathematical representation of networks. Therefore, it is worthwhile to briefly review this representation and some of the mathematical properties of networks at the foundation of the algorithms. The reader should also keep in mind the differences between the richer representations available through GIS-T data models and the more austere representations required by these procedures. This means that, at some level, data models such as dynamic segmentation and lane-based navigable data models must be translated into a model more suitable for network analysis.

### Mathematical representation of graphs and networks

Chapter 3 discussed the underlying basis of most transportation data models, namely, graph theory and network analysis. Recall that graph theory assesses topological properties (that is, properties associated with connections) among members of a set. Vertices represent members of the set and edges represent physical or logical connections among those members. An undirected graph means that relationships hold in both directions, while a directed graph specifies relationships that hold only in the directions specified. Graphs can be planar (embedded in a mathematical plane, so that a vertex exists at every edge intersection) or non-planar (independent of any mathematical plane).

We can represent a graph as a set. For example, a graph can be defined by

$G = (V, E)$ where $V = \{1, 2, 3, 4,\}$ and $E = \{\{1, 2\}, \{2, 3\}, \{1, 4\}\}$. This defines a graph with four vertices (1, 2, 3, 4) and edges between vertices 1 and 2, between vertices 2 and 3 and between vertices 1 and 4. Since we did not specify a directed graph, it assumed that the relationships are bidirectional (e.g., the relationship from vertex 1 to vertex 2 also holds for vertex 2 to vertex 1). The equivalent representation using a directed graph would require $E = \{(1, 2), (2, 1), (2, 3), (3, 2), (1, 4),(4, 1)\}$. (In mathematical notation *braces* or { } indicate a set in which the ordering of the elements is irrelevant. Conversely, *parentheses* or ( ) indicates a sequence in which ordering is relevant.)

A graph that has one or more real numbers associated with each edge is called a *weighted graph* or *network* (Balakrishnan 1995). In this case, we often refer to the vertices as *nodes* and the edges as *arcs*. Arc weights typically represent the unit flow cost for that arc, i.e., the cost for one unit of flow to traverse that arc. Arc weights can be *fixed* (independent of flow) or *variable* (related to the current level of flow within the arc). We discuss variable flow cost functions appropriate for transportation modeling in chapter 6.

Also recall from chapter 3 that network analysis extends graph theory to explicitly model flow or movement. Nodes are locations where flow originates, terminates or is relayed. Arcs represent conduits for movement between nodes. Arcs can correspond to a physical entity (e.g., a road segment) or a logical relationship (e.g., airline service between two cities) in the real world. Arcs can be *directed* (allow flow only in the direction specified) or *undirected* (allow flow in either direction). Similar to graphs, networks can be planar or nonplanar. Most currently available GIS software require planar networks, although controlled relaxation of topological consistency is possible (e.g., not inserting a node at every arc "crossing"). Object-orientation (see chapter 2) can allow the user to specify topological rules for particular object classes, but this requires careful specification by the user.

The set notation discussed above for graphs can also be used to represent networks. However, an alternative approach is to use *incident matrices* that maintain information on the connectivity relationships within the network. This is often useful when analyzing networks using shortest path and other network algorithms. Incident matrices can also be used to calculate accessibility properties within networks. For example, repeatedly powering an adjacency matrix (see below) can show the number of paths between all node pairs in a network (see Taaffe, Gauthier, and O'Kelly 1996 for an accessible introduction). Some GIS software (e.g., TransCAD) allows these type of matrix operations; these operations can be easily programmed for extensible GIS software.

Assume we have a network $G = (\mathbf{N}, \mathbf{A})$ where $\mathbf{N} = \{1, 2, \ldots, n\}$ is the set of nodes and $\mathbf{A} = \{(i,j)\}$ is the set of (directed) arcs. We can form an *adjacency matrix* $\mathbf{X} = \{x_{ij}\}$ by specifying a row and a column for each node in the network. Then, an element $x_{ij}$ equals one if an arc exists between node $i$ and node $j$, zero otherwise. Elements along the main diagonal (i.e., $x_{ii}$) are zero by convention. In the case of a directed network, this matrix is asymmetric, i.e., $x_{ij}$ may not equal $x_{ji}$.

Other types of incident matrices are possible (Bell and Iida 1997). In a *node-arc incident matrix*, the matrix rows correspond to nodes while the matrix columns correspond to arcs. An element $z_{ik} \in \mathbf{Z} = +1$ if node $i$ is the start node of arc $k$, −1 if node $i$ is the end node of arc $k$, 0 otherwise. Another important incidence matrix is the *arc-path incident matrix*. Rows correspond to network arcs while columns correspond to network *paths* or "routes". An element $r_{kp} \in \mathbf{R} = 1$ if path $p$ uses arc $k$, 0 otherwise. We can form this matrix after performing a shortest path analysis to determine network paths.

A *cost matrix* maintains the flow cost information for each arc in the network. We form a cost matrix $\mathbf{C} = \{c_{ij}\}$ also by specifying a row and a column for each node in the network. We define its elements as:

$$c_{ij} = \begin{cases} l_{ij}, & (i,j) \in A \\ \infty, & (i,j) \notin A \end{cases} \tag{5-1}$$

where $l_{ij}$ is the weight for arc $(i, j)$ and the symbol $\in$ indicates "element" or "member of" and $\notin$ indicates "not a member of." Elements along the main diagonal are zero and the matrix may be asymmetric if the network is directed.

Incident matrices are logical data models for transportation networks; a matrix is rarely if ever used as a physical data model to store data in primary or secondary memory. A matrix representation is only efficient when the network is *highly connected*, meaning that each node is connected to many of the other nodes in the network. In contrast, most transportation networks are *sparsely connected*: Nodes are connected only to a few neighbors that are proximal in geographic space. Storing a sparsely connected network in memory using a matrix structure means that superfluous empty cells occupy most memory locations, wasting valuable storage space.

### Connectivity in a network

Of particular interest in transportation analysis are properties related to connectivity and potential movement within a network. First, we discuss some basic terminology to describe direct connectivity among arcs and nodes in a network. An arc that is "connected" to a node is *incident* to that node. The *degree* of a node is the number of arcs that are incident to it. Two nodes that have an arc between them are *adjacent*; the arc is said to *join* the two nodes (Balakrishnan 1995).

In transportation analysis, we are often required to find the subnetworks of a network that have certain properties. A network $G' = (\mathbf{N}', \mathbf{A}')$ is a *subnetwork* of a network $G = (\mathbf{N}, \mathbf{A})$ if $\mathbf{N}' \subseteq \mathbf{N}$ and $\mathbf{A}' \subseteq \mathbf{A}$ (i.e., every node in $\mathbf{N}'$ is also a node in $\mathbf{N}$ and every arc in $\mathbf{A}'$ is also an arc in $\mathbf{A}$. Important subnetworks include walks, paths, and cycles. A *walk* is a sequence of nodes $n_1, n_2, \ldots, n_i, \ldots, n_k$ such that $n_i$ and $n_{i+1}$ are adjacent for every $1 \le i \le k - 1$. If all nodes in the walk are unique (i.e., no nodes are repeated), we refer to the walk as a *path*. If all nodes are unique except that $n_k = n_1$ (i.e., the last node in the sequence is the first node; the path returns to its beginning), we refer to the walk as a *cycle*. If the network is directed, then all arcs contained in the walk, path, or cycle must be oriented consistently in the direction of travel.

If two nodes in a network have at least one path between them, we refer to the nodes as being *connected*. The overall network is connected if a path exists between all node pairs. We call a network *acyclic* if it does not contain any cycles.

Two important types of paths and cycles are the *Hamiltonian path* and the *Hamiltonian cycle*, named after Irish mathematician W. R. Hamilton (1805–1865). The former is a path that includes every node in the network (or a specified subnetwork) and the latter is a cycle that includes every node and returns to the "start" node (Biggs 1985). Hamiltonian cycles are sometimes also referred to as *tours*. As we will see later in the text, Hamiltonian cycles are critical for solving routing problems such as the traveling salesman problem and the vehicle routing problem.

Another important subnetwork is a *tree*. A *tree* is a subnetwork in which each node can only be visited once (i.e., there is a unique path between any two nodes in the tree). More specifically, a subnetwork is a tree if it has the following properties:

- The subnetwork is *connected* (i.e., it is possible to find at least one path between any node pair).
- The subnetwork has no cycles.
- The number of arcs is $k - 1$, where $k$ is the number of nodes in the subnetwork.

The *root* of the tree is the node that only has arcs directed outward from it; we typically represent a tree graphically with the root node at the top. The *children* of a node are the nodes directly below it in the tree. We also refer to that node as the *parent* of the children nodes. A *spanning tree* is a tree that includes every node in the network (Bell and Iida 1997; Phillips and Garcia-Diaz 1981).

Trees are critical to shortest path calculation and network flow modeling. When we conduct a shortest path analysis, the result is usually a spanning tree rooted at the origin of interest. In other words, a *shortest path tree* is a spanning tree that provides the shortest paths from an origin node to all nodes in the network. Figure 5-1 illustrates a shortest path tree. We often use shortest path trees to allocate flows from an origin among destinations in the study area of interest.

## Fundamental properties of algorithms

In the remainder of this chapter and in the next chapter, we will be discussing procedures or *algorithms* for solving network problems such as shortest path, vehicle routing and facility location. Before presenting the algorithms themselves, it is worthwhile to discuss some fundamental characteristics of algorithms. This background knowledge is essential for evaluating and choosing algorithms and software for solving transportation problems.

A good starting point is to define the term "algorithm," at least informally. An algorithm is a sequence of well-defined rules or instructions for solving a problem in a finite number of steps (Biggs 1985; Illingworth and Pyle 1996). "Well-defined" means that each step or procedure is precise, straightforward and does not require interpretation or judgement. One can think of an algorithm as a computer program since these also must consist of finite instructions that can be implemented without human interpretation. Therefore, much of our intuition

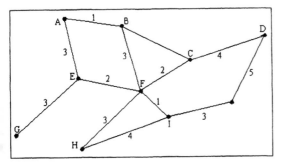

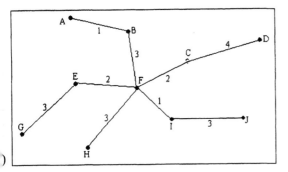

Figure 5-1: A shortest path tree. (a) Input network. (b) Shortest path tree rooted at node F

about computer programs applies to algorithms and vice-versa (Illingworth and Pyle 1996). (Incidentally, in 1936, Alonzo Church and Alan Turing independently discovered a deep correspondence between algorithms and computation: roughly, anything that can be stated as an algorithm can be computed and vice versa. For more on this subject, see Sipser 1997.)

Algorithm efficiency

When solving a network problem such as shortest path routing, we generally want the algorithm to be correct and cheap. "Correct" means that the answer is accurate; for example, the computed shortest path is in fact the true shortest path between the origin and destination. "Cheap" means that we want the algorithm to use the least amount of computing resources possible, i.e., it should be *efficient*. This is particularly critical given the size and complexity of many transportation applications.

Although efficiency can be interpreted in a broad sense to include all computing resources, an algorithm's efficiency is measured most often in terms of the time required for the calculation. Although the exact reasons are technical, generally the time requirements dominate the "space" (memory) requirements of an algorithm, i.e., it usually requires more time than space to run an algorithm. Therefore, the fastest algorithm for a problem is likely to be the most efficient overall (Garey and Johnson 1979; Sipser 1997).

Measuring the time requirements of an algorithm is not as straightforward as it might appear at first glance. Simply coding the algorithm, executing it on a computer, and measuring execution time for different problem sizes does not provide a definitive answer. Actual execution time can be affected by many factors. These can include the computer language, the skill of the programmer in translating the algorithm to the language, the computer's hardware configuration, its operating system, whether the computer is networked, what else was going on in the computer and network when the code executed, and so on. Therefore, a run-time analysis does not provide definitive results unless performed within very carefully controlled experimental settings.

Instead of run-time analyses, we can achieve more general (although more vague) answers. *Complexity analysis* attempts to derive a rough but general assessment of the time required for an algorithm. Rather than estimate the precise time, a complexity analysis identifies a general classification based on the time required. We accomplish this by ignoring machine-dependent details such as the time required for a given operation. Instead, we classify the algorithm by the general number of operations required as a function of input size. Although we lose detail and precision, we achieve a statement that can be generalized to any computational platform (including computers that do not exist yet but are conceivable).

The number of operations required for a problem of a given size can vary depending on the characteristics of the input data. For example, the operations required by a shortest path routing algorithm for a network of size $n$ (where $n$ is the number of nodes) can vary depending on the connectivity in the network and the arrangement of the input data. Therefore, the *time complexity* of an algorithm for a given problem size is actually a distribution of times rather than a single number (Sedgewick 1992).

We usually summarize the distribution of run-times for an algorithm in two ways. We can conduct an *average-case analysis* to determine the expected number of operations for a given problem size. Another approach is *asymptotic* or *worse-case analysis*. In this approach, we determine an *upper bound* or the guaranteed maximum number of operations required (Sipser 1997). To be more precise, we determine the general class of the upper bound and not the exact upper bound. This is sometimes misleading. For example, since we are determining only the general class of the upper bound, two algorithms that are in the same class can have very different run-times in practice. Also, the worse-case we are describing with an upper bound may only occur very rarely in practice. Nevertheless, worse-case analysis is more common than average-case analysis since it provides more definitive statements and is often easier to determine (Nemhauser and Wolsey 1988; Sedgewick 1992).

We use a system called *Big-O notation* to indicate the asymptotic time class of an algorithm. If we state that an algorithm has an asymptotic time complexity of $O[g(n)]$, we are saying that the upper bound on the number of operations for a problem size of $n$ is in the class $g(n)$. A bit more precisely, we are saying that the number of operations is no greater than $kg(n)$ for an input size $n$, where $k$ is some constant (see Sedgewick 1992 or Sipser 1997 for a more rigorous definition). We ignore the constant $k$ since that is subject to implementation details that vary by machine and therefore are not of interest in this general analysis.

An example will help (see Sipser 1997 for more details). Let's say we have a mathematical function $f(n) = 2n^3 + 20n^2 + 35n + 200$. Although this is not an algorithm, we can conduct a similar complexity analysis. The asymptotic complexity of this function is determined by choosing the highest order term ($n^3$) and ignoring its constant (2). We choose the highest-order term in this case since it *dominates* or it is much greater than the other terms in a general sense (e.g., $bn^3$ is much greater than $cn^2$ for most values of the constants $b$ and $c$). We ignore the constant associated with $2n^3$ since constants can be viewed as implementation-specific details that vary by machine. Although these can generate substantial run-time differences in practice, the complexity effects of these constants are relatively small compared to differences in the function class (e.g., linear versus exponential; see table 5-1). These qualitative differences are the major *complexity classes* we are trying to capture in our analysis. Therefore, we say $f(n)$ has a worse-case or asymptotic time complexity of $O(n^3)$.

Table 5-1 provides a general classification of asymptotic complexity classes. Although not strictly in the spirit of complexity analysis, table 5-1 also provides simulated run times based on an assumed time required for each operation. Note that complexity classes can be placed into two general groups: those that require *polynomial time* ("constant" through "cubic") and those that require *exponential time*. In general, polynomial time algorithms are suitable for applied problems while exponential time algorithms are not, for reasons illustrated in the table. However, quadratic and cubic time algorithms, while polynomial, may require a large amount of time for a large problem. Keep in mind that the complexity classes are very general. Therefore, even if two algorithms are linear (for example), one may take much longer in practice if each operation requires a large amount of time. This is some of the detail we lose when achieving general statements about the algorithm complexity.

### Exact versus heuristic algorithms

Table 5-1 illustrates another important point about algorithms. Sometimes, finding the *exact* or *optimal* answer to a problem requires an unreasonable amount of time. A famous example is the traveling salesman problem (discussed in detail later in this chapter). This requires finding the shortest Hamiltonian cycle in a network or subnetwork. The asymptotic time complexity for computing an exact solution to a TSP is $O(2^n)$. This is obviously unworkable in practice. Yet, TSP shows up in many real-world problems, such as routing a delivery truck from a depot to several stops and back to the depot, designating work orders in factories and manufacturing computer chips (Kolata 1991).

If an exact algorithm is too computationally burdensome, we often must resort to an approximate or *heuristic* algorithm. Heuristic algorithms provide "good" (near-optimal) answers within reasonable amounts of time. Therefore, they are very useful in practice.

Even though determining optimal solutions to a general problem is too difficult in practice, this does not mean that any heuristic algorithm is acceptable.

Table 5-1: Algorithm Complexity Classes and Examples

| Complexity Class | $O[g(n)]$ | Simulated Run Times[1] | | | | | | Comments |
|---|---|---|---|---|---|---|---|---|
| | | $n=10$ | $n=20$ | $n=30$ | $n=40$ | $n=50$ | $n=60$ | |
| Constant | $1$ | 0.000001 s | 0.000001 s | 0.000001 s | 0.000001 s | 0.000001 s | 0.000001 s | Most of the procedures in the algorithm are executed once or just a few times; ideal situation, but rarely achieved for sufficiently interesting problems |
| Logarithmic[2] | $\log n$ | 0.000001 s | 0.000013 s | 0.000015 s | 0.000016 s | 0.000017 s | 0.000018 s | Algorithm solves a "big" problem by cutting its size by some constant factor and transforming it into a "small" problem |
| Linear | $n$ | 0.00001 s | 0.00002 s | 0.00003 s | 0.00004 s | 0.00005 s | 0.00006 s | Each input element is subjected only to a small amount of processing |
| "$n \log n$"[2] | $n \log n$ | 0.00001 s | 0.000026 s | 0.000044 s | 0.000064 s | 0.000085 s | 0.000107 s | Algorithm solves "big" problem by breaking it down to a set of "small" problems, solving each one independently, and then assembling the overall solution from the smaller solutions |
| Quadratic | $n^2$ | 0.0001 s | 0.0004 s | 0.0009 s | 0.0016 s | 0.0025 s | 0.0036 s | Algorithm must process all pairings of input elements; practical only for small problems |
| Cubic | $n^3$ | 0.001 s | 0.008 s | 0.027 s | 0.064 s | 0.125 s | 0.216 s | Algorithm must process triples of input elements; practical only for small problems |
| Exponential | $2^n$ | 0.001 s | 1.0 s | 17.9 min | 12.7 day | 35.7 y | 366 centuries | Characteristic of "brute-force" (simple, straightforward, and not very clever) algorithms applied to complex problems; practical only for "toy" (very small) problems |
| | $3^n$ | 0.059 s | 58 min | 6.5 y | 3,855 centuries | $2 \times 10^8$ centuries | $1.3 \times 10^{13}$ centuries | |

*Source:* Based on Garey and Johnson 1979; Sedgewick 1992; and authors' calculations.

[1] Assumes that each operation requires 1 μs or 0.000001 s.
[2] Simulated run times for logarithmic and $n \log n$ assumes base 10 logarithm. The logarithm base is normally irrelevant in complexity analysis since these differ only by a constant factor.

Before the scientific community accepts a heuristic algorithm, its performance characteristics must be determined, particularly with respect to how close it comes to optimal solutions and how often this occurs. These properties can be determined by theoretical analysis (e.g., the behavior of the algorithm under certain, limited conditions) or by running the algorithm against published test datasets with known optimal solutions (e.g., TSPLIB, an online library of solved TSP and related problems).

## Shortest path algorithms

In this section of the chapter, we will discuss algorithms for solving the shortest path for one unit of flow through the network. The "one unit flow" condition is critical. The algorithms discussed in this section do not take into account the effects of multiple flows in the network. When routing multiple flows, we must consider *route choice externalities* such as congestion, in other words, the effect of route choices on each other. Shortest path algorithms treat each arc weight as static and not a function of all routes that use that arc, that is, the amount of flow on that arc. Nevertheless, shortest path algorithms are useful for many application problems. They are also embedded within algorithms that solve more complex network routing and flow problems.

### Shortest path properties

Before presenting actual shortest path algorithms, it is worthwhile to discuss two properties that a shortest path must obey. This will help in understanding the operation of shortest path algorithms. The first property is that a shortest path from a given origin to a given destination is composed of shortest paths between all intermediate locations on that path. In other words, a shortest path from node $i$ to node $j$ that contains an intermediate node $k$ must consist of: (i) a shortest path from node $i$ to node $k$ and (ii) a shortest path from node $k$ to node $j$. See Nemhauser and Wolsey (1988, p. 56) for a proof.

Although this sounds like a trite observation, the "intermediate shortest paths" property provides the basic mechanism for computing shortest paths. It implies that the shortest path from an origin to a destination can be derived by computing the shortest paths from that origin to the intermediate points and then assembling the overall shortest path from the intermediate shortest paths. As we will see, shortest path algorithms exploit this property by building larger shortest paths from an intermediate node once the shortest path from the origin to that intermediate node is known.

A second fundamental property is that shortest paths cannot be computed using a "greedy" approach. In other words, in general a shortest path cannot be found by simply choosing the arc with the minimum weight (least cost) from each node. Choosing the minimum weight arc may result in the path extending to a node that is in a disadvantaged situation with respect to the network; for example, the node is only connected to "high-weight" exiting arcs. Since we are trying to minimize the cost of the *entire* path, we may have to incur higher short-term costs to achieve lower costs overall. The basic message here is that a network shortest routing algorithm must have *foresight* or "look-ahead" to find a minimum cost route. Shortest path algorithms reflect this property by treating the estimated shortest path from an origin to each intermediate node as a temporary guess, subject to updating until the algorithm has looked ahead sufficiently to know that the current shortest path estimate is correct.

### Shortest paths through networks

In this section, we consider algorithms that compute the shortest path from an origin node to a destination node in a network. Because of the properties discussed above, most shortest path algorithms actually solve a more general problem, namely, the *shortest path tree* (SPT) or the shortest paths from an origin node to *all* destination nodes in the network (although some algorithms can be terminated when the desired destination node is reached). The result is a spanning tree rooted at an origin $r$ consisting of shortest paths to all nodes of the network represented by the leaves of the tree. We call this the SPT rooted at $r$ or $T^*(r)$.

The original SPT algorithms are due to Bellman (1958) and Dijkstra (1959). Since that pioneering work, many researchers have modified the basic shortest path algorithm. All perform the same fundamental operations and differ with respect to their implementation. Physical level implementation (e.g., data storage structures, processing rules for moving data between structures) determines the complexity of the algorithm and its suitability for different network types. Therefore, understanding both the algorithm and implementation strategies is important.

We will first present a generic SPT algorithm. We will then provide an example calculation using a common implementation strategy. Next, we will discuss critical implementation factors, including network data storage structures, "labeling methods" and decision rules/data processing. Finally, we will identify several shortest path implementations that work well on transportation networks.

#### *Generic shortest path algorithm*

Assume we have a directed network $G = (N, A)$ and a nonnegative arc weight $w_{ij}$ for each $(i, j) \in A$. The generic *shortest path tree* (SPT) algorithm is (Gallo and Pallottino 1988):

Algorithm SPT

Step 1. Initialize a directed tree rooted at node $r$, $T(r)$. For each $v \in N$, let $l(v)$ be the length of the path from $r$ to $v$ using subnetwork $T(r)$ and let $p(v)$ be the parent node of $v$ in $T(r)$.

Step 2. Find an arc $(i, j) \in A$ such that $l(i) + w_{ij} < l(j)$, set $l(j) = l(i) + w_{ij}$ and update $T(r)$ by setting $p(j) = i$.

Step 3. Repeat step 2 until $l(i) + w_{ij} \geq l(j)$ for every $(i, j) \in A$.

The directed tree initialized in step 1 provides an initial guess at the directed shortest path tree rooted at origin $r$. This initial directed tree is arbitrary. A common strategy is to specify a "null" tree (i.e., set $p(v) = 0$ for every $v \in N$). The node labels $l_i$ or estimated shortest path distances from node $r$ to each node $i \in N$ are initially set equal to $\infty$ (or some very large constant) except for the node label for $r$, which is set to zero. This pseudo-tree is sufficient as an initial guess; stage 2 updates this tree as it scans the network and extends the true shortest path tree from $r$.

Step 2 of SPT updates the shortest path tree. The algorithm scans the network and chooses an arc if the node label for its to-node ($l(j)$) is greater than the node label of its from-node plus the cost of traversing that arc ($l(i) + w_{ij}$). The node labels are the current estimate of the shortest path cost from $r$ to the node. Therefore, if $l(i) + w_{ij} < l(j)$ then a lower cost path can be found by going through node $i$ and arc$(i, j)$ to reach node $j$ instead through any previously estimated route to node $j$. This reflects a shortest path property discussed previously, namely, a "global" shortest path consists of "local" shortest paths. SPT updates node label $l(j)$ with the new estimate of the shortest path distance and updates the tree by replacing the arc that enters node $j$ with the arc$(i, j)$ in the tree $T$ by setting $p(j) = i$.

Stage 3 of the SPT algorithm is an optimality test. When this condition is met, it is no longer possible to update the shortest path cost estimates in $T(r)$. Therefore, it must consist of the shortest paths from $r$ to all network nodes. We report this tree as $T^*(r)$, the shortest path tree rooted at $r$.

*Example shortest path calculation*

Figure 5-2 illustrates the shortest path algorithm as applied to an example network. In this example, we generate $T^*(1)$, the shortest path tree rooted at node 1. The first row of the table is the initialization process in step 1. Following the convention discussed previously, we set $l(j) = \infty$ and $p(j) = 0$ for all nodes except for the origin.

Step 2 of the SPT algorithm tells us to check for network arcs that meet the condition $l(i) + w_{ij} < l(j)$. However, it does not specify an ordering for "scanning" or considering the arcs. This is a critical implementation decision that affects the SPT algorithm's efficiency and appropriateness for different network types. Although we discuss implementations in detail later, for now we use the strategy of choosing the node with the minimum label during each iteration and considering arcs directly connected to this node. This is sometimes referred to as a *best-first search* of the network (Gallo and Pallottino 1988).

A best-first strategy starts with the origin node (since it has the minimum label after initialization) and extends a SPT from the origin based on direct connectivity. Choosing the node with minimum label actually extends the SPT; in other words, the minimum label reflects a true shortest path cumulative cost. At any stage of the algorithm, only nodes that are directly connected to the current SPT can have noninfinite labels. Therefore, the minimum of the labels must be an extension of the current SPT since this node reflects a minimal cost extension of a known shortest path from the origin. (We encourage the reader to confirm this by sketching the evolution of $T(1)$).

Since its shortest path from the origin is known, we no longer need to

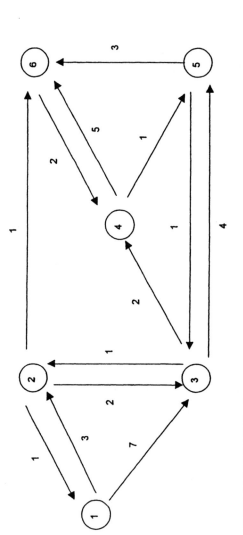

|  | Nodes | | | | | | | | | | | |
|---|---|---|---|---|---|---|---|---|---|---|---|---|
|  | 1 | | 2 | | 3 | | 4 | | 5 | | 6 | |
| Action | l(1) | p(1) | l(2) | p(2) | l(3) | p(3) | l(4) | p(4) | l(5) | p(5) | l(6) | p(6) |
| initialization | 0 | - | ∞ | 0 | ∞ | 0 | ∞ | 0 | ∞ | 0 | ∞ | 0 |
| scan from 1 | 0 | - | 0+3 < ∞? | 0 | 0+7 < ∞? | 0 | ∞ | 0 | ∞ | 0 | ∞ | 0 |
| update labels and T | 0 | - | 3 | 1 | 7 | 1 | ∞ | 0 | ∞ | 0 | ∞ | 0 |
| scan from 2 | 0 | - | 3 | 1 | 3+2 < 7? | 1 | ∞ | 0 | ∞ | 0 | 3+1 < ∞? | 0 |
| update labels and T | 0 | - | 3 | 1 | 5 | 2 | ∞ | 0 | ∞ | 0 | 4 | 2 |
| scan from 6 | 0 | - | 3 | 1 | 5 | 2 | 4+2 < ∞? | 0 | ∞ | 0 | 4 | 2 |
| update labels and T | 0 | - | 3 | 1 | 5 | 2 | 6 | 6 | ∞ | 0 | 4 | 2 |
| scan from 3 | 0 | - | 3 | 1 | 5 | 2 | 5+2 < 6? | 6 | 5+4 < ∞? | 0 | 4 | 2 |
| update labels and T | 0 | - | 3 | 1 | 5 | 2 | 6 | 6 | 9 | 3 | 4 | 2 |
| scan from 4 | 0 | - | 3 | 1 | 5 | 2 | 6 | 6 | 6+1 < 9? | 3 | 4 | 2 |
| update labels and T | 0 | - | 3 | 1 | 5 | 2 | 6 | 6 | 7 | 4 | 4 | 2 |
| STOP | 0 | - | 3 | 1 | 5 | 2 | 6 | 6 | 7 | 4 | 4 | 2 |

Note: Bold lines indicate nodes that are "frozen." See text for details.

consider additional updates to a chosen node. The shading in the table of figure 5-2 indicates nodes that are frozen and no longer considered. The algorithm terminates when the condition in step 3 is met, i.e., all nodes are frozen and we can no longer extend the SPT. This occurs in the last row of the table in figure 5-2; the entries in this row provides $T^*(1)$.

*Implementing the shortest path algorithm*

The SPT algorithm described above does not specify important implementation details that can affect the algorithm's performance. Given the increasing size of digital network databases and the complexity of problems being addressed using GIS-T software, seemingly small differences in implementation can make huge differences in practice. Even if you are not directly involved in software development, being an intelligent consumer of GIS-T software requires a basic understanding of these implementation issues.

The critical SPT implementation issues are (Gallo and Palottino 1988; Zhan 1997): (i) *network storage structure*, (ii) *labeling method*, and (iii) *selection rule/node processing structure*. Network storage structure refers to the physical data model for the transportation network. This includes the relative memory locations of network components and the memory *pointers* among locations that allow efficient retrieval of related information. The labeling method keeps track of which nodes are candidates for being scanned. The decision rule/node processing structure determines which candidate node should be scanned next. The decision rule is closely linked to the memory structure used to keep track of the candidate nodes identified by the labeling method.

*Network storage structure* Many shortest path implementations share an efficient network data storage structure known as the *forward star structure* (FSS). As suggested by the example SPT calculation, examining arcs according to their from-nodes is effective. The FSS supports this strategy through organizing network data by nodes and the arcs leaving each node.

A FSS consists of two data arrays, namely, an *arc array* $a(.)$ and an *arc weight array* $n(.)$. Also included is a *pointer array* $p(.)$ for accessing the two data arrays. Figure 5-3 provides an example for the network in figure 5-2. The array $a(.)$ contains the network arcs ordered according to their from-nodes. For example, in figure 5-3, the array $a(.)$ first lists all arcs leaving node 1, then all arcs leaving node 2, and so on. The ordering of multiple arcs incident to the same from-node is arbitrary. The set of arcs leaving a given node is called the *forward star* of that node. The array $w(.)$ contains the arc weights in the same order as $a(.)$. The pointer array $p(.)$ provides access to the information in $a(.)$ and $w(.)$. The $i$th element of $p(.)$ is the sequential location in $a(.)$ corresponding to the first arc leaving node $i$, i.e., the beginning of the forward star for that node. We also include an $n + 1$ element in $p(.)$ where $n$ is the number of nodes; this "points" to the $m + 1$ element of $a(.)$, i.e., a "null" location, indicating the end of the node list. If the forward star for node $i$ is empty, by convention we set $p(i) = p(i + 1)$; in other words, the next element of the pointer array.

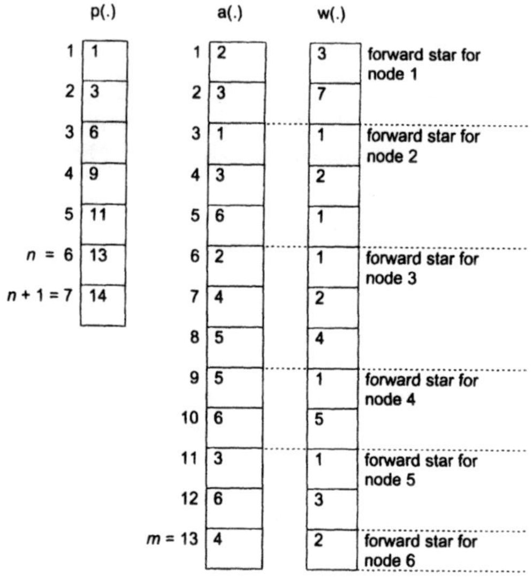

| p(.) | | a(.) | | w(.) | | |
|---|---|---|---|---|---|---|
| 1 | 1 | 1 | 2 | | 3 | forward star for node 1 |
| 2 | 3 | 2 | 3 | | 7 | |
| 3 | 6 | 3 | 1 | | 1 | forward star for node 2 |
| 4 | 9 | 4 | 3 | | 2 | |
| 5 | 11 | 5 | 6 | | 1 | |
| n = 6 | 13 | 6 | 2 | | 1 | forward star for node 3 |
| n + 1 = 7 | 14 | 7 | 4 | | 2 | |
| | | 8 | 5 | | 4 | |
| | | 9 | 5 | | 1 | forward star for node 4 |
| | | 10 | 6 | | 5 | |
| | | 11 | 3 | | 1 | forward star for node 5 |
| | | 12 | 6 | | 3 | |
| | | m = 13 | 4 | | 2 | forward star for node 6 |

Figure 5-3: Forward star structure for the network in Figure 5-2

*Labeling method* The labeling method supports the selection and updating of node labels $l(i)$ and parent nodes in the shortest path tree $T(r)$ during step 2 of the generic SPT algorithm. Efficient implementation of this step requires identifying a set of *candidate nodes* $Q$ for possible selection.

In the SPT example calculation (figure 5-2), the best-first search of the network treated nodes differently based on their current *status* in the execution. A node could be in one of the following states:

- *unreached*, meaning that the algorithm has not yet found a possible extension of the current SPT to the node; its label is the same as assigned initially
- *temporarily labeled*, meaning that the algorithm has found a possible extension of the SPT but it is tentative and is subject to further updating
- *permanently labeled*, meaning that the node's position in the SPT is known and not subject to further updating

In the best-first search example, unreached nodes were not candidates for the next node processed (the next base node for scanning), since not enough is known about these with respect to the current SPT. Permanent nodes have their SPT position known and are not subject to further processing; therefore, they are no longer candidates for possible selection. The temporarily labeled nodes were the only candidates for becoming the next node to be processed. This is common across all SPT implementations: temporarily labeled nodes comprise the candidate set $Q$.

*Selection rule and node processing* The SPT example discussed previously chooses the node with the minimum label in the current candidate set $Q$. This is known as a *best first* (BF) selection rule. Other possible selection rules include *first-in first-out* (FIFO), *last-in first-out* (LIFO) and *hybrid FIFO-LIFO*. These selection rules are the major source of performance differences among SPT implementations. As we will see, implementing these rules is more directly a matter of choosing a memory structure for processing the nodes in a desired order. We will first discuss the FIFO and LIFO rules since their memory structures are more fundamental. We will then discuss the BF implementations.

FIFO chooses the oldest node in $Q$ as the next node to be processed. This generates a *breadth-first search* of the network in which all nodes directly connected to the current base node are scanned before scanning other nodes (Gallo and Pallotino 1988). This is a sweep-like strategy that searches the network in concentric "rings" of increasing distance from the origin (see Sedgewick 1992, ch. 29). LIFO chooses the most recent node to enter $Q$ as the next node to be processed. This generates a *depth-first search* where the algorithm processes a node adjacent to the current base node, then processes a node adjacent to that node and so on. When continued expansion of the branch is no longer possible it returns to a neighbor of the original base node and repeats the process to that branch's maximum depth. This is a probe-like exploration of the network that looks for nodes at increasing distance from the origin, only returning after reaching a "dead-end" (Sedgewick 1992, ch. 29).

FIFO and LIFO selection rules require the members of $Q$ to be ordered according to the time of entry to the set. A *list* is a generic data structure that can accommodate an ordered set. A list is some sequence of elements with a *head* corresponding to the first element and a *tail* corresponding to the last element. Elements can be removed or added only to the head and tail of the list (Aho, Hopcraft, and Ullman 1974; Gallo and Pallottino 1988).

SPT implementations use several types of lists. A *queue* is a list that only allows additions at its tail and removals at its head. A queue typically implements the FIFO selection rule: since elements enter the tail but leave the head, only the oldest element can be removed. A *stack* is a list that allows an addition (a *push*) and a removal (a *pop*) at its head (or *top*). (A common analogy is the spring-loaded tray dispenser at many cafeterias.) A stack typically implements the LIFO rule since only the most recent element can be removed. Figure 5-4 illustrates the basic structure of queues and stacks.

As the name implies, the hybrid FIFO-LIFO selection rule uses both FIFO and LIFO temporal selection rules in concert depending on the current status of the element in $Q$. The first time a node enters $Q$, it is subject to FIFO processing. After removal from $Q$ and subsequent re-entry, the node is subject to LIFO processing. The rationale is that initial scanning of unknown portions of the network should be breadth-first. If a node is processed subsequently, it is likely that its current SPT information is more accurate; therefore, the algorithm should also try to update its successors through a depth-first search.

Two types of hybrid lists can implement the hybrid FIFO-LIFO selection rule. A *deque* or "double-ended" queue can allow additions and removals at both ends. A special type of deque used in SPT algorithms allows additions at both ends but removals from only one end. Figure 5-4 illustrates the deque structure. A node is added at the end of the deque on its first entry to $Q$ and therefore is subject to FIFO processing. During subsequent processing, the node is added at

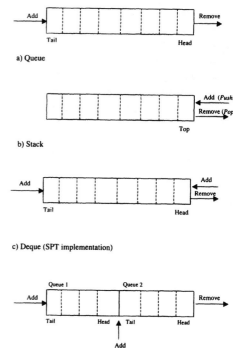

Figure 5-4: Types of list structures used in SPT algorithms

the head of the deque and is subject to LIFO processing. We can interpret this as a LIFO stack and a FIFO queue connected in series so that the tail of the stack points to the head of the queue. See Pallottino (1984) for additional details.

The *two-queue* is another type of list that can implement the hybrid structure. As the name implies, a two-queue is two queues linked together. Elements can be added to the tail of either queue but only removed from the head of the second queue (see figure 5-4). Initially scanned nodes are added to the tail of the first queue while previously scanned nodes are added to the tail of the second. Although both are FIFO structures, previously scanned nodes have higher priority than initially scanned nodes, resulting in a network search similar to the deque (Cherkassky et al. 1993; Gallo and Pallottino 1998).

Insert and delete operations on queues and stacks are very efficient. Both operations require only constant time (i.e., $O(1)$; refer to table 5-1) for all of the structures discussed above (Corman, Leiserson, and Rivest 1998). However, SPT implementations that use queues tend to perform poorly. Poor performance results from the relative inefficiency of FIFO and LIFO selection rules in the SPT search. For example, the breadth-first "sweeping" of the remaining network from each node in a FIFO search could result in repeated scanning of the same node from every node in the network. This implies an SPT algorithm with an asymptotic time complexity of $O(nm)$ where $n$ is the number of nodes and $m$ is the number of arcs. Hybrid FIFO-LIFO using the deque and two-queue both have much higher worse-case bounds ($O(n2^n)$ and $O(n^2m)$, respectively; the former being an exponential bound) (Cherkassky et al. 1993). (However, as we will see in the next section, performance can be good in practice, illustrating the occasionally misleading nature of complexity analysis.)

The shortest path properties discussed earlier suggest that a BF search is a more intelligent search rule and may perform better than LIFO and FIFO. However, BF requires additional processing with respect to maintaining the candidate set $Q$ as well as operations manipulating elements of the set.

The BF search strategy requires a data structure that can manipulate and update elements based on their current numeric values (i.e., the node labels) rather than when they entered the candidate set. A *priority queue* is a data structure that supports efficient sorting of elements based on their numeric values (sometimes referred to as their *priorities*), updating of the numeric values and removing the element with the minimum value. Naïve BF implementations use a type of priority queue known as a *linked list*. This strategy defines a sequence by defining a memory pointer between each data element and the next element in the sequence. The last element points to a "dummy" element that in turns points to the first element in the sequence. Finding the element with the minimum value, a key operation in BF search strategies, could require scanning the entire list (Gallo and Pallottino 1988; Horowitz and Sahni 1976).

More efficient priority queues for SPT implementation include *buckets* and *heaps*. In the bucket strategy, we store data elements with similar values into an aggregate memory unit, i.e., a "bucket." If we need to find an element with the minimum value, we only need to look a single bucket or relatively few buckets rather than through all of the elements. In the heap strategy, we store elements in a quasi-ordering in a "heap" or "pile"-like memory structure that has high values at the bottom, medium values in the middle and the minimum value at the top. Retrieving the minimum value is efficient, only requiring choosing the element at the top of the heap.

SPT implementations define buckets based on node labels falling within specified intervals, with the buckets arranged in increasing order based on their intervals (Denardo and Fox 1979). Figure 5-5 provides a (somewhat simplified) representation of a bucket data structure. The top half of the figure shows an input data vector with eleven node labels. The bottom half shows the same data stored in a bucket structure. The candidate set $Q$ corresponds to an array of pointers. Each element points to the first element of a linked list that contains node labels within a specified range. These pointers can be easily redirected if a node label is updated and must be moved to another bucket. The numbers to the left of $Q$ indicate the ranges for each bucket. Note that two buckets ("100–199" and "300–399") are empty; therefore these do not point to any list. (More precisely, they point to a null element.) See Aho, Hopcraft, and Ullman (1974), Cormen, Leiserson, and Rivest (1998), or Gallo and Pallottino (1988) for more details.

The bucket strategy trades fast retrieval operations for less efficient use of memory. The asymptotic time complexity of inserting or deleting an element into a bucket is linear (Cherkassky et al. 1993). While this is fast, the total time complexity can nevertheless be substantial since these operations may need to be performed across all buckets (in the worse-case). Therefore, it is critical to use

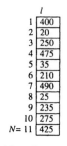

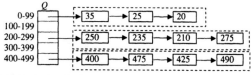

**Figure 5-5:** Bucket data structure

a) Input data vector

b) Data stored using a bucket structure

as few buckets as possible in a SPT implementation. A straightforward method is to set the number of buckets as $nC$, where $n$ is the number of nodes and $C$ is the maximum arc length in the network expressed as an integer (Dial 1969). (For ease of processing, we treat arc lengths as integers by scaling appropriately, such as multiplying arc lengths by a factor of 100.) However, it can be shown that at most $C + 1$ buckets are occupied at any stage in the SPT algorithm; exploiting this can reduce memory requirements (Ahuja, Magnanti, and Orlin 1993). We can reduce this number even more by using more-sophisticated bucket strategies such as overflow bag, approximate buckets, and double buckets.

The *overflow bag* approach requires the user to specify the number of buckets as $\alpha < C + 1$. This may be less than the number of buckets required. We get by with the smaller number of buckets by placing into buckets only the temporary node labels below a specified threshold. A special memory unit known as the overflow bag contains the node labels that are greater than or equal to the threshold. Overall asymptotic time complexity is $O[m + n(C/\alpha + \alpha)]$. In the *approximate buckets strategy*, the user specifies an integer $\Delta$ that sets a fixed interval for each bucket. We start with a single bucket containing node labels from 0 to $\Delta - 1$. Each stage of the algorithm adds a new bucket with the range $\Delta$, with the $i$th stage adding the bucket that contains temporary node labels within the range $[i\Delta, (i + 1)\Delta - 1]$. Nodes within each bucket are processed in FIFO order. This strategy sacrifices speed for reduced memory since each node may be scanned more than once but never more than $\Delta$ times. The worse-case complexity for this implementation is $O[m\Delta + n(\Delta + C/\Delta)]$. The *double bucket* strategy integrates ideas from the overflow bag and approximate buckets strategy. This approach designates two types of buckets, namely, *low-level buckets* and *high-level buckets*. The user specifies the number of low-level buckets as the integer $\Delta$. Similar to the overflow bag strategy, the algorithm stores the smallest node labels in the low-level buckets and searches these for the next minimum label to process. Larger node labels are placed into high-level buckets for distribution to the low-level buckets at later stages. After the low-level buckets are examined, their range shifts and nodes in the high-level buckets are moved to the corresponding low-level buckets. Asymptotic time complexity is $O[m + n(\Delta + C/\Delta)]$ (Cherkassky et al. 1993).

A *heap* is a tree structure that stores a set of elements or aggregate memory locations (e.g., buckets) in *heap-order* based on their current values. "Heap-order" means that the children of a node (if they exist) have numeric values that are greater than or equal to the parent's element. This restriction means that the root of the tree has the element with the minimum numeric value. Therefore, it is easy to find the extreme element of the set.

The simplest type of heap is a *binary heap*: this is a heap structured as a *binary tree* or a tree where each node has at most two children. Binary heaps are suitable for SPT algorithms only if the network is sparse (that is, each node is connected to only a few of the other nodes in the network; a condition satisfied by real-world transportation networks). Figure 5-6 illustrates a binary heap for the input data vector in figure 5-5. The top half of figure 5-6 shows a tree representation of the binary heap while the bottom half shows its representation as the array $Q$. Since we are maintaining node labels for a SPT algorithm, we require that the values stored at the children nodes are greater than or equal to their parent. We index the nodes such that a parent with index $i$ (indicated by the number to the left of each box) has children indexed as $2i$ (the left child) and $2i + 1$ (the right child). Also note that node labels within any given level of the tree are ordered arbitrarily; for example, the nodes labels stored in positions 4, 5, 6, and 7 are 235, 210, 250, and 275. Finally, note that the root of the tree points to the minimum label; this is the value we need to retrieve in a best-first search.

The major operations on heaps required for the SPT algorithm include creating the heap, removing the minimum element from the heap and inserting a new element to the heap. Another operation often used is to decrease the value of a node label in the heap; the label updating phase of the algorithm uses this operation directly. Building a binary heap requires at most $O(n)$ time, where $n$ is the number of nodes in the network. Other manipulations required for SPT algorithms such as removing the node with the minimum label and updating node labels both require at most $O(\log n)$ time for each operation. Since the BF search only scans nodes once, there are at most $n + m$ of these operations. Therefore, a binary tree implementation requires at most $O[(n + m)\log n]$ time (Cormen, Leiserson, and Rivest 1998).

Other types of heaps can improve the efficiency of the operations used in the SPT algorithms. Some of these heaps include *d*-heaps, Fibonacci heaps, and radix heaps. *d-heaps* are a generalization of a binary heap in which each parent has no more than $d$ children. If $d = \lceil 2 + m/n \rceil$, where $\lceil 2 + m/n \rceil$ indicates the largest integer less than or equal to $2 + m/n$, the worse-case time complexity for a *d*-heap SPT implementation is $O(m \log n)$ (Tarjan 1983). A *Fibonacci heap* is a collection of heap-ordered trees. Unlike binary or *d*-heaps, there is no restriction on the number of nodes or structure of the trees in a Fibonacci heap. Rather, the restric-

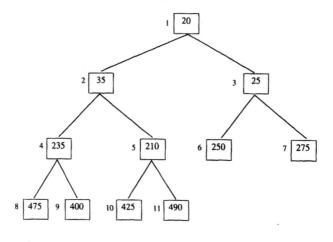

a) Binary heap – tree representation

b) Binary heap – array representation

Figure 5-6: Heap data structure for input vector in Figure 5-5

tions are on how the trees are manipulated in operations such as inserting and deleting nodes or linking trees. The term "Fibonacci" derives from the resulting heap displaying properties of the Fibonacci sequence (1, 2, 4, 7, 12, ...); see Fredman and Trajan (1987) or Cormen, Leiserson, and Rivest (1998) for more detail. Fibbonacci heaps result in a SPT algorithm that requires $O(n \log n + m)$ time in the worse-case (Fredman and Trajan 1987). A *radix heap* maintains a set of buckets in heap-order. The size of each bucket increases according to its index in the tree, with the root bucket containing at most one node. The range of node labels maintained in each bucket varies as the algorithm executes. A type of priority queue known as a *doubly linked list* maintains the set of node labels within each bucket. Total time complexity for the resulting SPT algorithm is $O(m + n \log C)$ where $C$ is the upper bound on the arc costs in the network. See Ahuja et al. (1990) for details.

*Choosing an SPT implementation: An evaluation using real-world networks*

Table 5-2 summarizes 15 SPT implementations with their asymptotic time complexity. As discussed earlier in this chapter, complexity analysis provides very

Table 5-2: Selected Shortest Path Tree Algorithms

| Implementation[a] | Selection Rule | Q Storage Structure | Complexity[a] | References |
|---|---|---|---|---|
| Bellman-Ford-Moore | FIFO | Queue | $O(nm)$ | Bellman (1958) |
| | | Queue with parent checking | $O(nm)$ | Ford and Fulkerson (1962) |
| | | | | Moore (1959) |
| Dijkstra | BF | Unordered list (naïve) | $O(n^2)$ | Dijkstra (1959) |
| | | Buckets | $O(m + nC)$ | Dial (1969) |
| | | Buckets with overflow bag | $O[m + n(C/\alpha + \alpha)]$ | Cherkassky et al. (1993) |
| | | Approximate buckets | $O[m\Delta + n(\Delta + C/\Delta)]$ | |
| | | Double buckets | $O[m + n(\Delta + C/\Delta)]$ | |
| | | Fibonacci heap | $O[n\log(n + m)]$ | Fredman and Tarjan (1987) |
| | | $d$-heap | $O[m\log(n)]$ | Tarjan (1983) |
| | | $R$-heap | $O[m + n\log(C)]$ | Ahuja et al. (1990) |
| | | | $O(n2^n)$ | Pape (1974) |
| Incremental graph | Hybrid FIFO-LIFO | Deque | $O(n^2 m)$ | Pallottino (1984) |
| | | Two-queue | $O(nm)$ | Glover et al. (1984, 1985) |
| Threshold | FIFO | Separate queue for nodes with labels less than a threshold value | $O(nm)$ | |
| Topological ordering | LIFO | Linearly ordered set based on topological distance from base node | $O(nm)$ | Goldberg and Radzik (1993) |
| | | Linearly ordered set based on topological distance with distance updating | | |

*Source:* Adapted from Cherkassky et al. 1993; Gallo and Pallottino 1988; Zhan and Noon 1998.

[a] $n$ = number of nodes; $m$ = number of arcs; $C$ = maximum arc length in network; $\alpha, \Delta$ = input parameters (see text).

general guidelines for selecting implementations. Run-time analyses are generally less reliable, but can provide more detailed answers if performed within a careful experimental design. Zhan and Noon (1998) conduct a careful run-time analysis of the 15 shortest path implementations. Unlike most computational evaluations of SPT algorithms, the Zhan/Noon experiments explicitly consider real-world transportation networks.

Real-world transportation networks have characteristics that are unique relative to most of the universe of possible networks. Transportation networks are *sparsely connected*: Each node is directly connected to only a tiny portion of the other nodes in the network (van Vliet 1978). For example, a street intersection is typically connected to only a few neighboring intersections. In abstract networks, a node may be connected to all other nodes in the network. Also, adjacent nodes in real-world networks tend to be proximal geographically. They have higher likelihood of being adjacent on the shortest path than in abstract networks where there is no such adherence to geography. Most run-time analyses of shortest path implementations use randomly generated networks; therefore, their connectivity patterns rarely reflect the situation faced by the transportation analyst. Zhan and Noon (1998) resolve this problem by using the U.S. National Highway Planning Networks (NHPN) and both low- and high-detail networks for ten states in the Midwest and Southeast United States.

The Zhan and Noon (1998) experiments lead to the following suggestions. First, implementations that should not be used for transportation networks are *Bellman-Ford-Moore*, *Bellman-Ford-Moore with parent checking* and *Dijkstra Naïve Implementation*. These implementations exhibited inconsistent run-time performance and performed very poorly on large networks. Recommended implementations for transportation networks are *Graph Growth by Palottino* when solving the one-to-all shortest paths problem (i.e., the spanning or "full" SPT for a network) and two Dijkstra implementations when solving the one-to-one and one-to-some shortest paths problems (i.e., terminating the algorithm after finding the destination(s) of interest). The *Dijkstra Approximate Bucket* implementation is best when the maximum arc weight is less than or equal to 1,500. Both *Dijkstra Approximate Bucket* and *Dijkstra Double Bucket* are best when the maximum is greater than or equal to 1,500.

*Generating shortest paths quickly: Parallel-processing and heuristic strategies*

Even with a judicious choice of SPT implementation, the vast size of some real-world transportation network databases can result in unacceptably long execution times, particularly for applications that demand answers in real-time or near real-time. Examples include intelligent transportation system (ITS) and emergency vehicle routing. We can follow one of two strategies in these and other time-critical applications. If a parallel platform is available, we can use a parallel-processing version of the SPT algorithm; this can result in much quicker processing times relative to a serial computer. If a parallel-processing platform is not available, or if we want to solve large problems on a modest computational platform, we can relax our requirements for an optimal shortest path and instead use a heuristic algorithm.

*Parallel-processing implementations* Parallel processing implementations involve two related issues. First is the type of parallel platform. The *single instruction multiple data stream* (SIMD) platform use less powerful but numerous processors, with each working with identical instructions on different data elements. In contrast, *multiple instruction multiple data stream* (MIMD) platforms use more powerful but less numerous processors, with each processor having its own instruction set and data elements. A second and related issue is the strategy for partitioning the overall problem into parallel computations. *Task parallelism* decomposes the overall problem into relatively independent processes; this strategy is more appropriate for MIMD platforms. *Data parallelism* decomposes the data set into smaller data elements that are distributed among the processors; this strategy is more appropriate for SIMD platforms. Since spatial databases tend to be very large and data are more easily decomposed than tasks, data parallelism tends to be used more often than task parallelism in spatial analysis problems (Xiong and Marble 1996).

Partitioning a network for parallel shortest path computations can be difficult. Databases can be easily partitioned if they are regular and homogeneous. Transportation networks rarely display these characteristics. Most real-world transportation networks have an irregular geometric structure with respect to the location of nodes and the connections among the nodes (a contrasting example of a regular network is a lattice). Also, the data elements in transportation networks tend to be heterogeneous; for example, a node can be an origin, a destination or neither. Another complicating factor is the domain of shortest paths: these can span the entire network and therefore can require global processing (Densham and Armstrong 1998).

Densham and Armstrong (1998) discuss two strategies for spatial data parallelism in shortest path calculations. The *nonoverlapping partitions* strategy divides the network into nonoverlapping partitions that are processed independently using an SPT algorithm. The procedure assigns each partition to a processor that builds shortest paths from every origin in that partition. These paths extend to the partition boundaries if the path length is unconstrained. After processing all partitions, the incomplete paths are assembled into full paths. Although quick, this strategy requires large amounts of memory to store the partial paths prior to assembly.

The *overlapping partitions* divides the network into partitions that overlap sufficiently to permit a full set of shortest paths to be generated for each. The degree of overlap depends on the maximum allowable path length. Therefore, this technique is more appropriate for shortest path problems with a restricted path length; otherwise, each partition may span most or the entire network. Densham and Armstrong (1998) use the *dynamic recursive alternating bisection* (DRAB) method to divide the irregular network into partitions with equal numbers of origins. Each partition is assigned to a processor that computes the shortest paths from the origins in that partition.

Karimi and Hwang (1997a, 1997b) discuss a method for enhancing the nonoverlapping partitions spatial data parallelism strategy. Their algorithm is:

Step 1. Partition the network into non-overlapping subnetworks.

Step 2. Identify the origin and destination nodes and the subnetworks that contain them.

Step 3. Run the SPT algorithm to compute shortest paths between the origin and all *boundary vertices* within its subnetwork (a "boundary vertex" is a node where two arcs from different subnetworks are incident).

Step 4. Run the SPT algorithm to compute shortest paths between the destination and all boundary vertices within its subnetwork.

Step 5. Run the SPT algorithm to compute the shortest path between all pairs of boundary vertices in the remaining subnetwork.

Step 6. Construct a hypernetwork or abstract network representing all of the above shortest paths.

Step 7. Run the SPT algorithm on the hypernetwork to determine the actual shortest path between the origin and destination.

Steps 3, 4, and 5 can be run in sequence with the task within each step executed in parallel. If enough processors are available, Steps 3, 4, and 5 can also be executed in parallel.

*The A\* algorithm* If a parallel processing platform is not available and shortest path calculations are required quickly, we must settle for a heuristic shortest path algorithm that will return "good" (near-optimal) answers quickly. A heuristic strategy that has good convergence properties and also performs well in practice is the $A^*$ algorithm. $A^*$ computes the shortest path between a specified origin-destination pair rather than the SPT rooted at a given origin. $A^*$ modifies the best-first processing strategy by using a heuristic "look-ahead" function to direct the search for the destination node. Under certain conditions, $A^*$ will return the exact optimal path.

Similar to the SPT algorithm, $A^*$ updates node labels based on current estimates of the shortest path and choosing the lowest node label for processing in its best-first selection. However, in contrast to the SPT algorithm, each node label is an additive function of two components:

$$l(i) = f(i) + h(i) \qquad (5\text{-}2)$$

where $l(i)$ is the node label, $f(i)$ is the estimated shortest path cost from the origin to node $i$, and $h(i)$ is the estimated shortest path cost from node $i$ to the destination node. In other words, each node's label is the current estimate of the shortest path cost from the origin to the destination constrained to go through that node. The $h(i)$ component of the node label helps to focus the search on the destination node rather than simply generate a SPT rooted at the origin node (Nilsson 1998).

The performance of $A^*$ depends critically on the specification of the "look-ahead" function $h(i)$. $A^*$ is guaranteed to return the optimal shortest path if the network does not have negative arc costs and if the look-ahead function never overestimates the true cost of the shortest path from the current node to the destination (Galperin 1977; Shekhar, Kohli, and Coyle 1993). Specifying a look-ahead function that meets this requirement involves a trade-off between the computational expense of computing that function and the savings enjoyed by narrowing the search in the overall algorithm. For example, we could define $h(i) \equiv 0$ for every node in the network. This requires no additional computational burden and meets the requirements on that function. However, the overall algorithm will behave like a standard Dijkstra-like search and offer no performance advantage. When applying $A*$ to transportation networks, an effective strategy is to use the Euclidean distance between the current node and the destination node as the look-ahead measure. This focuses the search geographically toward the destination node (Nilsson 1998).

Shekhar, Kohli, and Coyle (1993) and Shekhar and Fetterer (1996) analyze the performance of $A*$ as a shortest path routing algorithm for ITS applications. They compare $A*$ with a breadth-first search strategy and the Dijkstra algorithm. Their experiments use a synthetic grid network as well as an actual road network (Minneapolis, a mid-sized city in the Midwestern United States). Euclidean and Manhattan distance functions serve as the look-ahead measure. Results suggest that $A*$ is promising for ITS and other time-demanding routing applications. With respect to the synthetic networks, $A*$ outperformed the breadth-first and Dijkstra algorithms if the origin-destination path cost is smaller than the *diameter* of the network (the maximum shortest path length in the network). It also performed better if distribution of arc costs is skewed such that arc costs between the origin and destination tend to be lower than average. With respect to the real road network, $A*$ greatly outperforms Dijkstra but only outperforms the breadth-first search strategy for short path lengths. Results indicate that the look-ahead function is critical: The Manhattan function outperforms the Euclidean function on the grid networks. The authors also make some suggestions regarding management of the node candidate set in the algorithm.

### Shortest paths through surfaces

Although most transportation analyses focus explicitly on transportation networks, we sometimes need to consider "off-network" shortest or least cost paths. For example, we may wish to model off-network movement through terrain, open water or even air depending on the type of conveyance (Goodchild 1998). We can view these media as cost *surfaces* and solve for the shortest or least cost paths through these surfaces. Another example is *corridor analysis*; this involves determining "corridors" (sites) for locating new transportation or telecommunication facilities. Analyzing alternative corridors involves solving optimal paths through a surface representing varying suitability for the facility (e.g., Huber and Church 1985; Lombard and Church 1993).

Mathematically, a *surface* is a 2-D graph within a 3-D space representing the function $z(\mathbf{x})$, where $\mathbf{x} \in \mathcal{R}^2$ are continuous locations in 2-D Cartesian. We can also generalize this concept to higher dimensions, but this restricted definition is sufficient for most transportation applications. We designate this as a *cost surface* $c(\mathbf{x})$ if the function is nonnegative for all locations $\mathbf{x} \in \mathcal{R}^2$.

Given a cost surface, we could determine the minimum cost path between locations $x_i$ and $x_j$ by solving the following *variational problem* (Angel and Hyman 1976; Goodchild 1977):

$$\min_{\{P_{ij} \in \Re^2\}} \int_{P_{ij}} c(\mathbf{x}) d\mathbf{x} \tag{5-3}$$

where $P_{ij}$ is a subset of 2-D space corresponding to a "path" between locations $x_i$ and $x_j$ (roughly, a set of contiguous locations that includes locations $x_i$ and $x_j$). In other words, we must choose the $P_{ij}$ with the minimal cumulative cost. The cumulative cost of each path involves integrating the costs at each location since we are dealing with a continuous surface and therefore must "sum" the infinite costs corresponding to the infinite locations within each path.

Solving equation (5-3) directly is difficult except for special cases (see Miller 2000). For example, a classic, ground-breaking study by Angel and Hyman (1976) analytically solve equation (5-3) for the *radially-symmetric* case where $c(\mathbf{x})$ is a function of distance from a single point (e.g., the central business district of an urban area). This is somewhat unrealistic, particularly given contemporary multicentric urban form. However, we can solve *approximations* of equation (5-3) corresponding to discrete representations of the continuous surface.

One method for solving equation (5-3) is to aggregate the continuous locations into a tessellation (i.e., mutually exclusive and exhaustive) of cost *polygons*. In this case, the problem of finding the minimum cost path reduces to finding the locations where the path crosses polygon boundaries since the path within a polygon is a straight line segment. Werner (1968) demonstrates that the boundary point location problem can be solved using the *law of transportation refraction*. This reformulation of Snell's law for the refraction of light provides the conditions for the polygon boundary crossing locations (see Werner 1968, 1985 for more details).

Smith, Peng, and Gahient (1989) and Mitchell and Papadimitriou (1991) develop procedures for the case where the polygons are a triangulation of the 2-D plane. Smith, Peng, and Gahient (1989) use a generalization of Snell's Law to construct a family of local, asynchronous and parallel algorithms that find least cost path when the cost polygons are a triangulation. *Local* means that each processor only requires information from neighboring processors. *Asynchronous* means that neighboring processors do not execute simultaneously. These properties are achieved by restricting crossings to selected locations within the boundary segments of a triangle. The algorithms are exact for the special case of a corridor of triangles along a single dimension. Globally optimal paths are not guaranteed for the 2-D triangulation, but numerical experiments are promising. Mitchell and Papadimitriou (1991) apply Snell's Law for the case where the tessellation is a constrained Delaunay triangulation. This is an exhaustive tiling of nearly equilateral triangles used in triangulated irregular network (TIN) terrain models. Their procedure generates a "map" of shortest paths from a given source location, allowing the user to determine a specific shortest path by querying from the database. The procedure requires $O(n^4)$ in the worse-case but is likely to be smaller in practice.

Another special case occurs when the polygons correspond explicitly to terrain in 3-D Euclidean space. In this case, the polygons are not weighted by cost; instead, the objective is to find the minimum Euclidean distance path on this surface. Mitchell, Mount, and Papadimitriou (1987) develop an $O(n^2 \log n)$ procedure for arbitrary (possibly nonconvex) polyhedral surfaces. De Berg and Van Kreveld (1997) develop procedures that determine shortest paths over polyhedral surfaces that are restricted to stay below a specified elevation or minimize total ascent. This is useful for cross-country movement planning in mountainous terrain.

Another possible approach is to restrict the cost locations to a discrete and finite regular square grid (RSG) or lattice within the plane and specify interaction costs only between "neighbors." We can define the "neighbors" based on the *rook's case* (nearest neighbors only in directions parallel to the axes) or the *queen's case* (rook's case plus nearest neighbors along the diagonals). This creates a network, but the regularity and density of locations better represent the properties of continuous surfaces. Therefore, a RSG network can represent terrain and other "physical" cost surfaces. In a GIS, we can treat each RSG point as the centroid of a small raster cell.

Goodchild (1977) examines the properties of the RSG approximation when solving equation (5-3). We can solve this special case using the SPT algorithms discussed previously in this chapter. However, in general multiple paths rather than a single path will correspond to the least cost. Three types of errors are introduced by using a lattice approximation of a continuous cost surface. These are elongation, deviation, and proximity distortions. *Elongation* errors occur since the lattice path will be longer than the continuous space path. *Deviation* errors occur when the lattice path differs in location from the continuous space path; this occurs when all "moves" in one direction are executed first. *Proximity distortions* occur since the cost measure for a raster cell normally does not consider neighboring cells. The resulting paths can be optimal with respect to their site but suboptimal with respect to their situation (Goodchild 1977; Huber and Church 1985; Lombard and Church 1993; van Bemmelen et al. 1993).

Intuition suggests that as the RSG becomes finer and approaches continuous space, the solution of the shortest path problem will become a better approximation of the continuous space solution. This is not the case: Elongation and deviation errors are independent of the RSG density (Goodchild 1977). These errors relate to the permutation of move directions and therefore cannot be eliminated through a finer mesh. Strategies for reducing these errors include *more connected rasters* and an *extended raster approach*. The more-connected raster approach means connecting each lattice point to more of its neighbors (e.g., connecting to the closest 16, 32, 64, ... lattice points instead of just four or eight in the rook's and queen's cases). This can reduce elongation and deviation errors but requires additional computational expense and can create nonintuitive intersecting paths. The extended raster approach configures the network so that a path enters and exits each cell at finite boundary locations rather than traveling between the centroids of each raster (i.e., the RSG points). The cost of each segment of the path is easier to calculate since it is contained within a single cell. These strategies can also be combined with a quadtree structure for reducing data storage and reducing computational requirements through efficient data access (van Bemmelen et al. 1993).

## Routing vehicles within networks

In this section of the chapter, we will consider certain types of network problems that arise frequently in *logistics*, or the procurement and/or distribution of goods and services. We will consider two closely related problems, namely, the *traveling salesman problem* (TSP) and the *vehicle routing problem* (VRP). Recall that earlier in this chapter, we defined a *Hamiltonian cycle* or a *tour* as a path that includes every node of a network or of a selected subnetwork and returns to the start node. We also mentioned that the TSP involves finding the shortest or lowest cost tour of the network or a subnetwork. This problem has obvious importance for transportation applications such as customer delivery or pick-up services. However, as we will see, the TSP also has tremendous theoretical importance as well.

The VRP can be viewed as a "fleet" version of the TSP. The problem is to find the best tours for a set of vehicles such that each node is "covered" (i.e., receives a stop by a vehicle) and the capacities of the vehicles are not violated. We will see that this involves two highly interrelated subproblems, namely, partitioning the nodes among vehicles subject to capacity constraints and finding the shortest tour for the nodes assigned to each vehicle. The tour for each vehicle starts and ends at the same location, often referred to as the *depot*. There are numerous variations of the basic VRP, including time windows for pick-up and delivery, stochastic demand, multiple vehicle types, and so on. Therefore, understanding the basic VRP is critical for entry into a vast literature on vehicle routing methods.

### The traveling salesman problem

#### Overview of traveling salesman problems

The *traveling salesman problem* (TSP) is probably one of the most famous (some might say infamous) optimization problems of all time. (Although "traveling sales*person*" would be more appropriate, we keep the traditional name for historical consistency.) This is for a number of reasons. First, it is easy to state yet exceedingly difficult to solve. Second, it has tremendous and far-reaching importance. TSP is important directly as a transportation problem and for a surprising number of non-transportation applications such as computer wiring, hole punching in manufacturing processes, job sequencing, and crystallography (see Kolata 1991 and Laporte 1992). It is also indirectly important since it is also equivalent to a large class of important optimization problems. In other words, an optimal solution procedure for the TSP would also solve exactly a large number of other important optimization problems. However, while there is no proof that a polynomial time optimal solution procedure does not exist for the TSP, it is very likely that this is the case.

The name "traveling salesman problem" comes from the analogy of a salesperson who must visit every city in a designated territory and return to the starting city. For modeling purposes, we assume a network where the locations to be visited (referred to hereafter as "demand locations") are nodes and arcs represent *direct* connections between the demand locations. If we are solving the TSP in planar (continuous) space, the arcs represent direct travel between any pair of demand locations and the arc weights correspond to the distance (or some function of distance) between the two demand locations. If we are solving the TSP for customers embedded within a transportation network, the directly connected network can be viewed as an abstract "pseudonetwork" whose arc weights correspond to the shortest path lengths between customers in the real network.

Although there are many ways to formally state the TSP, a convenient statement that fits in well with the discussion in this chapter is the following *constrained optimization problem* (Dantzig, Fullerson, and Johnson 1954; Fisher and Jaikumar 1981; Nemhauser and Wolsey 1988). Assuming a directly connected network $G = [N, A]$ where $|N| = n$ (in this context, the symbol " $||$ " indicates the *cardinality* or size of the set, i.e., $n$ is the number of nodes):

$$\min_{\{x_{ij}\}} \sum_{i=1}^{n} \sum_{j=1}^{n} c_{ij} x_{ij} \tag{5-4}$$

subject to:

$$\sum_{i=1}^{n} x_{ij} = 1, \; j = 1, \ldots, n \tag{5-5}$$

$$\sum_{j=1}^{n} x_{ij} = 1, \; i = 1, \ldots, n \tag{5-6}$$

$$\sum_{\substack{(i,j) \in A \\ i \in N' \\ j \in N \setminus N'}} x_{ij} \geq 1, \; N' \subseteq N, 2 \leq |N'| \leq |N| - 2 \tag{5-7}$$

$$x_{ij} \in \{0, 1\} \tag{5-8}$$

where:

$$x_{ij} = \begin{cases} 1 & \text{if } j \text{ immediately follows } i \text{ on the tour} \\ 0 & \text{else} \end{cases}$$

$c_{ij}$ = cost of travel from $i$ to $j$

Equations (5-4)–(5-8) are referred to as a "constrained optimization problem" since the task is to optimize an *objective function* (equation (5-4)) subject to a set of *constraints* (equations (5-5)–(5-8)) that dictate which solutions are admissible or *feasible*. We often refer to the set of solutions allowed by the constraints as the *feasible region* of the problem.

Equation (5-4) requires minimizing the tour cost. The decision variables $\{x_{ij}\}$ indicate the ordering of stops and therefore the tour. Equations (5-5)–(5-8) state the mathematical properties of a tour and therefore which values of the decision variables are feasible. Equation (5-5) requires each node to be entered exactly once and equation (5-6) requires each node to be exited exactly once. Equation (5-7) prevents *subtours* that only include a subset of nodes (this will be explained shortly). Equation (5-8) requires the decision variables to be binary; that is, $x_{ij} = 1$ means that arc $(i, j)$ is included in the tour while $x_{ij} = 0$ means that the arc is not included.

By themselves, equations (5-5) and (5-6) are not sufficient as a constraint set for the TSP since they allow subtours that include only a subset of nodes. In other words, equations (5-5) and (5-6) would allow a solution consisting of independent and unconnected tours through subsets of nodes since these would meet their "single entry-single exit" requirements. Equation (5-7) eliminates subtours. Although the equation looks complex, it is based on a simple observation. If we partition a feasible TSP solution into any two subsets of nontrivial size, then there must be a tour link that goes from the first subset to the second subset and a tour link that goes from the second subset back to the first subset. If not, the two sets must correspond to unconnected subtours. The term $N' \subseteq N, 2 \leq |N'| \leq |N| - 2$ defines a series of subsets $N'$ ranging in size from 2 to $|N| - 2$; these are the nontrivial partitions for which the above observation must hold. The summation in equation (5-7) is over all arcs whose from-node is in the subset $N'$ and whose to-node is in the remaining subset ($N'\backslash N$; "\" indicates the set difference operator). At least one tour link ($x_{ij} \geq 1$) must exist between all of the subsets of the node set (Nemhauser and Wolsey 1988). Another equivalent statement of equation (5-7) is (Fisher 1995; Fisher and Jaikumar 1981):

$$\sum_{(i,j) \in N \times N} x_{ij} \leq |N| - 1, 2 \leq |N| \leq n \tag{5-9}$$

where $N \times N$ indicates the Cartesian product of the set of nodes with itself (i.e., all possible pairings of nodes with themselves).

### Solving the TSP

*The complexity of the TSP* The only known exact or optimal solution procedure for the general TSP is the "brute-force" method of complete enumeration of all possible solutions and then choosing the best. Unfortunately, the solution space for the TSP is exceedingly large. Given a TSP with $n$ stops and asymmetric distances, the number of possible tours is $(n - 1)!$; a number that becomes very large rapidly as $n$ increases. If distances are symmetric, then the above number reduces by half but is still very large for even modest problems.

Many authors cannot resist inserting startling calculations when discussing the complexity of the TSP. We are among them. If $n = 100$, then the number of possible tours is $10^{200}$. To put this number in perspective, if every electron in the universe was a computer that could do one billion calculations per second, it would take on the order of $10^{11}$ *years* to check all of the tours (Jon Bentley of Bell Laboratories; reported in Kolata 1991).

TSP belongs to a special class of problems known as "nondeterministic polynomial time complete" or *NP-complete*. NP-complete means two things. First, the problem is in a class in which there is no known polynomial time algorithm for finding a solution, meaning that finding a solution requires exponential time. (The example in the previous paragraph supports this; also refer back to table 5-1.) However, and somewhat intriguingly, a hypothesized answer can be *verified* in polynomial time. In other words, while we cannot easily find a solution to the problem we can easily check if a guessed answer is optimal. The phrase "nondeterministic polynomial time" refers to this property, with "nondeterministic" roughly meaning "randomly guessing." This is not very helpful in practice since blindly searching for the optimal solution by guessing could take a very, very long time even for a modest problem. Second, the term "complete" means that the problem is in a sense equivalent to every other NP problem: If we solve TSP optimally in polynomial time we could solve all NP problems optimally in polynomial time. (More precisely, every NP problem can be converted efficiently to the TSP, with "efficiently" meaning "in polynomial time.")

NP is a large class of problems with many important members (see Garey and Johnson 1979 for an extensive list of NP problems). If we could solve the TSP exactly in polynomial time we could solve all of these problems as well. To add the final wrinkle to this tale, there is no proof that NP is even a real class of problems. There is no mathematical proof that some yet undiscovered procedure does not exist that can solve NP problems (although there is an enormous amount of evidence that this is the case). These properties reveal some of the fundamental limits of logic and computation. For more information on the fascinating theory of NP-completeness, see the classic reference by Garey and Johnson (1979), a technical but accessible introduction by Sipser (1997) or a popular account by Poundstone (1988).

Equations (5-4)–(5-8) is a general formulation of the TSP since it does not place restrictions on the set of travel costs $\{c_{ij}\}$. This problem is sometimes referred to as the *asymmetric TSP* since it may be the case that $c_{ij} \neq c_{ji}$. In contrast, the *symmetric TSP* requires $c_{ij} = c_{ji}$. The *Euclidean TSP* assumes that the customers are located in continuous space and the $\{c_{ij}\}$ are the Euclidean distances between customers. This means that the travel costs obey the metric properties of *symmetry* ($c_{ij} = c_{ji}$), *non-negativity* ($c_{ij} \geq 0$), and *triangular inequality* ($c_{ik} + c_{kj} \geq c_{ij}$) (Laporte 1992). Unlike the general TSP, some of these special case TSPs, can be solved easily.

Despite the difficulty in solving the general TSP, there is a large amount of research activity directed towards developing solution procedures that produce heuristic (suboptimal but fast) solutions. Supporting this effort are several online repositories of TSP problem incidences with known solutions (the most famous is TSPLIB). Recent years have witnessed astounding breakthroughs in the size of problems solved within a few percentages of optimality. We will first discuss *optimization-based approaches* that use techniques from traditional optimization theory. We will then discuss *TSP heuristics*. These include solution construction and improvement procedures, computational intelligence-based approaches, and transforming the general problem to easily solved special cases.

*Optimization-based approaches* Although the TSP is NP-complete, optimization-based algorithms and solution procedures have been proposed. Some of these algorithms are remarkably successful and have solved quite large TSPs to optimality or near-optimality. These algorithms involve some very sophisticated techniques in integer or combinatorial optimization ("integer" optimization problems are problems where decision variables can only have integer values while "combinatorial" refers to optimizing the arrangement of objects). We will provide an overview of these techniques and point the reader to key references in the literature.

A powerful optimization technique for integer and combinatorial optimization is *branch and bound* (BB), originally applied to the TSP by Dantzig, Fullerson, and Johnson (1954) and Little et al. (1963) and continually refined over years. BB is a directed enumeration method. Instead of blindly enumerating all possible solutions to the problem, BB attempts to zero-in on an optimal or near-optimal solution through a "divide and conquer" strategy. A dividing or *branching* process partitions the solution space into increasingly smaller subsets in attempt to find the subset that contains an optimal or near-optimal solution. For each subset, we calculate a *bound* that estimates the best possible solution in the subset. This is often based on a simpler approximation of the TSP such as ignoring the requirement that the subtours connect to form an overall tour (i.e., ignoring equation (5-7) in the problem above). If the bound for a subset indicates that it cannot contain an optimal or near-optimal solution, we discard or *fathom* that subset and continue the partitioning process with another subset. The algorithm terminates when there are no subsets remaining or sometimes after a prespecified length of time (Cook et al. 1998; Hillier and Lieberman 1990). Phillips and Garcia-Diaz (1981) provide a detailed, introductory example of BB as applied to the TSP while Laporte (1992) provides a higher level, rigorous review. Carpaneto and Toth (1980), Balas and Christofides (1981), and Miller and Pekny (1991) develop promising BB algorithms that solve TSPs as large as $n = 240$, 325, and 5,000 (respectively) in reasonable computation time.

Another optimization-based approach to solving the TSP is through *cutting planes*. This approach involves solving a "relaxed" version of the problem, checking to see if the solution is feasible and (if not) introducing new constraints to the problem and re-solving. A *cut* is a new problem constraint that eliminates a feasible solution for the relaxed version of the problem without eliminating any feasible solution to the original problem. After solving the relaxed problem and finding that it is not feasible for the original problem, we selectively "cut off" that part of solution space so that the relaxed problem is a better approximation of the original problem. This "tightening" process increases the chances that the next solution of the relaxed problem will also be an optimal solution to the original problem. Hillier and Lieberman (1990), Cook et al. (1998), and Nemhauser and Wolsey (1988) provide general discussions of the cutting plane technique. See Grötschel and Padberg (1985) and Padberg and Grötschel (1985) for a rigorous discussion of cutting plane theory as applied to the TSP. Laporte (1992) provides a summary of these algorithms as applied to the TSP.

The cutting plane approach has proven to be very successful at solving TSPs to optimality or near-optimality. Grötschel (1980) reports solving an $n = 120$ symmetric TSP while Padberg and Hong (1980) report solving an $n = 318$ symmetric TSP within 0.26% of optimality; both were stunning breakthroughs for that time. The cutting plane approach continues to be improved through parallel implementations: in 1995, an $n = 7{,}397$ symmetric TSP was solved using 60 workstations in parallel (reported in Usami and Kitaoka 1997).

Fleischmann (1985) presents a cutting-plane solution procedure for solving the TSP on road networks. Traditional TSP formulations assume a completely connected network. As mentioned early in this chapter, road networks exhibit much less connectivity; typically, each node in a road network is only connected to a few close neighbors. One way to resolve this mismatch is to add all of the missing arcs to the network and designate a very large artificial weight for each of these arcs so that none will appear in the final solution. However, this approach can greatly increase the complexity of TSP solution procedures. Another property of TSPs on road networks is that not all of the network nodes require visits; in fact, usually only a small subset of the nodes are required stops for the TSP tour. Fleischmann (1985) designates these vertices as *visiting vertices* (using the graph theoretic terminology) and reformulates the TSP problem as:

> **R-TSP**: Find the shortest tour in the network that meets all visiting vertices at lease once.

The resulting tour is not Hamiltonian since it may contain cycles that meet a vertex more than once or pass through an arc more than once. Computational results for problems up to $n = 292$ (in this case, $n$ is the number of visiting vertices) are promising.

*TSP heuristics* Despite the stunning breakthroughs in optimization-based methods, we also require simpler heuristics for solving TSP on modest computational platforms. Classic heuristics for the TSP include *construction procedures* and *improvement procedures*. Construction procedures start with an arbitrary node and sequentially add nodes to the tour based on some selection rule until all nodes are included. Adding an arc from the last node to the starting node completes the tour. Criteria for selecting the next node on the tour include:

- *Arbitrary insertion*: select an unused node at random
- *Nearest neighbor*: select the closest node in planar space to the last node added
- *Nearest insertion*: select the closest node in planar space to any node currently in the tour
- *Farthest insertion*: select the farthest node in planar space to any node currently in the tour
- *Cheapest insertion*: select the node that results in the lowest additional cost.

All of the insertion procedures have an asymptotic time complexity for the Euclidean space problem of $O(n^2)$ except cheapest which requires $O(n^2 \log n)$ time. Nearest and cheapest are guaranteed to find a solution that is no worse than twice the cost of the optimal solution. The performances of arbitrary, nearest neighbor and farthest insertion in the Euclidean space problem degrade with the problem size, with arbitrary and farthest insertion guaranteeing a solution no worse than $2\ln(n) + 0.16$ of optimality. Nearest neighbor guarantees a solution no worse than $(1/2)\lceil \log_2 n \rceil + 1/2$ of optimality, where the symbol $\lceil n \rceil$ indicating the greatest integer $\leq n$ (see Golden et al. 1980; Rosenkrantz, Stearns, and Lewis 1977).

Other tour construction heuristics include Clarke and Wright (1964) (discussed below for the vehicle routing problem), a procedure by Christofides (1976) and a procedure based on *convex hulls* or the smallest convex polygon enclosing a set of points. Clarke-Wright and the convex hull procedure both require $O(n^2 \log n)$ while Christofides requires $O(n^3)$ in the worse-case for the Euclidean problem. The guaranteed solution quality is unknown for the Clarke-Wright and convex hull procedures while the Christofides procedure guarantees no worse than one-and-one-half times optimal for the Euclidean problem (Golden et al. 1980).

Improvement procedures start with an arbitrary feasible solution and attempt to repeatedly improve that solution through some perturbation until no further improvement is possible. A well-known improvement procedure is the *r-opt strategy* (Lin 1965). In this strategy, we perturb a current solution by changing $r$ components. For example, if $r = 3$ at each step, we randomly swap three arcs that are in the solution with three arcs that are not in the solution. We accept the new solution if it is feasible and has lower cost than the current solution. The algorithm stops when perturbations no longer improve the solution; we refer to the resulting solution as *r-optimal*. In practice, $r$ usually equals 2 or 3; a larger $r$ can lead to better solutions but at high computational costs (Golden and Stewart 1985). Lin and Keringhan (1973) develop a procedure that computes $r$ dynamically for each step of the solution algorithm.

Golden et al. (1980) suggest a composite solution procedure for the Euclidean TSP. This involves the following steps:

Step 1.  Obtain an initial tour using a tour construction procedure.

Step 2.  Apply a 2-opt procedure to the tour obtained in step 1.

Step 3.  Apply a 3-opt procedure to the output obtained in step 2.

The idea behind this heuristic is to use a simple procedure to obtain a good initial solution and then use procedures that are increasingly computationally expensive to fine-tune the solution.

Golden and Stewart (1985) summarize extensive computational experiments conducted with the TSP heuristics by Golden et al. (1980) and Adrabiński and Syslo (1983). Golden et al. (1980) use Euclidean TSPs in their experiments while Adrabiński and Syslo (1983) use non-Euclidean problem formulations. Golden et al. (1980) also test the performance of heuristics under increasing network sparseness by systematically deleting arcs and rerunning the heuristics. Some of their conclusions are:

- Many of the tour construction heuristics will find a TSP tour to within 5–7% of optimality but generally cannot do much better. Tour construction procedures are therefore appropriate when a reasonable solution is required with minimal computational effort.
- The 2-opt and 3-opt procedures perform roughly the same as the tour construction procedures.
- The three-step composite procedure can find tours within 2–3% of optimality with a high degree of regularity but require more computation time.
- The heuristics perform well when the network is 60% connected (i.e., 60% of the fully connected network arcs are present) but deteriorate in performance with lower connectivity.

- For non-Euclidean problems, the farthest insertion procedure seems to perform best.

A recent trend in applying heuristics to solving the TSP is using *computational intelligent* (CI) methods. CI methods are similar to the more well-known methods of *artificial intelligence* (AI). However, CI methods represent "intelligence" at a lower level than AI. Instead of representing explicit knowledge through rule structures, CI technique exploit intelligence properties such as robustness, adaptivity and tolerance for imprecision and uncertainty (Fischer 1997). CI methods such as *simulated annealing* (Dowsland 1993), *tabu search* (Glover and Laguna 1993), *genetic algorithms* (Reeves 1993) and *artificial neural networks* (Peterson and Söderberg 1993) have been applied to the TSP with varying degrees of success. (The above references are introductory in nature and provide references to TSP applications of each technique. Also see Ansari and Hou 1997 for a discussion of CI applications to the TSP.)

Another strategy for solving the TSP is recognizing that a particular problem either corresponds to or can be transformed to a special case that can be easily solved. Some well-solvable special cases of the TSP include problems whose optimal solutions are *pyramidal tours*. A pyramidal tour visits a sequence of locations in increasing order based on labels, reaches location $n$, and then visits the remaining locations in decreasing order. The trick in exploiting this property is recognizing that the optimal tour is pyramidal from the cost matrix $\{c_{ij}\}$. Also easily solved are certain *Euclidean special cases* such as all customers being arranged along (or near) parallel lines and the *graphical TSP* where the tour is allowed to visit each location and use each tour arc more than once. Burkhard et al. (1998) discuss these and other well-solvable special TSP cases.

## The vehicle routing problem

As mentioned previously, the vehicle routing problem (VRP) is essentially a "fleet" version of the TSP. The problem is to route a fixed number of delivery or pickup vehicles through a number of demand locations such that the total cost is minimized and vehicle capacity constraints are not violated. Typically, there is a designated location known as the *depot* where all vehicles must start and end their tours. The depot is usually not a member of the set of demand locations.

The VRP literature is very large and there are many formal statements of the basic problem (see Fisher 1995; Magnanti 1981). We will use a VRP statement by Fisher and Jaikumar (1981) since it clearly illustrates the structure of the problem as well as the linkage to the TSP. Again, we assume a network $G = [N, A]$ with nodes corresponding to vehicle stops and arcs representing direct travel between the locations. (Recall from the TSP discussion that we can solve both planar space and transportation network-based versions of this problem using this representation.) In addition to the $1, \ldots, n$ vehicle stops, we indicate the depot by a "0". There are $K$ vehicles available for servicing the locations.

The basic VRP is:

$$\min_{\{y_k\}} \sum_{k=1}^{K} f(y_k) \qquad (5\text{-}10)$$

subject to:

$$\sum_{i=1}^{n} a_i y_{ik} \leq b_k, k = 1, \ldots, K \quad (5\text{-}11)$$

$$\sum_{k=1}^{K} y_{ik} = \begin{cases} K & i = 0 \\ 1 & i = 1, \ldots, n \end{cases} \quad (5\text{-}12)$$

$$y_{ik} \in \{0, 1\}, i = 0, \ldots, n \quad k = 1, \ldots, K \quad (5\text{-}13)$$

where:

$y_{ik} = \begin{cases} 1 & \text{if location } i \text{ (customer or depot) is visited by vehicle } k \\ 0 & \text{otherwise} \end{cases}$

$y_k = (y_{0k}, y_{1k}, \ldots, y_{nk})$
$b_k$ = the capacity (in weight or volume) of vehicle $k$
$a_i$ = the requirements (in weight or volume) for pickup or delivery at location $i$
$f(y_k)$ = the cost of an optimal TSP tour of the locations assigned to vehicle $k$ (starting and stopping at the depot)

Equation (5-11) requires the load assigned to each vehicle to be less than or equal to its capacity. Equation (5-12) requires each stop to be visited by exactly one vehicle except for the depot which is visited by all $K$ vehicles.

Note that the function $f(y_k)$ requires finding the optimal tour of the locations assigned to vehicle $k$ (starting and stopping at the depot). This is essentially a TSP for each vehicle. Therefore, the VRP can be viewed as two interlinked problems: (i) finding an optimal assignment of customers to vehicles and (ii) solving the TSP for each vehicle based on its assigned customers. These two problems are highly interrelated since the solution to the assignment problem depends on the vehicle tours and vice versa (see Fisher 1995; Fisher and Jaikumar 1981; or Magnanti 1981).

*Solving the vehicle routing problem*

Similar to the TSP, the VRP is a NP problem; in this case, a *NP-hard* problem (Lenstra and Rinnooy Kan 1981). Recall that NP-complete means that the problem can only be solved in a nondeterministic polynomial (NP) time and every other NP problem can be efficiently converted to it. In contrast, a problem is NP-hard if it is not NP but every NP problem can be converted to it. "Not NP" means that the problem is harder than NP problems; i.e., we cannot even check an hypothesized answer in polynomial time. This means that NP-hard problems are as hard or harder than any NP problem, that is, there is no known polynomial time exact solution algorithm (Garey and Johnson 1979). This is not surprising since the TSP is a special case of the VRP with one vehicle and unlimited capacity.

*Optimization-based approaches* Due to the similarity between the TSP and VRP, several of the optimization strategies applied to the TSP have also been adopted for the VRP. Christofides, Mingozzi, and Toth (1981) develop branch and bound (BB) approaches based on shortest paths and minimal spanning trees to calculate the bounds. Christofides, Mingozzi, and Toth (1980) use another approach to calculate bounds in the BB using principles from *dynamic programming* (a technique for optimizing a sequence of linked decisions; see Hillier and Lieberman 1990 for a concise introduction.) Several branching techniques are available, with the simplest being choosing a current unserved customer to include in an existing route (see Christofides 1985). Laporte, Nobert, and Desrochers (1985) use cutting plane methods for the VRP with distance and capacity constraints.

*Heuristics for the VRP* VRP heuristics include strategies such as cluster first-route second, route first-cluster second, construction procedures and improvement procedures. *Cluster first-route second* first groups or clusters demand nodes and then designs optimal routes independently for each cluster. An example is the Gillett and Miller (1974) "sweep" procedure. This strategy starts with a random customer and "sweeps" either clockwise or counterclockwise using the depot as the origin for the sweeping ray. As they are swept, customers are assigned to a vehicle until its capacity is reached. A new vehicle is then selected and the sweep continues until all customers are assigned. A TSP heuristic routes each vehicle through its assigned customers. *Route first-cluster second* is the reverse strategy: A large (usually infeasible) route is constructed first and then the route is partitioned into smaller, feasible routes. Bodin and Berman (1979) and Newton and Thomas (1974) apply this strategy in school bus routing problems.

Similar to the TSP heuristics discussed previously, *construction* procedures build solutions incrementally while improvement procedures start with an initial (sometimes infeasible) solution and attempt to improve it. Bodin and Golden (1981) refer to construction procedures as *savings/insertion procedures* since they typically build a solution by inserting the tour component that leads to the greatest cost savings. For example, the well-known procedure by Clark and Wright (1964) starts with every customer served by its own vehicle. It then calculates alternative configurations obtained by combining routes (so that two customers are now served by a single vehicle). The savings achieved by combining customers $i$ and $j$ into a single route are:

$$s_{ij} = c_{0i} + c_{0j} - c_{ij} \tag{5-14}$$

where $c_{0i}$ is the cost of travel from the dept to customer $i$. The procedure calculates the savings for all customer pairs and chooses the combination that results in the greatest savings, subject to feasibility. This continues as long as positive savings can be realized.

*Improvement procedures* start with an initial feasible solution. At each step, one component of the solution is exchanged with another component not in the solution if this generates a feasible solution with lower cost. An example is Chistofides and Eilon (1969) who adopt the $r$-opt TSP solution procedure for the VRP.

Another heuristic is due to Fisher and Jaikumar (1981). The heuristic first approximates the vehicle assignment sub problem by identifying $K$ "seed locations" (one for each vehicle) to approximate the routing costs for that vehicle. This is either a subset of the existing demand locations suitably dispersed in space or new "cluster points" at appropriate places in continuous space. Fisher and Jaikumar (1981) and Fisher (1995) discuss several techniques for identifying these seed locations. Given these seed customers, we can solve the problem of assigning vehicle to customers efficiently using either optimal or heuristic techniques (see Fisher 1995). A heuristic technique is to assign customers to the seed locations based on the added cost of including the customer on the route between the depot and the seed location. Each cluster of seed location and its assigned customer locations provides a routing problem that can be solved using a TSP solution procedure.

Other heuristics for the VRP include *set-partitioning procedures*. The set-partitioning problem groups elements into sets in a manner that minimizes some cost measure. For the VRP, the elements are candidate routes that must be grouped into $K$ sets corresponding to the vehicles that must execute these routes. The trick to this technique is determining a good set of candidate routes. If the set of candidate routes encompasses all feasible routes, then the set-partitioning approach will (theoretically) find the optimal vehicle routing solution. However, this candidate set will often be too large to solve. Instead, we must determine a candidate set that is large enough to generate an optimal or near-optimal solution but not so large as to make the problem unsolvable. Problem parameters and constraints are helpful for this. For example, if the customer demands are large relative to the vehicle capacities, then the number of feasible route will be small. Similarly, constraints such as time delivery windows can make the candidate set manageable (Fisher 1990). Desrochers, Desrosiers, and Solomon (1992) use set-partitioning for a VRP algorithm whereas Cullen, Jarvis, and Ratliff (1981) use set-partitioning within a decision support system that allows the analyst to specify candidate routes interactively.

Finally, similar to the TSP, CI-based methods such as simulated annealing and tabu search have also been applied to the VRP. Several of the CI references provided above in conjunction with the TSP also discuss application to the VRP. Also see Fisher (1995) for a summary discussion. Schulze and Fahle (1999) present a parallel implementation of the tabu search procedure that appears promising for solving large-scale VRPs.

*Extensions to the basic VRP*

We have discussed only the basic VRP. There are many extensions to this basic problem; most attempt to introduce in a controlled manner more realistic characteristics. Some of these characteristics include different types of vehicles, maximum vehicle route-times, and different types of objectives (e.g., minimizing the number of vehicles required). Bodin and Golden (1981) and Assad (1988) provide general, synthetic reviews of VRP extensions. Perhaps one of the more important extensions concerns *time windows* or time intervals corresponding to when stops can occur for each customer. Time windows can be "hard" or "soft." In the former case, a customer's time window cannot be violated. In the latter case, windows can be violated but these are penalized in the objective function. For a comprehensive review of time-constrained routing problems, see Solomon and Desrosiers (1988).

## The geometry of traveling salesman and vehicle routing tours in zonal systems

Often, traveling salesman or vehicle routing tours occur within pre-designated zonal systems such as service districts, sales territories or municipalities. These zonal systems can be exploited in a simple but powerful heuristic strategy for solving the TSP or VRP. This strategy involves approximating the tour within a zone with a "swath" or path based on the customer density and zonal geometry. This path can be computed using a GIS or any software system that can handle basic computational geometry routines such as a point-in-line segment test. The order in which the swath intersects customers within a zone provides the stop sequence for the vehicle assigned to that zone. This approximation can also be used to estimate the expected cost of the tour within each zone prior to using one of the optimization or heuristic techniques discussed above. The geometric solution can also be fine-tuned directly using an interactive or heuristic method (Daganzo 1984a, 1984b; Newell 1986; Newell and Daganzo 1986a, 1986b; Robusté, Daganzo, and Souleyrette 1990).

If the distribution of customers within zones is random with density $\delta = n/A$, where $A$ is the area of the zone, the optimal width of the swath covering the zone is:

$$w_1 \approx \left(\frac{3}{\delta}\right)^{\frac{1}{2}} \text{ for the Manhattan distance metric} \quad (5\text{-}15)$$

$$w_2 \approx \left(\frac{2.95}{\delta}\right)^{\frac{1}{2}} \text{ for the Euclidean distance metric} \quad (5\text{-}16)$$

where the Manhattan metric restricts travel to path segments parallel to the axes. (For more information on Manhattan, Euclidean and other distance metrics, see Love, Morris, and Wesolowsky 1988, ch. 10.) To determine a TSP tour, we could cover the zone with parallel laps and record when the swath intersects a customer location. For large $n$, the expected TSP tour lengths with the zone are:

$$l_1 \approx 0.95(nA)^{\frac{1}{2}} \text{ for the Manhattan distance metric} \quad (5\text{-}17)$$

$$l_2 \approx 0.75(nA)^{\frac{1}{2}} \text{ for the Euclidean distance metric} \quad (5\text{-}18)$$

A more-sophisticated strategy could adjust the path and width based on the local density of customer locations. These swaths need to be modified for irregular geometry such as long and narrow zones (see Robusté, Daganzo, and Souleyrette 1990).

For the VRP, the expected tour lengths for a vehicle in its service area are:

$$l_1 \approx 2h\left(\frac{n}{c}\right) + 0.73\left(n\delta^{-\frac{1}{2}}\right) \quad \text{for the Manhattan distance metric} \quad (5\text{-}19)$$

$$l_2 \approx 2h\left(\frac{n}{c}\right) + 0.57\left(n\delta^{-\frac{1}{2}}\right) \quad \text{for the Euclidean distance metric} \quad (5\text{-}20)$$

where $c$ is the maximum number of stops per vehicle and $h$ is the average distance between customers and the depot. This assumes that the service areas for the vehicles are narrow, elongated toward the depot with a width of $z = w\sqrt{2}$ and do not overlap. If the service areas cannot be elongated and narrow, the equations are still valid if the depot is near the center of the vehicle service area and if $7 < c < 1.5(n/c)$. If the second half of the inequality is violated, meaning that the total number of tours is not much larger than the number of stops per tour, then other length approximations are available. For Euclidean distance metric:

$$l_2 \approx \left[0.9 + \frac{kn}{c^2}\right]\sqrt{nA} \quad (5\text{-}21)$$

where $k$ is a parameter whose value depends on the ratio $n/c^2$ and the zonal geometry. Example parameter values include:

$k = 0.40$ if $n/c^2 > 1.5^{-1}$ and the zone is circular;
$k = 0.45$ if $n/c^2 = 1.5^{-1}$ and the zone is square;
$k = 0.55$ if $n/c^2 = 1.5^{-1}$ and the zone is a six by ten rectangle.

(see Robusté, Daganzo, and Souleyrette 1990). These calculations are analytical; an open research question is deriving versions based on arbitrary computational representations of zonal geometry.

The geometric strategy discussed above could be used directly to determine the ordering of stops on TSP and VRP tours through a manual or computer-aided tour construction process. These approximations can also be improved through fine-tuning using some heuristic optimization method. Robusté, Daganzo, and Souleyrette (1990) achieve good results using a hybrid strategy that manually constructs tours using the geometric strategies discussed above and then fine-tunes the solutions using simulated annealing.

## Conclusion

In this chapter, we discussed several important network shortest path and routing problems. After reviewing some basic network properties, we discussed in some detail the fundamental network problem of finding shortest paths. This provided a forum for reviewing basic physical processing structures in a computer as well as clearly illustrating that all shortest path routines are not alike. We finished our shortest path discussion with a review of parallel processing and heuristic methods for quick calculation of paths and methods for solving shortest paths through terrain and other cost surfaces. The remainder of the chapter reviewed some common network routing problems, including the traveling salesman problem and the vehicle routing problem.

In the next chapter, we will complete our "tour" of network algorithms relevant to GIS-T by examining procedures for routing multiple flows and for locating facilities in a network.

# 6

# Network Flows and Facility Location

This chapter continues the discussion in the previous chapter on network algorithms for GIS-T. In the previous chapter, we reviewed shortest path algorithms, their implementations and methods for solving tours and vehicle routings within networks. In this chapter, we complete our discussion by reviewing methods for routing multiple flows and locating facilities within a network.

We begin the discussion in this chapter with network flow modeling. There are two broad classes of problems to consider. The first case is when the network is *uncongested*. This means that each unit of flow does not affect other flows; in other words, we do not need to worry about interaction and congestion among flows. As we will see, this case often corresponds to "abstract" networks that represent problems in *logistics* or the procurement and distribution of materials and goods across space. These abstract networks represent possible shipments, schedules, or other logical relationships that do not necessarily correspond to physical transportation networks. We will discuss several classic network flow problems for this case.

The second case involves *congested* networks. In this case, the network is saturated to the point where flows affect each other by creating delays and other congestion-related costs. This usually corresponds to flows through physical networks, in particular, flow through street and other transportation networks. We will discuss two different modeling approaches within this general problem class. The first is the *equilibrium* approach that attempts to solve for a realistic flow pattern analytically by assuming that the flow achieves some type of balance between users' travel demands and the performance of the network in meeting this demand. The second is the *simulation* approach that uses computational methods that require fewer assumptions but is also tied less strongly into a theoretical framework.

We end the chapter by discussing methods for locating facilities within a network. Facility location methods are important for configuring logistics networks. Facility locations are sensitive to the amount of demand that can be serviced: this involves movement to or from the facility sites. The new transportation demands created by facilities also should be considered when evaluating new land-use proposals; facility location models can be helpful for this. We can also use facility location methods to solve flow interception and sampling problems such as locating vehicle inspection stations and traffic counters.

## Flow through uncongested networks

### The minimum cost flow problem

The *minimum cost flow problem* (MCFP) is a fundamental network flow problem. The MCFP determines the minimum cost routes through a network for a set of fixed origin-destination flows. As we will see very shortly, this problem is central to an entire class of uncongested network flow problems. Several important flow problems are special cases of this generic problem. These flow problems are increasingly being incorporated into GIS and related transportation software, particularly software oriented towards logistics and other business applications.

Assume a directed network $G = [N, A]$ with the following attributes (Hillier and Lieberman 1990):

$c_{ij}$ = cost per unit flow through directed arc $(i, j)$; this is set equal to infinity if $(i, j) \notin A$

$u_{ij}$ = flow capacity of directed arc $(i, j)$

$b_i$ = net flow generated at node $i$. This depends on the type of node:
   $b_i > 0$ if node $i$ is an origin or *supply* node
   $b_i = 0$ if node $i$ is a *transshipment* node
   $b_i < 0$ if node $i$ is a destination or *demand* node

Supply nodes generate flows while demand nodes absorb flows. The amount of flow supplied and demanded at these nodes (respectively) is known and exogenous to the problem. Transshipment nodes do not generate nor absorb flows: All flow that enters these nodes also leaves.

Given the above notation, we can formulate the minimum cost flow problem as (Hillier and Lieberman 1990):

$$\min_{\{f_{ij}\}} \sum_{i=1}^{n} \sum_{j=1}^{n} c_{ij} f_{ij} \quad (6\text{-}1)$$

subject to:

$$\sum_{j=1}^{n} f_{ij} - \sum_{j=1}^{n} f_{ji} = b_i, \quad i = 1, \ldots, n \quad (6\text{-}2)$$

$$0 \leq f_{ij} \leq u_{ij} \tag{6-3}$$

where $|N| = n$ (i.e., $n$ is the number of nodes or the *cardinality* of set $N$). The objective function requires minimization of the total flow costs in the network. The constraints corresponding to equation (6-2) are *flow conservation requirements*. These dictate that the total outflow from node $i$ $\left(\sum_{j=1}^{n} f_{ij}\right)$ minus the total inflow to node $i$ $\left(\sum_{j=1}^{n} f_{ji}\right)$ must equal the known net flow generated at that node ($b_i$). The constraints corresponding to equation (6-3) require that the flow on each arc must be nonnegative and less than or equal to the arc's capacity. Figure 6-1a illustrates the structure of this problem for a simple network.

The minimum cost flow problem has two critical properties that also have implications for its special cases (Hillier and Lieberman 1990). The first is the *feasible solutions property*. A feasible solution to the minimum cost flow problem only exists if the following condition is true:

$$\sum_{i=1}^{n} b_i = 0 \tag{6-4}$$

In other words, the total outflow from the supply nodes exactly equals the total flow being absorbed at the demand nodes. (Since $b_i = 0$ for transshipment nodes, the above condition can only be met if total supply exactly balances with total demand.) The second property is the *integer solutions property*. If the arrays $\{b_i\}$, $\{u_{ij}\}$ all have integer values, then the optimal solution will also have integer values. The integer solutions property is very attractive if we are trying to solve a special case of the minimum cost flow problem where we are only concerned with integer solutions (e.g., if flow corresponds to package shipments or other indivisible units). This means that we do not need to impose explicit integer constraints on these problems that would require complex integer and combinatorial optimization techniques.

## Special cases of the minimum cost flow problem

As mentioned above, the MCFP is central since several important network flow problems are special cases of this more-general problem. Some of these special cases follow (Hillier and Lieberman 1990).

### The transportation problem

In the transportation problem, there are $m$ supply nodes that can ship to $n$ demand nodes. We designate the known total outflows from the supply nodes $\{s_i\}$ and the known total inflows to demand nodes as $\{d_j\}$. Each supply center is directly connected to every demand center with uncapacitated arcs (that is, $\{u_{ij}\} = \infty$). The arc weights $\{c_{ij}\}$ represent the cost of shipping one unit from supply center $i$ to demand center $j$. There are no transshipment nodes. Figure 6-1b illustrates this problem.

Network Flows and Facility Location 175

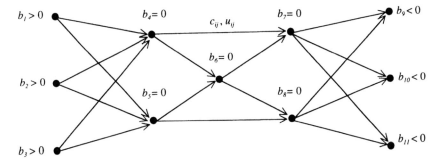

a) minimum cost flow problem

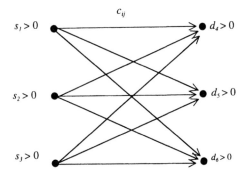

b) transportation problem

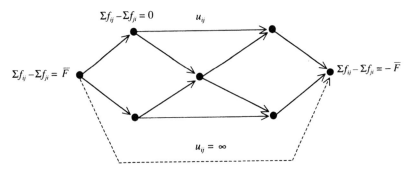

c) maximum flow problem

Figure 6-1: The minimum cost flow problem and some special cases

A standard interpretation of the problem views the supply centers as factories or warehouses and the demand centers as warehouses or retail outlets. For example, the transportation problem could solve for the optimal shipment patterns from factories to warehouses, factories to retail outlets, or warehouses to retail outlets. The transportation problem would determine from which (for example) factory each store should receive its supplies such that total cost is minimized, each store receives all of the supplies it needs and the supply at each factory is used completely.

We can state the transportation problem as the following optimization problem (Phillips and Garcia-Diaz 1981):

$$\min_{\{f_{ij}\}} \sum_{i=1}^{m} \sum_{j=1}^{n} c_{ij} f_{ij} \qquad (6\text{-}5)$$

subject to:

$$\sum_{j=1}^{n} f_{ij} = s_i, \quad i = 1, \ldots, m \qquad (6\text{-}6)$$

$$\sum_{i=1}^{m} f_{ij} = d_j, \quad j = 1, \ldots, n \qquad (6\text{-}7)$$

$$f_{ij} \geq 0, \quad i = 1, \ldots, m; j = 1, \ldots, n \qquad (6\text{-}8)$$

The feasible solutions property in this case requires $\sum_{i=1}^{m} s_i = \sum_{j=1}^{n} d_j$ for any basic feasible solution to exist. If this condition is not met by a particular problem instance, we can add "dummy" or fake supply or demand nodes to make up the difference (see Hillier and Liberman 1990, pp. 214–219, for examples). The dummy supply node could represent shipment purchases from outside the system (e.g., another company) while the dummy demand node could represent putting shipments into inventory (although inventory costs should be handled carefully). The solution to the transportation problem will be integer valued if $\{s_i\}$, $\{d_j\}$ are all integer valued.

A generalization of the transportation problem is the *transshipment problem*. The transshipment problem allows flow from a supply node to a destination node to pass through intermediate nodes. These intermediate nodes can be transshipment nodes or other supply and demand nodes. Unlike the transportation problem, the transshipment problem allows for simultaneous solution of the amount to ship from each source to each destination and the routing for these shipments (Hillier and Lieberman 1990). Routing decisions can be important in more complex logistical systems such as three-tier systems with, for example, factories (supply nodes), warehouses (transshipment nodes), and retail outlets (demand nodes). Mathematically, the transshipment problem is identical to the MCFP (equations (6-1)–(6-3)) except that the constraints corresponding to the arc capacity restrictions (equation (6-3)) are replaced with nonnegativity constraints (equation (6-8)).

*The maximum flow problem*

The maximum flow problem concerns finding the most amount of flow that can travel from a designated origin to a designated destination. Deriving this from the MCFP requires the following modifications (Hillier and Lieberman 1990). We designate one supply node and one demand node with the remaining serving as transshipment nodes. We set $\{c_{ij}\} = 0$ since we are trying to push as much flow as possible through the network regardless of cost. We add a dummy overflow arc that directly connects the supply node to the demand node; this arc has an arbitrarily large arc flow cost to discourage its use and an unlimited flow capacity. Finally, we designate an upper bound $\bar{F}$ on the maximum amount of flow that can conceivably pass through the network and assign this as the supply at the origin and the demand at the destination. Figure 6-1c illustrates this problem.

Designating the origin node with "1" and the destination node as $n$, the maximum flow problem is (Phillips and Garcia-Diaz 1981):

$$\max_{\{f_{ij}\}} \sum_{i=1}^{n} \sum_{j=1}^{n} f_{ij} \tag{6-9}$$

subject to:

$$\sum_{j=1}^{n} f_{ij} - \sum_{j=1}^{n} f_{ji} = \begin{cases} \bar{F}, & i = 1 \\ 0, & i = 2, \ldots, n-1 \\ -\bar{F}, & i = n \end{cases} \tag{6-10}$$

$$0 \le f_{ij} \le u_{ij} \tag{6-11}$$

At optimality, the solution will pass the maximum amount possible of $\bar{F}$ through the regular network and send the overflow through the costly dummy arc.

### Solving the minimum cost flow problem and its special cases

Note that MCFP, transportation/transshipment and maximum flow problems all consist exclusively of linear equations. In fact, all of these problems as stated above as *linear programs* (LPs); these are a special type of constrained optimization problem that consist of a linear objective function (e.g., equation (6-9)) and linear constraints (e.g., equations (6-10) and (6-11)). We can solve LPs using a very powerful solution technique known as the *simplex method*. This method can solve very large LPs quickly in the average case (see Hillier and Lieberman 1990 or any other operations research textbook). We can modify the general-purpose simplex method to make it even more efficient for the special problems of interest here.

The *network simplex method* and the *transportation simplex method* are tailored for the MCFP and the transportation/transshipment problems (respectively). The network simplex method exploits the property that arcs with positive flow in a feasible solution comprise a spanning tree (see chapter 5). We can calculate the spanning tree more efficiently than the matrix inversion required for the general-purpose simplex method (see Hillier and Lieberman 1990). The transportation simplex method exploits the feasible solution property and quick methods for calculating cost differentials among possible flow assignments to determine the next solution to examine from the current solution. See Balakrishnan (1995), Hillier and Liberman (1990), or Phillips and Garcia-Diaz (1981) for technical discussions of this procedure. Also see Taaffe, Gauthier, and O'Kelly (1996) for a very accessible introduction to the transportation simplex method.

### Flow through congested networks

When solving for the simultaneous shortest paths for multiple flows in a congested network, we must recognize the influence of each unit flow on travel costs. Multiple flows can create congestion effects, with the result that arc costs depend on the amount of flow currently within the arc. Thus, the shortest paths used by the flow units influence the current network flow costs that in turn influence the shortest paths, and so on. We can slice into this "Gordian knot" by determining an *equilibrium flow pattern*. By "equilibrium" we mean (roughly) a stable flow pattern where no flow has any incentive to switch routes unless some disturbance occurs. Different types of equilibrium exist depending on the behavioral principles assumed for the route choice mechanism. While equilibrium flow models have weaknesses, particularly from a behavioral perspective, they allow access to rigorous microeconomic theory and related techniques for policy evaluation (see the discussion below).

We will first present some concepts and notation for flow problems. We then discuss some fundamental network flow properties that must be met for the flow pattern to be realistic. Next we discuss common *arc flow cost functions*; these relate the flow cost for an arc as a function of its current flow. We will then discuss different types of network flow equilibrium. We end this section with a brief discussion of microeconomic strengths and behavioral weaknesses of equilibrium flow models.

Our discussion of network flow equilibrium will focus on fundamental models. The "state-of-the-art" (although not necessarily the "state-of-the-*practice*") is well beyond the introductory material presented here. For clarity, we will present general network flow conditions for multiple modes (e.g., automobile, bus) but restrict our discussion of solution algorithms to single mode flows. Again, more advanced multimodal flow methods can be found in the literature. We encourage the interested reader to pursue the citations of more-advanced material provided below.

### Congested network flow concepts

#### Network flow notation

We will use the following notation in network flow models. The transportation network is a directed graph $G = [N, A]$. $I \subseteq N$ is the set of origins and $J \subseteq N$ is the set of destinations. A network arc a is defined as $a \equiv (n_l, n_m)$ where $n_l, n_m \in N$; in other words, as a "from-node, to-node" tuple. A network path $r$ is defined as a sequence of incident arcs $r \equiv \{(n_i, n_k), (n_k, n_l), \ldots, (n_l, n_m),(n_m, n_j)\}$. The set of paths in the network is $R$ while the set of paths that connect origin-destination (O-D) pair $i, j$ is $R_{ij}$.

To find equilibrium flow patterns within networks we must keep track of the flows loaded on paths and the resulting flow levels within arcs. Recall from chapter 5 that sequences of network arcs that comprise network paths. We use an arc-path incidence variable (see chapter 5) to keep track of the relationship between individual arcs and paths for mode-specific flow:

$$\delta_{ar}^k = \begin{cases} 1 & \text{if arc } a \text{ belongs to path } r \text{ and allows flow by mode } k \\ 0 & \text{else} \end{cases} \quad (6\text{-}12)$$

We will use the following variables to indicate arc, path and origin-destination flows in the network:

$f_a^k$ = mode $k$ flow on arc $a$
$F$ = set of all arc flows in the network
$h_r^k$ = mode $k$ flow on path $r$
$H$ = set of all path flows in the network
$D_{ij}^k$ = total mode $k$ flow between origin-destination pair $i, j$
$D_{ijr}$ = total flow on path $r$ between origin-destination pair $i, j$

The following flow conditions that must be met if the generated flow pattern is to be a realistic depiction of transportation flows. All path flows must be non-negative since negative flows obviously do not make sense when modeling transportation networks:

$$h_r^k \geq 0 \quad \forall\, r \in R, k \in K \quad (6\text{-}13)$$

where the symbol "$\forall$" means "for every" (i.e., "$\forall\, r \in R$" means "for every path in the set of network paths"; this type of set notation is convenient in this class of models). The mode $k$ flow on an arc must equal the sum of the mode $k$ flows on all paths that use that arc:

$$f_a^k = \sum_{i \in I} \sum_{j \in J} \sum_{r \in R_{ij}} \delta_{ar}^k h_r^k = \sum_{r \in R} \delta_{ar}^k h_r^k \quad (6\text{-}14)$$

The mode $k$ flow on path $r$ must equal the sum of mode $k$ flows on all arcs that comprise that path:

$$h_r^k = \sum_{a \in A} \delta_{ar}^k f_a^k \quad (6\text{-}15)$$

Finally, the total mode $k$ flow between an O-D pair equals the sum of mode $k$ flows on all paths between that pair:

$$D_{ij}^k = \sum_{r \in R_{ij}} h_r^k \quad (6\text{-}16)$$

The restrictions imposed by equations (6-13)–(6-16) may seem obvious. However, in mathematical models of congested network flows we cannot be assured that these realism and consistency requirements will be met unless we explicitly require these in our models. Therefore, we will see varying combinations of these conditions in equilibrium flow models.

Although network equilibrium models require consistency between flows at the arc and path level, an interesting theoretical result is that at equilibrium only arc flows and aggregate travel demands are unique: Path flows are not unique (see

Fernandez and Friesz 1983; Sheffi 1985, pp. 66–69). This means that any set of path flows that are consistent with the equilibrium arc flows is allowable; in theory, this is an infinite set. This is not a major problem from a practical perspective. We are primarily concerned with flow levels within arcs not along paths when analyzing properties such as congestion. However, keep in mind that path flow estimates from these models are not suitable for analysis since they are arbitrary. (See Bell and Iida 1997 for a discussion of this property from an entropy perspective.)

*Arc and path costs*

We will use the following variables to indicate arc and path costs in the network:

$c_a^k$ = average travel cost for a mode $k$ traveler on arc $a$;
$C_r^k$ = average travel cost for a mode $k$ traveler on path $r$;
$C_{ij*}^k$ = minimum average travel cost for a mode $k$ traveler between O-D pair $i, j$.
$C_*$ = the set of minimum travel costs for all modes and all O-D pairs

The relationship between arc costs and paths costs is:

$$C_r^k = \sum_{a \in A} \delta_{ar}^k c_a^k \qquad (6\text{-}17)$$

that is, the total mode $k$ cost for path $r$ is the sum of mode $k$ costs for arcs that comprise that path.

To capture congestion effects, travel costs should reflect the current flows within the network. These costs can be either the mode specific flow only on that arc:

$$c_a^k = c_a^k(f_a^k) \qquad (6\text{-}18)$$

or the flows in all arcs in the network:

$$c_a^k = c_a^k(F) \qquad (6\text{-}19)$$

Equation (6-18) is a *separable* flow cost function. This implies that the flows across different modes and different arcs can be meaningfully separated, e.g., we can model flow costs for automobiles as a function to the current automobile flow on that arc only. Equation (6-19) is a *nonseparable* flow cost function. This implies that flows across different arcs cannot be separated meaningfully into independent flows. This can capture the effects of neighboring arcs (e.g., cross-street traffic) on an arc. Nonseparable cost functions are much more realistic than separable cost functions. However, solving for the network flow equilibrium is much more difficult with nonseparable cost function (see Fernandez and Friesz 1983 for a good discussion).

In addition to being dependent on flow, arc cost functions often must also represent any out-of-pocket expenses associated with travel through an arc (e.g., tolls). For separable cost functions, a typically invoked specification is:

$$c_a^k = d_a^k + \omega s_a^k(f_a^k) \qquad (6\text{-}20)$$

where $d_a^k$ is any out-of-pocket expense required for using mode $k$ on arc $a$, $s_a^k(f_a^k)$ is the travel time on arc $a$ associated with flow level $f_a^k$, the mode $k$ flow on arc

$a$, and ω is a value-of-time (VOT) parameter that translates travel time into equivalent monetary units, that is, travelers' time cost. Although stated as a nonflow dependent cost in equation (6-20), out-of-pocket expenses can also be a function of flow to model systems such as congestion pricing schemes.

Equilibrium flow models require the arc cost functions to behave in particular ways with respect to flow. Typically, they require the flow cost function to be nonnegative, continuous and nondecreasing with respect to arc flows (i.e., an increase in flow cannot result in a decrease in cost). Other desirable features include the existence of an overload region; in other words, the function should not return an infinite travel cost even if flow is near or above capacity (see Ortúzar and Willumsen 1994).

Given the format of equation (6-20) we can ensure the conditions mentioned above by choosing the correct specification of $s_a^k(f_a^k)$. A typical specification that meets these requirements is the U.S. Bureau of Public Roads (BPR) flow cost function (Branston 1976):

$$s_a^k(f_a^k) = \bar{s}_a^k \left[ 1 + \beta_1 \left( \frac{f_a^k}{B_a^k} \right)^{\beta_2} \right] \quad (6\text{-}21)$$

where $\bar{s}_a^k$ is the mode $k$ free-flow travel time, $B_a^k$ is the mode $k$ capacity of arc $a$, and $\beta_1$, $\beta_2$ are parameters estimated through observations of traffic flows and travel times. Sheffi (1985, ch. 13) provides a basic discussion of estimation procedures while Branston (1976) provides (often-borrowed) BPR parameter estimates for different road types in Winnepeg, Canada. Also see Ortúzar and Willumsen (1994) for a discussion of other cost-flow functions.

### Network flow equilibrium

In this section, we discuss several common types of network flow equilibrium used in travel demand modeling and transportation forecasting. As mentioned in the introduction to this section, "equilibrium" refers to a "stable" network flow pattern. This pattern results from a multitude of individual route choices and the interactions among these choices in the form of congestion. Since route choices are central, different types of equilibria can be formulated based on postulated decision processes.

In this section, we will discuss the major types of behavioral equilibria in congested network flow modeling. We will focus mainly on the so-called "user-optimal" equilibrium; this is the most commonly applied behavioral equilibrium. We will also discuss, albeit more briefly, other variations on this principle. This section is based on Miller (1997).

Another dimension to consider when formulating network equilibrium models is *time*. Most network equilibrium models are *static* and assume that the network flow pattern converges to a steady state with respect to time. However, intuition and observation tells us that congested network flow is a *dynamic* phenomenon. Congestion builds up and dissipates over time. Congestion also propagates through the network from localized incidences. Static network equilibrium models, while useful for long-term, strategic planning, are less useful for tactical operations such as traffic management and intelligent transportation systems. Although several dynamic network equilibrium models are available, we will restrict our discussion to a pragmatic model developed by Janson (1991a, 1991b). In addition to its practicality, the Janson (1991a, 1991b) model is also a good introduction to the complex modeling issues involved in modeling dynamic network flows.

### User optimal equilibrium

*Equilibrium conditions*  The user optimal (UO) network flow equilibrium, originally due to Wardrop (1952), is the most common type of network equilibria analyzed. The traditional UO condition is:

> (UO) At network equilibrium, no traveler can reduce his or her travel costs by unilaterally changing routes (i.e., independently change routes without other users' route changes). *Alternatively*: All used routes between an O-D pair have the same, minimal cost and no unused route has a lower cost.

UO implies the following network flow characteristics. First, positive mode $k$ flow on a route implies that it must have a travel cost equal to the minimum mode $k$ cost for that O-D pair:

$$h_r^k > 0 \Rightarrow C_r^k = C_{ij\cdot}^k \; \forall \; i \in I, j \in J, k \in K, r \in R_{ij} \qquad (6\text{-}22)$$

where the "open arrow" ($\Rightarrow$) indicates "implies" (i.e., the condition on the right-hand side of the arrow follows from the conditions on the left-hand side.) Second, any route with a cost greater than the minimum implies that the flow level is zero on that route:

$$C_r^k > C_{ij\cdot}^k \Rightarrow h_r^k = 0 \; \forall \; i \in I, j \in J, k \in K, r \in R_{ij} \qquad (6\text{-}23)$$

Equations (6-22)–(6-23) require flow for each mode to only occur on the minimum cost routes between each O-D pair. No traveler has a less costly alternative route unless some outside disturbance occurs (Smith 1979).

The behavioral *motivations* assumed in the UO conditions are reasonable: travelers follow the "cheapest" available route for their user class. However, UO also requires strong assumptions about travelers' behavioral *capabilities*. The UO conditions imply that travelers have perfect decision-making capabilities and perfect knowledge about network conditions. In other words, travelers know the exact cost on each available route and react to these costs with perfect accuracy.

Despite the strong behavioral assumptions required, there is a strong underlying belief in the literature that the UO equilibrium conditions are the natural stable flow patterns that occur in real-world settings. However, there have been few attempts to validate empirically the UO equilibrium conditions (Fernandez and Friesz 1983). One of the few attempts is by Florian and Nguyen (1976). Using empirical data from Winnipeg, they discover good correspondence between the UO predicted flows and observed flows. They note that the procedure tended to overpredict arc and route travel times. They also comment on the sensitivity of the results to the arc performance function parameters and the network details (i.e., the level of network aggregation).

*Solving for the user optimal flows* Models that solve for the UO flows use the *equivalent optimization* (EO) strategy pioneered by Martin Beckmann and colleagues (Beckmann, McGuire, and Winsten 1956). The EO strategy requires specification of a travel model, e.g., a network assignment model such as UO (other travel demand components can be added; see chapter 8). We then form an EO problem whose solution corresponds to the equilibrium conditions stated in the travel model.

If we wish to interpret the EO solution as a travel demand pattern (in our present case, a UO network equilibrium), we must be sure that its solution is unique and equivalent to the desired equilibrium conditions. Equivalency can be assured by comparing the "first-order" (first-derivative) conditions for optima of the EO program to the theoretical conditions for equilibrium (equations (6-22)–(6-23)). In the interest of space, we will not review the well-known UO conditions. See Sheffi (1985) for an accessible introduction or Bell and Iida (1997), Boyce (1984), Boyce (1990), and Boyce, LeBlanc, and Chon (1988) for advanced treatments. We can be assured that the solution is unique if the objective function to be minimized is convex. This can be visualized roughly by imagining a "u-shape" in 2-D or "bowl-shape" in 3-D. (For a more rigorous definition of convexity, see Varian 1992.) To achieve this, we impose constraints on the components of the travel model. We require the arc flow cost functions in the network to be separable (equation (6-18)), nonnegative, continuous, and increasing with respect to arc flows. Equations (6-20)–(6-21) together form a cost function that meets these requirements.

Sheffi (1985) provides an excellent, accessible introductory discussion UO equivalent optimization problem. For single-mode flow, the required constrained convex programming problem:

$$\min_{\{f_a\}} \sum_a \int_0^{f_a} c_a(x)dx \qquad (6\text{-}24)$$

subject to:

$$f_a = \sum_{r \in R} \delta_{ar} h_r \qquad \forall \ a \in A \qquad (6\text{-}25)$$

$$\sum_{r \in R_{ij}} h_r = D_{ij} \qquad \forall \ i \in I, j \in J \qquad (6\text{-}26)$$

$$h_r \geq 0 \qquad \forall \ r \in R \qquad (6\text{-}27)$$

The objective function (equation (6-24)) is the summed cumulative costs of each arc cost function in the network given its current flow. Note that we integrate the cost function $c_a(x)$ for each arc from 0 to $f_a$ (the current flow level): this is the area under the flow cost curve up to the current flow level. We sum this integral over all arcs in the network. The decision variables in the optimization problem are the flow levels on each arc; the objective is to choose these flows so that the objective function is minimized.

The arc flow levels that we find to minimize equation (6-24) must meet the following conditions. First, the total flow on arc must equal to the summed flow for all paths that use that arc (equation (6-25)). Second, the flow on all routes between an O-D pair must sum to the aggregate travel demand for that pair (equation (6-26)). Finally, all path flows must be non-negative (equation (6-27)).

Solving the UO problem requires searching though the vast numbers of different combinations of arc flows that satisfy conditions (6-25)–(6-27) for the set of arc flows that minimizes equation (6-24). Heuristic search methods include the *capacity restraint method* and the *incremental assignment method*. Capacity restraint involves performing a sequence of all-or-nothing assignments. All flow between an origin-destination pair is assigned to the shortest path between that pair calculated using a SPT algorithm (see chapter 5). The previous assignment's travel costs are used for the SPT calculation in the current iteration, with the first iteration using free-flow travel costs. This continues until convergence or until a preset number of iterations. A problem with this approach is that the algorithm can get trapped in cycles where flow changes "bounce" back and forth for a subset of the network while other subsets are ignored. Smoothing procedures that combine the last two iterations' travel costs in a weighted average solves this cycling problem but still does not guarantee that we find the optimal solution.

The incremental assignment method divides the O-D flow matrix into portions and performs an all-or-nothing assignment for each portion. After each portion of flow loads on the network, the travel costs are updated and the next portion loads based on these updated costs. Although an improvement over the capacity restraint method, the procedure also does not guarantee finding an optimal solution (Sheffi 1985).

An exact method that guarantees finding an optimal solution for user optimal EO problem is the *convex combinations* method (also known as the Frank-Wolfe algorithm). This procedure is a *feasible direction method*. It starts with some feasible solution; this is any solution that only satisfies the constraints (equations (6-25)–(6-27)). This can be an all-or-nothing assignment to the shortest path trees based on free-flow travel costs. At each iteration, the algorithm determines the *direction* and *step-length* or move size within the solution space that best improves (i.e., reduces the current value of) the objective function. This continues until the solution can no longer be improved (Sheffi 1985). Figure 6-2 illustrates this method.

The direction-finding step of the convex combinations method requires a linear approximation of the objective function at the current solution. We do this by computing an *auxiliary* or alternative solution using the current solution as a base. The auxiliary solution is an all-or-nothing assignment using the current travel costs within the network. The current and auxiliary solutions form a line that determines which direction to move, i.e., which network flows should be adjusted. We then refer back to the original objective function and determine the optimal step size. This requires solving for the move-size that minimizes the objective function in that direction. Since this is a single parameter, we can easily solve for this value using a one-dimensional search algorithm such as the bisection method (see Sheffi 1985 for an elementary discussion or Gill, Murray, and Wright 1981 for an advanced discussion of these search methods).

Each adjustment of flows in the network during the convex combinations

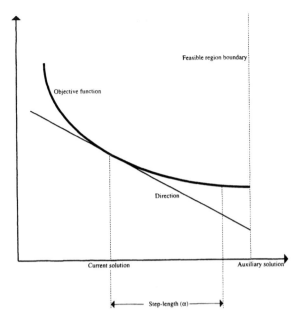

Figure 6-2: The convex combinations method (after Sheffi 1985; Miller 1997)

methods requires solving the SPT problem from each origin and loading its O-D flows on the resulting tree using an all-or-nothing assignment strategy. Asymptotic time complexity for the convex combinations method is therefore $lmO[g(n)] = O[lmg(n)]$, where $l$ is the number of iterations, $m$ is the number of origins, $n$ is the number of nodes in the network and $O[g(n)]$ is the asymptotic complexity of the SPT implementation used.

In practice, the convex combinations method tends to converge slowly and therefore require a large number of iterations. Several methods are available for accelerating its convergence; see Bell and Iida (1997) and Ortúzar and Willimsen (1994) for general discussions or Jayakrishnan et al. (1994) for a specific example. Also available is a more general version of the UO equivalent optimization that allows nonseparable cost functions; see Bell and Iida (1997) or Fernandez and Friesz (1983) for general discussions. Dafermos (1980) provides an early statement of the modeling approach and Smith (1979) provides the earliest statement of the theoretical conditions. Also see Dial (1996) for a model that allows the perceived value of time to vary among travelers (i.e., the VOT parameter in equation (6-20)).

### Other static equilibrium principles

*System optimal* The UO conditions require travel costs to be minimal for each individual. This does not minimize *total* (system-wide) cost for travelers. If the network is congested, each traveler's route choice influences the costs of other travelers; for example, a car entering a congested street not only experiences delay but also imposes additional delays on other cars within that street. If a driver does not consider these *externalities* of his or her route choice, the aggregate cost for the system as a whole can be higher although each driver is minimizing their individual costs.

To accommodate the possibility that individuals either choose or can be directed to choose routes that minimize total rather than their individual travel costs, Wardrop (1952) formulated a second, *system optimal* (SO) principle:

(SO) At network equilibrium, the total travel cost is minimum.

A flow pattern that satisfies this principal is appealing from a society-wide perspective. A SO flow minimizes the total operating cost of the network, implying efficiency (Fernandez and Friesz 1983). Also, if total cost correlates with system-wide use of energy resources and output of pollution, this pattern would minimize these negative impacts. However, this flow pattern is not likely to occur in practice since it requires travelers to make joint decisions to minimize total cost rather than their individual cost. At SO, some travelers could switch routes and reduce their individual costs, meaning that the pattern will be difficult to sustain without some external control mechanism (Fernandez and Friesz 1983; Sheffi 1985).

The difference between UO and SO is clear when one considers the type of information required for travelers to achieve each pattern. UO postulates that travelers consider the average cost on routes: travelers choose the route between an O-D pair that has the minimum average cost, equations (6-22)–(6-23). In contrast, the SO conditions imply that travelers only consider the *marginal costs* for routes, that is, the *added cost of their entry* into a route. Formally, these conditions are

$$h_r > 0 \Rightarrow M_r = M_{ij^*} \; \forall \; i \in I, j \in J, r \in R_{ij} \tag{6-28}$$

$$M_r > M_{ij^*} \Rightarrow h_r = 0 \; \forall \; i \in I, j \in J, r \in R_{ij} \tag{6-29}$$

where: $M_r = \dfrac{\partial C_r}{\partial h_r}$ is the marginal travel cost on path $r$ and $M_{ij^*}$ is the minimum marginal travel cost between O-D pair $ij$. Equations (6-28)–(6-29) state that, at equilibrium, flow only occurs along routes whose marginal cost for the mode is minimum for that O-D pair. Thus, travelers will only choose routes that minimize their impact on total travel cost.

SO-based model formulations have two valuable features. First, SO flow patterns provide a valuable benchmark for assessing the efficiency of other flow patterns (Sheffi 1985). In addition, while the SO principle has traditionally been viewed as an unrealistic ideal, the increasing sophistication of congestion pricing policies and ITS in general can make these conditions an obtainable goal.

We can solve for the SO flows through a simple extension of the UO equivalent optimization program and the convex combinations solution algorithm (Ortúzar and Willumsen 1994). This requires replacing the arc flow cost in equation (6-24) with the marginal cost of traveling on each arc. The convex combinations algorithm is the same as before, except that we use marginal rather than total arc costs in the step length calculation (see Ortúzar and Willumsen 1994 for additional details).

*Stochastic user optimal* The stochastic user optimal (SUO) relaxes UO's strict behavioral assumptions. SUO assumes that travelers are motivated by cost minimization but allows cost perceptions to vary. The SUO principle is (Daganzo and Sheffi 1977):

> (SUO) At network equilibrium, no traveler can reduce his or her perceived travel costs by unilaterally changing routes. *Alternatively*: No traveler believes he or she can reduce costs by unilaterally changing routes.

This principle assumes that the route travel costs include random components that reflect variations in travelers' perceptions. Randomness can result from behavioral factors such as limited information, inaccurate decision making or nonmeasured route attributes (Daganzo and Sheffi 1977). This means that route flows are a function of perceived cost of each route:

$$h_r^k = D_{ij}^k \pi_r^k \ \forall \ i \in I, j \in J, k \in K, r \in R_{ij} \tag{6-30}$$

where:

$$\pi_r^k = \text{Prob}\lfloor P_r^k \le P_s^k \ \forall \ r \ne s \in R_{ij} \rfloor \tag{6-31}$$

Prob[·] indicates the probability operator and $P_r^k$ is the perceived mode $k$ travel cost on path $r$. Equation (6-31) states the probability of path selection as the probability that the path is perceived as the cheapest for mode $k$ travelers between O-D pair $i, j$. Equation (6-30) distributes the aggregate mode $k$ flow between the possible routes based on the probabilities. Different assumed probability distributions for the error component lead to different analytical models for calculating the route choice probabilities (e.g., Sheffi and Powell 1981, 1982). However, at equilibrium the actual route costs for used routes will not be equal and minimal as in the UO case (Sheffi 1985). In general, under SUO each route between an O-D pair will have a nonzero flow level, although it may be small in some cases.

There are several formulations of the SUO EO problem. They differ mainly with respect to the route choice probability mechanism. Two common route probability mechanisms are *logit based* and *probit based*. Logit-based SUO assumes that the route choice error terms are independently and identically Gumbel distributed. *Probit*-based network loading assumes a very general error structure, meaning that route choice utilities can be correlated (see Ben-Akiva and Lerman 1985).

Logit-based route choice has some behavioral weaknesses. First, it can load too much flow to overlapping routes since it is insensitive to network topology. This results from the logit assumption that choices are independent and does not share unmeasured attributes. (This is sometimes referred to the *independence of irrelevant alternatives* (IIA) property of the Luce choice axiom framework; see Ben-Akiva and Lerman 1985 for a good discussion.) A second problem is the logit's reliance on travel cost differences only. This implies that the *magnitude* of the route length is ignored, e.g., a 5-minute travel time difference has the same effect whether the route lengths are 10 versus 15 minutes or 120 versus 125 minutes in length (see Sheffi 1985, pp. 302–305 for a clear illustration).

Probit-based network loading takes into account network topology and route length magnitudes. However, behavioral realism is gained at the expense of more difficult path probability calculations. Calculations must often be obtained through Monte Carlo simulation or other, computationally intensive methods (although progress continues; see, e.g., Bolduc 1999 or Maher and Hughes 1997). Consequently, logit-based SUO remains popular despite its behavioral weaknesses (Bell and Iida 1997). Fisk (1980) presents probably the most common EO problem statement for a single mode, logit-based SUO. Vovsha and Bekhor (1998) develop a link-nested logit route choice model that resolves the route overlap problem. Their model is a generalization of the Fisk (1980) formulation.

As noted above, in general, the SUO problem will generate a positive flow level for each network route regardless of its travel cost, although many of these flow levels can be quite small. Since the number of routes can be quite large, a solution algorithm must work directly with arc flow levels rather than route flow levels (Damberg, Lundgren, and Patriksson 1996) or define a set of plausible or *efficient* routes. Fisk (1980) discusses two methods for identifying a set of efficient paths when solving the SUO problem: (i) shortest path assignment and (ii) Dial's (1971) STOCH algorithm. The former method loads flow onto the shortest path identified during an iteration. The latter method uses a logit function directly to load flows onto the set of efficient paths during an iteration. In this case, the "efficient paths" are those that only include links that bring the traveler closer to the destination and farther from the origin (i.e., a path is not efficient if it brings the traveler closer to origin during any of its subpaths).

Several solution methods have been proposed for the SUO problem. One of the earliest is the *method of successive averages* (MSA). MSA can be used with any stochastic network loading routine, including both logit and probit (Sheffi 1985; Sheffi and Powell 1982; Slavin 1995). MSA is similar to the convex combinations method. In this case, the step-size along the feasible direction is not determined during an iteration. Instead, we prespecify a sequence of step-sizes prior to algorithm execution. The move direction is determined through the stochastic network loading model. MSA's fixed step length means that the algorithm requires a large number of iterations to converge (Chen and Alfa 1991; Huang 1995). Other solution methods specifically tailored for the Fisk (1980) SUO problem include Chen and Alfa (1991), Damberg, Lundgren and, Patriksson (1996) and Leurent (1995).

*Dynamic equilibrium*

Overview   Static equilibrium formulations assume that the network flow pattern converges to a "steady-state" condition where temporal fluctuations do not occur. This is a convenient simplification that makes the mathematics and computations relatively easy. In reality, network flow often fluctuates over time. Travel demand patterns in space and time have become more dispersed and complex due to factors such as increases in suburb-to-suburb commuting, service sector employment and nonwork trips (see Cervero 1986; Hanson 1995). Also, since many urban transportation networks operate at near-saturation levels, an isolated incident (such as traffic accident) can propagate and result in widespread delays.

Traditional transportation modeling is oriented towards strategic infrastructure planning and policy evaluation. Therefore, these models can ignore minor temporal fluctuations. However, emerging applications of transportation modeling and GIS-T technology require capturing the dynamic properties of network flows. These include intelligent transportation systems, *just-in-time* (JIT) logistics, environmental impacts assessment and other tactical problems that require sensitive control or predictions of network flows and travel times.

There is an extensive body of literature on dynamic network flow modeling. We can broadly classify these models into vehicle-based and flow-based approaches (Ran and Boyce 1996). Vehicle-based approaches are microscopic and model individual vehicle movements based on a set of rules and interactions with other vehicles. This generally requires simulation although optimization methods have also been used (e.g., Ghali and Smith 1993). Flow-based approaches disregard individual vehicle movements and instead focus on macroscopic properties governed by dynamic flow equations and "first-in/first-out" (FIFO) restrictions.

*Equilibrium conditions* "Equilibrium" is a much broader concept when we consider temporal dynamics in network flows. First, time can be viewed as *discrete* (i.e., divided into finite intervals) or *continuous*. Second, equilibrium conditions can be stated for "within-day" or *intra-periodic*, "day-to-day" or *inter-periodic* or combined intra/inter-periodic dynamics. Within-day dynamics capture daily fluctuations in network flows with respect to inherent fluctuations and unplanned disturbances such as road closings, accidents, and so on. Within-day dynamics also allows modeling timing decisions for travel; this is important for discretionary travel as well as flexible commuting in congested networks. Day-to-day dynamics capture the slower learning process of travelers as they acquire information about the travel environment. Third, the existence of traditional network equilibria is not guaranteed, particularly with respect to continuous time dynamics. The system may converge to different attractors and display complex behavior as with dynamical systems in general (Cantrella and Cascetta 1995).

Another consideration is the assumed reaction dynamics of travelers. We can assume a *reactive* dynamic user optimal (DUO) in which travelers react to instantaneous travel times and continuously update their route choices according to prevailing traffic conditions. However, this may not lead to equilibrium and can result in unrealistic flow patterns (Jayakrishnan, Tsai, and Chen 1995). Another possibility is a *predictive* DUO in which travelers determine a priori route choices on anticipated network conditions (Janson 1991a, 1991b; Jayakrishnan, Tsai, and Chen 1995). A third possibility is an *ideal* DUO where during each unit of time the actual travel times experienced by travelers departing at the same time are equal and minimal (Ran and Boyce 1996).

Continuous-time dynamic flow models can be stated as variational inequality (VI) or continuous optimal control problems (see Chen 1999; Ran and Boyce 1996). A VI is a general mathematical problem that requires finding a solution vector that makes a vector of cost functions orthogonal to a feasible region. Originally developed to study dynamical systems, VI techniques are also effective for studying network flow and optimization problems in economics (see Nagurney 1993; Ran and Boyce 1996). The UO conditions discussed above are actually a special case of more general UO conditions formed as a VI (Dafermos 1980; Smith

1979). VI equilibrium formulations allow more general and realistic flow cost functions (e.g., nonseparable rather than separable cost functions) as well as dynamic equilibrium formulations. Examples of VI-based dynamic flow models include Fisk and Boyce (1983) and Friesz et al. (1993). Optimal control problems involve determining a set of control variables such that a dynamic system maximizes a performance criterion subject to physical constraints. Examples of optimal control-based formulations include Wie et al. (1990) and Friesz et al. (1989).

Janson (1991a, 1991b) and Janson and Robles (1995) formulate an intra-periodic, discrete-time predictive DUO principle. We will focus our discussion on this DUO principle and a basic equivalent optimization problem for two reasons. First, it is a direct extension of the Wardrop UO principal to the dynamic case and is therefore more easily understood given the previous material in this chapter. Second, it is *practical* in the sense that it can be executed for large, urban-scale transportation networks on modest computational platforms. In contrast, most of the optimal control/variational inequality formulations are computationally complex and require substantial computational platforms even for fairly moderate networks.

Given a set of discrete time intervals spanning a model time horizon, we can state the Janson (1991a, 1991b) dynamic user optimal equilibrium conditions based on departure or arrival times:

> (DUO) At network equilibrium, no traveler who departed (arrived) during the same time interval can reduce his or her travel costs by unilaterally changing routes. *Alternatively*: All used routes between an O-D pair have the same, minimal cost and no unused route has a lower cost for travelers that departed (arrived) during the same time interval.

Since travel times are variable, we cannot fix both departure and arrival times within the equilibrium conditions. Therefore, the DUO conditions assume either a known (fixed) departure or arrival time interval for flows and require equivalent minimal travel costs for all flows "scheduled" to depart or arrive during each interval (Wu, Miller, and Hung 2001).

The DUO principle implies the following network flow characteristics (Janson 1991b). First, positive flow on a route for users who departed (arrived) during a given time interval implies that it must have a travel cost equal to the minimum cost for those users between the particular O-D pair:

$$h_r^{t'} > 0 \Rightarrow C_r^{t'} = C_{ij*}^{t'} \ \forall \ t' \in T, r \in R_{ij}, i \in I, j \in J \quad (6\text{-}32)$$

where $t'$ is a (discrete-time) departure or arrival interval, $T$ is the set of time intervals in the analysis, $C_r^{t'}$ is the travel time on path $r$ for flow that departed (arrived) during $t'$ and $C_{ij*}^{t'}$ is the minimal travel time between origin $i$ and destination $j$ for flow that departed (arrived) during $t'$. Second, any route with a cost greater than the minimum for users who departed during a given time interval implies that the flow level for those users is zero:

$$C_r^{t'} \geq C_{ij*}^{t'} \Rightarrow h_r^{t'} = 0 \ \forall \ t' \in T, r \in R_{ij}, i \in I, j \in J \quad (6\text{-}33)$$

Note that these conditions are a direct extension of the UO conditions: UO is a special case of the DUO principles (Janson 1991b).

*Solving for the dynamic user optimal flows* Janson and colleagues have formulated several EO problems to approximate or solve exactly for the dynamic network flows that meet the equilibrium conditions stated in equations (6-32)–(6-33) (Janson 1991a, 1991b; Janson and Robles 1995). To keep the discussion accessible, we will first discuss in detail the original DUO formulation in Janson (1991b). For more detail on implementing this algorithm, see Wu, Miller, and Hung (2001).

The departure-based DUO problem is (Janson 1991b):

$$\min_{\{f_a^t\}} \sum_{a \in A} \sum_{t \in T} \int_0^{f_a^t} s_a(x) dx \tag{6-34}$$

subject to:
(static constraints)

$$f_a^t = \sum_{r \in R} \sum_{d \in T} h_r^d \delta_{ra}^{dt} \quad \forall\, a \in A, t \in T \tag{6-35}$$

$$D_{ij}^d = \sum_{r \in R_{ij}} h_r^d \quad \forall\, i \in I, j \in J, d \in T \tag{6-36}$$

$$h_r^d \geq 0 \quad \forall\, r \in R, d \in T \tag{6-37}$$

(dynamic constraints)

$$\sum_{t \in T} \delta_{ra}^{dt} = 1 \quad \forall\, r \in R, a \in A_r, d \in T, t \in T \tag{6-38}$$

$$b_{rn}^t = \sum_{t \in T} \sum_{a \in A_{rn}} s_a(f_a^t)\delta_{ra}^{dt} \quad \forall\, r \in R, n \in N, d \in T, t \in T \tag{6-39}$$

$$\lfloor b_{rn}^t - t\Delta t \rfloor \delta_{ra}^{dt} \leq 0 \quad \forall\, r \in R, n \in N, d \in T, t \in T, a \in A_n \tag{6-40}$$

$$\lfloor b_{rn}^t - (t-1)\Delta t \rfloor \delta_{ra}^{dt} \geq 0 \quad \forall\, r \in R, n \in N, d \in T, t \in T, a \in A_n \tag{6-41}$$

where:

- $s_a(f_a^t)$ = arc travel time function (returns travel time on arc $a$ as a function of flow (e.g., equation (6-21)))
- $f_a^t$ = flow on arc $a$ at time $t$
- $t$ = discrete time interval
- $d$ = origin departure time interval
- $\Delta t$ = length of each time interval
- $T$ = total number of time intervals
- $h_r^d$ = flow on path $r$ that departed during time interval $d$
- $D_{ij}^d$ = total flow from $i$ to $j$ departing in time interval $d$
- $\delta_{ra}^{dt}$ = temporal arc-path incidence variable; equal to one if trips departing during time interval $d$ and assigned to path $r$ use arc $a$ during time interval $t$, zero otherwise
- $b_{rn}^d$ = travel time of path $r$ from its origin to node $n$ for travelers departing in time interval $d$
- $A_n$ = set of all arcs incident to node $n$
- $A_{rn}$ = set of all arcs on path $r$ prior to node $n$

Note that the objective function (equation (6-34)) and static constraints (equations (6-35)–(6-37)) are the same as the UO problem except for the addition of the discrete-time temporal dimension. This essentially extends the UO problem across multiple, discrete time periods.

The dynamic constraints (equations (6-38)–(6-41)) ensure temporal flow consistency. The temporal arc-path incidence variable $\delta_{ra}^d$ maintains relationships between arcs and paths across time intervals for flows that depart during the interval. Note that this is a temporal extension of the static arc-path incidence variable. A critical difference is that the temporal arc-path incidence is an endogenous decision variable solved within the problem. In DUO, the arc composition of paths for flows that departed during a given time period cannot be predetermined since the time interval of arc usage is affected by travel costs which in turn are affected by flow loadings (Janson 1991b).

The endogenous nature of arc-path incidence in DUO requires the problem to have nonlinear dynamic flow constraints to ensure flow consistency. First, we require flows to only use each arc on a given path only once during each time interval (equation (6-38)). Second, we require each path to be consistent with respect to the required travel times to reach each arc within the path. To ensure this, we measure the total travel time on a path from the origin to a given node for trips departing in a given time interval (equation (6-39)). Then, we force flow to use the arcs in a path in a temporally consistent manner. If the cumulative travel time to the from-node is greater than or less than the current interval then the temporal arc-path incidence variable is forced to zero and the path cannot use that arc (equations (6-40)–(6-41)). This is clearer if we combine equations (6-40)–(6-41) for the case $\delta_{ra}^d = 1$ (i.e., a trip that departed in time interval $d$ and uses arc $a$ of path $r$ in time interval $t$) and substitute $b_{rn}^d = \sum_{a \in A_{rn}} s_a(f_a^t)$ where $A_{rn}$ is the set of arcs on path $r$ prior to node $n$:

$$(t-1)\Delta t \leq \sum_{a \in A_{rn}} s_a(f_a^t) \leq t\Delta t \tag{6-42}$$

(see Miller, Wu, and Hung 1999). Equation (6-42) ensures that paths use links in the proper time "window."

Two solution strategies are available for the basic DUO problem. A *dynamic traffic assignment* (DTA) procedure generates heuristic solutions with reasonable computational times (Janson 1991b). The *convergent dynamic algorithm* (CDA) is an exact (optimal) algorithm (Janson 1991a).

DTA incrementally assigns the known flows departing during each interval to shortest paths while anticipating future link volumes. Note that in a static NA problem all flow assignment occurs at the same "time." Therefore, the procedure can simply compute the shortest paths from an origin to each destination based on the current levels of arc flows and costs (although these flows can be readjusted until convergence to equilibrium). In the dynamic realm, it is unknown how current and future arc volumes will be affected by assignments from other origins. After each assignment, the DTA procedure must project current arc flow assignments into future time intervals. DTA projects future arc flows based on current arc flow levels, ratios of future (not yet assigned) travel demands and flows assigned in previous intervals. DTA uses these projections only to calculate the shortest paths; these flows are only assigned during their appropriate time intervals. (See Janson 1991b; or Wu, Miller, and Hung 2001 for more details.)

The CDA strategy combines the convex combinations method with linear programming. The convex combinations method solves for a UO equilibrium with fixed node time intervals. This solution is then passed to a linear program to update node time intervals. These updated node time intervals are then passed back to the convex combinations routine for flow updating. This continues until convergence (See Janson 1991a for more details.)

The major data requirement for the DUO problem is a time-specific origin-destination (O-D) flow matrix. Ideally, this requires O-D flow data tagged with the time of day when each trip occurred. These data can be aggregated to the discrete time intervals of the DUO model. Since the DUO model specifies a dynamic equilibrium for flows departing within the same time interval, the critical "time stamp" is the departure time of each trip although the arrival times can also be used for model validation. In contrast, an alternative, arrival-time DUO formulation requires time stamps corresponding to arrival times (Janson 1991a).

If a temporal O-D matrix cannot be obtained directly from primary data it must be estimated. A range of techniques are available for estimating dynamic O-D matrices; however, most of these methods address the situation where origins and destinations are the entry and exit points for a limited portion of the network such as highway interchanges (e.g., see Cremer and Keller 1987; Sherali et al. 1997). Janson and Southworth (1992) discuss a more-general method that uses the dynamic traffic assignment procedure to estimate origin departure times from observed link traffic counts; these data are often readily available. A less-sophisticated option is to temporally disaggregate a daily O-D flow matrix. The simplest method is to divide the O-D matrix equally into the $T$ daily intervals implied by the specified time duration. However, since O-D flows typically exhibit morning and daily peaks this approach is crude. Daily O-D flows could be distributed over the time period of interest by using daily peak profile curves; this would provide a more-realistic estimate of the time-dependent O-D flows.

*Extensions to dynamic user optimal* Although innovative, the basic DUO formulation has some weaknesses. The major problem is the use of binary variables ($\delta_{ra}^{dt}$) to allocate flows across the network over time. This is a crude approximation of a continuous flow process in reality. The "chunking" of flows across time into discrete packets creates difficulties with respect to maintaining aggregate FIFO conditions. Binary flow allocation also creates cumulative inaccuracies in defining node travel times, poor handling of spillback queuing effects resulting from link congestion and difficulties in defining dynamic link capacities (Carey 1992; Janson and Robles 1995). Its also requires specification of the discrete time intervals that are much greater than the average or longest link travel time in the network. This exacerbates the flow-chunking problem and creates substantial and somewhat unrealistic variations in flow levels between discrete time intervals (Janson and Robles 1995; Jayakrishnan, Tsai, and Chen 1995).

To reconcile problems associated with binary flow allocation across discrete time intervals, Janson and Robles (1995) develop a quasi-continuous DUO model that allows fractional allocation of flows within discrete time intervals. Fractional assignment better approximates continuous time conditions and improves handling of and speed transitions between time intervals. This consists of a two level optimization problem. The upper level is a multi-interval UO assignment problem similar to the basic DUO discussed above. The lower level is an optimization that determines the "time reachability" of each node and enforces temporal constraints on flows. A CDA-like procedure iterates back and forth between the upper and lower problems until convergence. Boyce et al. (1997) apply a modified version of this model to a large portion of the Chicago street network. Although the data required for external validation were not available, the model converged to a set of dynamic flows that were internally consistent and intuitive.

Jayakrishnan, Tsai, and Chen (1995) further refine the dynamic equilibrium approach to achieve smaller discrete time intervals that better approximates continuous traffic dynamics. Their approach is based on three modifications: (i) modeling traffic *load* instead of traffic flow, (ii) explicit treatment of the space-time network, and (iii) a modified link travel time function.

Traffic *flow* and traffic *load* are related but distinct concepts. The former is the number of vehicles crossing a given location in unit time (e.g., vehicles per hour). This can be different depending on the measured location within a link (e.g., "upstream" or the entry point, midpoint, "downstream," or the exit point). In contrast, traffic load is the number of vehicles present in a link at an instant in time or during some discrete time period. While flow is a suitable measure for static network flow modeling, it is less well suited for dynamic flow modeling. This can be seen clearly by examining the state equation describing the time-dependent behavior of flow on each link:

$$\frac{df_a(t)}{dt} = f_a^{in}(t) - f_a^{out}(t) \qquad (6\text{-}43)$$

where $f_a(t)$ is the traffic load at time $t$ within link $a$, $f_a^{in}(t)$ is the number of vehicles crossing the upstream end of link $a$ per unit time (i.e., "inflow") and $f_a^{out}(t)$ is the number of vehicles crossing the downstream end of link $a$ per unit time (i.e., "outflow"). Static network flow models assume $df_a(t)/dt = 0$; in other words, flow is constant with respect to time and $f_a^{in}(t) = f_a^{out}(t)$. Conversely, the more realistic situation that dynamic network flow models attempt to capture is $df_a(t)/dt \neq 0$, implying $f_a^{in}(t) \neq f_a^{out}(t)$. This suggests that flow does not adequately characterize a network link since it depends on the measurement location within the link. Load is more adequate since it is a spatial measure expressed over the entire link.

The second, closely related, concept is the idea of the transportation network existing in space-time. In static network flow, the flow assigned to a path exists simultaneously on all of its constitute arcs. This provides computational convenience since it allows calculating network flows by computing shortest path trees and loading flow onto these paths. In the dynamic case, flow propagates through the network over time, meaning that the load on a network link in one time period can exist on a different arc during another time period. We can reconcile this by treating the network as existing in the 2-D spatial plane plus a third dimension corresponding to time. In this expanded view, traffic load can exist simultaneously on all links in a 3-D path. See Jayakrishnan, Tsai, and Chen (1995) or Pallottino and Scutellá (1998) for discussions of shortest path calculations and enforcing dynamic flow constraints in space-time networks.

The third concept relates to the arc flow cost functions and their treatment of arc travel times. Although useful for modeling static network flows, the BPR and related flow-based functions are less useful for modeling dynamic network flows. Equation (6-21) describes a convex relationship in which travel times increase at an increasing rate as flow increases. In reality, after reaching a saturation level flow will begin to decrease as travel times continue to increase since link congestion reduces throughput. This implies a nonconvex relationship (see Jayakrishnan, Tsai, and Chen 1995; or Ortúzar and Willumsen 1994 for more detail).

As we noted above, arc flow cost convexity is required for many static flow equilibrium models; this produces a convex objective function in the equivalent optimization problem. This assumption is not a major problem in static flow models since they do not intend to capture traffic flow dynamics. In dynamic flow models, these functions result in unrealistic flow behavior. To reconcile these problems, Jayakrishnan, Tsai, and Chen (1995) develop the following arc cost function:

$$s_a(f_a^t) = \frac{L_a B_a}{u_{min} B_a + (u_{max} - u_{min})(B_a - f_a^t)}, \quad f_a^t \leq B_a \quad (6\text{-}44)$$

where:

$S_a(f_a^t)$ = travel time on arc $a$ in time step $t$ (seconds);
$L_a$ = length of arc $a$ (feet);
$B_a$ = jam density (vehicles/mile) of link $a$;
$u_{min}$ = minimum speed at jam density (feet/second)
$u_{max}$ = free flow speed (feet/second)

Equation (6-44) is a density-based (as opposed to flow-based) arc cost function. It is nondecreasing and convex with respect to density and continuous, implying desirable properties for the equivalent optimization objective function. However, equation (6-44) is best suited for freeway and other limited access facilities rather than surface streets since it does not capture queuing effects and interactions (see Jayakrishnan, Tsai, and Chen 1995 for more detail).

### Network flow equilibrium: A behavioral critique

Network flow equilibrium is a valuable theoretical perspective for analyzing travel demand and flows within congested networks. Network equilibrium methods view transportation systems as *markets*. In transportation networks, "demand" corresponds to the origin-destination flows while "supply" corresponds to the levels of transportation service provided by the network measured by the arc flow cost functions. Individual route choices involve demand *externalities* in the sense that each individual's choice imposes costs on other individuals and therefore influences their choices. Equilibrium approaches assume that these markets can achieve, at least in the short-run, a "balance" between demand and supply. See Fernandez and Friesz (1983) for an excellent discussion of market concepts in travel demand modeling.

Accepting a view of flows within congested transportation networks as a type of market equilibrium allows us to use the powerful theoretical frameworks available in microeconomics and transportation economics in particular. This theoretical apparatus supports the derivation of clear and logically consistent statements about the costs and benefits of transportation systems. For example, benefit measures such as *consumer surplus* can be applied to transportation systems to evaluate travel demand management policy, congestion pricing schemes or infrastructure changes (see Button 1993; Jara-Diaz 1990; Jara-Diaz and Friesz 1982; or Niskanen 1987 for discussions and applications of these concepts). This can provide a solid foundation for guiding transportation planning and policy.

Despite the benefits of microeconomic theory, network equilibria are less defensible from a behavioral perspective. As we noted above, there have been few empirical tests of their validity. Goodwin (1998) argues quite forcefully that the behavioral properties of congested networks are at odds with the equilibrium assumptions of tending toward a steady state, even if this steady state is dynamically defined. In particular, network equilibria mask the multiple dynamic behavioral processes at work at different time-scales when travelers react to network conditions and adjust their route choices. For example, changes in transport infrastructure, tolls, and land use have qualitatively different effects over time due to factors such as habit, response lags and different thresholds required for triggering responses. This can introduce bias into benefit measures such as consumer surplus (see Dargay and Goodwin 1995). Exacerbating this situation is the difficulty in capturing equilibrium conditions (if they exist) with cross-sectional data (i.e., data captured at one point in time).

An alternative approach to modeling congested network flow is through *network flow simulation*. Network flow simulation models can handle very realistic components and properties of a transportation system. These include signalized intersections with varying signal timing policies; different vehicle types such as cars, buses, trucks, and carpools; multiple lanes; and detailed geometry at intersections to capture turning behavior (see Taori and Rathi 1996). Consequently, traffic engineers and others interested in traffic management systems have used simulation models for years. Growing interest in dynamic properties of traffic flows and congestion and modeling requirements associated with intelligent transportation systems is creating greater interest in flow simulation methods. Continued increases in the power of computational platforms, improvements in visualization methods and near-real-time traffic data collection are also enabling the wider use of these methods. In particular, recent modeling efforts such as TRANSIMS are shifting the domain of flow simulation models from transportation corridors and limited portions of the transportation network (such as limited access highways and city centers) to comprehensive, urban-scale applications.

Network flow simulation

Simulation refers attempting to depict or imitate a real-world system as realistically as possible. There is a fuzzy boundary between "simulation" and more-traditional modeling. Some distinctions include the types of techniques used and the type of solution sought. Traditional modeling tends to use algebraic and calculus based techniques to derive analytical solutions, that is, solutions that can be described as a mathematical equation or system of equations (e.g., the UO equivalent optimization problem discussed earlier). Although we often must use numeric techniques to find the solution (such as the convex combinations method), the solution properties are prespecified using analytical methods. In contrast, simulation approaches tend to specify the behavior of system components and then use computational or numeric methods to allow these components to behave and interact. The solution characteristics are not pre-specified; rather, they occur as a consequence of the components' behaviors. Simulation allows more realistic representations than traditional modeling since we do not need to impose strict assumptions to ensure the existence and uniqueness of a solution.

There are generally three classes of flow simulation models depending on the level at which vehicle behavior is modeled. *Macroscopic models* use traffic flow theoretical relationships and treat vehicles as aggregate streams. Although similar to equilibrium approaches in this respect, macroscopic models can capture more sophisticated traffic flow dynamics (see Daganzo 1997). *Microscopic flow models* represent vehicle behavior at the individual level. They simulate each vehicle's movement relative to the positions of other vehicles, incorporate behavioral principles such as car-following behavior and capture individual vehicle responses to factors such as traffic control devices. Flow properties emerge from the interactions of the individual vehicle behaviors. *Mesoscopic models* are "halfway" between macro- and microscopic simulation models: They do not treat flow as an aggregate stream nor do they model vehicles at the individual level. Mesoscopic models are sometimes based on methods from statistical physics that describe a complex system based on the *distribution* of states among the system's components (see Kulkarni, Stough, and Haynes 1996).

There is a large number of network flow simulation models and this number seems to grow almost on a daily basis (Weiss 1999). An example of a macroscopic model is TRANSYT (*Traffic Network Study Tool*); this treats flow as a distribution that propagates from arc to arc in the network. A platoon dispersion parameter controls inter-vehicle spacing at an aggregate level (Chard and Lines 1987; Manar and Baass 1996; Rakha and van Aerde 1996). DYNASMART is another example of a macroscopic simulation model. DYNASMART uses continuous flow dynamics to dictate traffic flow behavior, but treats vehicles as "macroparticles" whose speeds are determined by these flow dynamics. Originally, macroparticles were aggregates of 5–20 vehicles, but more recent versions use macroparticles of one vehicle. This allows individual vehicle tracking and multiple user classes but cannot capture individual behaviors such as car following (Center for Transportation Research 1994; Mahmassani et al. 1992).

NETFLO II and NETFLO I are macro- and meso-level (respectively) flow simulation models developed in the late 1970s under an U.S. Federal Highway Administration contract. NETFLO II is a derivation of TRANSYT with simplified inputs but better handling of time and dynamic flow properties. Although NETFLO I models vehicles at the individual-level, it moves a vehicle only when triggered by an event or some change in status. This event may correspond to a large "jump" downstream in the traffic flow. Therefore, it is mesoscopic since it does not provide detailed vehicle movement trajectories. Turn movements, free-flow speed and other movement attributes are stochastic (Taori and Rathi 1996).

NETSIM and INTEGRATION are examples of microscopic models. NETSIM represents the position of individual vehicles across one-second discrete time intervals. Vehicles move based on their relative positions with respect to other vehicles and according to car-following logic, traffic controls, and other vehicles' behaviors. Turn movements, free-flow speed and other attributes are stochastic (Rathi and Santiago 1990; Taori and Rathi 1996). INTEGRATION is a microsimulation model specifically geared towards evaluating a combined network of freeway (limited access) and signalized arterial roads (Rakha and van Aerde 1996; van Aerde 1992). Rakha et al. (1998) describe experience with implementation and calibration of INTEGRATION in a mid-size metropolitan area in the United States.

DRACULA (*Dynamic Route Assignment Combining User Learning and Microsimulation*) is one of the few traffic simulation models that captures day-to-day learning processes in route choice processes. DRACULA is a flexible modeling system that supports flow modeling at varying levels of detail (macroscopic and microscopic) as well as different assumptions regarding route choice behavior (such as fixed routes or route choices changing enroute based on traffic conditions). The model updates travelers' perceived average travel costs using a learning mechanism for the next day's route choice process (Liu and van Vliet 1996; Marcotte and Nguyen 1998).

The flow simulation models discussed above treat flow as moving explicitly through a traditional node-arc network (although the network is often expanded to handle features such as individual lanes). An alternative approach is to treat the street network as a tessellation of cells with varying states over time that implicitly generate movement. At the macroscopic level, Daganzo (1994) develops the *cell transmission model* that uses hydrodynamic theory to capture flow propagation between the spatial units. At the microscopic level, the TRANSIMS project of Los Alamos National Laboratory in the United States models network flows using *cellular automata* (CA). A CA specifies the next state of each cell based on the current states of its neighboring cells. In the case of traffic flow, the state of a cell is the presence or absence of a vehicle. As the CA iterates over discrete time intervals, consecutive "vehicle present" states in contiguous cells over adjacent time intervals simulates the movement of a vehicle. Nagel (1998) shows that CA rules are consistent with car-following and traffic flow theory can generate realistic traffic pattern simulations. TRANSIMS has been applied at the urban-scale for Dallas, Texas, and is currently under development for Portland, Oregon (Weiss 1999).

One of the benefits of flow simulation models is the ability to test and calibrate advanced traffic management systems and intelligent transportation systems prior to deployment. The ability to capture sophisticated traffic dynamics, multiple user classes and traffic control devices make these models better suited for pre-deployment testbeds than most equilibrium approaches (although see Ran and Boyce 1996). See Summers and Southworth (1998) for a discussion of the functional requirements of these testbed environments and Ben-Akiva et al.(1997) for a description of an operational simulation laboratory system.

## Facility location within networks

In this final section of the chapter, we discuss facility location methods. These methods attempt to find optimal or near-optimal locations for activities such as retail stores, public sector facilities, emergency facilities, and so on that service some type of demand distributed in space.

### Facility location on a network: Terminology and properties

There are three general types of facility location problems based on their objectives (Larson and Odoni 1981). *Median problems* attempt to locate facilities so as to minimize the total (or, equivalently, average) travel cost between demand locations and facilities. This is the most commonly applied objective. It is appropriate for private sector facilities such as retail outlets and public sector facilities such as government services. *Center problems* (also known as minimax problems) locate facilities so as to minimize the maximum travel cost that any customer will travel to a facility. This is typically applied in emergency service location where we are concerned with minimizing the worse response time or cost in the system. *Requirements problems* locate facilities according to some prespecified performance standard. For example, we may define a demand location as serviced only if it is within a prespecified distance of a facility. This is a more general type of location problem that is suitable both for emergency and nonemergency facilities.

Median and center location problems can be further subdivided based on the distribution of demand within the network and the restrictions on facility locations. Table 6-1 provides a summary of these subcategories. The unqualified terms "median" and "center" refers to the case where both demand locations and facilities are restricted to the vertices of the network (we use the term *vertex* rather than *node* for consistency with the facility location literature and its graph theoretic orientation). *Absolute* medians or centers are the case where demand locations are located at vertices but facilities can locate anywhere. *General* medians or centers are the case where facilities can locate at vertices only but demand locations are anywhere in the network. Finally, *general absolute* medians or centers can locate anywhere in the network to serve demand locations anywhere in the network (Evans and Minieka 1992; Tansel, Francis, and Lowe 1983a, 1983b).

A very important theoretical result regarding centers and medians in a network is due to Hakimi (1964). Hakimi (1964) proves that the median

Table 6-1: Classification of Network Location Problems

| Spatial Pattern of Demand | Objective | Facility Sites | |
|---|---|---|---|
| | | Vertices Only | Unrestricted |
| Vertices | Minimize maximum cost | Center | Absolute center |
| | Minimize total cost | Median | Absolute median |
| Anywhere | Minimize maximum cost | General center | General absolute center |
| | Minimize total cost | General median | General absolute median |

*Source*: After Evans and Minieka 1992.

locations within a network will always be at vertices if customers are located at vertices. In other words, the medians of the network are equivalent to the absolute medians of the network. This theorem has tremendous practical importance. Hakimi's theorem implies that we can restrict our search for optimal medians to the finite set of nodes rather than explore the infinite number of locations anywhere in the network. This has resulted in some powerful solution procedures that can tackle very large median location problems with reasonable computational time.

### Some selected facility location problems

In the next few subsections, we will discuss in some detail some common facility location problems. Our discussion is far from complete: For more comprehensive reviews, see Tansel, Francis, and Lowe (1983a, 1983b), ReVelle (1987), or Hansen et al. (1987). In the discussion below, we will use very general statements of discrete space location problems with the understanding that the travel costs can be measured within a transportation network. Where necessary, we will point out properties associated specifically with network location problems.

### The p-median location problem

*Formal statement and characteristics*   Perhaps the most commonly applied facility location problem is the *p-median problem*. In discrete space, this problem requires siting $p$ facilities among a set of $m$ candidate sites to serve $n$ demand sites, where $p$ is a user-supplied integer such that $p < m \leq n$. When solving the *p*-median problem in a network, the candidate sites and demand sites are restricted to the network nodes. Not all of the network nodes may be candidate sites nor may all of the network nodes have demand.

The *p*-median problem is (Hillsman 1984; ReVelle 1987):

$$\min_{\{x_{ij}\}} \sum_{i=1}^{n} \sum_{j=1}^{m} s_{ij} x_{ij} \qquad (6\text{-}45)$$

subject to:

$$\sum_{j=1}^{m} x_{ij} = 1, \quad i = 1, \ldots, n \quad (6\text{-}46)$$

$$x_{jj} - x_{ij} \geq 0, \quad i = 1, \ldots, n; j = 1, \ldots, m; i \neq j \quad (6\text{-}47)$$

$$\sum_{j=1}^{m} x_{jj} = p \quad (6\text{-}48)$$

where:

$s_{ij} = w_i c_{ij}$
$w_i$ = amount of demand at node $i$; $w_i \geq 0$
$c_{ij}$ = shortest path travel cost from node $i$ to node $j$
$x_{ij} = \begin{cases} 1 & \text{if demand locaion } i \text{ is assigned to facility site } j \\ 0 & \text{otherwise} \end{cases}$
$x_{jj} = \begin{cases} 1 & \text{if a facility is opened at site } j \\ 0 & \text{otherwise} \end{cases}$

The standard p-median problem assumes that customers will patronize the nearest (minimal travel cost) facility. Equation (6-46) requires each demand location to be assigned to only one facility site; this must be the nearest site for equation (6-45) to be minimized. Equation (6-47) only allows a demand location to be assigned to a facility site if a facility is located at that site. (Equation (6-47) is equivalent to the requirement $x_{ij} \leq x_{jj}$, but isolating the decision variables on the left-hand side and constants on the right-hand side is the preferred format in optimization problems.) Equation (6-48) allows exactly p facilities to be sited.

The problem (6-45)–(6-48) is sometimes referred to as the *location-allocation problem* since it requires simultaneous location of facilities (the $\{x_{jj}\}$) and the allocation of the customers to facilities (the $\{x_{ij}\}, i \neq j$).

*Solving the p-median problem* As Hakimi's theorem implies, since demand locations are restricted to vertices we only need to search the vertices for the optimal locations (Hakimi 1964). Nevertheless, the p-median problem is NP-complete (Kariv and Hakimi 1979b), meaning that an exact or optimal solution technique is not feasible for the general problem.

Solution procedures for the p-median problem include *enumeration, graph-theoretic, mathematical programming techniques* and *heuristic procedures* (Mirchandi 1990). Enumeration is only feasible for very small problems. Graph-theoretic techniques attempt to exploit the structure of the underlying network. However, this is only effective if the network has a special structure such as a tree; transportation networks do not typically meet these conditions. Mathematical programming techniques including linear programming relaxation (ReVelle and Swain 1970), branch-and-bound strategies (Khumawala 1972), Lagrangian relaxation (e.g., Narula, Ogbu, and Samuelsson 1977), linear programming duality (Erlenkotter 1978) and decomposition strategies (Swain 1974). See Mirchandi (1990) for an excellent summary discussion.

Heuristics for the p-median problem have received a great deal of attention since they can be applied to general networks and provide very good solutions at reasonable computational expense even for the large problems typically encountered in real-world applications. Some commonly applied heuristics for the $p$-median problem are the add algorithm, drop algorithm, the alternating algorithm and the interchange algorithm (Densham and Rushton 1992a). The *add algorithm* begins with no facilities sited. The procedure searches the $m$ candidate sites for the location that will result in the greatest decrease in the objective function. The facility is located, and the algorithm repeats the search over the remaining sites. This continues until $p$ facilities are sited. The *drop algorithm* is the inverse strategy: the algorithm begins with facilities located at all $m$ sites and removes facilities one at a time such that the value of the objective function is increased minimally. This continues until only $p$ facilities remain.

The *alternating algorithm*, originally due to Maranazana (1964), starts with $p$ facilities located. These can be arbitrary, a set of existing facilities or just the analyst's guess of a good solution. The procedure assigns customers to their closest facility and then relocates each facility independently by solving a single ($p = 1$) weighted median problem based on its assigned customers. (The single median problem can be solved in polynomial time; see Hakimi 1964). The procedure then reallocates customers and resolves the single median problems. The procedure alternates between allocation and single median location until convergence.

The *interchange* or *vertex substitution algorithm*, originally due to Teitz and Bart (1968), has several valuable properties that make it amenable for implementation on modest computational platforms and as part of general-purpose software such as a GIS. First, it often converges to an optimal solution (see Rosing, Hillsman, and Rosing-Vogelaar 1979). Second, irrespective of problem size, it is typically close to a solution within two iterations and often within four, although the final stages of convergence can be slow. Finally, although it is not tied to any particular data structure, some simple data editing techniques can allow it to be applied to a wide range of facility location problems (Densham and Rushton 1992b).

The interchange heuristic begins with an initial, arbitrary solution. Starting with the first of the $(m - p)$ unoccupied facility sites, it attempts to swap the facility at each of the occupied sites with the current open candidate site. The swap that most decreases the objective function, if any, is chosen and the procedure moves on to the next unoccupied candidate site. The iteration completes when all open sites have been considered for a swap. The algorithm terminates when no swap occurs during an iteration. At convergence, the solution will exhibit the three necessary conditions for an optimal solution. These are: (i) all facilities are local medians for their assigned customers, (ii) all customers are allocated to their closest facility, and (iii) removing a facility in the solution and exchanging it with an open site will increase the objective function. Note, however, that although these conditions are necessary they are not *sufficient* and an optimal solution is not guaranteed (Densham and Rushton 1992a).

A hybrid heuristic technique for the $p$-median problem is the *global/regional interchange algorithm* (GRIA) developed by Densham and Rushton (1992a). GRIA modifies the interchange heuristic through a more intelligent search for possible swaps between occupied and unoccupied facility sites. GRIA first looks at a "global" level for swaps by finding the best site to drop from the current solution through the drop heuristic and the best unoccupied site to include in the solution through the add heuristic. This can result in a radical reconfiguration of facility locations and assignments. The algorithm then fine-tunes the solution regionally by considering each facility in turn and attempts to swap it with candidate sites in its service area (i.e., assigned demand sites). At both levels, the swap that best improves the objective function is made permanent. The algorithm terminates when no swaps occur at either the global or regional level.

Densham and Ruhston (1992a) argue that GRIA requires evaluation of fewer swaps and therefore achieves greater computational efficiency than other $p$-median heuristics with comparable solution quality. Numerical experiments that compare GRIA with the interchange heuristic and a modified interchange procedure developed by Goodchild and Noronha (1983) support this claim. However, analysis by Horn (1996) suggests that GRIA can overlook some swaps and therefore terminate with a local optimum with inferior quality compared to the interchange procedure. Whether the conditions under which this occurs is of practical importance is unknown. Required are additional numerical experiments.

*Data structures to support* p-*median heuristics* As we noted in the previous chapter with respect to the shortest path tree algorithm, the computational implementation of an algorithm can have a nontrivial impact on its efficiency. This is also true with respect to $p$-median heuristics. Algorithms such as the interchange heuristic and GRIA require swapping of facilities among candidate sites and the allocation of demand sites to facilities. Proper storage of these data can greatly enhance an algorithm's efficiency.

The key data in a $p$-median algorithm are the node-to-node travel costs. The most straightforward way of storing these data is using an $n \times m$ travel cost matrix. However, this can be quite large for realistic problem sizes. It is also inefficient: since demand locations are assigned to the closest open facility, we are more concerned about the travel costs from a demand location to a subset of proximal candidate sites than to all candidate sites. A clever method for maintaining these data more efficiently are *distance strings* (Hillsman 1980).

A distance string is an ordered set that maintains the travel costs from a base node to other nodes in increasing order of travel cost. In practice, these are usually implemented as two separate lists, where with one list maintaining the travel costs and the other maintaining the corresponding node labels. Figure 6-3 illustrates the distance strings for a sample network. Since the relevant travel costs are sorted in increasing order rather than stored arbitrarily, demand strings can greatly increase the efficiency of a $p$-median solution algorithm (Sorenson and Church 1995, 1996).

Storing all of the available travel costs in a distance string will not result in substantial savings in storage: in fact, it can increase the storage due to overhead requirements. Instead, we need to store only the travel costs for candidate sites that are proximal to the demand node since these are the likely assignments. The problem is defining a way to "cut-off" the portion of the distance string that will

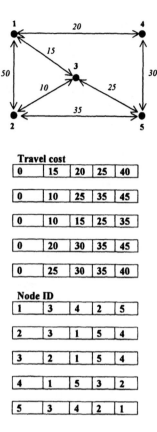

Figure 6-3: Distance strings

be irrelevant when solving the *p*-median problem. Cutting off too much of the distance string can degrade the quality of the solution from the *p*-median heuristic by eliminating potentially optimal assignments of demand sites to facilities.

Strategies for truncating distance strings include defining a *fixed service distance* (FSD), a *fixed-string length* (FSL) and a *demand-varied string length* (DVSL). FSD sets a fixed distance beyond which sites and travel costs are not recorded. FSL only records a fixed number of sites in each distance string. DVSL allows the length of a string to be shorter for base nodes with higher levels of demand since it is likely that a facility will be opened near these nodes. The FSD strategy is the most commonly implemented in *p*-median software. Sorenson and Church (1995, 1996) compare these strategies for some test datasets and find that FSL and DVSL both outperform FSD. However, there is no clear relationship between the size of the problem (with respect to *n* and *p*) and the amount of information that should be stored in the distance strings. This makes it difficult to pre-specify the parameters for the FSD, FSL and DVSL strategies for a given problem. Instead, Sorenson and Church (1995, 1996) suggest a hybrid strategy that sets a fixed number of entries for each string individually.

We can also define two types of distance strings, namely, demand strings and candidate strings (Densham and Rushton 1992b). *Demand strings* are distance strings corresponding to base nodes with positive demands. These list the candidate sites and their corresponding travel costs in increasing order of cost. *Candidate strings* are distance strings corresponding to candidate locations. These list the demand locations and their corresponding travel costs in increasing order of cost. Demand strings optimize the retrieval of the candidate sites that can serve a particular demand site. Candidate strings optimize the retrieval of demand sites that can be served by the particular candidate site.

Usually, a *p*-median heuristic uses either demand strings or candidate strings. For example, the *p*-median software developed by Hillsman (1980) uses candidate strings while the PLACE suite developed by Goodchild and Noronha (1983) uses demand strings. Densham and Rushton (1992b) suggest that using both types of strings combined with an *allocation table* structure can improve algorithmic efficiency to a degree that compensates for the increased data storage. It accomplishes this by requiring less data access from secondary memory (disk) and less data storage in primary memory (RAM) during iterations. For more details, see Densham and Rushton (1992b).

*Variations on the* p-*median problem: The Unified Linear Model (ULM)*

Besides being widely applicable, the *p*-median problem is central to many other types of facility location problems. Hillsman (1984) defines the *p*-median problem as a general mathematical structure he terms the *unified linear model* (ULM) since it consists of a linear objective function and linear constraints. The version of the *p*-median problem specified above (equations (6-45)–(6-48)) is in the ULM format. Editing the coefficients of this problem allows the derivation of *p*-median special cases.

p-*median problem with maximum distance constraints* The *p*-median problem with maximum distance constraints treats demand sites as unserved if they fall outside a set distance or travel cost of a facility. See Khumawala (1973) or Rushton and Kohler (1973) for more details on this problem. To derive this model within the ULM, set:

$$s_{ij} = \begin{cases} w_i c_{ij}, & c_{ij} \leq C \\ M, & c_{ij} > C \end{cases} \quad (6\text{-}49)$$

where $C$ is the maximum service distance or cost and $M$ is an arbitrary large number.

*Covering problems* Covering problems attempt to locate facilities so that demand locations are within a specified "covering" distance, time or travel cost of at least one facility. There are two major types of covering problems. The *set covering location problem* (SCLP) finds the locations and number of facilities such that all demand sites are within a set distance or travel cost of facility. Unlike the problem above, travel costs to facilities are not considered beyond this covering requirement. The *maximal covering location problem* (MCLP) finds the locations of a fixed number of facilities such that the maximum number of demand sites are covered.

We will discuss the SCLP and MCLP in more detail below, for now we will show how to derive these within the ULM. We can specify the SCLP by ignoring the $p$ facility constraints (equation (6-48)) and setting:

$$s_{ij} = \begin{cases} 1, & i = j \\ 0, & c_{ij} \leq C \\ M, & c_{ij} > C \end{cases} \quad (6\text{-}50)$$

We specify the MCLP by setting:

$$s_{ij} = \begin{cases} 0, & c_{ij} \leq C \\ M, & c_{ij} > C \end{cases} \quad (6\text{-}51)$$

and enforcing the $p$ facilities constraints (equation (6-48)). Again, we will examine these problems in more detail below using the traditional formulations.

*Maximum attendance facility location problem* Rather than binary ("all-or-nothing") assignment of demand to facilities, the maximum attendance location problem assumes that customers patronize their nearest facility but patronage declines with increasing distance. See Holmes, Williams, and Brown (1972) for more detail. To derive this within the ULM, set:

$$s_{ij} = w_i(1 - \beta c_{ij}) \quad (6\text{-}52)$$

and change equation (6-42) to

$$\sum_{j=1}^{m} x_{ij} \leq 1, \quad i = 1, \ldots, n \quad (6\text{-}53)$$

where $\beta$ is a "distance decay parameter" showing the effect of travel costs on facility patronage. Equation (6-52) assumes that all of the demand at location $i$ will patronize the facility if it is sited at that location (i.e., $c_{ij} = 0$). As travel costs increase, demand will decrease in a linear manner at a rate determined by $\beta$. Holmes, Williams, and Brown (1972) suggest setting $\beta = 1/C$, where $C$ is the maximum travel cost a person will incur to patronize the facility.

*p-median with powered distance* The $p$-median problem with powered distances is an attempt to build equity considerations into the $p$-median problem. The traditional $p$-median problem attempts to maximize efficiency by minimizing the total travel cost in the system. However, this can result in substantial variations in travel costs among demand sites, with some sites poorly served by a facility. Raising the distance or travel cost to an exponential power (usually squared) can improve the equity of the resulting location pattern by penalizing higher travel costs or distances between demand sites and facilities. See Morrill and Symons (1977) for more details. Within the ULM, this occurs by setting

$$s_{ij} = w_i c_{ij}^{\beta} \quad (6\text{-}54)$$

where β is the distance decay parameter.

Other problems, such as the simple plant location problem, can also be derived from the ULM (see Hillsman 1984). This has practical significance beyond its theoretical insights: it implies that a $p$-median solution algorithm can also be used to solve a very wide range of facility location problems. Of course, this may not be the most efficient strategy since specially tailored solution algorithms can exploit the structure of a particular facility location problem. Also see Church and ReVelle (1976) for a related, earlier discussion regarding the $p$-median, set-covering, and maximal covering location problems.

*Set covering and maximal covering problems*

The set covering and maximal covering location problems have emerged as very important facility location problems, particularly for siting emergency services, telecommunication facilities and other activities where a maximum response time, travel cost or distance is the overwhelming consideration. Therefore, it is worth considering in detail beyond the ULM discussion above.

The set covering and maximal covering problems are a type of requirements objective in facility location. As mentioned above, the *set covering location problem* (SCLP) finds the number and location of facilities such that all customers are within a specified distance or travel cost of a facility. As noted above, we can formulate the set covering problem as a special case of the ULM. The traditional formulation is (ReVelle 1987; Toregas et al. 1971):

$$\min_{\{x_j\}} \sum_{j=1}^{m} x_j \qquad (6\text{-}55)$$

subject to:

$$\sum_{j \in N_i} x_j >= 1, \qquad i = 1, \ldots, n \qquad (6\text{-}56)$$

where:

$$x_j = \begin{cases} 1 & \text{if a facility is opened at site } j \\ 0 & \text{otherwise} \end{cases}$$

$N_i = \{j \mid c_{ij} \leq C\}$

$C$ = maximum service distance or travel cost

The set $N_i$ contains the list of all facility sites that are within $C$ units of distance or travel cost of customer $i$. Equation (6-56) requires at least one facility to be within $C$ units of each customer. The objective function locates the minimum number of facilities such that these conditions hold.

The SCLP is *NP*-hard and therefore cannot be solved optimally for the general case. However, several very effective heuristics are available. One method is to ignore the restriction that the decision variables $\{x_{ij}\}$ be binary (i.e., each customer being allocated to one facility only with no fractional allocations). The resulting problem can be solved using linear programming techniques. Under certain conditions, the LP-derived solutions will be binary and therefore solve the original problem. If not, the binary conditions must be enforced. Branch-and-bound with cutting-plane techniques can be effective for this. It is also sometimes possible to reduce the complexity of the problem prior to solution. If the problem is structured such that the $\{c_{ij}\}$ are mostly zero, then the coefficient matrix can be examined to eliminate many of the constraints implied by equation (6-56), thus simplifying the problem and making it more tractable (see Hansen et al. 1987).

The SCLP has some "looseness" in the sense that multiple solutions can correspond to the same minimal value for the objective function. This "looseness" can allow for multiple criteria to be accommodated within the problem. For example, secondary objectives such as maximizing redundant coverage (i.e., demand points covered by more than one facility) in addition to the set covering criterion. This could be important for backup coverage; for example, we may want a second emergency response unit to respond to demand points within $C$ if the primary unit is busy. See ReVelle (1987) for a summary discussion and Daskin and Stern (1981) and Plane and Hendricks (1977) for some problem formulations.

The SCLP assumes that the budget for opening facilities is unlimited: the system can open as many facilities as required to cover all demand points. This may not be possible or even efficient given the available budget. A problem is that the marginal benefit of adding facilities decreases as the number of facilities increase. In other words, the additional number of demand points covered within $C$ travel cost units by $p$ facilities may only be slightly more than the number covered by $p - 1$ facilities. Therefore, it may not be cost effective to open all of the facilities required to cover all demand points (ReVelle 1987).

An alternative to opening sufficient facilities to cover all demand points is to prespecify the number of facilities to be opened and maximize the number of demand points covered by those facilities. This is the *maximal covering location problem* (MCLP), originally due to Church and ReVelle (1976):

$$\max_{\{x_j\}} \sum_{i=1}^{n} w_i y_i \qquad (6\text{-}57)$$

subject to:

$$y_i - \sum_{j \in N_i} x_j \leq 0, \qquad i = 1, \ldots, n \qquad (6\text{-}58)$$

$$\sum_{j=1}^{m} x_j = p \qquad (6\text{-}59)$$

where:

$$y_i = \begin{cases} 1, & \text{if customer } i \text{ is covered within } C \\ 0, & \text{else} \end{cases}$$

and the other variables as defined previously. Equation (6-58) requires the $\{y_i\}$ to equal one only if the demand point $i$ is covered within $C$ travel cost units. These are the demand weights that are counted in the objective function (6-57) when trying to maximize the covered demand. Similar to the set-covering problem, we can solve the maximal covering problem by ignoring the binary restrictions on the decision variables, and solving the relaxed problem using linear programming techniques, and then using branch-and-bound or cutting-plane techniques to restore the binary solution requirements.

The SCLP and MCLP are only the basic formulations of a very large and active literature on covering problems. Variations on the basic problems include considerations of facility reliability, probabilistic demand, capacitated facilities, and hierarchical facilities. See Schilling, Jayaraman, and Barkhi (1993) for a comprehensive review.

*The p-centers problem*

As mentioned above, the $p$-centers problem attempts to locate $p$ facilities so as to minimize the maximum transportation cost between clients and facilities (Hakimi 1965). This is sometimes referred to as the "minimax" facility location problem (Hansen et al. 1987):

$$\min_{\{x_{ij}\}} \max s_{ij} x_{ij} \qquad (6\text{-}60)$$

subject to equations (6-46)–(6-48).

The $p$-centers problem is *NP*-hard and therefore difficult to solve optimality in the general case (Kariv and Hakimi 1979a), although special cases such as tree networks and unweighted demand locations are more tractable (Tansel, Francis, and Lowe 1983b). Halpern and Maimon (1982) provide a survey of solution techniques within a unified framework.

An effective heuristic for the $p$-centers problem is to solve it as a series of SCLPs with varying covering distances. Suppose the number of facilities sited for a SCLP with covering distance $C$ is $p$. If the number of facilities increases when solving the SCLP for any $C' < C$, then the solution to the SCLP with covering distance $C$ is also a solution to the $p$-centers problem. This location pattern also minimizes the maximum distance ($C$) that anyone must travel given $p$ facilities (Tansel, Francis, and Lowe 1983a). Therefore, we can solve the $p$-centers problem by solving the SCLP with increasing values of $C$ until we find the smallest $C$ that locates $p$ facilities. Christofides and Viola (1971) and Toregas et al. (1971) provide examples of this strategy for the absolute $p$-center and $p$-center problem, respectively. Using this strategy requires determining a reasonable lower bound for $C$ to start the search as well as a step length for incrementing $C$. This involves a trade-off between solution accuracy and computational expense.

*Flow-intercepting location problems*

The facility location problems discussed above only consider demand that is geographically "fixed" (i.e., stationary in space), typically at nodes. However, there are often good reasons to locate facilities within a transportation network that explicitly considers the flow within the network arcs. Examples of flow dependent facilities include retailing activities such as gasoline stations, convenience stores, bank machines and signage such as billboards (see Ghosh and McLaferty 1987; Goodchild and Noronha 1987; Hodgson 1990). Other applications include locating vehicle inspection stations, for example, hazardous material inspection stations or driving under the influence of alcohol (DUI) interception points (Hodgson 1990; Hodgson, Rosing, and Zhang 1996). Finally, locating traffic counters for estimating network flows and travel demands also require consideration of network flows to generate unbiased estimates (see Miller 1999b).

A wide variety of flow interception location problems exist; problem dimensions can include deterministic versus probabilistic flows, allowing deviations from shortest paths, capacity constraints and customer queuing, and single versus multiple interceptions. See Berman, Hodgson, and Krass (1995) for a rigorous review of these problems.

In this section, we will focus on two deterministic, single-interception flow interception problems. The *flow capturing location problem* (FCLP) locates facilities in order to intercept flow at any point in their journeys. This is useful for locating convenience-retailing activities, signage, or "punitive" vehicle inspection stations that attempt to deter violations such as DUI. The *preventative inspection model* (PIM) locates facilities to intercept flow as early as possible during their trips. This is useful for locating facilities designed to protect the network (such as hazardous material inspection stations) or competitive retail location. Both formulations assume that the paths between origin and destinations in the network are known.

The formulation of the FCLP is very similar to the MCLP (Berman, Hodgson, and Krass 1995; Hodgson 1990):

$$\max_{\{x_j\}} \sum_{r \in R} f_r y_r \quad (6\text{-}61)$$

subject to:

$$y_r - \sum_{j \in r} x_j \leq 0, \quad \forall r \in R \quad (6\text{-}62)$$

$$\sum_{j=1}^{m} x_j = p \quad (6\text{-}63)$$

where:

$r$ = a network path

$R$ = set of all paths in the network

$f_r$ = flow on path $r$

$y_r = \begin{cases} 1, & \text{if at least one facility is located on path } r \\ 0, & \text{otherwise} \end{cases}$

$x_j = \begin{cases} 1, & \text{if a facility is located at node } j \\ 0, & \text{otherwise} \end{cases}$

Equation (6-62) requires $y_r$ to be zero if no facility is located on path $r$. This prevents $f_r$ from being included in the objective function's maximization.

Similar to the MCLP, the FCLP is *NP*-hard. Hodgson (1990) develops two heuristic solution procedures. A *cannibalizing* solution procedure does not take into account multiple capturing of flows, that is, more than one facility capturing the same flow. This procedure simply locates the facilities at the $p$ most heavily traveled nodes. A *noncannibalizing* solution procedure considers multiple flow capture. The procedure locates a facility at the most heavily traveled node, removes that flow from consideration, and then locates the next facility at the most heavily traveled node based on the remaining flow. This continues until $p$ facilities are sited. Hodgson (1990) reports numerical experience with these procedures.

The PIM locates facilities so as to intercept flows as early as possible on their paths between origins and destinations. The PIM formulation is (Hodgson, Rosing, and Zhang 1996):

$$\max_{\{x_j\}} \sum_{r \in R} \sum_{j \in r} \rho_{rj} y_{rj} \qquad (6\text{-}64)$$

subject to:

$$\sum_{j \in r} y_{rj} \leq 1 \text{ for all } r \in R \qquad (6\text{-}65)$$

$$y_{rj} - x_j \leq 0 \text{ for all } r \in R, j \in r \qquad (6\text{-}66)$$

$$\sum_{j=1}^{m} x_j = p \qquad (6\text{-}67)$$

where:

$\rho_{rj}$ = protection available to path $r$ at node $j$; calculated as the product of the flow on path $r$ and the persons at risk on path $r$ between node $j$ and the destination

$y_{rj}$ = proportion of the total protection to path $r$ available at node $j$ that is used

with the $\{x_j\}$ as defined previously. The $\{y_{rj}\}$ are no longer binary; instead, these are proportions calculated based on the relative position of node $j$ on path $r$. The value of $y_{rj}$ equals one if node $j$ is the origin node of path $r$; a facility located here protects the entire path by intercepting flow before it traverses any portion of the arc. Conversely, $y_{rj}$ is zero if node $j$ is the destination node of path $r$; a facility located here provides zero protection for path $r$ since the flow has traversed the entire path before being intercepted. Other values of $y_{rj}$ depend on the proportional location of node $j$ along the (directed) path $r$.

Equation (6-66) ensures that a node cannot provide protection unless a facility is located there. Equation (6-65) ensures that a path receives no greater than 100% protection that one facility can provide. If there are two or more inspection stations on a path, the objective function only counts the protection provided by one. The maximization requires that this will be the one with the largest $\rho_{rj}$.

The PIM is a mixed integer program, meaning that it is a linear optimization problem where some of the variables are required to have integer values (specifically, the $\{x_j\}$ are required to be binary). Hodgson, Rosing, and Zhang (1996) report solving PIM for modest networks using mathematical programming software. They initially solve the problem by relaxing the binary requirements on the $\{x_j\}$ and then resolving nonbinary solutions using branch-and-bound.

## Spatial aggregation in network routing and location problems

When solving network routing and location problems, we often must work with spatially aggregated data. Often, travel demand and facility demand data are only available in spatially aggregated form such as traffic analysis zones, census tracts, enumeration districts, postal codes, and so on. Even if individual level data are available, confidentiality concerns may require some form of aggregation. Finally, since the computational burden of many routing and location problems depends on the size of the input data (e.g., the number of origins and destinations), we are often forced to work with spatially aggregated data even if data collection efforts and confidentiality issues are not concerns.

We have already discussed general issues surrounding spatial aggregation and the *modifiable areal unit problem* (MAUP) in chapter 4. We will also discuss techniques for designing traffic analysis zoning systems with respect to travel demand modeling in chapter 8. At present, we will briefly discuss spatial aggregation issues in facility location problems. Most of this literature concerns facility location problems in continuous space rather than networks. However, some of the results are general and translate to the network domain.

Hillsman and Rhoda (1978) identified three types of error resulting from spatial aggregation in facility location problems. *Source A* errors result when the measured travel cost between a facility and an aggregated demand point (e.g., a zonal centroid) does not accurately reflect the total travel cost between that facility and the set of disaggregated demand locations that the aggregated demand location represents. These errors can be negative, zero, or positive. *Source B* errors result when the facility location is coincident with the aggregate demand location: in this case, the measured travel cost is zero even though it is positive in reality since the disaggregated customer locations are dispersed geographically. *Source C* errors result when an aggregated demand point is allocated to a facility when some of the disaggregated demand locations are closer to another facility (see Miller 1996 for a summary discussion).

Several authors (e.g., Casillas 1987; Current and Schilling 1987, 1990; Francis and Lowe 1992; Francis, Lowe, and Rayco 1996; Goodchild 1979; Murray and Gottsegen 1997; Rayco, Francis, and Lowe 1996) analyze errors in the estimated costs and facility locations resulting from spatial aggregation and (in some cases) suggest spatial aggregation strategies for minimizing these errors. Francis et al. (1999) carefully and comprehensively review this literature. They note that aggregation error is related to the ratio $q/p$, where $q$ is number of aggregate demand locations and $p$ is the number of facilities. For fixed $p$, both absolute error and relative error decrease as $q$ increases (i.e., the underlying demand is more disaggregate). For fixed $q$, the absolute error decreases while the relative error increases as $p$ increases. A minimum threshold for $p$-median problems appears to be $q/p \geq 10$. They suggest using one or both of the following strategies: (i) use a method to define the spatially aggregated entities that attempt to minimize aggregation error, and (ii) increase the ratio $q/p$. These results are based on Euclidean distance measurements; a similar analysis that considers the transportation network and its varying representation at different levels of aggregation is still an open research question.

## Conclusion

This chapter completes our review of network algorithms that are relevant to GIS-T. We began the chapter by discussing methods for calculating flows within uncongested networks; these problems tend to correspond to transportation logistics applications where we are moving multiple shipments through abstract networks that represent procurement and/or distribution systems. We then discussed modeling flow through congested networks; this corresponds to traffic flow through physical transportation networks such as streets. We ended the chapter reviewing methods for locating facilities within transportation networks, including facilities that explicitly attempt to intercept network flows. In all of these cases, particularly the dynamic network flow problem, we have only touched on the basic methods in the literature. We encourage the interested reader to use this chapter as a foundation to explore the more advanced treatments in the literature.

In the next chapter, we turn our attention directly to GIS software and discuss analytical capabilities relevant to transportation analysis.

# 7

# GIS-Based Spatial Analysis and Modeling

Why GIS-T? In other words, why do we care so much about using geographic information systems in transportation analysis? An obvious answer is that transportation systems are distributed across geographic space. Indeed, the objective of transportation systems is to *overcome* geographic space. But what is it about GIS that enhances analysis, planning and decision making for geographic phenomena such as transportation systems?

Eastman et al. (1995) suggest that GIS can provide solutions in three generic roles, namely, GIS as an *information base*, GIS as an *analytical tool*, and GIS as a *decision support system*. As an information base and a decision support system, GIS can aid the "front-end" (data collection and management) and "back-end" (communication of alternatives) of the transportation decision process. We will discuss more about the role of GIS in decision support in chapter 8. The objective of this present chapter is to discuss the analytical capabilities of GIS.

The history of GIS can be viewed as an evolution from a mapping software to a geographic analysis and a decision support toolkit (although the traditional mapping base is still critical and in fact being enhanced, as we will see below). The evolution of GIS has important implications to both GIS software vendors as well as GIS users. In the last two decades GIS software vendors have been continuously expanding their software functionality in order to support spatial analysis and modeling needs, albeit with varying degrees of success. To be an appropriate tool, GIS software must provide analytical functions customized to a wide range of application domains.

McCormack and Nyerges (1997) identify three levels of GIS functions to support travel demand modeling (see chapter 8). These are *information man-*

*agement*, *information manipulation*, and *information analysis*. At the lowest level, GIS information management functions are used to retrieve and store land use and network data and to assist in zone and network construction. At the next level, an increasing number of analysts are using GIS information manipulation functions to prepare geographic data for transportation modeling software. At the highest level, transportation analysts use GIS as an information analysis tool. At this level, GIS spatial analytical functions are used to explicitly scrutinize the geographic dimension of the process being modeled or the problem being solved.

In this chapter, we will discuss GIS spatial analysis and modeling capabilities with special reference to transportation. To provide a context, we will begin with a discussion of general issues surrounding GIS and spatial analysis. We will then discuss GIS analytical functions, including basic functions common to most GIS software and transportation-specific tools that have been integrated into many commercial GIS products.

Since it is unlikely that many of the sophisticated tools required for transportation analysis will be built into GIS software, we will then discuss coupling GIS and transportation analysis software. We will discuss both general strategies and specific examples. We will also identify some recent developments in GIS interoperability and componentware that support tight software integration. We follow this with a review of software tools such as macrolanguages and object-oriented programming that allow users to customize GIS software.

Advanced transportation applications, particularly disaggregate travel demand modeling approaches and intelligent transportation systems, require representation of complex and abstract transportation features that are not well supported by the node-arc data model. We discussed some of these in chapter 3. We will expand on this toward the end of this current chapter to set a context for the use of GIS in some of the advanced transportation applications discussed in the second part of this book.

We end this chapter with a discussion of geographic visualization, an integration of scientific visualization and cartography. We will discuss several modes and strategies for geographic visualization, including virtual reality. These information exploration and communication tools will become increasingly important in the emerging data-rich environment for GIS-T.

### GIS and spatial analysis

*Spatial analysis* refers to the scientific study of properties that vary with geographic location. Spatial analysis encompasses questions of extent, pattern, association, interaction and change in geographic space (Nyerges 1991b). These questions are highly complementary with issues surrounding the capture and processing of geographic information. Indeed, some have suggested that GIS is to spatial analysis what the microscope is to biology and the telescope is to astronomy, that is, watershed events that dramatically change the course of the science (Abler 1987). Consequently, there are a number of research programs that explore and strengthen the linkages between spatial analysis and GIS. These include the U.S. National Center for Geographic Information and Analysis

(NCGIA) I-14 research initiative and the European GISDATA scientific program (see Bailey and Gatrell 1995; Fischer and Nijkamp 1993; Fischer et al. 1996; Fotheringham and Rogerson 1994; Goodchild, Parks, and Steyaert 1993; Longley and Batty 1996).

GIS is often distinguished from other computer mapping and graphics systems by its analytical capabilities with respect to geographically referenced data (Goodchild 1987). Cowen (1988) suggests that, unlike computer mapping, GIS creates new geographic information rather than just retrieving previous encoded information. Unfortunately, even with the wide spread of GIS applications today, spatial analytical capabilities of GIS are still somewhat limited. Longley and Batty (1996) argue that "GIS provides the new context for spatial analysis: it provides a new and more accessible basis for representing spatial systems but, to date, it does not provide very clear or easy handles for the development of new functionality which embrace spatial analysis" (p. 2). Openshaw and Clarke (1996) are more critical and argue that "(D)espite many promisingly titled books on the subject of spatial analysis and GIS or vice versa, most of the problems have not been solved although there is now a much improved awareness of some of the issues" (p. 22). Fotheringham (2000) suggests that coupling spatial analysis and GIS may be a partial step backwards since it can make poor and inappropriate tools accessible. (We discuss the four-step approach, a transportation example of the last case, in chapter 8.)

Fischer et al. (1996) argue that both *statistical spatial data analysis* and *spatial modeling* have the potential to enrich the analytical functionality of GIS. They identify four ways that statistical spatial data analysis can enhance GIS. The first is related to *spatial sampling methods*. For example, spatial statistical methods can be used to help GIS-T users determine where and how densely to sample for a household travel survey. A second enrichment is in *spatially integrating and comparing data* that are collected for the same study area but with incompatible sets of zones or incompatible spatial referencing systems. In transportation analysis, we often need to analyze data that are referenced in different spatial reference systems (e.g., UTM versus "route and milepost") or analyze socioeconomic, demographic, and transportation data that are collected by different geographic units (e.g., census tracts, ZIP codes, traffic analysis zones) in the same study area. We noted in chapter 4 that spatial analytical problems (such as boundary effects, scale, the modifiable areal unit problem) and tools (such as areal interpolation methods) can be addressed using GIS functions.

A third spatial data analysis enhancement to GIS is *exploratory spatial data analysis* (ESDA) for data-rich but theory-poor questions and problems such as exploring flow, interaction and activity data. ESDA tools employ techniques such as scatterplots, nearest neighbor, outlier detection methods, spatial interpolation methods, spatiotemporal interactive data displays, autocorrelation tests, among others, to help users identify patterns in their data sets (e.g., see Anselin 1996; Gunnink and Burrough; 1996; Unwin 1996). With large data volumes collected in the transportation community, ESDA will be critical to GIS-T in the coming data-rich era.

A fourth enhancement to GIS is *confirmatory spatial data analysis* that involves systematic analysis and hypothesis testing of spatially referenced data. This requires the development of spatial statistical analysis software linked with GIS (e.g., see Anselin 1992; Anselin and Getis 1993).

"Modeling" refers to a simplified representation of an object under investigation for purposes of description, explanation, forecasting or decision making. A *spatial model* is a model that studies objects existing in bispace (geographic space, attribute) and possibly trispace (geographic space, time, attribute). Spatial modeling encompasses a large range of diversified models that include atmospheric, hydrological, land surface, ecological, economic, and sociological modeling (Wegener 2000).

GIS and related tools have great potential for improving the detail and realism of the geographic space being modeled (Miller 2000). However, the modeling tools available in a GIS are determined by the data models employed. For example, a transportation problem must be conceptualized into a representation framework that is capable of performing the intended analyses. As Goodchild (1998) suggests: "the versatility of a GIS is ultimately determined by the set of data models it enables" (p. 25).

Current GIS data models are often designed to facilitate the inventory of geographically referenced data. Consequently, they are weak in dealing with the abstract spatial modeling and entities such as travel path, trip chain, zonal centroid, and origin-destination matrix that are frequently encountered in transportation analysis. It is apparent that "the focus of research attention now needs to move on from geographic information *handling* to geographic information *using*" (Openshaw, 1996, p. 56).

In a review of various data models needed to support transportation analysis, Goodchild (1998) points out that we have seen many extensions to the basic GIS data model, such as turn tables, dynamic segmentation, route and milepost schemes, and traffic lanes over the years. However, current GIS are still inadequate in handling data analysis of flows, complex paths, and temporal changes. Goodchild (2000) suggests that the evolution of GIS-T can be characterized by three views.

The *map view* is static in nature and has a focus on data inventory and data description. Linear transportation networks are often represented as a collection of nodes and links such as the topological data model used in the U.S. Census Bureau's TIGER files and many commercial GIS software. This topological data model however has its disadvantages. For example, the planar enforcement of topological data model is inadequate for handling overpasses and underpasses. In addition, the topology of a network must be re-generated when new intersections are created. Attribute values must be associated with the fixed segmentation of network links. Dynamic segmentation therefore is required to extend the node-arc topological data model to handle the inventory of linearly referenced data that do not necessarily change their values at network nodes.

The *navigation view* raises concerns with respect to network connectivity and planarity. Although the node-arc structure offers a convenient representation for applying algorithms such as finding the optimal path between two locations on a network, it is insufficient to support many navigational needs. At the very minimum, navigation through a street network requires information such as one-way streets and turn restrictions. Some commercial GIS software products consequently add a variable to flag one-way streets and a turn table to handle turn movements. Furthermore, certain intelligent transportation systems (ITS) applications requires the representation of individual traffic lanes in order to give navigational instructions such as lane shifts or no U-turn on a divided highway. Traditional centerline representations of a network are therefore inappropriate for such applications. Even with the many extensions to the basic GIS data model over the years, current GIS are still inadequate in handling data analysis of flows, complex paths, and temporal changes. The map view is essentially a static perspective and the navigation view attempts to represent the information of a dynamic nature on a static network geometry.

The third view, *behavioral view*, deals explicitly with the behavior of discrete objects to represent moving geometry. This view is closely related to the concept of space-time path proposed by Hägerstrand (1970). A space-time path traces the activity locations of an individual through time in 3-D with time as the third dimension. The activity-based approach in transportation planning studies is a good example of analysis with such data sets (e.g., see Carpenter and Jones 1983; Janelle et al. 1988, 1998; Jones 1990). The behavioral view therefore requires a new set of representation methods that are beyond those developed under the map view and the navigation view. Some research efforts have been taken to address these needs (e.g., Kwan and Golledge 1995; Shaw and Wang 2000); however, there still remain many research challenges to fully implement the behavioral view in GIS.

## GIS analytical functions

There are several basic GIS analysis functions that exist in almost all GIS software. These basic GIS analysis functions are useful for many GIS applications, including GIS-T applications. This section briefly reviews basic GIS analysis functions such as query, map overlay, dynamic segmentation overlay, buffer, and spatial join operations. Other analysis functions that are more specific to GIS-T applications will be discussed in the following sections.

### Basic GIS analytical functions

*Query*

One fundamental function to support GIS-T analysis is the ability of querying GIS databases according to user-specified criteria. An example of attribute-based Boolean query could be "find all street segments whose traffic volume is greater than 10,000 vehicle trips per day and whose volume/capacity ratio is greater than 0.8." Boolean logical operators such as AND (i.e., intersection), OR (i.e., union), NOT (i.e., negation), and XOR (i.e., exclusive OR) can be used in various possible combinations to derive a result set that meets the user-specified criteria.

Boolean logical operators are part of the *Structured Query Language* (SQL) in *relational database management systems* (RDBMS; see chapter 2). An extension of the Boolean operations is to query geographic features based on their locations instead of their attribute values. One effort is the development of SQL/MM, which is part of the SQL3 standard for multimedia use (i.e., graphics, image, voice, and video).

We discussed some of the shortcomings of SQL for geographic data in chapter 2. Worboys (1995) comments on the pros and cons of extending SQL for spatial queries. He indicates that the traditional SQL can extend its database retrieval functionality only "by embedding SQL in a computationally complete programming language, such as C" (p. 297). The new standards for SQL, on the other hand, will allow users to define their own operations and embed them in SQL commands. Spatial query operations, such as "within a distance of $x$ feet," can be programmed and used in an SQL to handle spatial queries.

Parallel efforts of extending relational database functionality to handling spatial data are also found among the RDBMS and GIS software vendors. Extensible DBMS (e.g., Informix® Dynamic Server Universal Data Option, IBM DB2 Universal Database, and Oracle™ Objects and Extensibility Option) allow users to define and add new data types and functions. Spatial data types (e.g., point, line, and polygon) and functions (e.g., "touch" and "overlap") therefore can be incorporated into a RDBMS. Users of Oracle Spatial, for example, can create and manipulate user-defined object types and user-defined components (Oracle Corporation 1999a). User-defined or *abstract data types* (see chapter 2) extend the relational database beyond the built-in types. Groups of objects can form a database component to provide domain-specific behavior. A spatial database component, for example, can store geometric data types and user-defined operators for handling geographic data. Oracle Spatial stores geometry of points, lines, arcs, circles, and polygons to render a digital map in the database and offers operations such as proximity and overlap to evaluate the relationships between geometric objects (Oracle 1999b, 1999c).

Some GIS software vendors also develop their own products to work with the leading commercial database products. This allows geographic data to migrate to existing legacy databases rather than the more difficult inverse task. The *Spatial Database Engine* (SDE) developed by Environmental Systems Research Institute (ESRI) is designed as an application server as well as a spatial data management server (ESRI 1998). SDE enables a standard RDBMS to store and manage geographic data by adding a spatial data type to a relational data model. SDE stores a geographic feature as a row of the database table. SDE stores the coordinates that represent the geometry of the feature as a single binary object value in a column of the table. In addition, SDE extends standard SQL queries to support spatial relationships and spatial searches. For example, SDE can evaluate the spatial relationships such as "equals," "touches," "within," "contains," "crosses," "overlaps" between geographic features.

Map overlay

Map overlay had been used as a map analysis tool long before the era of GIS. An example is the classic *Design with Nature* by McHarg (1971) that describes an analog overlay and suitability assessment methodology. GIS now makes the task easier, faster and less error-prone (although not error free, of course).

Map overlay functions are handled differently in raster GIS and in vector GIS. A raster GIS divides each map layer into a regular grid. When two map layers are overlaid to derive new information, the values of corresponding grid cells in the two map layers are evaluated. Vector GIS, on the other hand, assumes a continuous coordinate system. It requires comparing the coordinates of map features in two map layers to evaluate topological relationships.

There are two types of vector map overlay functions, namely, topological overlay and dynamic segmentation event overlay. Figure 7-1 illustrates three common topological overlay operations: (i) polygon-on-polygon, (ii) line-on-polygon, and (iii) point-on-polygon overlays. They are known as topological overlay operations because they evaluate the connectivity relationships of map features between the two GIS layers and generate a new GIS layer with the combined topology of the original two GIS layers. These procedures combine attribute data from both GIS layers into a single attribute table. Users therefore can retrieve and analyze the attribute data based on the combined topology. Boolean logical operators discussed above are often embedded in the map

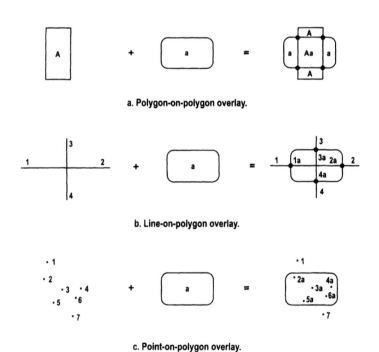

Figure 7-1: Vector topological overlays

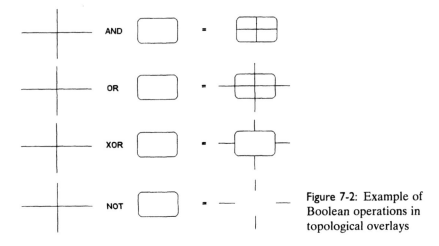

Figure 7-2: Example of Boolean operations in topological overlays

overlay functions to indicate how the different map layers should be combined; see Figure 7-2.

GIS-T applications often deal with linear features and point features located on networks. To find the spatial coincidences between these linear and point features, *dynamic segmentation overlays* are required. Recall from chapter 3 that in the dynamic segmentation data model, line segments are grouped into *routes* and the locations of both linear and point features are referenced by a *linear referencing system* (LRS). For example, a bus stop is located at 5.7 mile from the beginning of a bus route and a divided 4-lane highway segment exists between 50.2 mile marker and 72.5 mile marker on an interstate highway route.

Each attribute that describes the characteristic at a particular location or along a segment on a network route is known as an *event*. There are three types of events that could occur on network routes. *Point events* are the attributes that occur only at point locations on a route. Examples include traffic accidents, bus stops, and toll booths. *Continuous events* represent the attributes that are present, but often varies, along an entire route, such as pavement conditions, number of traffic lanes, posted speed limits, among others. *Linear events*, represent the attributes that are present on selected linear segments on a route only. For example, construction zones along a highway route and tunnels along a railroad track are associated with discontinuous linear segments of a route.

Each event must contain at least two pieces of data: the route identification number and the linear measurements along the route. In general, point and continuous events are referenced by a single linear measure and linear events are referenced by linear measures at both the "begin" and "end" locations where the event attribute value changes. Additional attributes describing the events can be added to the event data file. In most cases, event data are stored in relational tables and can be linked to the GIS network layer through the route ID and the linear referencing measures.

We use dynamic segmentation overlays to analyze the spatial coincidence between different events on network routes. The Boolean intersection operation splits all linear events where there is a change of attribute value on any of the included events and then creates an output file of overlapping events only. Records in an event table that do not overlap with any record in other event tables will be excluded in the output file. The union operation also splits the linear events at all locations with a change of attribute value, but it reports both overlapping and nonoverlapping events in the output file. Figures 7-3 and 7-4 show examples of a line-on-line overlay with the union operation and a point-on-line overlay with the intersection operation (respectively).

## Buffer

We apply the GIS buffer operation when we need to find the features that are located within a specific distance from another set of map features. For example, analysts working on a proposed highway expansion project need to identify all of the land parcels that are located within 100 ft of the existing

Resurface Event Table

| Route ID | From Measure | To Measure | Last Resurface Month |
|---|---|---|---|
| 12 | 1.2 | 3.0 | 07/1997 |
| 12 | 4.5 | 8.2 | 05/1993 |
| 12 | 12.7 | 15.6 | 03/1996 |

Pavement Condition Event Table

| Route ID | From Measure | To Measure | Pavement Condition |
|---|---|---|---|
| 12 | 0.0 | 2.5 | Good |
| 12 | 2.5 | 7.6 | Fair |
| 12 | 7.6 | 9.3 | Poor |
| 12 | 9.3 | 12.5 | Fair |
| 12 | 12.5 | 15.6 | Good |

Union of Resurface Event and Pavement Condition Event Overlay

| Route ID | From Measure | To Measure | Last Resurface Month | Pavement Condition |
|---|---|---|---|---|
| 12 | 0.0 | 1.2 | - | Good |
| 12 | 1.2 | 2.5 | 07/1997 | Good |
| 12 | 2.5 | 3.0 | 07/1997 | Fair |
| 12 | 3.0 | 4.5 | - | Fair |
| 12 | 4.5 | 7.6 | 05/1993 | Fair |
| 12 | 7.6 | 8.2 | 05/1993 | Poor |
| 12 | 8.2 | 9.3 | - | Poor |
| 12 | 9.3 | 12.5 | - | Fair |
| 12 | 12.5 | 12.7 | - | Good |
| 12 | 12.7 | 15.6 | 03/1996 | Good |

Figure 7-3: Line-on-line overlay with the union operator

## GIS-Based Spatial Analysis and Modeling

Subway Stations Event Table

| Route ID | Location Measure | Station Name | Parking |
|---|---|---|---|
| 3 | 0.0 | NE 35th Av. | Y |
| 3 | 1.8 | Green St. | Y |
| 3 | 2.7 | Red St. | N |
| 3 | 3.8 | Downtown | N |
| 3 | 5.1 | Blue St. | N |
| 3 | 6.8 | SW 25th Av. | Y |
| 3 | 8.2 | Airport | Y |

Subway Route Event Table

| Route ID | From Measure | To Measure | Structure Type |
|---|---|---|---|
| 3 | 0.0 | 2.5 | Ground |
| 3 | 2.5 | 4.7 | Underground |
| 3 | 4.7 | 6.5 | Ground |
| 3 | 6.5 | 8.2 | Elevated |

Intersection of Subway Stations Event and Subway Route Event Overlay

| Route ID | Location Measure | Station Name | Parking | Structure Type |
|---|---|---|---|---|
| 3 | 0.0 | NE 35th Av. | Y | Ground |
| 3 | 1.8 | Green St. | Y | Ground |
| 3 | 2.7 | Red St. | N | Underground |
| 3 | 3.8 | Downtown | N | Underground |
| 3 | 5.1 | Blue St. | N | Ground |
| 3 | 6.8 | SW 25th Av. | Y | Elevated |
| 3 | 8.2 | Airport | Y | Elevated |

Figure 7-4: Point-on-line overlay with the intersection operator

right-of-way. In this case, a 100 ft buffer zone will be generated around the existing right-of-way. A topological overlay operation can then be performed between the land parcel layer and the buffer zone layer to derive the desired information.

GIS buffer functions can be performed on point, line, or polygon features. A buffer operation on a set of point features creates a circle, with a radius equal to the user-specified buffer distance, around each point feature (figure 7-5a). For line features, buffer operation will create a polygon extending to the user-specified buffer distance around each line feature (figure 7-5b). For polygon features, the user has a choice to create a buffer zone extending outward or inward from each polygon feature, or to create buffer zones in both directions (figure 7-5c). In all three situations, the buffer operation can automatically dissolve the boundary lines in the overlapped areas between the adjacent buffer zones. The polygons created from buffer operations become separate GIS layers that can be used with other existing GIS layers for additional analyses such as the topological overlay operation.

In some GIS software, a *spatial search* function is a substitute for the two-step process of performing buffer operation and topological overlay operation separately. For example, a spatial search function can identify all of the traffic accident sites that are located within 100 ft around the street intersections. Therefore,

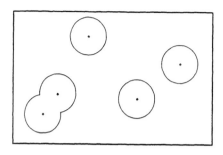

a. Buffer operation on point features.

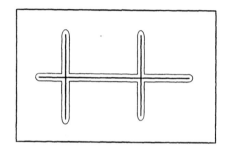

b. Buffer operation on line features.

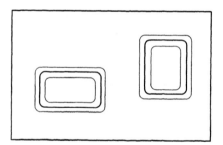

c. Buffer operation on polygon features.

Figure 7-5: Buffer operations around point, line, and polygon features

it does not require the creation of a buffer GIS layer before performing a topological overlay operation.

*Spatial join*

*Spatial join* is an operation based on the spatial relationships between the features in two different thematic layers. For example, we have a thematic layer of retail store locations and another layer of warehouse locations. A logistics manager needs to devise a plan for allocating each retail store to its closest warehouse based on the distance measures between these two sets of point locations. A spatial join operation will compute these distance measures and find the nearest warehouse to each retail store. Spatial join operations also can be applied to the analysis of spatial relationships between different types of map features (e.g., points and lines, points and polygons, lines and lines, lines and polygons, polygons and polygons).

For point-point, point-line, and line-point spatial relationships, the spatial join function finds the nearest point or line feature in the second map layer to each feature in the first map layer. One limitation of these types of spatial joins is the use of the Euclidean distance measures (i.e., straight-line distances) between the two sets of features. For GIS-T applications that require network distance measures between locations, the spatial join function can be inadequate.

Spatial joins based on other pairings of geometric primitives are based on topological relationships. The spatial join function for line-line determines if a line feature in one map layer is "part of (i.e., substring)" of any line feature in another map layer and joins the matched records together. When the spatial join function is applied to point-polygon, line-polygon, and polygon-polygon spatial relationships, features in the first map layer that are located "completely inside" the individual polygon features of the second map layer will be identified. This type of spatial join function is similar to the map overlay function; however, the spatial join function does not perform a topological overlay that splits line features or polygon features in the first map layer where they intersect with the polygon boundaries in the second map layer.

## Built-in transportation analysis functions

Market demand is the basic driving force behind most GIS software development (Goodchild et al. 1992). A rise in market demand for GIS-based transportation analysis and modeling tools have resulted in expansion from the basic network analysis functions in early commercial GIS to more specialized and advanced model tools in contemporary releases. However, as we will see below and in the remaining chapters of this book, it is often the case that commercial GIS software implements the most basic, default versions of many transportation analytic tools. Useful ancillary tools and capabilities are often missing. GIS data models require modifications to handle some of these modeling constructs. Finally, we often do not know how these tools are implemented, a concern particularly with respect to problems that require heuristic solution procedures.

### Shortest path and routing

We discussed the fundamental shortest path and routing problems and their solution algorithms in chapter 5. Since they are compatible with the node-arc topological data model, shortest path tree (SPT) procedures were among the first transport analysis functions available in many commercial GIS software products. The basic SPT procedures were soon extended to take into account the restrictions and penalties associated with the turn movements at street intersections through the use of ancillary data models such as turn tables (see chapter 3). In chapter 5, we noted that there are several ways to implement the SPT algorithm in a computational platform. We also noted the possibility of SPT heuristics such as A* and parallel implementations.

Some leading commercial GIS software products also include functions for solving routing problems such as vehicle routing problems (VRP) and arc routing problems. Recall from chapter 5 that there are numerous variations of the basic VRP, including multiple-depot, time windows for pick-up and delivery, time windows for depot hours and driver work hours, stochastic demand, multiple vehicle types, mixed products and so on.

Current commercial GIS software products have limitations in handling some variations of the VRP. In general, tools for handling multiple-depot and time windows are readily available in the commercial software. Capabilities of handling other variations of the VRP vary between the hard GIS software products. Also recall from chapter 5 that the VRP is *NP*-hard and therefore cannot be solved optimally for realistic size problems. Therefore, GIS vendors must use one or more heuristic or approximate solution algorithms. There are many heuristic procedures available and their effectiveness depends on the particular VRP being solved.

While the VRP is to find efficient routes for visiting a set of nodes in a network, the *arc routing problem* (ARP) is to find efficient routes of traversing a set of links in a network. Applications of the ARP include, for example, garbage collection, mail delivery, water/electricity meter reading, street sweeping and snow plow routing. The ARP attempts to minimize the amount of deadheading, which is the part of traversed route with no demand for the service. It therefore requires a different algorithm from the VRP to find the efficient routes. Again, there are many variations of the ARP, including multiple depots, multiple work shifts, multiple passes of the same link, single pass of a link to serve both sides of the link, specific schedule of serving certain links, among others. Similar to the VRP, ARP cannot be solved optimally for realistic-sized problems. Commercial GIS software normally employs heuristic algorithms to solve selected variations of the ARP only.

We encourage the reader to seek out answers from their GIS software vendors with respect to the computational implementations and algorithm choice for the shortest path and routing tools discussed above. While online documentation, web pages and help desk facilities can be useful, we have found that public questions at user conferences and vendor booths are far more entertaining (although usually less effective).

*Spatial interaction models*

In chapter 8, we will discuss *spatial interaction* models in the context of travel demand forecasting. SI models estimate the amount of interaction (movement of people, material, energy or information) between geographic locations. SI models consider trip generation factors at the origins, destination attractiveness, and the cost of traveling between origin-destination (O-D) pairs.

a. Relational table.

| Feature ID | Item #1 | Item #2 | Item #3 | Item #4 |
|---|---|---|---|---|
| 1 | ... | ... | ... | ... |
| 2 | ... | ... | ... | ... |
| 3 | ... | ... | ... | ... |
| 4 | ... | ... | ... | ... |

b. Matrix data.

Figure 7-6: Relational table and matrix data

SI models are central to trip distribution forecasting (i.e., where trips are going) and are also embedded in more comprehensive travel demand and land-use/-transportation models. Consequently, several commercial GIS software products have implemented SI models, albeit not always in a completely satisfying manner. For example, parameter calibration procedures are usually poor (see chapter 8).

A unique characteristic of SI models is that interaction data correspond to object pairs (i.e., O-D pairs). An effective logical data model for representing these data is the flow matrix. Unfortunately, relational database management systems (RDBMS) implemented in commercial GIS software are not designed to handle data associated with object pairs. Each row in a relational table corresponds to a particular object and the columns represent the different data items associated with the objects (see figure 7-6). Accommodating O-D within the standard relational data model requires storing an "abstract" completely connected network with $m^2$ arcs (tuples). This is inefficient and reflects poor relational database design (see chapter 2). Some commercial GIS packages modify this by including an extension of relational tables to accommodate these paired data.

The centrality of O-D pairs to many types of transportation modeling and analysis suggests the need for a matrix data type in GIS-T software (Shaw 1993). Some GIS allow a special matrix file type. Although not entirely satisfactory from a database design perspective, it does support efficient implementation of matrix algebra operations (e.g., matrix addition and matrix multiplication). These can be used to compute high-level connectivity properties for the interaction data (see chapter 5).

*Network flow models*

In chapter 6, we reviewed a set of basic network flow models such as the transportation problem, the minimum cost flow problem, and the maximum flow problems in uncongested networks. We also discussed congested network flow models such as the user optimal and dynamic user optimal equilibrium models that can be used to forecast aggregate traffic flow through street networks. Disaggregate, simulation approaches are also available for modeling traffic flow.

Network flow models in general have received limited attention by commercial GIS software vendors, with some exceptions (e.g., TransCAD). As we noted in chapter 6, the minimum cost flow problem (MCFP) and its special cases can be solved using linear programming tools and specialized procedures. These procedures are easily embedded within GIS and other transportation software. However, the LP solvers in GIS and other general-purpose software tend to be less powerful and flexible than the tools available in special-purpose LP software. A particularly useful functionality provided by these specialized packages is sensitivity analysis. Special-purpose LP software allow the analyst to compute the *dual* or inverse of the original problem. Solving the dual generates solution values known as *shadow prices*. These show the value (i.e., the improvement in the objective function) that can be obtained by relaxing a resource or capacity constraint (see Hillier and Lieberman 1990). In chapter 11, we will see an example of a transportation logistics application where shadow prices help to guide resource and capacity planning for the system.

*Facility location models*

In their most general form, facility location models simultaneously determine the number of facilities, their locations and the allocation of demands among facilities. Special cases of this general model fix one or more parameters (e.g., the number of facilities) and require different objectives (e.g., minimize the summed travel cost, cover all demand points within a set distance).

As discussed in chapter 6, the number of possible solutions to a facility location problem can be astronomical and it is often impractical to solve within a reasonable amount of time and computing resources. Heuristic algorithms find "good" (near-optimal) solutions within a reasonable amount of time and computing resources. Facility location solution algorithms, particularly those for the $p$-median or "location-allocation" problem, perform very well and can solve very large problems. Nevertheless, similar to many heuristics, their performance degrades as the problem size increases.

Many off-the-shelf GIS software do not support a very simple yet powerful method for improving performance of some heuristics for large facility location problems. Since interchange heuristics require an arbitrary (random) starting configuration, we can improve solution quality simply by *restarting* the heuristic several times with different random starting solutions and then choosing the best of the results across all restarts.

Church and Sorenson (1996) performed an experimental analysis of the restarting strategy for the interchange and GRIA heuristics for the $p$-median problem (see chapter 6). Their analysis determined a very useful stopping rule for the restarting strategy. This rule is:

> *Stopping rule.* Continue to restart the interchange or GRIA heuristic with additional random starting configurations until the best-generated solution has been identified $k$ times without finding a better configuration, where $k$ is a user-specified integer.

The remarkable result of their experiments is that it appears that the optimal solution can be found with very low values for $k$. For example, if $k = 3$, the optimal solution was identified 87% of the time on average across their test datasets. At $k = 5$, the success rate was 96% on average while at $k = 8$ the success rate was 100% on average. These rates varied with the size of the dataset, with the success rate being lower for larger datasets. Clearly, restarting heuristics is a powerful strategy that does not require a major increase in computational time.

## Coupling transportation analysis and modeling with GIS

### Coupling strategies

Instead of "reinventing the wheel" within a GIS, often a better strategy is to integrate the GIS with another software with analytical tools tailored for a particular domain. This allows complementary linkages between general-purpose geographic information processing tools and special-purpose tools. As suggested above, special-purpose analytical software is often more powerful and flexible than the necessarily limited tools in a general-purpose GIS. Software integration, however, is not always simple (see Burrough et al. 1988; Kessell 1990; Lewis 1990; Nyerges 1992).

Lewis (1990) identifies different options for integrating GIS and land-use/transportation (LUT) models. One (poor) option is independent GIS and LUT software. Another option is to integrate separate GIS and LUT software using an interface module linking the two software systems. These strategies maintain GIS and LUTs as separate software products. Another strategy is complete LUT functionality with partial GIS capabilities embedded in the LUT software. The inverse strategy is to develop a full GIS with partial LUT modeling capabilities. The most powerful and perhaps satisfying strategy from a user's perspective is an integrated system with essential capabilities for both GIS and LUT models.

More generally, Goodchild et al. (1992) discuss three alternatives to coupling GIS with spatial data analysis tools. The *full integration* approach embeds spatial analysis procedures fully within a GIS environment. This requires a software design that shares data at the physical level to support data storage, management, analysis, and display needs for different applications. Due to the different data models and analysis/modeling tools required to support not only GIS-T but also a wide range of other GIS applications, vendors are unlikely to offer a full integration solution until the market demand is high enough to justify the investment.

An alternative is the *close coupling* approach that offers GIS application developers direct access to the standard user interfaces and data in the GIS software. Under this approach, data can be passed between GIS software and the spatial analysis software without loss of high-level information such as topology and object identity. An example is the development of a standard spatial query language that allows users to access the spatial data without having to know the specific data structures used in the GIS software.

A third alternative is the *loose coupling* approach. This approach allows the developments of GIS software and spatial analysis procedures as independent products. Data are passed between GIS and spatial analysis procedures through data export and data import functions that convert the data formats between them. This is the easiest and the most widely adopted approach in practice. However, this approach often involves information loss (e.g., topology and object identity) and introduces additional overhead of data processing.

Nyerges (1992) also proposes a similar conceptual framework with four categories of coupling environments for GIS and spatial analytical modeling. The *isolated* approach deals with different data models that require off-line manual data construct conversion to transfer GIS data to models. *Loose coupling* requires manual resolution of data construct differences, but the transfer of data between GIS and models is an on-line process. It therefore demands less effort on the user than the isolated approach. *Tight coupling* relies on interoperable interface services that understand data schemas in all software components. It provides seamless software integration through an application integration framework. The *integrated* approach is built on a single design and is not really a coupling. Integration requires a common data model and a single-user interface for GIS and other software; therefore, there is no need for data construct resolution.

### GIS and model integration: Some examples

Most GIS modeling applications are developed under either the loose coupling approach or the tight coupling approach. In this subsection, we will review some GIS-model integration projects to provide some concrete examples. We will discuss some of these projects from an application perspective in the second part of this book.

Willer (1990) uses structured query language (SQL) in TransCAD GIS software to export data to a suite of location-allocation modeling programs and import the results back into TransCAD for map displays and GIS queries.

Densham (1996) develops a location-allocation modeling toolkit, the *Locational Analysis Decision Support System* (LADSS), that can be coupled with a number of commercial GIS, mapping, computer-aided design (CAD), and database software, including ArcInfo, TransCAD and GISPlus, System 9, Atlas Graphics, Map Viewer, AutoCAD, dBase, and Paradox. The heart of LADSS is a *model-base management system* (MBMS) that implements a number of heuristic location-allocation algorithms and a suite of supporting capabilities accessed through an object-oriented hypertext interface. Hooks in the interface and the MBMS allow a linkage to GIS, mapping, and database software.

Bennett (1997) develops a geoprocessing framework that incorporates GIS, model-base management, computer simulation, and *spatial decision support systems* (SDSS) into an integrated environment to support geographical models.

Johnston and de la Barra (2000) use a sequential linking of travel and land use projections from an integrated urban model (TRANUS) with a GIS-based land allocation model that allocates different land uses within each zone according to simple accessibility rules.

Wu (1998) develops a prototype simulation model that integrates cellular automata and multicriteria evaluation with GIS for the simulation of land conversion. Both the cellular automata and the multicriteria evaluation modules are written in the C programming language and built within ArcInfo GIS. The tightly coupled GIS and modeling environment offers several advantages, such as easier access to spatial information, more realistic definition of transition rules in cellular automata module, and visualization of decision-making.

Anderson and Souleyrette (1996) implement a GIS-based transportation forecast model for small urbanized areas through data conversions between TRANPLAN (Urban Analysis Group, Daville, California) and MAPINFO (MAPINFO Corporation, Troy, New York).

To assess the impacts of large developments on traffic management, Chung and Goulias (1996) develop an interface written in the C programming language to link a regional model of travel demand forecast with a local model of traffic simulation. The regional model is based on the four-step travel demand models in TranCAD, while the local model uses a stochastic microscopic traffic simulation model (Traf-NETSIM) developed by the U.S. Federal Highway Administration.

Hallmark and O'Neill (1996) discuss using GIS for microscale air quality analysis. GIS functions such as contour generation, classification, and map overlay are used to improve the air quality analysis. Due to the incompatible data formats and data models between the GIS software (TransCAD) and the air quality models (CALINE3 and CAL3QHC), they use a loose coupling approach that passes data between the GIS and the air quality models through data conversions.

*Knowledge-based GIS* (KGIS) is another example of extending the basic GIS functions for transportation applications (see chapter 10 for a more-detailed discussion). Spring and Hummer (1995) link a knowledge base with the MAPINFO software to identify hazardous highway locations. Panchanathan and Faghri (1995) develop a knowledge-based expert system (KBES) for identifying hazards and suggesting improvement actions at rail-highway crossings. This set of external procedures is incorporated into the TransCAD user interface as additional menu options that allow data passing between the KBES and the GIS environment. Sarasua and Jia (1995), on the other hand, use the ArcInfo *arc macro language* (AML) and the C programming language to develop a graphic user interface that integrates a KBES design into a GIS environment for pavement management.

### Interoperability standards and component-based software

Some recent developments in the GIS and other software environments, such as data interoperability standards and component-based software, have important implications to the integration of transportation analysis and modeling with GIS. Members of the Open GIS Consortium, Inc. (OGC), including most major GIS and database software vendors, have been working on the development of specifications for interoperable geoprocessing. *Interoperable geoprocessing* refers to "the ability of digital systems to: (i) freely exchange all kinds of spatial information about the Earth and about objects and phenomena on, above, and below the Earth's surface and (ii) cooperatively, over networks, run software capable of manipulating such information" (OGC Technical Committee 1998, p. xi).

The OpenGIS specification, as a comprehensive specification of a software framework for distributed access to geographic data and geoprocessing resources, can provide software developers with a detailed common interface template for software development that will interoperate with other OpenGIS-conformant software written by other software developers. Figure 7-7 shows how the OpenGIS specification can be used to develop integrated systems for different applications, including transportation and logistics. An interoperable environment provides both GIS software vendors and GIS users with a platform to share their data transparently and to better integrate analysis and modeling procedures into this open environment.

Leading commercial GIS software vendors also have adopted the component GIS approach in their software design. A component GIS is based on code that is organized into different software components. These software components are modular and can be assembled together to create custom software. In addition, these software components can be made available to the GIS users who write their own programs. This offers the GIS user community a more flexible environment to build their own analysis and modeling tools and integrate them into a GIS environment.

### Customizing GIS

Since it is unrealistic to expect that general-purpose commercial software to support the wide range of GIS applications with the built-in functions, GIS and other vendors are increasing their support for user customization or modification of their software. At the simplest level, users can modify the default user interface by adding/removing tools or turning on/off particular user interface windows. This is similar to, for example, the customization of user interface in

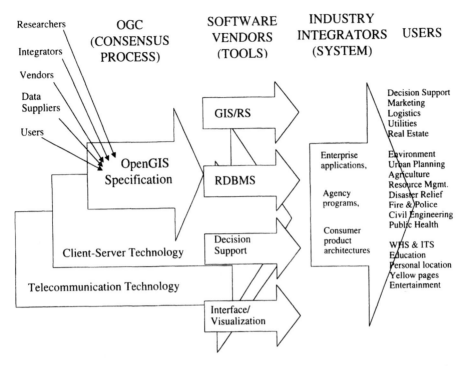

Figure 7-7: The role of the OpenGIS Specification in structuring the geoprocessing channels (after OGC Technical Committee 1998)

Microsoft Word by adding/removing toolbars (e.g., drawing toolbar, tables, and boarders toolbar) and/or commands (e.g., cut and paste). This simple customization of graphical user interface (GUI) makes it possible to create different application GUIs to support the needs of, say, the mapping division versus the right of way division in a transportation organization. At a more-advanced level, users can write their own programs to develop custom tools for specialized analysis and modeling needs and integrate these custom tools into the GUI of a commercial GIS software product. With the trends of moving toward component-based and interoperable GIS environments, GIS users now have a greater flexibility of developing custom applications that are less limited by the built-in analysis and modeling functions in commercial GIS software products.

### Macro languages

Customization capabilities in the early generation of commercial GIS software were rather limited. Macro programming languages were the first tools available to GIS users to write programs within a commercial GIS package. A macro programming language can access the GIS functions in a given software and offers a set of programming controls such as variable declaration and flow control statements. It allows users to create customized procedures for repeated GIS tasks and to develop menu-driven interfaces for casual GIS users.

Macro languages are often unsatisfying as a software customization tool. Shaw (1989) used the ArcInfo arc macro language (AML) to simulate some simple network assignment routines and found that the approach was inefficient for tasks that are computational intensive and/or require frequent data updates during the computation. Like all macro languages, AML is an interpreted language and it is much slower than compiled languages. Streit and Wiesmann (1996) also indicate that the power of GIS-specific macro languages is usually not sufficient for purposes of advanced modeling.

## Object-oriented programming

Ralston (1999) distinguishes the differences between procedural programming, structural programming, event-driven programming, and object-oriented programming. *Procedural programming* is based on a predetermined path that a program follows. For example, an AML script will execute a set of ArcInfo commands, along with the programming controls embedded in the script, in a particular sequence. *Structural programming* places an emphasis on modularization. Each module will take care of a specific subtask and together accomplish the overall task. *Event-driven programming*, which is closely related to object-oriented programming, does not follow a particular path. Instead, the flow is dependent on the interactions with the user. For example, users could choose different menu options or click various tool icons in a Windows-based GIS software to carry out their tasks in the sequence of their choice. Each menu choice or icon click represents an event that triggers the program to perform a particular function requested by the user.

*Object-oriented programming* (OOP) is now a dominant programming paradigm. The *object-oriented* (OO) approach "arises out of a desire to treat not just the static data-oriented aspect of information, as with the relational model, but also the dynamic behavior of the system" (Worboys 1995, p. 85). The OO approach can be applied to several aspects of computer-based systems, such as *object-oriented programming* (OOP), *object-oriented database management system* (OODBMS), or *object-oriented design methodologies* (OODM). These applications are based on the same fundamental concepts; see chapter 2 for a discussion.

Objects are the building blocks of OOP. Conceptually, an *object* can correspond to a real world entity (e.g., road, railroad crossing, or traffic analysis zone) in a problem domain or a programming component (e.g., a menu item, a toolbar icon, or a display window) in a programming environment. Each object is defined by its *properties* (e.g., the name and the width of a road, or the size and position of a display window), along with the *methods* that can be applied to the object (e.g., change a road name, widen a road or move or minimize a display window). The set of methods defined with an object represents the permissible dynamic behavior of the object. As suggested by Worboys (1995, p. 85), the key notion of object-oriented approach is:

$$object = state + functionality$$

In this relation, *state* refers to the location and the attributes of an object and *functionality* refers to the methods that are permissible to the manipulations of the object.

Different OOP languages (e.g., Visual C++, Visual Basic, Java, and SmallTalk) have varying level of OO enforcement (Ralston 1999). For example, programs written in SmallTalk must be entirely in objects, while we can write a perfectly acceptable C++ program without references to any objects. In addition, there are languages with OOP implementations (e.g., VBA: Visual Basic for Applications and Avenue, ArcView's programming language) that are not full OOP languages. For example, although Avenue implements many OOP concepts, it does not allow users to create new objects.

Avenue represents an earlier version of GIS OOP environment that is designed to work under the vendor's proprietary software only. This programming environment makes it easier for users to develop extensions of modeling and analysis functions and to design custom GUIs. However, users are limited to the objects and the methods available in the programming language for application development. Demands from the GIS user community for a more interoperable environment between the hardware platforms and software packages have led to GIS OO programming toolboxes that do not require proprietary GIS software.

### Componentware

Due to the popularity of Microsoft's Windows operating system, several leading commercial GIS software vendors have developed their products to be compliant with the Microsoft's Component Object Model. The *Component Object Model* (COM) consists of objects and interfaces to the objects. Each interface consists of one or more related properties, methods, and/or events. *Properties* are the attributes of an object (e.g., *Name* or *Color*). *Methods* define the actions that an object can take (e.g., "Draw" or "Move"). *Events* communicate with the rest of the system when something happens to an object (e.g., a "Mouse Click"). An object can interact with other objects through their interfaces. In addition, the interfaces define what types of manipulations are allowed for the objects.

With a COM-compliant commercial GIS software, users can choose among a set of object-oriented programming environments (e.g., Visual Basic for Applications, Visual Basic, Visual C++, and PowerBuilder) to manipulate ActiveX controls to develop their custom applications. *ActiveX* controls are reusable components that can be interconnected for custom application developments. Because ActiveX components expose objects, the calling program can access methods, properties, constants, and enumeration of the objects. The ActiveX acts like a server (thus called *Object Linking and Embedding* (OLE) Automation

Server) taking requests from a client program and returning the desired result (Ralston 1999).

ESRI's MapObjects is an example of a collection of ActiveX controls that correspond to the selected functionality in ArcInfo. User of ArcInfo 8 also have access to similar components, known as *ArcObjects*, to develop their own custom GIS applications with an object-oriented programming language. This is also known as the *component GIS* approach. Some commercial GIS software are packaged with the Visual Basic for Applications (VBA) that offers a run-time, integrated programming environment for users to create custom tools within a GIS environment. In this case, users can create and test their custom tools completely within the GIS environment without having to use another development environment to create the custom tools. However, one shortcoming of VBA is the scripts are accessible to the users and, therefore, the source codes are not protected (ESRI 1999b). An alternative is to use other OOP languages such as Visual Basic or Visual C++ to access the GIS components in a custom application program.

### Supporting advanced transportation analysis in GIS

Transportation analysis, particularly for advanced applications such as disaggregate modeling and real-time navigation, offers some unique GIS integration and customization challenges (see Lewis 1990; McCormack and Nyerges 1997; Prastacos 1992; Shaw 1993; Spear and Lakshmanan 1998; Sutton and Gillingwater 1997). A particular challenge is the limitation of GIS data models for transportation modeling.

Goodchild (1998) identifies several data model improvements required to support advanced transportation analysis. These include non-planar networks, turn tables, dynamic segmentation, route and milepost schemes, traffic lanes, off-network travel, traffic flows, complex paths, and temporal changes. Chapter 3 provides detailed discussions of the data models for planar and non-planar networks, turn tables, dynamic segmentation, route and milepost schemes. Most leading commercial GIS software vendors have extended their basic GIS data model over the last two decades to incorporate these innovations. In Chapter 3, we also reviewed navigable transportation network models such as lane-based relational models and 3-D object models.

Travel activities also take place off of a recognized transportation network. Boats can navigate along many possible paths in open water and aircraft sometimes deviate from fixed flight paths. Even vehicles traveling on a highway network often enter parking lots that are not part of a linear network representation. Goodchild (1998) suggests the use of a digital representation of continuous 2-D spaces in conjunction with a linear network representation to model such situations. This approach will require functional linkages between the 2-D continuous surface representation and the network representation.

Complex paths are common features in many transportation studies. Transit routes, garbage collection routes, delivery routes, and our daily trips are examples of travel paths on a transportation network. Although a path could be represented as an ordered list of network links on which the path traverses, GIS users should be able to directly deal with paths instead of a collection of individual network arcs. With the recent trend of extending the GIS data models to handling objects in some commercial GIS software (e.g., the geodatabase data model in ArcInfo 8), users can define their own objects, the object behaviors, and the relationships between objects (Zeiler 1999). This development enables users to model their problems in a GIS environment that is closer to their own conceptualization of the problems (e.g., modeling the transit routes instead of an order list of geometric representations of network links).

Temporal change is the last, but not the least important, transportation data model need discussed by Goodchild (1998). Incorporating the temporal dimension into GIS (known as spatiotemporal GIS) has received extensive research attention in the literature since the 1990s (e.g., Kemp and Groom 1994; Langran 1992; Miller 1991; Peuquet 1994; Peuquet and Duan 1995, Peuquet and Wentz 1994; Worboys 1992; Yuan 1999). We reviewed spatiotemporal GIS data concepts in chapter 2 of this book. Most transportation related data (e.g., land use patterns, travel activities, and traffic accidents) have a temporal dimension. Shaw (2000) provides examples of incorporating temporal data into a GIS environment for three types of transport data sets: historical traffic count data, travel diary data, and simulated GPS-based vehicle tracking data. Kwan (1998) and Miller (1999a) develop methods of measuring individual accessibility within a space-time context using GIS. Chen et al. (1997) discuss a spatial/temporal data model that can integrate both historical and real-time data to solve dynamic network routing problems for ITS applications. Although there have been many efforts of incorporating temporal data into a GIS environment, currently available GIS products provide very limited capabilities of managing and modeling spatiotemporal data.

In order to overcome the above weaknesses of handling more advanced transport analysis and modeling needs, current GIS must be enhanced to better represent different transportation phenomena. Egenhofer et al. (1999) review the progress and status of research in the area of computational methods for representing geographic information. They suggest that past GIS research concerned the development of fast and efficient implementations of data storage, retrieval, and analysis of traditional cartographic concepts. As the GIS user community demands a more-intuitive and user-friendly environment to support a wide range of applications, we need to address "the linkage between human thought about geographical space and the mechanisms of computational models." (Egenhofer et al. 1999, p. 776)

Egenhofer et al. (1999) suggest several research and development directions, some of which overlap with the discussions in Goodchild (1998). First of all, they indicate that a true interoperating GIS is needed to overcome the current limitations of performing analysis between discrete (vector) and continuous (raster) geographic phenomena. This is consistent with functional linkages between a 2-D continuous surface representation and a linear network representation for handling off-network travel activities suggested by Goodchild (1998). In addition, GIS should be able to represent qualitative spatial information that matches human cognition and spatial reasoning. For example, driving directions with reference to landmarks are more intuitive to most people than the location references by Cartesian coordinates used in GIS. Therefore, it is necessary to better integrate the semantics of natural language and the models for quantitative spatial information. Temporal capabilities is another future enhancement area to deal with space-time behavior as well as continuous moving, point-like objects. The massive amounts of spatial data collected through technologies such as remote sensing, GPS, and ITS require more effective spatial data mining methods to help us discover the patterns embedded in large data sets. Finally, Egenhofer et al. (1999) suggest GIS software should become the spatial data management components of other information system architectures. The recent trend of extending existing commercial RDBMS to include spatial functionality shows a first step toward this direction. It is anticipated that future GIS software will provide better interfaces with other commercial software.

## Geographic visualization

### Geographic visualization, spatial analysis, and GIS

Cartography is the ancient science and art of storing and communicating information about the surface of the Earth using graphical methods. Mapping is a fundamental spatial analysis method, summarizing large amounts of information organized by geographic location (MacEachren 1994b; Philbrick 1953). GIS has greatly expanded the ease and (arguably) the power of cartographic display and presentation in spatial analysis. The role of geographic visualization will only increase in importance as we face major challenges in developing effective tools to make sense of the exploding volume of geographic data being collected and geographic information generated from these data (MacEachren et al. 1999).

Visualization is central to *exploratory data analysis* (EDA). EDA uses graphical techniques such as histograms, scatterplots, scatterplot matrices, density ellipses, spin diagrams, and cluster diagrams to search for and to understand patterns, trends, and outliers in data sets (MacDougall 1992). Research on the use of GIS in EDA, however, did not receive much attention until the late 1980s (e.g., Goodchild 1987; Monmonier 1990; Openshaw et al. 1990; Walker and Moore 1988; Wills et al. 1990). Traditionally, cartography and GIS use abstract representations of spatial phenomena to simplify the complexity of real world. EDA for geographic data employs methods such as stereo views, shading, and geographic brushing to develop more realistic rendering for depicting complex data relationships in abstract data spaces (MacEachren et al. 1994).

DiBiase (1990) develops a framework that places *geographic visualization* (GVis) in the context of *scientific visualization*. A U.S. National Science Foundation report on *Visualization in Scientific Computing* defines scientific visualization as research that "studies those mechanisms in humans and computers which allow them in concert to perceive, use and communicate visual information" (McCornick et al. 1987, p. 3). In general, scientific visualization places an emphasis on the use of information technology to help scientists explore data, identify patterns, develop hypotheses, and gain insights. Scientists employ visualization in two types of related activities: visual thinking and visual communication (DiBiase et al. 1992). *Visual thinking* is exploratory in nature and helps scientists investigate a research problem. *Visual communication*, on the other hand, offers scientists a tool to explain and distribute their findings in graphics forms. The framework proposed by DiBiase (1990) defines "map-based scientific visualization as including all aspects of map use in science, from initial data exploration and hypothesis formulation through to the final presentation of results." (quoted in MacEachren 1994b, p. 2) Therefore, maps can support both private visual thinking in a research process as well as public visual communication of research results.

Taylor (1991) places visualization at the center of a triangular framework that consists of *cognition* (i.e., analysis and applications), *communication* (i.e., new display techniques) and *formalism* (i.e., new technologies) as the three sides of the triangle. MacEachren (1994b) comments that the framework proposed by Taylor places an emphasis on computer graphics technology for visualization, while the DiBiase's framework focuses more on the use of visualization. Nevertheless, both frameworks include an analysis/visual thinking component and a communication/presentation component in visualization. MacEachren (1994b) further proposes a representation of cartography as a 3-D space defined by three continua. One dimension is a continuum from *private map use* (i.e., generate a map for personal use) to *public map use* (i.e., make maps available to the public). Another dimension represents a continuum of map use for *revealing unknowns* (i.e., data exploration) to *presenting knowns*. The third dimension is a continuum from *low human-map interaction* (i.e., no or limited user interactions with map displays) to *high human-map interaction* (i.e., user can substantially interact with map displays). This 3-D representation offers a framework to the GIS community to consider both visualization and communication aspects of map use. As a visualization tool, GIS need to provide users with capabilities of a high human-map interaction level for revealing unknowns in addition to the presentation of knowns with a lower level of human-map interaction.

MacEachren and Monmonier (1992, p. 197) suggest that computer-assisted geographic visualization "facilitates direct depiction of movement and change, multiple views of the same data, user interaction with maps, realism (through 3-D stereo views and other techniques), false realism (through fractal generation of landscapes), and the mixing of maps with other graphics, text, and sound." Due to the interaction and animation capabilities of computer display technology, we are no longer limited to a single static map view of our data. Several commercial GIS software products allow users to visualize their data in multiple views that include, for example, a *map view* (for spatial data), a *graphics view* (for bar charts, scatterplots, etc.) and a *table view* (for nonspatial data). In addition, more interactive visualization tools such as fly-over are available in some commercial GIS software.

Visualization has been an integral part of GIS from its origin. Map displays are critical elements in almost all GIS applications for displaying spatial patterns and for communicating the results of "what if" analysis. Scientific visualization provides a number of innovative ways that take advantages of the developments in computer GUIs and the use of interactive visual programming. For example, a network editor built in an interactive visual programming environment can enable users to build their applications by selecting and connecting data and functions represented in iconic form (Visvalingam 1994). This allows users to assemble data flows in a flexible manner and to visualize multiple perspectives of the data.

Visualization in GIS traditionally was limited to conventional maps and 3-D views of terrain or statistical surfaces (Turk 1994). Recent research in GIS calls for more interactive and dynamic ways to visualize data and to represent processes. Spatial decision support systems (SDSS, discussed in chapter 8), for example, require interactive and dynamic data manipulation and visual displays of different what-if scenarios, especially in a group-based decision-making environment.

*Visual realism*

Although abstract representation has been a tradition of cartography and GIS, it may not be an intuitive way to communicate with the GIS users. Many GIS users want a user-friendly system that minimizes abstract representations and maximizes information content. *Visual realism* is an approach that allows users to explore data sets with their natural sensory perceptions (Bishop 1994). Realistic simulation is one way to provide direct experience of alternative realities of the consequences of different what-if scenarios. For example, visual simulations of projected traffic flows on a street network due to different proposed land use change plans provide more intuitive representations than the projected traffic flow data presented in a tabular form or as static maps. Video log systems that have been developed at many transportation agencies to capture visual displays of roadway conditions are another example of visual realism. Transportation engineers and planners can perform a virtual drive-through of highway segments to gain a direct visual experience of pavement conditions, roadside traffic signs, among other roadway information.

To support more realistic visualization of GIS data in both spatial and temporal terms, Bishop (1994, pp. 63-64) suggests the following enabling technologies:

- development of dynamic modeling tools within the GIS environment
- use of photographs or video as textual maps to increase the realism of GIS-generated perspectives
- use of animations
- interactive exploration of highly realistic imagery
- incorporation of other display options such as stereo projection, head-up display, and data gloves

- retrieval of numerical values from simulated objects using an interactive probe in 3-D space

With the rapid development of computer technology such as virtual reality, we can expect that visual realism will increasingly become part of the GIS environment.

*Dynamic and temporal visualization*

Spatial analysis concerns identifying patterns as well as underlying processes in the data sets. Processes are inevitably related to the time dimension. As time changes, a process can lead to changes of spatial patterns. DiBiase et al. (1992) identified three types of changes. The first type of change involves change in position of the *observer in relation to the geographic space*. An observer can change his/her location and/or scale in space to view spatial patterns (e.g., fly-by). Displays of such changes are therefore spatially dependent, but do not require a particular temporal sequence to display the changes (i.e., they are temporally independent). A second type of change is related to change in position and/or attributes of an *object within geographic space*. This is often shown as time series that the sequence of displays is clearly time dependent. The third type of change deals with change in *position within attribute space* (also known as *re-expression*). Displays of these changes are based on the values of a selected attribute rather than on either spatial or temporal orders. The first two types of change are good candidates for visual realism (Bishop 1994) and for animation of time driven map sequences (MacEachren 1994a).

Pan and zoom functions available in most commercial GIS software are rudimentary tools to animate the change in position of an observer in geographic space. Some commercial GIS software (e.g., ArcView 3D Analyst extension) also provide fly-by capability for users to navigate in an animating space. Time-series changes, on the other hand, place an emphasis on the temporal order of displays to show the changes in temporal space (MacEachren 1994a). Display frames are ordered to match with the duration of discrete time samples.

With an increasing number of digital geographic databases that are referenced in both space and time, "We need to devise an exploratory technology to assist in searching for patterns and processes in GIS databases which would not have been immediately obvious if mapped in conventional ways" (Openshaw et al. 1994, p. 132). Openshaw et al. (1994) suggest computer animation of time driven map sequences as a spatial analysis tool. Tobler's (1970) movie of visualizing time-dependent data of urban population change and Moellering's (1976) computer animated film of traffic accidents are early examples of creating computer movies as visualization tools.

Openshaw et al. (1994) propose three alternatives of ordering map displays in an animated movie. The first alternative is to speed up time, while retaining the temporal structure of the data, in search for time-driven spatial patterns. The second alternative reorders the data by a finer temporal resolution (e.g., hours and minutes) and ignores the longer temporal effects (e.g., day, month, and year).

This allows us to examine, for example, time-of-day-dependent patterns. The third alternative uses an attribute rather than time to order map displays in an animated movie. This approach, which focuses on the patterns of variables that are not necessarily functions of time, is similar to re-expression (DiBiase et al. 1992).

*Integrating geographic visualization and GIS*

Scientific visualization and GIS have developed in parallel but not in concert. Spatial data standards do not give much consideration to data visualization and visualization software do not recognize GIS data models (Rhyne 1997). As a result, we encounter difficulties in registering spatial data within visualization software, producing animation sequences within a GIS environment, and connecting GIS databases with the visualization environment to support various display, query, and analysis functions (Hearnshaw and Unwin 1994; Rhyne 1997).

Rhyne (1997) defines four levels of GIS and scientific visualization integration methods. At the rudimentary level, a minimum amount of data can be shared and exchanged between the two technologies. For example, data in DXF and DEM formats now can be shared and exchanged between some GIS and visualization software. Certain GIS software also can export data into Virtual Reality Modeling Language (VRML) format for visualization on the Internet. At the operational level, visualization software is able to directly access the GIS databases to remove data redundancies between the two technologies. For example, Cook et al. (1997) use ArcView and Xgobi software to provide exploratory data analysis tools that are dynamically linked to a GIS. Currently, data access is often limited to displays and does not allow GIS functions from the visualization environment. The functional level of integration requires open GIS data standards to support visualization tools accessing spatial data analysis functions. The highest level of integration will require fundamental changes to the GIS and visualization software development processes in order to design a merged GIS and scientific visualization system.

Geographic Visualization (GVis) can be very complex, arguably more complex than traditional cartography. Effective GVis requires identifying a formal mapping of multiple geographic attributes to a 2-D, 3-D, or possibly 4-D *scenegraph* (a computational description of objects to be rendered). This should preserve or can be easily related to geographic space by the viewer. The number of possible ways to visualize data is exponential with respect to the data attributes, creating an enormous set of alternative visualizations even for a small number of attributes. Displaying multiple information themes at once can create complex interactive effects that are poorly understood and confound the signal or message in the visualization. These interactions are understood poorly and require additional research (Gahegan 1999).

Effective GVis also requires very fast rendering speeds, particularly for dynamic and interactive visualizations. The scenegraph can be several orders of magnitude larger than the data being visualized. Rendering performance depends on the software environment and hardware platform. Software environments include visual programming interfaces such as IBM Data Explorer or toolkits such as ENVI/IDL. Toolkits generally provide better performance and hence faster rendering speeds, although this requires a sacrifice with respect to functionality, ease of use and sometimes quality. Computational platforms with hardware components specially designed for graphics may be required for large geographic visualization tasks (Gahegan 1999).

### Virtual reality

*Overview*

Jaron Lanier coined the term of *virtual reality* (VR) in 1988 to refer to a computer generated 3-D interactive environment (Williams 1999). Fairbairn and Parsley (1997, p. 476) suggest that "VR is a non-deterministic method of presenting spatial information in graphical form." They indicate the following characteristics of a VR system:

- provides a user-navigation capability to access the data in a free and random manner.
- offers an interactive environment for users to dynamically explore the data.
- permits modification of the viewing position and/or modification of the data.
- creates an environment for users to make assumptions and to test hypothesis.
- enables the creation of user-defined views and maps.
- reduces dependence on professional cartographers to produce maps.

There are several types of VR (Fairbairn and Parsley 1997; Verbree et al. 1999). *Full* (also known as *immersive*) VR requires equipment such as head-mounted displays (HMD), audio speakers, moving platforms, and tactile gloves that subject the user to a simulated virtual reality environment. *Transparent* VR uses a simulated real world as a backdrop to present the spatial information in a virtual environment to the user (e.g., a flight simulator that displays a fishnet digital terrain model on the cockpit window). Transparent VR employs a virtual workbench with a 3-D-scale model that allows users to visualize the 3-D image from above the virtual workbench. *Projection* VR uses large format graphical displays to provide a virtual reality environment for multiple participants. This creates a stereoscopic surround projection around the viewer who wears special glasses to visualize the simulated 3-D environment. *Desktop* VR creates artificial worlds on standard computer monitors using conventional PC software. All VR systems keep track of the head position and the viewing direction of the user to constantly update the stereoscopic and perspective display of the virtual environment perceived by the user. For additional information about the virtual reality technology, readers can access a number of books on VR, such as Earnshaw et al. (1993), Burdea and Coiffet (1994), and Kalawsky (1994).

*Virtual reality GIS*

The immersive and interactive visualization environments available through VR have prompted the GIS community to investigate the potential of linking VR with GIS (e.g., Kraak et al. 1998; Neves et al. 1997; Raper et al. 1993, 1997; Rhyne 1997). Williams (1999), in a discussion of developing 4-D *virtual reality GIS* (VRGIS), argues that the conventional 2-D map displays (e.g., contour map) in GIS are too far removed from reality. By adding a third dimension (altitude; e.g., draping contours on a 3-D surface map) and a fourth dimension (time; e.g., time-based animation of surface changes) into GIS data, displays can be placed into a more-realistic context. VRGIS is "a multi-dimensional, computer-based environment for the storage, modelling, analysis, interrogation of, and interaction with geo-referenced spatio temporal data and processes" (Williams 1999, p. 209).

In order to integrate VR with GIS, Williams (1999) suggests further research at the user interface level, at the system architecture level, and at the data level. Research at the *user interface* level needs to identify the metaphors that will allow users to maximize their cognition of data and problem at hand through interactions with a VRGIS. At the *system architecture* level, a fundamental issue is the different system architectures used in existing GIS, virtual reality, and 3-D graphics engines. A specific example is the Yon Clipping (this defines the point beyond which objects will not be rendered in the display) used in many VR systems versus the "zoom to extent" function that displays the entire map in most GIS. This fundamental difference at the system architecture level between VR and GIS can present major problems when a VRGIS application needs to view the entire map rather than a limited virtual universe defined by Yon Clipping. At the *data level*, a common problem is that most GIS data are 2-D and are unsuitable for VR. In addition, GIS and VR normally employ incompatible file formats and require file conversions.

One method of passing 3-D display data from GIS to VR is through the *Virtual Reality Modeling Language* (VRML). VRML is a standard file format approved by the International Standards Organization (ISO) for describing interactive 3D objects and worlds that can be used on the Internet, intranets, and local client systems (Carey et al. 1997). It can create 3-D virtual world files to be viewed on the Internet through the web browsers that support VRML. Each VRML file composes a set of 3-D and multimedia objects, defines object behaviors, and implicitly establishes a world coordinate space for all objects defined in the file as well as those included by the file. Similar to the hyper text markup language (HTML), VRML can define hyperlinks within an image that allows users to create a walk-through of the graphic image and link to other connected images, movies, sounds, and text. Fairbairn and Parsley (1997) implemented a web site using VRML for the campus of the University of Newcastle upon Tyne. Some commercial GIS software (e.g., ArcView 3D Analyst extension) provide a utility for converting a 3-D GIS display into the VRML format.

In 1998, the VRML Consortium (renamed the Web3D Consortium in December 1998) established a GeoVRML Working Group to study the standards needed for representing and visualizing geographic data using a standard VRML browser. GeoVRML 1.0 includes several key capabilities of handling geographic data. First of all, it can embed latitude/longitude or UTM coordinates directly into a VRML file such that a browser can transparently fuse the geographic data into a global context for visualization. Second, GeoVRML extends the single-precision floating-point values in VRML to enable a submillimeter positional accuracy level. Third, GeoVRML provides various scalability features to manage the visualization of large, multiresolution models over the web. Fourth, GeoVRML can specify a subset of metadata describing geographic objects and link it to a full metadata description. Fifth, it supports an ability to interpolate within the geographic coordinate system such that animations can be defined with respect to key points on the Earth's surface. Sixth, users can query a GeoVRML scene and retrieve the geographic coordinate of any georeferenced point. Finally, GeoVRML supports basic navigation schemes for geographic applications. For example, an appropriate navigation velocity of a fly-over will be computed according to a linear relationship between velocity and the user's elevation above the terrain. Although the GeoVRML is still at its early development stage, the potential of visualizing geographic data in a virtual reality environment could be beneficial to many transportation applications such as corridor analysis, environmental impact studies, and traffic safety assessment.

## Conclusion

This chapter discusses the capabilities and limitations of GIS for transportation analysis and modeling. Various transportation applications can benefit from the basic GIS analysis functions as well as other network analysis and facility location modeling functions available in currently available commercial GIS software products. Due to the recent trends toward object-oriented and component GIS, users are experiencing greater flexibility in developing custom applications. However, the analysis and modeling capabilities are closely related to the data models implemented in a GIS. Although a significant progress has been made to extend the GIS data models in supporting transportation analysis and modeling needs, current GIS fall short of providing adequate functions for handling transportation phenomena such as off-network travel, traffic flows, complex paths, and temporal changes.

Geographic visualization is another important function to support transportation analysis and modeling. Recent technological advancements in remote sensing, GPS, and ITS allow the transportation community to collect more detailed data (e.g., high-resolution images, disaggregate travel data, and dynamic real-time data) at lower costs. We are facing major challenges in developing more effective tools to efficiently identify the patterns and trends in a data-rich environment. Information visualization and exploration technologies such as fly-by, simulation, and virtual reality have received increasing attention in the GIS community as enhancements to analysis and modeling.

GIS software development is an evolutionary process. As the GIS-T user community continues to develop more sophisticated applications, both the GIS research community and the GIS software vendors are responding to their demands by pushing the forefront of GIS analysis and modeling capabilities toward the next generation of GIS. In remaining chapters of this book, we will examine the role of GIScience and GIS tools in transportation analysis and problem solving in four areas, namely, *transportation planning*, *intelligent transportation systems*, *environmental and hazards analysis*, and *logistics*. In these chapters, we will see many efforts to push the envelope of GIScience principles and GIS tools to provide better understanding, prediction, and decision making.

# 8

# Transportation Planning

This chapter is the first of several on applications of geographic information science and systems in transportation. In this chapter, we discuss the role of GIS and spatial analysis in transportation planning. The objective of transportation planning is to guide development of a land-use/transportation system to achieve beneficial economic, social and environmental outcomes. This can range from tactical decisions such as planning new right-of-ways or public transit routes to long-term, strategic planning of the entire system. (Issues concerning operational or day-to-day transportation management will be discussed in the next chapter on intelligent transportation systems.)

GIS and spatial analysis offer much to the transportation planning process. Most obvious is support at the "front-end" as a spatial database management system and at the "back-end" to produce graphic and cartographic visualizations of present and future scenarios. In the middle are tools for processing geographic data into geographic information. As saying goes, "spatial is special": Geographic processes such as land-use/transportation systems have properties that non-spatial analysis techniques do not capture. In particular, geographic processes often exhibit properties of *spatial dependency* and *spatial heterogeneity*. The former property reflects interrelationships among geographic entities while the latter refers to an inherent degree of uniqueness for each location in a geographic system. Modeling techniques that do not take these properties into account do not exploit the full range of information implicit in geographic data. In fact, many "standard" techniques (particularly statistical and parameter estimation methods) are based on the assumptions that dependency and heterogeneity do not exist. Since their properties do not meet these assumptions, the results from standard techniques may not be reliable when applied to geographic data.

GIS software tools, as well as spatial analysis techniques, support processing of very large quantities of land-use/transportation data and the analysis of these data to produce reliable geographic information. This can help find effective solutions to transportation problems, identify the factors that influence the land-use/transportation system, determine desirable future states of the system and the paths to achieve the desirable states. GIS can also greatly enhance the quantity and quality of information flows among all components of transportation planning process, potentially enhancing decision-making within this process.

We will begin this chapter with a discussion of methods for analyzing geographic information to assess, forecast and design future transportation systems at several scales. We will first review the *traffic analysis zone design* problem and techniques for resolving this problem. Traffic analysis zones are central to traditional travel demand techniques but are also commonly used in other transportation planning activities. We will then review tools for *transportation analysis and planning*. This will include traditional and emerging methods for travel demand analysis, combined land use/transportation modeling and route planning (including public transit and transportation facilities). Finally, we will discuss the role GIS in *supporting decision-making* in transportation planning. This will include an overview of decision making within the transportation planning process and the role of spatial decision support systems in facilitating this process.

## Transportation analysis zone design

As discussed previously, traffic analysis zones (TAZs) have historically been central to travel demand analysis. This is partly due to past limits on computational power and data availability. Working with aggregate spatial units allows the analyst to reduce the input size to travel demand models. TAZs also provide a basis for integrating aggregate secondary data such as census data into the modeling process. Although both restrictions are easing, it is likely that TAZs will be common in transportation planning for some time.

Figure 8-1 provides an example of the traffic analysis zones in Dade County, Florida. Defining a TAZ system requires demarcating an external boundary encompassing the study area. The external boundary must be determined such that most trips have their origins and destinations within this boundary. This study region must also be partitioned into mutually exclusive and exhaustive zones that serve as aggregate trip origin and destination locations.

When partitioning the study region into TAZs, the overwhelming criterion is to "capture" as many trips as possible. Any trip that does not cross a TAZ boundary is lost to the travel demand analysis. TAZ size is an obvious factor: the smaller the TAZ, the more likely for that a trip will cross its boundary (see Ord and Cliff 1976; Rogerson 1990; and Kirby 1997 for careful geometric analyses of this property). Another way to maximize interzonal flow is to maximize homogeneity of

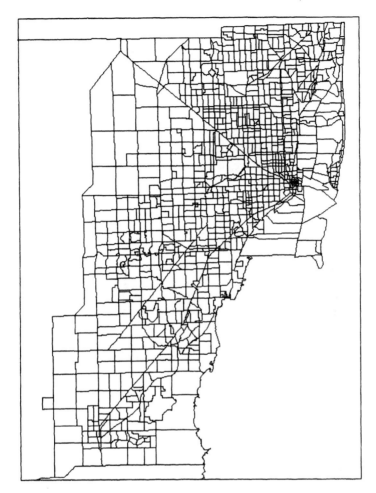

Figure 8-1: Traffic analysis zones in Dade County (Miami), FL

each TAZ with respect to socioeconomic characteristics and/or land use. Travel demand results from moving to a different location that has resources necessary to carry out some desired activity (such as work or shopping). Therefore, making TAZs as homogeneous as possible maximizes the likelihood that a person will need to cross boundaries to go somewhere different.

Other TAZ design criteria strive for simplicity and ease of spatial data integration. One criterion is requiring *simple polygons*: Roughly, these are polygons with no "holes" or "islands" (see Worboys 1995 for a precise definition). In the transportation literature, this appears as the requirement of *spatial contiguity* (not allowing "fragments" or "islands" that are not connected to the rest of the zone) and not allowing zones that are surrounded by only one other zone. Other criteria include *spatial compactness* and *compatibility with existing linear features* (e.g., rivers, railroads, and roads) and *compatibility with census geography* (see Ding 1998; Ortúzar and Willumsen 1994; You, Nedović-Budić, and Kim 1997a).

As discussed in chapter 4, we must be extremely careful when working with aggregate spatial units. In this case, any observed or estimated process may be an artifact of the analysis units rather reflecting something in reality (the so-called *modifiable areal unit problem* or MAUP). One strategy is to configure spatial units that are optimal for the given model or application. Heuristic optimization techniques such as simulated annealing, tabu search and genetic algorithms have been applied to diverse zonal configuration problems such as census unit design and political redistricting (see chapter 4).

O'Neill (1991) suggests the technique of *cluster analysis* for the TAZ design problem. Cluster analysis is actually a suite of techniques for grouping similar entities into a small number of sets based on their attributes. Cluster analysis techniques measure differences among entities through distance in attribute space and/or correlation (see You, Nedović-Budić, and Kim 1997a, 1997b). The objective is to make the differences among entities within each cluster smaller than the differences among entities assigned to different clusters. Some techniques determine a *representative point* that serves as a type of "centroid" for each cluster. In this case, the clustering problem is to simultaneously find the representative points and assign entities to these points (this is very similar to the location-allocation problem discussed in chapter 6; see Murray and Estivill-Castro 1998). In the TAZ problem, the entities are smaller spatial units such as census enumeration districts and each cluster corresponds to a TAZ aggregated from these smaller spatial units. If we wish to maintain spatial properties such as contiguity and compactness, we must often impose side conditions or use specially-tailored techniques since standard cluster analysis does not guarantee these conditions.

You, Nedović-Budić, and Kim (1997a, 1997b) use cluster analysis in a GIS-based toolkit that configures TAZs to maximize zonal homogeneity and maintain spatial contiguity. Figure 8-2 illustrates the conceptual design of their system. *Iterative clustering analysis* uses agglomerative clustering to derive the initial TAZs and iterative partitioning to refine the initial zoning system. "Agglomerative clustering" is a cluster analysis technique that is especially well-suited for grouping spatial units (see Plane and Rogerson 1994). Multiple attributes can be used in the clustering, with the user supplying relative weights for each attribute that reflects their importance. Iterative partitioning involves individual reassignment of entities among clusters based on their relative differences to cluster centroids. The user can visualize and statistically evaluate solution properties between iterations. Adjacency measures enforce contiguity constraints by not allowing reassignments that violate these constraints.

Another feature of the You, Nedović-Budić, and Kim (1997a, 1997b) system is the use of a *spatial autocorrelation* measure for evaluating TAZ system performance. Spatial autocorrelation is the degree of association among entities in space; it is essentially an extension of standard correlation analysis that allows association to be weighted with respect to geographic proximity measures such as distance, connectivity or the degree of shared borders. The system uses the Geary's $c$ and Moran's $I$ spatial autocorrelation measures: these are two of several available spatial autocorrelation statistics. In the TAZ design problem, spatial autocorrelation is a measure for zonal homogeneity; that is, a relatively

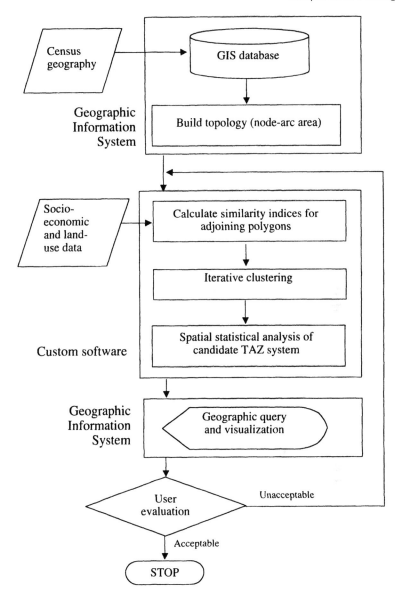

Figure 8-2: Conceptual model of a GIS traffic analysis zone design system (after You, Nedović-Budić, and Kim 1997a)

homogeneous zone should be composed of smaller spatial units that have positive spatial autocorrelation with each other.

Ding (1998) describes an interactive TAZ design module incorporated into a software system that links a GIS with the TRANPLAN travel demand software. The module measures properties of a candidate TAZ system, including the observed trip generation from each zone, boundary consistency, compactness, contiguity, and zonal homogeneity. The user can specify relative weights as well as minimum threshold values for these measures. A simple algorithm determines the TAZ system based on selecting $m$ "seed units" that score high on an attribute (e.g., population density) and are as separate as possible, where $m$ is the number of TAZs desired. The user can visualize the resulting system, measure system properties and adjust the weights and thresholds to generate a new solution.

An extreme but increasingly viable solution to the TAZ design problem is to eliminate the zonal system and use a continuous representation of trip origin and destinations (Miller 1999b). Angel and Hyman (1976) pioneered this approach in a classic but widely overlooked work that develops field-based representations of travel times and spatial interaction in an urban area. However, analytical solutions to these models are difficult without strict and unrealistic assumptions regarding variations in travel times by location. More recently, Spiekerman and Wegener (2000) use GIS tools to disaggregate zonal data into a discrete (raster-based) approximation of continuous trip generation and attraction variables such as residences and workplaces. The GIS allows the raster approximations to be linked with a detailed transportation network for microsimulation models of travel behavior and accessibility.

Disaggregate representation of travel demand factors can also be used in conjunction with traditional network flow models. Daganzo (1980a, 1980b) develops a method for incorporating continuous representations of origins and destinations in user equilibrium network flow assignment (see chapter 6). Daganzo (1980a) adopts a modified Frank-Wolfe algorithm developed by LeBlanc, Morlok, and Pierskalla (1975) for handling large numbers of origins and destinations. Dagnazo (1980b) uses the modified procedure in conjunction with mean, variance and covariance distance measures to approximate the equilibrium flow for the continuous trip origins and destinations. The procedure requires polynomial time and is therefore tractable for urban-scale problems. GIS technology can make this procedure accessible to a wide range of analysts and practitioners.

### Travel demand analysis

#### Overview

Travel demand analysis is an important foundation for transportation planning efforts. Planners must evaluate the effect of proposed changes in transportation infrastructure, service or policy on different parts of a transportation system before these are implemented. This can involve estimating some or all of four travel demand components in a study area. These components are *trip generation* (TG; where trips are coming from), *trip distribution* (TD; where trips are going to), *modal split* (MS; which modes are used, e.g., auto versus bus), and *network assignment* (NA; which routes are used within each mode). This is typically accomplished for a particular type of trip such as home-based trips to work.

There are two traditional approaches to estimating travel demand, namely, the sequential approach or equilibrium modeling. The *sequential or four-step approach* conceptually separates the four travel demand components using separate models for each component. These separate components are related to each other by feeding answers between models in a sequence such as TG → TD → MS → NA. Feedback loops are sometimes included to account for the influence of latter stages in the sequence on estimates from earlier ones. The four-step approach is very popular in GIS-T software due to its historical adoption in the *urban transportation modeling system* (UTMS) formalism supported by the U.S. Federal Highway Administration and its subsequent implementation in commercial software such as TRANPLAN. The components of the four-step approach are greatly enhanced through the spatial data management and manipulation functionality of GIS software packages (see McCormack and Nyerges 1997; Shaw 1993). However, as we will see, while each component's model may be viable as an independent "snapshot" of one travel demand facet, trying to link these models to derive the overall combined travel demand is fundamentally flawed.

The *equilibrium approach* views travel demand as a type of economic market that achieves a balance between the demand for travel and the supply of transportation services. In other words, it assumes that trip makers employ a simultaneous choice process of which all attributes of all alternatives are considered at the same time. Since there could be many possible interactions (known as *cross-elasticities* in microeconomics) among the different travel demand components, an equilibrium model is more complex mathematically. Nevertheless, the data and computational requirements are not much greater than the sequential approach and therefore there is no real barrier for its widespread adoption in GIS-T.

Although the equilibrium approach is a substantial improvement over the sequential approach, it still suffers from some weaknesses. Similar to the network flow equilibrium models discussed in chapter 6, equilibrium travel demand models assume that a transportation system achieves a balance between supply and demand. Whether or not this occurs in reality is debatable. Emerging alternatives include *activity-based analysis* and *computational intelligence approaches* such as neural networks. Also, a modeling approach that complements (but does not replace) travel demand forecasting is *accessibility modeling*. All of these emerging approaches are greatly enhanced by GIS tools and software since they critically depend on individual-level data on movement and activity in geographic space.

## The four-step approach to travel demand modeling

The four-step approach uses independent models for each travel demand component and attempts to link these solutions by feeding answers from one model to the next one in a pre-specified sequence or loop. Several software systems have been developed that link four-step modeling software with a GIS for supporting for spatial data management and cartographic visualization of estimated travel demands (see Anderson and Souleyrette 1996). Although linking independent models in the four-step approach has some inherent weaknesses, these models are valid when used in isolation for analysis of one travel demand component. In this section, we will give a brief overview of the models used for each component and the GIS software tools that support and enhance each model.

*Trip generation*

Trip generation models attempt to predict and sometimes explain the number of trips originating in TAZs (at the aggregate level), household (at the disaggregate level) or a similar spatial unit. Two common methods are cross-classification analysis and regression analysis. *Cross-classification analysis* develops meaningful categories along one or more attribute dimensions in a travel survey database and classifies each observation based on its membership in the multiple categories (e.g., "two-person households with two or more cars"). Each category combination defines a group for which a trip generation rate can be measured or estimated using statistical methods. Multiplying the number of households falling in each group for a TAZ by its estimated trip generation rate predicts the number of trips from that zone. Methods are available for defining categories and testing their significance (see Otrúzar and Willumsen 1994).

Regression analysis involves the specification and estimation of a linear equation that relates explanatory or *independent* variables to the amount of trip generation. The latter is known as the *dependent* variable. We can estimate regression equations at the TAZ or household level using the method of *ordinary least squares* (OLS) that finds the parameters that minimizes the sum of squared deviations between the predicted and observed trip generation from a calibration dataset.

Standard linear regression requires strict statistical assumptions. These include the observations being statistically independent and normally distributed, a linear relationship between each explanatory variable and the dependent variable and equal variability or dispersion of data with respect to the regression line. These assumptions are often violated in practice. For example, nonlinear relationships, non-normality, and *heteroscedasticity* (changes in the variance about the regression line) are common. *Multicollinearity* (correlation among the independent variables) is especially troublesome in trip generation modeling since many of its measurable antecedent factors are related (e.g., income and car ownership). Multicollinearity results in unstable regression parameters; at the extreme, the regression parameters cannot be calculated since a unique solution will not exist (i.e., any parameter values will fit the data).

Specifying and estimating regression models for a geographic phenomenon such as trip generation creates additional difficulties. Geographic phenomena almost always exhibit the properties of *spatial dependence* and *spatial heterogeneity*. As mentioned previously, spatial dependency refers to entities being related across space. This is often measured as the degree of *spatial autocorrelation* among observations in the geographic data. Spatial heterogeneity refers to the estimated parameters of a spatial model being an inadequate description of the phenomenon in a particular location. This occurs since each location in space has some degree of uniqueness with respect to the spatial system (Miller 1999b).

A regression model that does not consider spatial dependency may have inefficient parameter estimates. This means that the estimated parameter errors are much smaller than the true errors. This can seriously limit the predictions from the estimated model. Also, the parameter significance and overall goodness-of-fit tests are no longer reliable, meaning that the battery of statistical tests associated with the regression is useless for assessing model performance (see Anselin and Griffith 1988).

Spatial dependency can be captured in a regression model in one of three ways. It can be an effect captured in the variables, part of the regression error term, or both simultaneously. We refer to the two separate cases as *substantive spatial dependence* and *spatial error dependence* respectively (Anselin 1993). A traditional test for spatial error dependence is to apply a spatial autocorrelation statistic to the regression residuals, although this approach can be sensitive to model specification errors. In other words, the test is not accurate if the model variables are incomplete or some included variables are not appropriate (see Anselin and Rey 1991; Cliff and Ord 1972). Burridge (1980) provides a direct test for spatial error dependence while Anselin (1988a) develops a related test for substantive spatial dependence. These tests should be used in conjunction for a complete assessment of spatial dependency in a regression model (Anselin 1993).

In addition to simply testing for spatial dependency, it is possible to reformulate the regression model to account for these effects. Models that take into account substantive spatial dependency include *spatial autoregression* and *mixed regression-spatial autoregression*. The former regresses the dependent variable against a spatially lag function of itself while the latter includes spatially lagged versions of both the dependent and independents. Spatial error dependence can be handled through a specification similar to the mixed regression-spatial autoregression model (see Anselin 1988b, 1993; Bivand 1984).

Spatial heterogeneity effects usually manifest themselves in a statistical model through "parameter drift," that is, changes in the parameter values when the model is re-estimated for different (geographic) subsets of the data (see Fotheringham, Charlton, and Brunsdon 1996, 1997). These effects are not often noticed since re-estimating with data subsets is not a common practice. Parameter drift makes model transferability across space and time difficult, meaning that a trip generation model estimated for a study area at one time cannot be readily used for that study area at a different time or in a different study area.

One way to account for spatial heterogeneity is to explicitly model parameter variation across space. The recently developed technique of *geographically weighted regression* (GWR) estimates parameters based on a location-based weighting function. The analyst must determine the appropriate function by testing hypothesized functions against the dataset (see Brundson, Fotheringham, and Charlton 1996; Fotheringham, Charlton, and Brunsdon 1997). Developing, testing, deriving predictions and assessing transferability for a GWR trip generation model is an open research question.

Regression-based trip generation models estimated on spatial data that do not take into account spatial dependency and spatial heterogeneity could be invalid and unreliable. Similar to the modifiable areal unit problem discussed in chapter

4, these problems have been long-recognized but seldom acted on in practice since the digital geographic databases and computational power were not available until recently. GIS and related software packages are increasingly incorporating methods for assessing spatial autocorrelation, spatial heterogeneity and accounting for their effects in statistical estimation. Examples, discussion and integration strategies include Anselin (1992), Anselin and Getis (1992), Anselin, Dodson, and Hudak (1993) and Fotheringham and Rogerson (1993).

*Trip distribution*

The most commonly applied methods for partitioning the estimated or observed trips from an origin among destinations are *random utility* and *spatial interaction* models. Random utility models are disaggregate and attempt to model destination choice at an individual level. Spatial interaction models are aggregate and attempt to determine flows between origin and destinations as a function of origin and destination attributes and the travel costs between them. Although seemingly distinct, random utility and spatial interaction models have deep structural equivalencies and can be derived from the same theoretical base (see Anas 1983; Fotheringham and O'Kelly 1989).

Random utility models assume that we can measure an individual's preferences (and therefore choices) only up to a random or error term. The random component results from unmeasurable psychological factors or changes in the individual's "state of mind" over time. Specifying different distributions for the random components results in different types of random utility models. The most common is the *multinomial logit model* (MNL) that assumes the unmeasured components of each destination are unrelated. As discussed in chapter 6 with respect to stochastic user equilibrium-based network flow models, although the logit model is tractable it requires strict behavioral assumptions that can create problems (see Ben-Akiva and Lerman 1985).

The spatial interaction (SI) or "gravity" model attempts to relate quantitatively the components of spatial interaction system. Using general function notation, we can state a highly general SI model as

$$t_{ij} = f(\mu v_i, \alpha w_j, \beta c_{ij}) \tag{8-1}$$

where $t_{ij}$ is the flow between origin $i$ and destination $j$, $v_i$ is a variable summarizing the attributes of origin $i$ that influence its trip outflow, $w_j$ is a variable summarizing the attributes of destination $j$ that influence its trip inflow, $c_{ij}$ is the travel cost from $i$ to $j$ and $\mu, \alpha, \beta$ are parameters that reflect the relative effects of origin attributes, destination attributes and travel costs on the flow between the origin-destination (O-D) pair. The origin "propulsiveness" variable $v_i$ and the destination "attractiveness" variable $w_j$ can be functions of several attributes that are calibrated as part of the modeling process. We calibrate the parameters from data on these variables and a complete O-D flow matrix (see Fotheringham and O'Kelly 1989 and comments below).

The origin and destination components in equation (8-1) usually enter as power functions, e.g., $f(\mu v_i) = v_i^\mu$. The travel cost component can enter either as a power function or as an exponential function, i.e., $f(\beta c_{ij}) = \exp(-\beta c_{ij})$; see equation (8-2) below. The choice between the travel cost functions involves four issues. The first consideration is whether one wishes to compare and re-use parameter values between studies. The exponential form is sensitive to travel cost scale so cannot be used for comparison or re-use between very different settings. A second consideration is the hypothesized relationship between changes in travel costs and flow changes. A third issue is whether any travel costs are near or at zero (this causes problems with the exponential function). The final consideration is the level of homogeneity among the travelers that comprise the flows. Fotheringham and O'Kelly (1989) provide a detailed discussion of these considerations. A rule-of-thumb is that the exponential form is more appropriate for urban scale analysis while the power function is more appropriate for regional and national scales.

In addition to providing information for calibration, the O-D flow matrix also allows one to constrain the model with known aggregate flow information. An *unconstrained* SI model is required only to match the total number of observed flows in the system. An unconstrained SI model with an exponential cost function is:

$$t_{ij} = v_i^\mu w_j^\alpha \exp(-\beta c_{ij}) \qquad (8\text{-}2)$$

A *singly constrained* SI model can be origin constrained or destination constrained. In the former case, we replace the attribute variable $v_i$ for an origin with its observed outflows $O_i = \sum_j t_{ij}$. (Note that this is the total for row $i$ of the O-D matrix.) An example is:

$$t_{ij} = A_i O_i w_j^\alpha \exp(-\beta c_{ij}) \qquad (8\text{-}3)$$

where:

$$A_i = \left[\sum_j w_j^\alpha \exp(-\beta c_{ij})\right]^{-1} \qquad (8\text{-}4)$$

Equation (8-4) is an origin-specific *balancing factor* that ensures the origin outflow total constraints are met; see Fotheringham and O'Kelly (1989) for an explanation of how these factors work. The destination-constrained model similarly replaces $w_j$ with the observed inflow $D_j = \sum_i t_{ij}$ (the sum for column $j$) and requires predicted inflows to match these totals using destination-specific balancing factors. An example is:

$$t_{ij} = v_i^\mu B_j D_j \exp(-\beta c_{ij}) \qquad (8\text{-}5)$$

with the destination-specific balancing factors:

$$B_j = \left[\sum_i v_i^\mu \exp(-\beta c_{ij})\right]^{-1} \qquad (8\text{-}6)$$

Finally, a *doubly constrained* SI model requires both predicted outflows from origins and inflows to destinations to match observed totals:

$$t_{ij} = A_i B_j O_i D_j \exp(-\beta c_{ij}) \qquad (8\text{-}7)$$

$$A_i = \left[ \sum_j B_j D_j \exp(-\beta c_{ij}) \right]^{-1} \tag{8-8}$$

$$B_j = \left[ \sum_i A_i O_i \exp(-\beta c_{ij}) \right]^{-1} \tag{8-9}$$

In a major contribution to this field, Alan Wilson (Wilson 1967) derived all four members of this "family" of SI models within an *entropy-maximizing framework*. This framework calculates the most likely distribution of *microstates* based on known *macrostate* information about the systems (see Webber 1977). In the trip distribution case, the microstates are the detailed O-D flows while macrostate information is the origin outflow and destination inflow totals. In addition to their behavioral foundation, this provides SI models with a solid mathematical basis for estimating origin-destination flows based on total flow data that are collected more easily.

There is a trade-off between quantity and quality of information among the four SI models. The unconstrained model generates a high quantity of poor-quality information about the system. The doubly constrained model generates a low quantity of high-quality information. The singly constrained models fall in the middle (see Fotheringham and O'Kelly 1989).

Despite solid theoretical foundations, the MNL and the family of standard SI models share a common weakness for modeling trip distribution. In the MNL, independence among error terms requires that we assume destinations do not share unmeasured characteristics that influence choice. Similarly, the family of standard SI models does not consider the influence of geographic proximity among destinations.

Trip destinations can share characteristics related to their geographic proximity that influence choice. For example, "downtown" versus "suburban" shopping destinations share unmeasured characteristics related to the different experiences of shopping in a city center versus a mall. Also, there is strong evidence that individuals deal with the complexity of the destination choice process through a hierarchical process. People mentally cluster destinations that are proximal to each other and first choose a cluster and then a destination within that cluster. For example, an individual may first choose to shop downtown and then choose a store within the city center. Finally, certain types of destinations have complementary or competitive interactions. For example, comparison shopping for retail goods (e.g., automobiles) or multipurpose shopping among complementary retail lines (e.g., clothes and shoes) means that stores that locate in proximity can generate more customer traffic than stores that locate in isolation. Conversely, for some retail goods (e.g., food) proximal locations can result in lower customer patronage relative to an isolated location.

Figure 8-3 illustrates two different cases of proximity among destinations. In figure 8-3a, three destinations are located equidistant from an origin but are geographically dispersed. In Figure 8-3b, the three destinations are at the same distance from the origin as in Figure 8-3a, but are geographically clustered. The MNL and standard SI models do not distinguish between these two cases: they only consider site-relevant attributes and the origin-destination distances (more

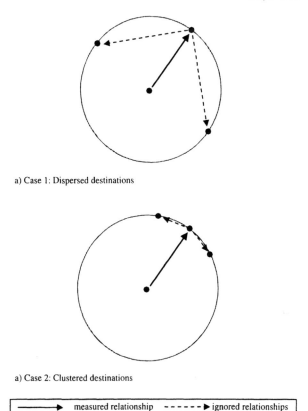

a) Case 1: Dispersed destinations

a) Case 2: Clustered destinations

——▶ measured relationship   - - - - -▶ ignored relationships

Figure 8-3: Proximity among destinations in a spatial interaction system

generally, travel costs; indicated by the solid arrows). These models do not consider possible interactions due to distance or travel costs among the destinations (indicated by the dashed arrows). For certain types of destination choice, both types of models are therefore misspecified since they do not consider these potential interactions.

Researchers have long-recognized these misspecification errors in the MNL and the family of standard SI models. As mentioned in chapter 6, there are alternatives to the MNL such as the probit model but these are not as tractable with respect to calculation and parameter estimation (although progress continues; see chapter 6). One tractable approach for trip distribution modeling is the *nested logit* (NL) model. The NL model requires the analyst to specify clusters of destinations. Logically, we can specify these clusters through a tree that shows the clusters as parent nodes and individual destinations that comprise each cluster as the children of the parent. Mathematically, we represent the tree by specifying a MNL model for each level, with an *expected maximum utility* term representing the predicted choice at the next lower level. Ben-Akiva and Lerman (1985) provide a general discussion of this approach while Kanaroglou and Ferguson (1996) for a recent and sophisticated application of these ideas in aggregate destination choice. A problem with the NL approach is that the analyst must prespecify and test rival tree structures to determine the appropriate specification. It is often not clear in advance which structure is appropriate. Many alternative structures may require testing (see Daly 1987).

SI models disaggregated by origin exhibit a strong "map pattern" effect where travel costs have an apparently less dampening effect on flows from origins that are relatively accessible within the spatial system. Attributing this pattern to some real variation in travel cost effects among locations can lead to some illogical conclusions. Rather, the map pattern effect is related to misspecification errors related to geographic proximity (Fotheringham 1981).

Fotheringham (1983, 1984) develops *competing destinations* versions of the unconstrained, singly constrained and doubly constrained SI models that resolve these errors and eliminate the map pattern effect. These models have an extra term that measures the relative accessibility of a destination to other destinations in the system:

$$P_j = \left( \frac{1}{m-1} \sum_{k \neq j} w_k c_{jk}^{-1} \right)^\theta \tag{8-10}$$

where $m$ is the number of destinations in the system. Equation (8-10) is a type of *geographic potential* function: It is the summed attractiveness of the other destinations weighted by the inverse of the travel cost to the given destination. The parameter associated with this term can be positive to reflect the enhancing effect of proximate location, negative to reflect a retarding effect or zero to reflect no effect due to proximity. This parameter is estimated as part of the model calibration process. An advantage of the competing destinations models relative to the NL model is that clusters are fuzzy and consequently do not require prespecification and testing.

Bhat, Govindarajan, and Pulugurta (1998) reconcile the MNL and SI approaches by developing a competing-destinations type model at the disaggregate level. Their model incorporates a destination accessibility measure into a multinomial logit model. Empirical results indicate superior fit relative to a comparable disaggregate model without the destination accessibility term.

Historically, SI models tend to be more widely used in travel demand analysis than MNL and NL models. This partly results from the latter models requiring individual-level data that can be expensive to collect (Shaw 1993). Consequently, SI models have been more widely incorporated into travel demand modeling and GIS software. The competing destinations formulations are not as widely available. However, the difficulty and expense associated with individual-level data collection is dropping due to GIS and related technologies (especially global positioning systems; see the section below on activity-based analysis). This will likely create renewed interest and applications of individual-level analysis.

Also lacking in current software are sophisticated tools for calibrating and performing diagnostics for SI models. Calibrating SI models requires a statistical method known as *maximum likelihood* (ML). Roughly, this requires determining the parameter that maximizes the likelihood of realizing the observed flows from an hypothesized theoretical distribution (e.g., normal or Poisson). Some

GIS software correctly specifies the condition that must be met but requires the user to adjust parameters until this condition is realized. This user-driven trial and error approach is unnecessary. Efficient computational techniques exist to find the ML parameter estimates (see Fotheringham and O'Kelly 1989). Diplock and Openshaw (1996) demonstrate the use of genetic algorithms and evolutionary computing to calibrate SI models with functional forms and data that create difficulties for ML estimation. It would be helpful if this type of functionality is built into GIS software that incorporate SI models, developed as an extension or as a standalone package that can share or translate GIS data files.

*Modal split*

Similar to trip distribution, we can model modal split at the aggregate or disaggregate level. Aggregate-level models include *diversion curves* that show the varying split among two competing modes based on their travel cost differences. Cross-classification and regression models, discussed above with respect to trip generation, have also been used for modal split modeling (see Ortúzar and Willumsen 1994; Shaw 1993).

The disaggregate level approach of random utility theory is the most popular modal split technique in practice. The MNL model discussed above can be calibrated using data on mode-specific travel costs, in-transit travel times and wait times as well as travel-specific characteristics such as income and household size. However, the MNL IIA assumption can create problems in modal split.

Certain modes have highly correlated characteristics that cannot be measured directly. For example, all public transit modes share the characteristic of being public transit. If we try to model these choices as equal with the choice of private transit (e.g., automobile) we will get nonsensical answers. A famous example is the blue bus/red bus choice paradox where a false choice (i.e., the color of the bus) still generates a "real" modal split (see Mayberry 1973) in an MNL modal split model. The NL model has emerged as an alternative, with (for example) the first-level choice between private and public transport and then the second level choice being competing alternatives within each mode (such as bus versus train). See Ben-Akiva and Lerman (1985) and Otrúzar and Willumsen (1994) for general discussions and Hunt (1990) and Otrúzar (1983) for specific examples.

Modal split models can also suffer from misspecification errors if spatial dependency and spatial heterogeneity effects are not considered. Due to spatial aggregation into TAZs, the same interzonal travel cost measure for each mode is assigned to all individuals with the same origin and destination zone. When applied at the individual-level in a mode choice model, this confounds the effects of *individual heterogeneity* (variations in travel costs across individuals with the same origin-destination pair) with *place heterogeneity* (variations in travel costs across origin-destination pairs). This is a type of ecological fallacy that must be untangled or the mode choice model is misspecified (Bhat 2000).

Problems with confounding place and individual heterogeneity problem in modal choice models can be addressed through *multilevel analysis*. This is a general modeling technique that explicitly considers hierarchical levels in the phenomenon being modeled. See Jones and Duncan (1996) for a general overview and applications in modeling geographic systems. Bhat (2000) formulates a multilevel discrete choice model that accounts for these effects in mode choice. The multilevel formulation captures spatial dependency in unobserved factors among individuals within the same origin and destination zones, removing potential parameter estimate problems such as underestimates of standard errors and inconsistencies among parameter estimates.

*Network assignment*

We discussed the major methods for network assignment in chapter 6. Common techniques are equilibrium-based models such as user-optimal, system-optimal, stochastic user-optimal and dynamic user-optimal. Breakthroughs in other modeling approaches such as variational inequality and optimal control theory are also providing powerful techniques for modeling dynamic network flows. Finally, simulation techniques can model dynamic network flows at varying levels of resolution (see chapter 6).

A difficult problem in travel demand analysis is flow assignment in public transportation networks. Public transportation trips are complex, involving components such as costs (time and money) for accessing the system, waiting time, in-transit time, transfer time and egress from the system to the final destination. Survey or patronage data can be used to estimate parameters that weight the different time components as well as fares into a generalized cost measure (see Ortúzar and Willumsen 1994; Talvitie 1992).

Adding to the measurement complexity in assignment to public transit networks is the *common lines problem*. This occurs when there are multiple public transit services exist between an origin-destination pair. In contrast with "private" (automobile) flow assignment where each traveler chooses a single route, in public transportation systems with parallel services a traveler can choose a strategy consisting of several possible routes. System performance (e.g., which vehicle arrives first at a station) determines the actual route executed (Ortúzar and Willumsen 1994; Speiss and Florian 1989).

Speiss and Florian (1989) formulate a general version of the public transit assignment problem that requires minimizing users' expected travel time (including wait time). Using service frequency as a measure of wait and in-transit time, they state a special case of this problem as a linear program. A polynomial time and therefore tractable algorithm solves this linear programming.

The weakness of the four-step approach

A travel demand pattern results from a multitude of individuals making interrelated choices about whether to travel (trip generation), where to travel (trip distribution), which mode to use (modal split) and what route to use (network assignment). These choices both affect and are affected by travel costs in the transportation system. When modeling travel demand, we must carefully link these complex interactions either based on empirical evidence or using a valid theoretic framework that can substitute for these evidence with theoretical expectations about hidden traveler behaviors (see Fernández and Friesz 1983).

The four-step approach implicitly assumes that all travel demand components balance to equilibrium. However, the four-step approach implements these principles in a deeply flawed manner. The sequence imposed by the modeler imposes the highly doubtful assumption that all travelers make their decisions according to the sequence specified by the model builder. It is highly doubtful that all travelers in a study area consider these choices in the same (or for that matter, *any*) order. Solving the models in sequence does not even guarantee convergence to any solution let alone the equilibrium. Travel costs that change in latter steps (e.g., due to congestion effects from the network assignment) must be accounted for in earlier steps (e.g., trip generation). "Feedback loops" are often used to cycle back to previous steps to re-estimate demand based on changed travel costs. In practice, solution oscillations frequently occur and the analyst simply stops after a few loops through the cycle and accepts the most recent answer.

These weaknesses of the four-step approach are not just theoretical: they have been confirmed by empirical evidence. As far back as the 1970s, Florian, Nguyen, and Ferland (1975) found that sequential estimation with feedback loops of TD → NA does not converge to a consistent solution. A study by COMSIS Corporation (COMSIS 1996) compared the four-step approach with feedback loops in mid-sized U.S. cities. The standard feedback method did not consistently converge to an equilibrium solution. Using the method of successive averages (see chapter 6) to "smooth" the feedback mechanism compares more favorably, although the study concludes that this may not perform well in larger cities with high levels of congestion. Boyce, Zhang, and Lupa (1994) compare the four-step procedure, with and without feedback loops, using data from Chicago. The four-step approach was poor at reproducing known data, particularly key variables such as automobile link flows and total automobile trips. Hansan and Al-Gadhi (1998) compare the four-step approach to an equilibrium model and find that the equilibrium model produces better traffic flow predictions. Clearly, the four-step approach is inferior to alternative methods and should be abandoned (Boyce 1998).

## Equilibrium travel demand modeling

As discussed in chapter 6, the basic idea in the equilibrium approach to travel demand modeling is a view of transportation systems as *markets*. The network equilibrium approach embeds the elements typically found in the traditional, four-step approach into a *market equilibrium framework*. At market equilibrium, the travel pattern should exhibit stability that *simultaneously* encompasses all four of the travel demand components. Since these components are tightly linked it is impossible to solve for each component in isolation without considering its effects on the other components (see Aashtiani and Magnanti 1981; Fernández and Friesz 1983).

An early and influential travel demand equilibrium model is Evans (1976): This solves for a combined TD/NA. The TD component is a doubly constrained SI model with an exponential cost function while the NA component is user optimal (UO) equilibrium (see chapter 6). Evans (1976) combines these into a constrained optimization problem solved using a *partial linearization algorithm* similar to the convex combinations method discussed in chapter 6. This is a powerful method that has been subsequently adapted for solving a wide-range of equilibrium travel demand models.

Safwat and Magnanti (1988) develop an equilibrium model that encompasses all four travel demand components. However, their model does not treat the MS component explicitly. Their model requires independent modal subnetworks linked through transfer arcs (see chapter 5), with the combined MS/NA pattern being UO equilibrium of route choices through this multimodal network. The TD component is a MNL consisting of two variables, namely, the minimum average travel cost between the O-D pair and the attractiveness of the destination. TG also consists of two components, namely, a variable summarizing the origin's propulsiveness factors and its relative accessibility. Safwat and Walton (1988) discuss two solution algorithms; one is similar to the convex combinations methods while the other is similar to the Evans (1976) partial linearization method. Also see Safwat (1987) for additional discussion of implementation issues.

Sheffi and Daganzo (1980) formulate travel demand equilibrium as a stochastic user optimal (SUO) equilibrium (see chapter 6) within a *hypernetwork*. The hypernetwork represents the multimodal network in the study area and includes abstract links to represent destination and generation choices. Determining the SUO equilibrium through this network provides a consistent SUO equilibrium among travel demands. We can estimate the parameters of this model and solve for equilibrium using standard network SUO methods (see chapter 6).

Oppenheim (1995) develops a "trip consumer" approach that integrates random utility theory with the microeconomic theory of consumer demand, providing consistency between behavioral choice theory and aggregate demands. At the individual level, the trip consumer approach uses nested logit to represent an hypothesized choice structure of the four components. The aggregate travel demands result from maximizing the utility of an aggregate "representative traveler" whose preference structure is consistent with utility function at the micro level. Both UO and SUO network equilibria can be accommodated.

Abrahamsson and Lundqvist (1999) test two nested choice model specifications and a simultaneous choice model of a combined TD/MS/NA equilibrium. In all three models, the trip distribution component is consistent with a doubly constrained SI model, mode choice is consistent with a multinomial logit model, and route choice is UO. They also develop and conduct specification testing and estimation using empirical data from Stockholm. Estimation and diagnostic tests suggest that a nested choice structure corresponding to MS-TD-NA provides the best performance with respect to consistent parameter estimates.

In chapter 6, we mentioned that the UO equilibrium conditions could be stated as a type of mathematical problem known as a *variational inequality* (VI). This allows more realistic network flow cost functions as well as dynamic formulations. Similarly, the travel demand equilibrium conditions are actually a special case of more general equilibrium conditions stated as a VI (Dafermos 1982).

Florian, Wu, and He (1999) specify a VI formulation of the travel demand equilibrium problem that is consistent with a nested logit choice mechanism at the individual level. Their formulation also encompasses multiple user classes and travel purposes in addition to multiple travel modes. They implement the model using the EMME/2 software for Santiago, Chile.

At the time of this writing, there are few available GIS or special-purpose software packages that implement equilibrium travel demand models (Boyce 1998). This is not due to any substantial impediments to developing this software or applying these methods in practice. Many travel demand equilibrium models do not require substantially greater data or computational power than the four-step approach. Since these models are unified and based on network equilibrium principles, they can be implemented as part of a network analysis toolkit in general-purpose and transportation-specific GIS software (see Miller and Storm 1996 for an example prototype design).

### Artificial neural networks

Continuing development and deployment of technologies for capturing geographic information such as vehicle-based GPS receivers and real-time traffic flow monitors are generating huge but error-prone datasets. To exploit this noisy information in transportation planning requires techniques that are robust (fault-tolerant) and tractable (process large databases in reasonable, perhaps even real, time). In chapter 5, we briefly mentioned *computational intelligence* (CI) methods in association with the traveling salesman and vehicle routing problems. CI methods attempt to exploit "low-level" intelligence to develop robust, adaptive and computationally-tractable methods for solving complex optimization problems and searching for patterns in very large datasets. CI-methods such as fuzzy logic, genetic algorithms, knowledge-based systems (to be discussed in chapter 10), rule induction and intelligent search techniques have been applied to transportation planning and traffic flow control. See the edited volume by Bielli, Ambrosino, and Boero (1994) for good overview chapters.

In this section, we will focus on *artificial neural networks* (ANNs), a technique that is appropriate for the voluminous and noisy information increasingly available from geographic information technologies. ANNs are an analog to biological neural networks such as the brain. ANNs adapt their structure based on subtle regularities in the input data. Similar to a brain, they are robust with respect to error and can find patterns in noisy data in a short amount of time (Dougherty, Kirby, and Boyle 1994). ANNs offer these advantages over "brittle" statistical methods that require minimal and well-behaved error. A disadvantage is that ANNs can be a "black-box" with limited ability to generate explanations in addition to forecasts. ANNs have been applied to a wide range of transportation problems, including driver behavior, pavement maintenance, vehicle detection and travel demand forecasting (Dougherty 1995).

An ANN replicates (on a very limited scale) the behavior and connectivity among biological neurons in a brain. Biological neurons adjust their firing frequencies over time to other neurons in response to the firing frequencies from their input neurons. Some of these neurons are connected to external sensors

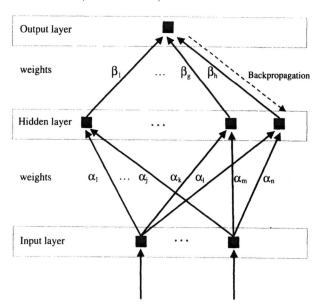

Figure 8-4: A feed-forward/backpropagation neural network

(such as eyes). Through a learning process, the biological neural network adjusts firing frequencies until an appropriate response is achieved (e.g., ideas, behavior). An ANN connects nodes corresponding to input devices, neurons, and output conditions in a manner that mimics a biological net. Through a training process, an ANN generates appropriate outcomes (predictions and even behavior if connected to a control device) by adjusting arc weights that affect signals transmitted between nodes (see Beale and Jackson 1990; Dougherty 1995; Peterson and Söderberg 1993).

Figure 8-4 provides a basic ANN architecture known as *feed-forward network* (Fischer and Gopal 1994). Two types of main architectures exist, namely, feed-forward and feed-back networks, with the feed-forward architecture more commonly applied for pattern recognition and the feed-back network more commonly applied in optimization problems (Peterson and Söderberg 1993). Each node in the input layer receives data or information directly (e.g., from monitoring devices) and sends a value to a hidden layer based on some mathematical function of its input. There is an input node for each input variable (or sensor). The number of nodes in the hidden layer is a design decision. Based on some mathematical function of its inputs, each node in the hidden layer similarly sends signals to the output nodes. Output nodes, one per variable being predicted, generate responses based on their received inputs. This is known as a feed-forward network since the information from the database travels in the forward direction (from input nodes to output nodes). In feedback architectures, the links are bidirectional (see Peterson and Söderberg 1993).

The arcs in an ANN have associated weights that modify the value of the transmitted signals. Initially, the weights are arbitrary. Training an ANN involves processing a set of input data and comparing the ANN prediction to the corresponding outcome in the database and outputs until the ANN generates similar responses. A measure of the prediction error is *backpropagated* through the network in the opposite direction of the feed-forward signals. Arc weights are adjusted based on moving in the direction within the parameter space that minimizes the error function (see Dougherty, Kirby, and Boyle 1994; Gopal and Fischer 1996). This is usually a local, heuristic search method for finding the optimal weights; other methods such as genetic algorithms can be used (see Fischer and Leung 1998).

Backpropagation is a type of *supervised* learning due to the comparison between estimated and empirical outcomes. Other training paradigms that have been used in the transportation literature include *reinforcement* that "reward" the ANN for good performance by strengthening weights for the most active components. *Self-organization* uses information only in the inputs to adjust weights so that its dimensionality is reduced. This is useful for classification problems where the classes are not known beforehand (see Dougherty 1995).

Although an ANN requires a large amount of data during training, a potential danger is overtraining. Similar to statistical overfitting, this occurs when the ANN is fit so well to a particular dataset that it conforms to its peculiar patterns and cannot generalize to other data (see Fischer and Gopal 1994 for some experience and discussion). Good practice requires dividing the test data into independent training, testing and validation subsets, with the testing subset used to determine when to stop training and the validation subset used to analyze the generality of the trained ANN (Dougherty and Schintler 1997).

A potential weakness of ANNs is that they tend to be a "black-box" modeling technique. The weights in a trained network cannot easily be interpreted from economic or behavioral perspectives (Nijkamp, Reggiani, and Tritapepe 1996). However, under certain conditions, an ANN can be interpreted as a set of conditional probabilities (Buntine 1996) or as a type of nonlinear regression (Fischer and Gopal 1994). Performing a sensitivity analysis by perturbing input data and assessing output changes could also provide insights for interpreting the trained ANN weights (see Dougherty, Kirby, and Boyle 1994).

ANNs have been applied with mixed experience in trip distribution, modal split and route choice analysis. Some researchers report good performance for ANNs relative to SI models for forecasting telecommunication (Fischer and Gopal 1994; Gopal and Fischer 1996) and commodity origin-destination flows (Black 1995), although the former benchmark is based on a low-quality unconstrained SI model. Mozolin, Thill, and Usery (2000) report inferior ANN performance in forecasting commuter trip distribution relative to a doubly constrained SI model calibrated using maximum likelihood methods. Nijkamp, Reggiani, and Tritapepe (1996) report mixed results from comparing three types of feedforward ANNs and a logit modal split model. Only one ANN outperformed the logit model in goodness-of-fit. Raju, Sikdar, and Dhingra (1996) report good fit to a training subset but poor generality based on a validation subset for an intraurban modal split dataset. Dougherty, Kirby, and Boyle (1994) argue that an ANN is an effective alternative to logit-based choice modeling of multidimensional survey data on route choice.

Experience with ANNs in network flow and traffic flow problems is similarly mixed. Chang and Su (1995), Kikuchi and Tanaka (2000), and Yang and Qiao (1998) report good ANN performance with respect to intersection delay, flows within an expressway system and traffic prediction for a highway segment (respectively). Dougherty, Kirby, and Boyle (1994) and Smith and Demetsky (1997) are less enthusiastic, both reporting that statistical methods outperform the feed-forward/back propagation ANN in traffic flow and congestion prediction.

More comprehensive and rigorous studies are required to resolve the mixed experience from applying ANN to travel demand analysis (Dougherty 1995). Varying experience may reflect difficulties in specifying an appropriate ANN architecture for a particular problem. Design decisions made by the analyst can affect performance. Design variables in the feed-forward/back propagation architecture include network topology, particularly with respect to the number of hidden nodes and number of layers. Increasing the size of the network can improve performance at the expense of greater computational and data requirements. In addition to supporting interpretation of trained ANN weights, sensitivity analysis can also be used to determine the importance of a particular node in the ANN; this can guide pruning the network to achieve a parsimonious configuration (Dougherty and Cobbett 1997).

There are numerous competing ANN architectures in addition to the feed-forward/back propagation architecture that seems to be the default ANN in travel demand analysis (Dougherty 1995). Yun et al. (1998) compares a feed-forward/back propagation with two other common ANN architectures, namely, a *finite impulse response* (FIR) network and a time-delayed recurrent network. They compare performance across three network flow datasets (limited access highways, surface highway, and urban intersection) with different dynamic patterns with respect to periodicity, volatility, and fluctuation. Results indicate that the time-recurrent network performed best on data with a high degree of randomness (urban intersection flow) while the FIR network performed best on the data with a high degree of periodicity (limited access highway flow). In a highly specialized design, Yang, Akiyama, and Sasaki (1998) transform the logical transportation network directly into an ANN and use back-propagation to derive real-time dynamic flow predictions with promising results.

Activity-based analysis

*Overview*

Most aggregate-level travel demand models treat transportation as something desired for its own sake. In fact, transportation is a *derived demand*: Except for some types of leisure travel, transportation exists so individuals can fulfil other activities, such as work, shopping, recreation and socializing. The increasing decentralization and locational specialization in contemporary economies means that individuals must consume transportation services to overcome space and conduct periodic activities at different locations (Button 1993). *Activity-based analysis* is a theory and modeling strategy that views travel demand (and, over the long-run, transportation services and land use) as emerging from the interactions of individuals fulfilling activity needs through travel choices.

Activity-based approaches to travel demand modeling are based on the following principles. Individuals have a set of periodic activities with varying levels of necessity and urgency. The resources that satisfy these activities are distributed in space and time, i.e., at few locations and for limited time intervals. This can include requirements that other individuals be contemporaneous in space, time or both (such as work or socializing). Individuals must distribute their limited time availability or *time budget* among these activities, using transportation to trade time for space when traveling to locations. It is also sometimes possible to substitute in-situ activities that do not consume transportation services. Aggregate-level outcomes such as transportation system performance, urban development, lifestyle decisions and long-term mobility both condition and are conditioned by these individual-level choices. See Ben-Akiva and Bowman (1998), Bhat and Koppelman (2000), Golledge and Stimson (1997) and Thill and Thomas (1987) for reviews of this literature.

In addition to explicit recognition that travel demand derives from activity participation, activity-based models can capture empirical trip complexities typically ignored by travel demand models. Travel demand models generally assume that each trip has a single purpose, consists of a single stop, and uses a single mode. In contrast, there is a large amount of evidence that trips are often multipurpose and multistop, such as stopping to shop or pick up a child at day care when commuting home from work (Kitamura 1984; O'Kelly and Miller 1984). Individuals also make multiple stops within a single purpose trip to acquire information when they are uncertain about alternatives; an example is comparison shopping (Miller 1993). Individuals may also frequent several locations to conduct a regular activity due to changing constraints and variety-seeking behavior (Hanson 1980). The transportation mode often changes with the type of activity being conducted (Ben-Akiva and Bowman 1998).

*Activity-based approaches to travel demand analysis*

Travel demand analysis using an activity-based approach is a difficult combinatorial problem. Decisions such as the number of activities within a time period, sequencing, timing, mode and route choice are interlinked, implying an information space that is exponential with respect to choice dimensions (Ben-Akiva and Bowman 1998). *Econometric and statistical approaches* attempt to determine regularities and patterns within this information space by specifying and testing multidimensional utility functions or summarizing inter-trip linkages using data derived from travel diaries or other related activity-recording data collection methods (see O'Kelly and Miller 1984). *Utility maximizing approaches* include microeconomic models for optimal allocation of time and money as well as behavioral models based on nested logit (see Kitamura 1984). *Rule-based reasoning systems* construct activity and travel schedules based on preference rules and decision heuristics derived from cognitive science and artificial intelligence (see Garling, Kwan and Golledge 1994; Hayes-Roth and Hayes-Roth 1979; Vause 1997).

*Simulation* approaches to activity-based travel demand modeling focus on deriving limited but plausible choice sets from the multiple decision dimensions and modeling individual choices from those sets (Ben-Akiva and Bowman 1998). The PESAP (Programme Evaluating the Set of Alternative Sample Path) model generates feasible activity and travel patterns based on space-time constraints (Lenntorp 1976, 1978). STARCHILD (Simulation of Travel/Activity Response to Complex Household Interactive Logistic Decisions) generates feasible activity patterns through a combinatorial scheduling algorithm and predicts choices based on utility maximizing behavior subject to random changes in the transportation environment (Recker, McNally, and Root 1986a, 1986b).

The TRANSIMS (Transportation Analysis and Simulation System) model is based on a cellular automata-based dynamic network assignment model (discussed in chapter 6). The travel costs derived through traffic dynamics and congestion patterns are linked with the other travel demand components through an activity scheduling simulation based on *representative activity patterns*. These are synthetic and generic activity patterns that reflect some cohort of the population in the study area. Since activity patterns are stable and can be generalized (Janelle et al. 1998; McNally 1998), it is possible to generate representative activity patterns and distribute these across synthetic households using stochastic modeling techniques estimated from travel diary data (see Kitamura, Chen, and Pendyala 1997).

*GIS support for activity-based modeling*

GIS and related geographic information technologies can support several phases of an activity-based modeling system, including data collection, data integration and management, spatiotemporal analysis and cartographic visualization of results (Greaves and Stopher 1998). *Global positioning systems* (GPS) combined with handheld recording devices such as *personal digital assistants* (PDA) can allow for more accurate and detailed recording of activities in space and time. Murakami and Wagner (1999) tested a GPS/PDA configuration in Lexington, Kentucky USA. Subjects used a menu in the PDA to record trip purposes for driver and passengers. The GPS recorded locations at 3-second intervals. This allows collection of detailed route data not usually collected in these diary methods. Self-reported distance and travel times were consistently larger than those derived through the GPS/PDA unit, indicating potential biases in the self-reported travel data. GPS receivers simultaneously collect network travel time information during the travel event, greatly improving the realism of the activity modeling process (see Guo and Poling 1995). Even without an activity-recording device, a GPS can be useful. Stopher and Wilmot (2000) found that during follow-up phone interviews, respondents were able to recall trip activities when prompted with the detailed time and location data collected using a GPS receiver mounted in their vehicles.

Since activity-based simulations focus strongly on using constraints to determine plausible activity patterns, they require realistic representations of the transportation environment. The Simulation Model for Activities, Resources and Travel (SMART) software system is a good example of integrating a household activity simulator with a commercial GIS software to support spatial data integration and management (Stopher, Hartgen, and Li 1996). The household activity simulator fixes mandatory activities (such as work) in time and space and schedules flexible activities around those fixed activities so as to minimize travel cost. The GIS maintains information on households and their characteristics, activity locations (from land-use coverages) and the transportation system. The system uses a network routing algorithm in the GIS to construct a stochastic network assignment procedure. The GIS also maintains network flow data and determines travel costs based on these flows. Future land-use configurations and their impacts on activity scheduling can also be simulated using this system.

A GIS also provides user interface and visualization tools that can provide decision support and facilitate exploratory spatial analysis of activity patterns. Van der Knaap (1997) develops a prototype software system that uses dynamic cartographic visualization techniques for exploring activity patterns within their geographic context. Users can visualize dynamic transportation network usage patterns generated by the activity patterns while controlling for the temporal dimension using an interactive slider bar. This and other types of dynamic cartographic visualization systems can provide powerful insights into the dynamic activity and travel demand patterns in a geographic environment.

Another useful feature of a GIS is the ability to calculate feasible activity space and locations given scheduling and other constraints. Miller (1991) develops a procedure for calculating accessible locations within a network given a travel origin, destination and a time budget for travel. Golledge et al. (1994) combine a rule-based activity-scheduling model with a GIS for calculating feasibility opportunity sets based on space-time constraints. Kwan and Hong (1998) develop a method for integrating cognitive constraints (e.g., limited information) into the space-time accessibility calculation within a network. Kwan (1997) extends these approaches to the dynamic case in the GIS-Interfaced Computational process model for Activity Scheduling (GISICAS) intended to provide decision support for advanced traveler information systems.

Huisman and Forer (1998) discuss methods for computing and visualizing potential activity regions in space-time and forecasting travel demand based on travel diary data. Using empirical activity data, they show how to construct a 3-D region that demarcates locations in space and time where individuals can conduct discretionary activities. Travel demand for specific activities can be estimated using intelligent agents who follow simple behavioral rules and are constrained by the empirically derived space-time activity region.

### Measuring accessibility

Travel demand forecasts are often critical for predicting transportation system performance in the future and under different "what-if?" planning and design scenarios. Models constructed properly from theory can also help identify the causal factors that determine travel demands. Travel demand models are often used (implicitly or explicitly) as a way to measure transportation system *performance*, that is, how well the transportation system is serving the study area. As discussed previously with respect to activity-based analysis, travel demand is a method for meeting required and desired activities; it is rarely demanded for its own sake. Using travel demand as a surrogate for system performance involves the implicit assumption that increasing system throughput will increase the system benefits to individuals. This may not be the case as the residents of most large cities can attest (Miller 1999a).

Since transportation systems exist to increase participation in activities distributed in space and time, a more sensitive approach to measuring system performance is to measure the *accessibility* it provides to individuals. We can measure "accessibility," or the freedom to participate in activities in the environment (Weibull 1980) in three major ways (Miller 1999a). First, we can assess the *spatio-temporal constraints* faced by individuals based on their fixed activities and the transport system's ability to trade time for space in movement. Second, we can derive a "potential interaction" or *attraction-accessibility* measure for individuals or locations that assesses the attractiveness versus travel cost for opportunities in the environment. Finally, we can measure the *benefits* to individuals from the feasible choice sets allowed by the transportation system. Accessibility measures are sensitive to geographic representation since they require accurate calculation of the costs and constraints facing individuals in complex activity systems. Consequently, there has been a considerable amount of research on using GIS to derive accessibility measures.

*Space-time accessibility*

An elegant and powerful perspective to derive constraints-based accessibility measures is the time geographic framework of Torsten Hägerstrand. The time geographic framework recognizes that time cannot be separated from space when examining activity participation and accessibility. See Hägerstrand (1970) for a classic reference on the time geographic framework. Burns (1979), Forer (1998), Miller (1991), and O'Sullivan, Morrison, and Shearer (2000) provide discussions of useful theoretical concepts and empirical measures from this framework.

Perhaps the most powerful theoretic concept from the time geographic framework is the *space-time prism*. Figure 8-5 illustrates the space-time prism. Figure 8-5a shows a set of *isochrones* (lines of equal travel time) based on travel from a given location. The vertical line beneath the isochrones indicates that the person is stationary in space (e.g., at a fixed location such as home). The isochrones start at the point in time when the individual can begin a travel episode. The spacing of the isochrones and the volume of the space-time prism (in this case, a cone) are determined by the allowable travel velocity, that is, the ability to trade time for space. Classic time geography assumes that travel velocity is constant across space. Figure 8-5b illustrates the space-time prism that results when anchoring the travel episode at two fixed locations (e.g., home and work) and imposing a time budget for a discretionary travel episode (e.g., shopping) that can occur during the trip. Even under simplistic assumptions regarding the travel environment (such as a constant travel velocity), the morphology of space-time prism can provide remarkable insights into accessi-

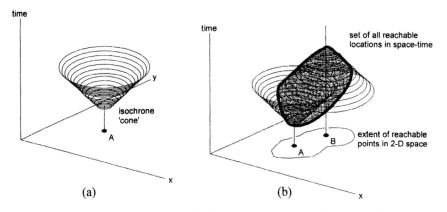

Figure 8-5: Space-time prisms (from O'Sullivan, Morrison, and Shearer 2000)

bility. See Burns (1979) for a rigorous and creative theoretical analysis of space-time prisms.

Although a powerful theoretical concept, the unrealistic constant travel velocity assumption limits the empirical relevance of the space-time prism. To enhance the space-time prism as an empirical measure, Miller (1991) develops an algorithm for computing space-time prisms within a transportation network where travel velocities and times can vary by arc. The *potential path area* is the locations in space accessible to an individual given fixed activity locations and durations, a time budget for travel and a constant travel velocity. In a geometric sense, it is the projection of the space-time prism to the plane representing geographic space (see figure 8-5). A network version of this concept can be used in activity modeling as well as a direct measure of network-based accessibility.

A generalized version of the *network potential path area algorithm* (N-PPA) is:

*Algorithm N-PPA*

Step 1. Compute the shortest path tree anchored at the first fixed activity location using network travel times, only extending the tree if arc travel times are less than $T$, where $T$ is the net time budget available (minus the required activity participation time, if any). Call this set of arcs the *candidate set*.

Step 2. Test each arc in the candidate set for inclusion in the N-PPA by computing the minimum travel time path to the second fixed activity. Include the arc only if $T - t_1 - t_2 \geq 0$, where $t_1$ is the travel time from the first fixed activity location to the arc and $t_2$ is the travel time from the arc to the second fixed activity location.

Empirical path estimates or another behavioral model can replace the shortest path calculations in both steps of the algorithm. The N-PPA can be used to query from address-indexed data files on activity locations using the address-matching capabilities of a GIS.

The N-PPA algorithm can be computationally expensive since it requires solving the shortest path from each candidate arc to the second activity location.

This can be a large set in detailed, urban-scale transportation networks. One possibility is to grow the shortest path tree in Step 1 to some limit less than the net time budget. Another possibility is to reduce the time budget through $T' = T - t_{12}$, where $t_{12}$ is the minimum travel time between the first and second fixed activities. Since travel between these locations will require at least $t_{12}$, there is no point in adding an arc to the candidate set in Step 1 unless it leaves at least this much residual in the time budget. Judicious choice of shortest path implementation, heuristic shortest path algorithms and using parallel implementations (particularly for Step 2) can reduce run-time for this algorithm (see chapter 5 for discussion of these issues and strategies).

Within a GIS, space-time prisms can provide effective support for querying about individual and aggregate accessibility. This can be useful both for analyzing a transportation system as well as providing travel decision support in an advanced traveler information system (also see Kwan 1998). Queries supported by a space-time prism include (the generic space-time queries are in parentheses):

- Which locations can an individual reach in 15 minutes? (What is the volume of the space-time prism for an individual at location $x$ at time $t$?)
- When and where does an event occur? (What volume of space-time does an event occupy?)
- Who can attend that event? (What is the relationship between the space-time volume of an event and individual space-time prisms?)
- Can I meet my friends this evening? (What is the relationship between several space-time prisms?)

Forer (1998) provides a good discussion of implementing these queries and other space-time prism concepts within a GIS.

O'Sullivan, Morrison, and Shearer (2000) customize off-the-shelf GIS software to develop a system for calculating space-time constraints for travel on public transportation. They calculate *isochrones* (lines of equal travel time) based on walking to and from a multimodal public transportation system. An algorithm determines the minimum time travel routes between locations based on street network and public transport service schedule data. Standard GIS tools such as buffering are used to perform required spatial analysis tasks such as determining service areas for public transit stops, intermodal connections and generating the isochrones.

Since limited information about the environment also affects accessibility, Kwan and Hong (1998) extend the spatiotemporal constraints approach to GIS-based accessibility modeling by incorporating cognitive constraints. Kwan and Hong (1998) frame the problem from the perspective of identifying a set of feasible activity locations based on space-time constraints and the individual's geographic familiarity and preferences. A GIS allows preference scores for polygons (in this case, grid cells) to be overlaid with the network to remove locations in the N-PPA that are unknown or undesirable to the individual. Kwan and Hong (1998) also make some valuable suggestions about improving the N-PPA calculations by including scheduling constraints.

## Attraction-accessibility

The attraction-accessibility approach to accessibility measurement computes a summary score for an individual or place based on the attractiveness of potential activity locations and their required travel costs or times. The most common attraction-accessibility measure is the Hansen (1959) potential measure:

$$A_i = \sum_j w_j f(\beta c_{ij}) \qquad (8\text{-}11)$$

where $A_i$ is the accessibility of travel origin $i$, $w_j$ is the attractiveness of destination $j$ and $f(\beta c_{ij})$ is a general function describing the travel cost between $i$ and $j$. (This can be power, exponential or some other function.) Weibull (1976, 1980) formulates a rigorous framework for deriving and applying these and related accessibility measures. Geertmann and van Eck (1995) develop GIS software for calculating and visualizing these measures within their geographic setting. The system calculates the potential measure for a regular lattice of locations in geographic space using network travel times and constant travel velocities with straight-line travel for off-network locations. Applying an interpolation routine to this lattice of accessibility measures allows a potential surface to be calculated and visualized with respect to the transportation network and other geographic data.

## User benefits

The problem with attraction-accessibility measures is they are difficult to interpret since they are not derived from a behavioral or economic framework. Although Weibull (1976, 1980) provides strict rules for calculating these measures, the framework still does not dictate what exactly we should be measuring when we measure accessibility. This makes comparisons difficult (Miller 1999a; Morris, Dumble, and Wigan 1979). The third major approach to accessibility measurement calculates the benefits that an individual receives from a choice set. Using random utility theory at the disaggregate level, we can calculate a *users' benefit* measure as the expected maximum utility of the choice set. Assuming a logit choice mechanism, this measure is (Ben-Akiva and Lerman 1979, 1985):

$$A_i = \ln \sum_{j \in \mathbf{K}_i} \exp(v_j) \qquad (8\text{-}12)$$

where $v_j$ is the net (minus travel costs) measured utility of activity location $j$ and $\mathbf{K}_i$ is the choice set available to person $i$ (or location $i$). At the aggregate level, we can rearrange the terms of a SI model and derive a location benefits accessibility measure (Wilson 1976):

$$A_i = \sum_{j \in \mathbf{K}_i} \frac{\alpha}{\beta} \ln a_j - c_{ij} \qquad (8\text{-}13)$$

where $\alpha$, $\beta$ are the destination attractiveness and distance decay parameters from the spatial interaction model, $a_j$ is the attractiveness of destination $j$ and

$c_{ij}$ is the required travel cost. Both measures are consistent with the classic economic concept of consumer surplus, although this is controversial (see Miller 1999a).

*Reconciling the three approaches: Space-time accessibility benefits*

Miller (1999a) develops accessibility measures that integrate the three major accessibility measurement approaches discussed above and computational procedures for calculating these within a transportation network. A space-time utility function includes destination attractiveness, travel times from and to fixed activities and the available activity participation time. The space-time utility function is at the basis of user benefits and locational benefits measures. All accessibility measures are also consistent with the Weibull (1976, 1980) framework for attraction-accessibility measures.

To calculate the space-time accessibility measures within a transportation network, Miller (1999a) adapts a procedure for calculating network-based market areas developed by Okabe and Kitamura (1996). The procedure calculates an extended shortest path tree from activity locations by extending the "leaves" of the standard shortest path tree until they meet at boundaries within arcs. Boundary nodes are inserted into the network at these locations. Travel times to and from locations within network arcs can now be easily interpolated based on travel times at nodes. Using these travel times within the space-time accessibility measures allows calculation of accessibility for given individuals and locations in the network. A simple algorithm also allows the identification and visualization of high or low accessibility regimes within the network. Miller and Wu (2000) discuss prototype GIS software that combines this functionality with user-friendly database and project management tools.

Figure 8-6 illustrates high-accessibility locations within the Salt Lake City transportation network based on one of the space-time accessibility measures in Miller (1999a). This shows highly accessible home locations based on a given activity schedule, namely, traveling from home to a given work location (the University of Utah in the upper right corner) and stopping to shop at a store on the way to work. Note that the high accessibility regimes are disconnected from each other. This reflects an important point about space-time accessibility measures. Most accessibility measures simply reflect urban structure and indicate high accessibility where activities are geographically concentrated. By including activity-related constraints, the space-time accessibility measures reflect the more realistic view that accessibility as the interface between urban form and how the city is used in a person's activity schedule (see Kwan 1998; Miller and Wu 2000).

## Land-use/transportation modeling

A legitimate criticism of many travel demand analysis methods is that they ignore the "fifth component" of travel demand, namely, *land use*. This weakness

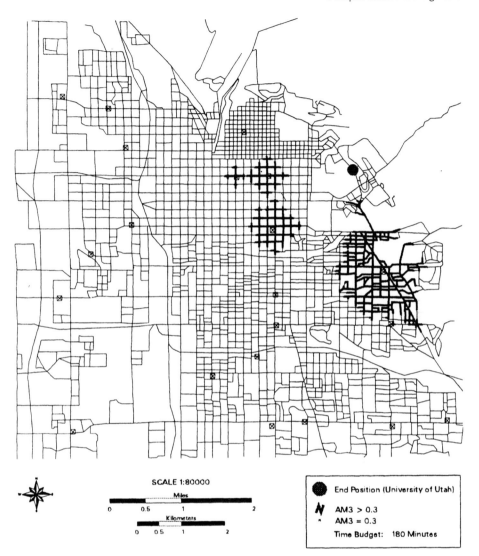

Figure 8-6: High space-time accessibility locations within a transportation network (from Miller and Wu 2000)

is especially critical for long-term forecasting and planning. Over multiple year time horizons, it is impossible to separate land use from travel demand. Land use influences the magnitude and pattern of travel demand and travel demand influences future land uses. The public and private sectors influence these complex relationships by providing transportation facilities and services that lower travel costs in particular places and times. Ignoring the mutual influences between transportation and land use in long-term forecasting and planning can lead to decisions with unintended and undesirable consequences.

There is a large literature on integrated *land-use/transportation* (LUT) models. This literature encompasses a wide range of theoretical perspectives. A detailed review would require a lengthy discourse beyond the scope of this book. Instead, we will provide an overview of the literature that will point the reader to key references. For more detailed reviews, see Boyce (1990), Kim (1989), Southworth (1995), or Wegener (1990). We will then discuss the role of GIScience tools and GIS software in enhancing this type of modeling.

Overview

Most LUT models encompass a range of urban subsystems that can be grouped according to their speed of change. *Slow-changing subsystems* (changing over multiple years to decades) include transportation, communication and utility networks and the land use distribution. *Medium-speed subsystems* (multiple years) include workplaces and housing. *Fast-changing subsystems* (less than one year) include employment and population. *Immediate-change* subsystems include individual travel and transport of goods. The urban physical environment can also be considered another subsystem but with more complex temporal dynamics (Wegener 1994).

Integrating the urban subsystems into a common modeling system is difficult and requires a strong theoretical framework to tie these components together. Adding a fifth step to the four-step approach and arbitrarily looping around these models is not sufficient. Southworth (1995) identifies five general types of LUT models based their major theoretical perspective. Wegener (1994) classifies models based on the subsystems modeled, underlying theory and policy variables encompassed. In the next few subsections, we will summarize Southworth (1995) and Wegener (1994); we refer the reader to these very good reviews for more detail.

Alternative approaches to land-use/transportation modeling

*Lowry and related models*

The pioneering *Lowry model* (Lowry 1964) captures population, employment, commercial and land-use through a system of equations that are equivalent to two interlinked SI models. One model allocates workers and related individuals to zones based on exogenous forecasts of *basic* ("industrial" as opposed to service) employment. The model allocates commercial activity to zones based on these predicted residential activities. Solution requires iterating the two models until convergence. Refinements of this approach include the *Disaggregate Residential Allocation Model* (DRAM) and the *Employment Allocation Model* (EMPAL). See Putman (1991) for the theory and model specification and Watterson (1990, 1993) for empirical applications.

*Mathematical programming*

*Mathematical programming* models are similar to the routing, flow and facility location optimization problems discussed in chapters 5 and 6. Wilson and colleagues (Coelho and Wilson 1976; Wilson et al. 1981) develop the underlying theory based on the entropy-maximizing SI framework. The objective function requires maximizing a measure of net locational benefits and an entropy component representing choice variation at the micro level. Constraints include limits on the supply of land and spatial interaction terms dictating travel demands among zones. An example is the Projective Optimization Land Use System (POLIS) that encompasses the employment, population, housing, land use and travel subsystems (Prastacos 1986a, 1986b). Policy variables include land use regulations and transportation improvements, although the network is implicit.

*Multisector models*

*Multisector* LUT models combine an input-output model of an economy with a spatial interaction component for allocating economic flows and implied travel demands among zones. Input-output models estimate economic activity across multiple industry, commercial, service, and household sectors based on an exogenous economic forecast and estimates of economic interlinkages among sectors. These technical coefficients represent feedback and loops in the economy that magnify direct economic impacts into greater economic activity (see Leontief 1967; Miller and Blair 1985). Multisector models extend this approach by spatially disaggregating economic sectors across zones. An interaction component distributes flows between economic sectors among zones based on transportation costs.

An example of a multisector LUT model is MEPLAN (Hunt and Simmonds 1993). MEPLAN encompasses all eight subsystems and also assumes that land use and network flows achieve separate equilibrium, with the latter based on Dial's (1971) stochastic assignment procedure. TRANUS (de la Barra 1989) is also comprehensive and assumes land use and network flow equilibrium. In both models, relevant policy variables are land use regulations, transportation infrastructure improvements and changes in transportation costs.

*Urban economic models*

*Urban economic* models treat transportation and land use as a market and determine a general equilibrium that encompasses both. A central mechanism is the *bid rent curve* (Alonso 1964). The bid rent curve shows the cost of land at varying distance to a desirable "center" (such as a central business district). Land uses compete to locate near the center and at equilibrium have sorted themselves out based on their ability to substitute transportation for land (e.g., locate further away for cheaper land but higher transport costs). An example is Kim (1989): This model solves for a general equilibrium of employment, population, transportation network, freight transport, and travel demands. Input-output coefficients capture inter-sectoral economic linkages that are translated into demands for travel and goods transport. Network flows achieve user-optimal equilibrium. Policy inputs to the modeling process focus on transportation improvements.

### Microsimulation and related models

The previous LUT models can be viewed as "top-down" since they solve for a combined land use and transportation system by imposing some central theoretical framework. *Microsimulation* models are "bottom up" in the sense that they specify behaviors for the individual entities in the system (possibly imposing some physical constraints at the macro level) and allow aggregate properties to emerge from the interaction of these entities. We discussed these approaches in chapter 6 with respect to network flow and earlier in this chapter with respect to activity-based modeling. Microsimulation models with explicit land use and transportation components include the *California Urban Futures* (CUF) model. Policy inputs for this model include land-use restrictions, environmental policies, public facilities, and transportation improvements.

Unlike many LUT models, CUF is site specific and spatially disaggregate: It does not require prespecified aggregate spatial units. A population submodel generates population growth trends for existing cities based on calibrated regression equations (but apparently not spatial regressions). A GIS database on undeveloped or underdeveloped areas, known as *developable land units* (DLU), combines with a spatial allocation submodel to distribute population growth among DLUs based on their physical site characteristics and accessibility. An annexation-incorporation submodel annexes DLUs to adjacent cities or creates new municipalities from DLU clusters. See Landis (1994) for detailed model discussion, data sources and experience with coupling the model with a commercial GIS. The second generation CUF includes multiple land uses. See Landis and Zhang (1998a) for description of CUF2 and Landis and Zhang (1998b) for calibration experience.

Another emerging approach to urban simulation modeling is *cellular automata* (CA). CA are discrete spatiotemporal dynamic systems based on local rules. CA can generate very complex spatiotemporal dynamics from relatively simple rule sets. CA are tractable and can address very large and detailed geographic applications. CA is also a natural fit with GIS, particularly raster GIS software tools. Commercial GIS software that provide map algebra or related macro languages or allow direct manipulation of the raster structure can easily accommodate CA.

Cellular automata partition space into discrete units (e.g., raster cells) and time into discrete steps. A rule set specifies the *state* of each cell in time $t + 1$ based on the states of its neighbors in time $t$. The system starts with an initial seed of cellular states and then iterates over discrete time steps, updating the state of each cell. For reviews of CA-based modeling of urban systems and pointers to the literature, see Batty and Xie (1997), Câmara, Ferreira, and Castro (1996), Couclelis (1997), Sanders (1996), or White and Engelen (1997).

In the LUT modeling context, the cell states in the CA encompasses different types of land use. The rule set specifying the land-use dynamics can be derived from urban land-use theory or empirical observations of the land-use conversion process in an urban setting. Although sometimes viewed as sterile computations, CA-based geographic modeling can be tied into basic geographic theory such as Waldo Tobler's cellular geography (Tobler 1979) and the *central place theory* of urban spatial systems (Clarke, Hoppen, and Gaydos 1997). Xie (1996) develops a generic and rigorous formalism for CA-based dynamic urban modeling that integrates processes and constraints at the macro and micro levels.

A criticism of CA-based LUT modeling is that it emphasizes local rules and bottom-up forces. One reason for LUT modeling is to assess the influences of government land-use and transportation policy on future urban development. Some of these policies can be difficult to state as local rules. Li and Yeh (2000) develop a CA model for modeling sustainable urban development that incorporates local, regional and global (system-wide) constraints using development suitability factors with varying degrees of restriction.

Another criticism of CA-based LUT modeling is the rigidity of the CA rule set, particularly over time. This can be unrealistic; for example, if all flat land is developed in an area, the local political process may relax the "rules" against developing steeper slopes. Clarke and Gaydos (1998) and Clarke, Hoppen, and Gaydos (1997) describe a self-modifying CA for simulating urban growth at a regional level (i.e., at relatively small map scales with highly generalized land-use categories). Four different types of urban growth processes are encompassed by the model (spontaneous, diffusion, road-induced, and "organic"). The rule set modifies if growth is too rapid or too slow. Using satellite-based remote sensing imagery and a commercial GIS, the authors implement and calibrate the model for San Francisco Bay area and Washington/Baltimore corridor in the United States.

## Enhancing land-use/transportation model through GIS

A GIS software system developed by Batty and Xie for modeling urban dynamics in Buffalo, NY, USA is a good (and detailed) example of the powerful enhancements to LUT modeling available through GIS tools (Batty 1996; Batty and Xie, 1994a, 1994b). Their system uses a basic spatial population density model adjusted for the discrete representation in the GIS. The GIS supports model calibration at varying levels of spatial aggregation to assess effects due to the modifiable areal unit problem (see chapter 4). Model results can be visualized as plots of attributes and spatial measurements (such as the fractal dimension) against distance from downtown. Animations allow the urban spatiotemporal dynamics to be visualized and explored at varying levels of spatial and temporal resolutions.

A rapidly emerging frontier that can greatly enhance the realistic quality of LUT model visualizations in 3-D GIS. The ability to create natural depictions of the urban environment at the human scale allows analysts and decision makers to assess and convey the human-scale experience in an urban future. "Fly-throughs" and other interactive animations can allow visual and cognitive summaries of current and future urban fabrics (see Sapeta and Barnell 1999; Wilson 1998). Users could manipulate *avatars* or virtual representations of themselves interacting and even interactively manipulate design elements in the urban environment. Developments in virtual reality will greatly enhance this experience as will WWW tools for disseminating these experiences to a wide range of stakeholders. One could also imagine *enhanced reality* systems that project visualizations of future urban landscapes onto the visual field of a real landscape through wearable eyeglasses or goggles (Mark 1999).

Köninger and Bartel (1998) develop a framework for a 3-D GIS to support urban visualization for design and planning. Their object-oriented design specifies an urban object space populated by the base-level objects buildings, streets, green areas, public places, and terrain surface. These are composites containing elements with varying levels of spatial resolution maintained as a level-of-detail object hierarchy. The level-of-detail when visualizing and manipulating urban objects depends on a combination of its size, the observer's distance and the visual angle. Objects can be combined into higher level structural elements corresponding to blocks, districts or the entire city.

Figure 8-7 provides the conceptual model for the urban 3-D GIS system (Köninger and Bartel 1998). Arrows indicate data and information flows. The system integrates data from traditional sources such as analog maps, GIS coverages, satellite imagery, aerial photography, and close-up images. Computer-aided Design and Drafting (CADD) software maintains 3-D architectural and urban design data. A semiautomatic classification process maps these data to the level-of-detail hierarchy. The output from a GIS-based LUT modeling system can be enhanced and visualized as a three-dimensional scene using CADD data representing current or imagined future architecture and urban design elements. This includes photo-realistic 3-D urban views with superimposed thematic data generating from analysis and modeling (Köninger and Bartel 1998).

## Route planning

### Public transit system planning

Planning a public transit system requires configuring a set of routes and levels of service along those routes that optimize some measure of performance, typically minimization of the travel time incurred by system users. Constraints on this problem usually include minimum service frequencies, maximum load factors for vehicles and restrictions on the total fleet size. We can state this problem as a constrained optimization problem solved using mathematical programming techniques and heuristic search. Practical guidelines and rules-of-thumb are also common (see Baaj and Mahmassani 1991).

An important determinant of public transportation usage is system access. Public transportation systems tend to serve populations that do not use automobiles due to choice, necessity or lack of resources. Consequently, a reasonable walking distance to the stops along a public transit route usually defines its service area. (Determining the service areas for other multimodal configurations such as "park-and-ride" stations is more complex). Constructing an appropriate buffer around transit stops and using overlay and areal interpolation methods allows the transportation analyst to determine the socioeconomic and demographic characteristics of service areas. This can be compared to target populations for the public transit service to assess how well the system is

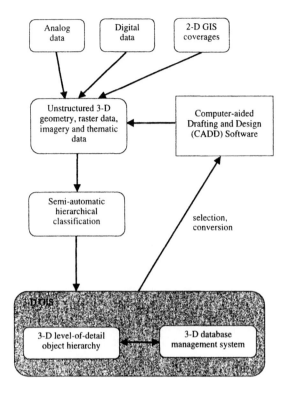

Figure 8-7: Conceptual design for a 3-D urban GIS (after Köninger and Bartel 1998)

performing (O'Neill 1995; O'Neill, Ramsey, and Chou 1992). Different proposed route configurations can be easily summarized and compared.

A GIS also allows integration of service areas with other geographically referenced data. For example, calculated service areas can be overlaid with georeferenced addresses to conduct marketing, public awareness campaigns and user needs surveys. This can also support provision of route information via dial-up services or the WWW as well as customized "dial-a-ride" services for disabled populations (Jia and Ford 2000; Souleyrette and Strauss 1999).

Vehicle-mounted GPS receivers and automatic passenger counting devices are creating a wealth of public transportation patronage data. A GIS can provide effective enterprise-wide database to integrate route, patronage and service area data to assess current route performance. A relational database can accommodate the primary data on route boarding and alighting activities by stop along each route (see Jia and Ford 2000). This can be georeferenced and integrated with the transit network, service areas, the multimodal transportation network and other geographic data through the dynamic segmentation data model (see chapter 3), supporting querying and visualizations for exploring and assessing system performance.

The data collected through automated passenger counters and GPS receivers can be integrated with other geographic data to conduct public transit passenger forecasting. A straightforward forecasting model is (Azar and Ferreira 1995):

$$B_i^d = P_i O_i^d L_i^d \tag{8-14}$$

where $B_i^d$ is the boarding count on route segment $i$ in direction $d$, $P_i$ is a "production factor" describing the potential public transit trip generation in the service area of segment $i$, $O_i^d$ is a "trip opportunity" variable describing the attractiveness of destinations downstream from segment $i$ in direction $d$, and $L_i^d$ is the level of service of segment $i$ in direction $d$. For calibration, a combined GPS/passenger counting system can generate the boarding count. The production factor can be estimated using the buffering, overlay and areal interpolation methods within a GIS. The trip opportunity can be similarly measured by buffering around transit stops downstream and using overlay and interpolation methods with secondary data on employment, retail or recreation activities. Service headway obtained through schedules or (more accurately) through the vehicle mounted GPS receiver. GIS software also allows visualization of model results (Azar and Ferreira 1992, 1995). See Ramirez and Seneviratne (1996) for another example of GIS-based public transit ridership forecasting.

### Right-of-way planning

Right-of-way (ROW) planning requires finding an optimal or satisfactory *corridor* through geographic space for a new transportation or telecommunication facility based on a set of criteria. By "corridor" we mean a 2-D region with a strong linear orientation that can be conceptualized as a route. Each location in geographic space is characterized by a set of attributes that affect the suitability of the corridor.

ROW planning is usually a multistage process. The *network design* phase identifies the need for new ROWs based on modifying the existing transportation network. This can result from the transportation planning process, including the results of the travel demand or accessibility analyses. This problem can also be stated mathematically as a constrained optimization problem that minimizes flow and construction costs subject to physical constraints. The physical constraints are usually limited to flow consistency and capacity constraints, meaning that the results are purely topological. See Current and Min (1986) and Magnanti and Wong (1984) for classic references and bibliographies and Bell and Iida (1997) for a more contemporary treatment and references.

Once we have determined the topological network modifications, we must determine a ROW among the infinite number of possible corridors between two locations in geographic space. This problem can be conceptualized differently based on map scale. At a small map scale, we can treat it as a *corridor location problem* that determines a good path or set of alternative paths in geographic space based on criteria such as land use, wetlands, floodplains, slope, soil stability and sensitive historical or archaeological sites. At a larger map scale, we can conceptualize it as a *corridor design problem* since we are concerned with the detailed configuration of a 2-D (possibly 3-D) region in geographic space based on design and engineering principles (Versenyi, Holdstock, and Fischer 1994). The different conceptualizations suggest a two-stage solution method. We first solve the corridor location problem at a small map scale and then perform detailed corridor evaluation and design of the solutions generated in the first stage. The location problem provides a first approximation while the second stage refines the approximation (Nicholson, Elms, and Williman 1976).

The network design, corridor location and corridor design phasing for ROW planning is somewhat arbitrary. Methods exist for combining these stages. For example, we can combine network design and corridor location by incorporating physical constraints on horizontal and vertical curvature for the new alignment into network design optimization models (see Trietsch 1987a, 1987b). Iterating between the stages may be necessary. A three-step with feedback loops ROW planning model makes sense since each phase in the sequence involves larger geographic scales requiring different techniques, technologies, and resources. We can manage the problem complexity by identifying a small set of alternatives before performing detailed design that requires greater resources with respect to data and person-hours.

*Corridor location problem*

*Decision tools for corridor location*  A difficult aspect of the corridor location problem is that it involves multiple, often conflicting, criteria. These can include construction cost, user (flow) cost, environmental impacts (e.g., wetlands, sensitive species, biodiversity) and physical factors such as slope and soil stability. *Multicriteria decision making* (MCDM) refers to the process of solving problems involving multiple, conflicting attributes.

Figure 8-8 shows a generalized version of the MCDM process (after Jankowski 1995; Jankowski and Richard 1994). We must first identify a set of criteria for evaluating alternatives. This supports identifying a set of feasible alternatives. To support comparison among solutions, we must calculate *criteria scores* that translate an attribute value into a score on some common scale. A function for converting an attribute measurement (e.g., population density, slope) into a criterion score is

$$c_k(x) = \omega_k a_k(x), \quad 0 < \omega_k < 1 \tag{8-15}$$

where $c_k(x)$ is the value of attribute $k$ at location $x$, $a_k(x)$ a measure of attribute $k$ at that location and $\omega_k$ is a scaling function for attribute $k$ (Hepner 1984; O'Bannion 1980). The measured attribute can be along nominal, ordinal, or interval/ratio scales but the weight(s) must be chosen so that the resulting value is a ratio measure. It is helpful to choose a scaling function that *normalizes* the data (i.e., puts all attributes on the same dimensionless scale, usually between 0 and 1). Linear and nonlinear methods are available; see Jankowski (1989).

After computing the criterion scores, we must then compare and contrast the alternatives. The objective is to identify relative importance of the criteria to decision-makers so we can identify a good solution or small set of solutions. One possible representation of alternatives and their criterion scores is the *decision table*. GIS can complement and support effective summary and comparison of alternatives through geographic data integration and visualization.

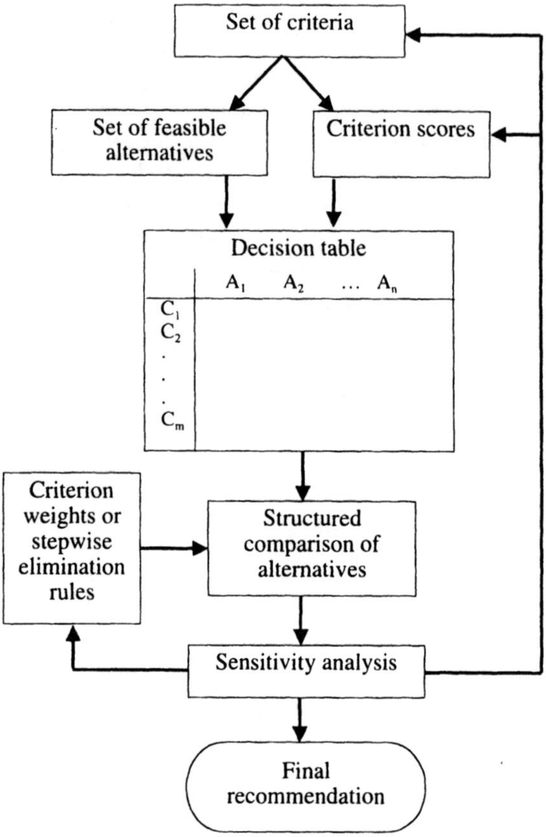

**Figure 8-8**: A generalized multi-criteria decision-making process (after Jankowski 1995)

Identifying preferences among criteria depends on the nature of the problem. Some decision problems are *compensatory*. This means that we can make meaningful comparisons among the decision criteria such that a poor value for one criterion can be compensated by a good value for another criterion. This is reflected in a set of *criterion weights* that allow the set of criterion scores to be combined into an overall value or score for an alternative. Determining criterion weights can be difficult and subjective. MCDM tools allow these scores and weights be elicited from experts and decision-makers; common techniques include weighted summation, concordance analysis, and multidimensional scaling. See Jankowski (1995) for a concise overview and pointers to the literature.

A technique for calculating the overall desirability of an alternative in a compensatory site selection problems is *suitability mapping*. Although greatly facilitated by GIS, this method has roots in analog overlay mapping (see McHarg 1971). The overall suitability for an activity at site $X$ (where this is a composite of individual locations $x$) is:

$$s(X) = \sum_{x \in X} \sum_{k=1}^{m} w_k c_k(x) \tag{8-16}$$

with the requirements:

$$w_k > 0, \sum_{k=1}^{m} w_k = 1 \tag{8-17}$$

where $s(X)$ is the overall suitability for the activity at site $X$, $w_k$ is weight indicating the relative importance of attribute $k$ among the $m$ attributes and $c_k(x)$ is the criterion score for attribute $k$ at location $x$ (Hepner 1984; O'Bannion 1980).

Some decision problems are *noncompensatory*, meaning that we cannot make meaningful comparisons between different attributes and generate an overall preference score for a solution. For example, there may be no way to trade off environmental impacts or aesthetics with construction or maintenance costs, particularly if diverse stakeholders are involved and consensus cannot be achieved. In this case, we must relax our requirement to obtain quantitative weights for the different criteria and eliminate solutions using less-demanding techniques.

Most noncompensatory MCDM techniques attempt to eliminate alternatives in a stepwise process until a single solution or a small set remains. A first step can be to eliminate *inferior solutions*. These are solutions that perform worse than at least one other solution on each criterion and therefore are not worthy of further consideration. We can then continue to reduce the candidate set by systematically adjusting acceptable thresholds for the criterion scores and examining the impact on this set. We can also ask experts or decision makers to make ordinal comparisons (e.g., rankings) among criteria and eliminate alternatives correspondingly (see Jankowski 1995).

GIS supports MCDM for spatial decision problems (such as corridor location) through geographic data integration and visualization tools. GIS, modeling software, or some combination can generate a set of solution alternatives. The alternatives can be integrated with geographic data through overlay and areal interpolation. *Criterion coverages* are thematic GIS layers showing the spatial distribution of the attributes in the study area expressed using criterion scores. We could construct geographic visualizations of elements, rows, and columns (or some combination) of the decision table. We could also summarize, overlay and visualize the original attribute data (Carver 1991; Jankowksi 1995; Jankowski and Richard 1994).

*Generating alternative corridors*   Generating alternative corridor locations for evaluation and comparison is a critical component of the ROW planning process. Alternative solutions are sometimes dictated by the political process or institutional constraints. A GIS allows selection and evaluation of alternative corridors using buffer, overlay, interpolation and visualization techniques for summarizing and communicating solution attributes (see Hartgen and Li 1994; Versenyi, Holdstock, and Fischer 1994).

A technique for generating alternative corridors is to solve the problem as a type of *surface shortest path problem*. Recall from chapter 5 that this solves for a minimum cost path through a surface that often represents geographic space. We can solve this problem at a computational level using polygons or the regular square grid (RSG) and triangulated irregular network (TIN) surface models. One strategy is to generate solutions based only on cost and use the MCDM techniques discussed above to compare these solutions based on other attributes. We could also incorporate other attributes at this stage by using suitability mapping with proper scaling to generate a pseudocost for each location (i.e., higher score implies lower suitability). Re-solving the surface shortest path problem based on a careful sensitivity analysis of the parameters can identify solutions that are *robust* or perform well under alternative but plausible scaling and weighting scenarios (see Lowry, Miller, and Hepner 1995).

It is often desirable that the set of alternative corridor locations be spatially dispersed. Otherwise, the alternatives may not be judged substantially different by decision makers and stakeholders. Methods for generating a good set of spatially dispersed alternative corridor locations using surface shortest path techniques include the *iterative penalty method* (IPM) and the *gateway shortest path problem* (GSPP). The IPM repeatedly solves the shortest path problem after penalizing arcs and nodes used in any previously generated solution (Huber and Church 1985).

The GSPP constrains the least cost path between an origin and destination to go through a prespecified location known as a *gateway* (Lombard and Church 1993). The gateway can be any node that is not a member of the optimal path. We can identify all gateway shortest paths in a network efficiently by solving the shortest path trees from the origin and the destination (see chapter 5). Adding the two labels for a node after solving the two shortest path problems provides the cost of the solution constrained to go through that gateway. A geometric procedure is also available for calculating the spatial difference between a given path and the optimal solution, allowing the analyst to identify a spatially dispersed set of alternatives (see Lombard and Church 1993). We can use flow, construction or both costs in these routing methods and then analyze other criteria using GIS buffer, overlay, and interpolation tools. We can also use a synthetic unsuitability score in the routing based on a reasonable parameter setting since we are generating multiple alternatives.

A potential problem with finding alternative corridors using basic surface shortest path techniques is that the resulting corridors may not be physically feasible. For example the geometric configuration of the corridor may prohibit vehicle travel due to sharp curves, steep inclines and so forth. While this can often be refined during the corridor design phase, methods are available for considering these factors in the location phase. Trietsch (1987b) provides a careful review of techniques for calculating minimum cost path through a RSG subject to constraints on its horizontal and vertical curvatures. Trietsch (1987b) also suggests a honeycomb lattice with 12 possible path directions from each node to reduce the type of distortion errors discussed in chapter 5. This is an analog to the more-connected rasters approach but is not compatible with most existing GIS software.

If we are simultaneously locating several corridors (i.e., a subnetwork), we can use special methods for *network design problem in geographic space*. This is an extension of the topological network design problem that explicitly considers terrain. The problem is to determine a network configuration between locations distributed within a surface representing terrain that minimizes costs (possibly partitioned into flow, construction, and maintenance components) subject to horizontal and vertical curvature constraints on the paths between locations. The computational version of this problem uses a lattice to represent the surface and the problem becomes one of selecting a subnetwork of the implied network describing the surface, possibly inserting extra nodes to allow finer approximations of the alignment curvature (similar to *shapenodes* in some GIS software). Trietsch (1987a) formulates this optimization problem and discusses some solution heuritics based on a honeycomb lattice.

Another technique for generating multiple solutions to the ROW location problem is *genetic algorithms* (GA). This technique generates a set of diverse but well-performing solutions to a problem rather than a single best solution. Guimarães Pereira (1996) combines GA with a multicriteria ranking technique to solve a highway ROW location problem. Jha and Schonfeld (2000) develop a GIS that combines GA with procedures for calculating routes through general (nonconvex) polygons. However, their system focuses on a single solution rather than a solution set since each polygon is characterized by a general cost summarizing several attributes.

In chapter 4, we discussed issues surrounding spatial data error and uncertainty. Spatial data error and uncertainty can have particularly complex effects in corridor location techniques that use information derived from *digital terrain models* (DTMs) such as a regular square grid (RSG) (including USGS digital elevation models (DEMs)) or a triangulated irregular network (TINs).

Ehlschlaeger, Shortridge, and Goodchild (1997) develop geographic visualization (GVis) techniques for assessing the effects of uncertainty in RSG terrain models on corridor location. They explore the effects on corridor location from using a coarse USGS DEM when higher resolution data are required. A stochastic simulation method generates different realizations of the required higher resolution DEM based on sampled elevation data from the coarse DEM. They visualize corridor variability under the lower resolution data through animations showing the optimal routes or calculated cost surfaces across the terrain realizations. An interpolation method allows smooth animation of these "snapshot" realizations. The animated variability in optimal route or cost surfaces is an easily understood indicator of the effects of uncertainty on the corridor location.

*Corridor design problem*

Once we have determined the ROW as a linear feature at a small map scale, we can zoom-in to a larger map scale and perform detailed design of the 2-D or possibly 3-D corridor. At this scale, transportation facility design involves complex interconnected decisions regarding the configuration of design elements with respect to each other and the geographic environment (which itself may be modified). This involves detailed construction, maintenance and safety engineering techniques as well as aesthetic design criteria beyond the scope of this book. We will discuss GIS tools for environmental design of transportation facilities in chapter 10.

Transportation engineering and design documents are usually developed through computer-aided drafting and design (CADD) software. CADD software is very versatile for graphic editing and design visualization but less capable for spatial analysis (see Cowen 1988). Integrating CADD with GIS can allow integrated representation of the facility design and geographic setting, capturing interconnections among design elements and the physical environment. This also allows generation of a common design document spanning construction, maintenance, bridge and traffic engineering. "As-built" CADD files and change orders generated during the subsequent construction and maintenance processes can be integrated into this representation (Lee and Clover 1995; Lee, Wong, and Clover 1991).

Most leading commercial GIS packages now provide functions for accessing CADD files in their native formats. In addition, data exchange programs are readily available for conversions between the popular CADD and GIS file formats. Good practice is to develop consistent data models and standards (including coordinate system, map projection, geodetic datum, and graphic symbols). Whenever a consistency is impossible (e.g., CADD does not have an equivalence of the dynamic segmentation data model), one of these two tools should be designated as the primary platform for these data.

## Decision support for transportation planning

In addition to the analytical and computational tools discussed previously in this chapter, GIS provides a conduit for building and sharing geographic information among different components of the planning process. Increasing the spectrum (range), fidelity (detail) and flexibility of information flows can profoundly change the decision-making process within transportation planning.

Figure 8-9 shows the information flows in traditional (pre-GIS) transportation planning (after Nielsen 1995). The expense and difficulty of data collection, analysis and communication makes information dear. This tends to create a rigid and linear planning process in which data and information are revised at a minimal rate. This encourages disconnection between planning process and other stakeholders including the public. Although metropolitan planning organizations (MPOs) or departments of transportation (DOTs) have traditions of public participation, this tends to be in narrowly circumscribed roles at the beginning (to identify preferences) and end (to select among a small set of alternatives) of the process. There is limited input to the processes that translate preferences into a small set of alternatives. This is not the result of malicious intent; rather, it is a byproduct of distributing scarce time and limited resources among competing needs, including traditional methods for geographic information processing that are expensive, slow and have limited spectrum, fidelity and flexibility.

Figure 8-10 shows the information flows in a planning process enhanced through GIS technology (after Nielsen 1995). Since most transportation planning data have a geographic dimension or can be related to geographic features, a GIS can support all activities. The relative ease of data sharing, integration, data

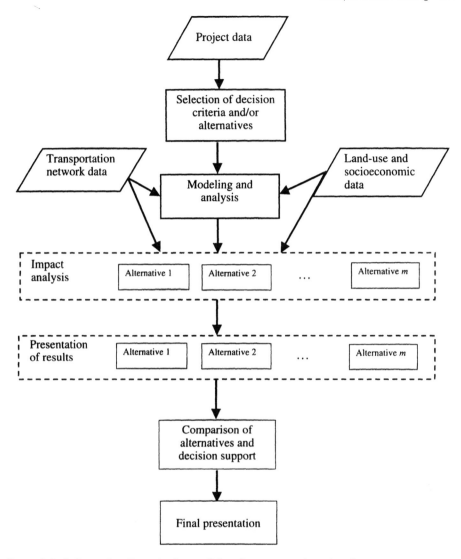

Figure 8-9: Information flows in the traditional transportation planning process

management and reporting encourages a more flexible structure where plans can be manipulated and revised at any point in the process. It is easier to involve other stakeholders and the public in technical decisions so they understand and provide input.

GIS is not a "magic bullet" for planning that guarantees only bright transportation futures. GIS can have negative effects on this process. The combination of sophisticated computer technology and powerful visual communication (and in the near future, virtual reality) can be very compelling. This could allow an analyst or planner to create an illusion of technical soundness for poorly conceived and constructed plans. Indeed, GIS can strengthen the rigid and linear

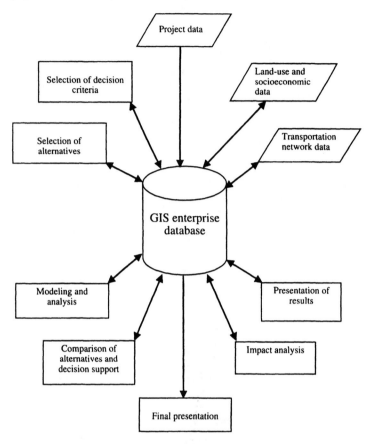

Figure 8-10: Information flows in the GIS-enhanced transportation planning process

planning process by giving the facade of a more flexible and democratic process. GIS only enables a nimble and responsive transportation planning; this must be actively pursued by the planning organization.

### Spatial decision support systems

As suggested by Figure 8-10, decision support plays a more central role in a flexible and democratic transportation planning process. A *spatial decision support system* (SDSS) is a set of linked technologies for generating and comparing alternative solutions to a geographic problem or alternative future geographies. This requires tools for spatial database management, spatial analysis and modeling, and communicating the alternatives to analysts, decision makers, and stakeholders. Consequently, GIS is a core technology (Armstrong et al. 1986). SDSS facilitate the MCDM process described in the section on corridor location modeling.

A SDSS can consist of several independent components, including separate or various software bundles of GIS, modeling and decision support tools as well as hardware platforms. The software components can be *loosely coupled* through simple file exchange mechanisms. *Tightly coupled* systems such as those achieved through OLE/COM and similar technologies share a common interface and database (Jankowski 1995; Nyerges 1993). OLE/COM can also support the development of user-configured special purpose software with small footprints (resource requirements). These can be constructed using visual programming language environments and software objects or components (independent objects) increasingly available in the software "toolbox" provided by GIS and other software vendors. Maximizing system performance may require a distributed or network-based computational platform such as client-server or Internet-based architectures. In chapter 2, we discussed some of the design principles involved in distributed geographic databases. In the next chapter, we will discuss client-server and Internet-based architectures within the context of intelligent transportation systems.

To date, many SDSS use a limited range of GVis techniques (see chapter 7), usually only the generic map products generated as standard output from an off-the-shelf GIS. This can diminish the effectiveness of the spatial decision support process (Armstrong et al. 1992). GVis techniques encompass a wide range of uses, including exploration, analysis, synthesis and presentation of geographic data (see MacEachren 1994b; MacEachren and Kraak 1997; and the discussion in chapter 7). All of these activities are relevant to the decision activities of identifying criteria, comparing solutions and conducting sensitivity analysis with decision parameters. Since standard GIS output functionality focuses mainly on presentation, it provides only limited support for the spatial decision process.

### Collaborative spatial decision support systems

As mentioned above, traditional planning typically involves a relatively small group of experts in the detailed decision process. The experts generate a small set of technically good alternatives for public review and comment. Discovering that their preferences are not in tune with public preferences at this late stage is usually an uncomfortable experience. A democratic transportation planning process requires input from as diverse a group of stakeholders as possible. However, this naturally increases the likelihood of conflicting objectives. Supporting the decision process in this case requires tools for facilitating negotiation and consensus building. *Collaborative SDSS* (CSDSS) are a set of integrated geographic information technologies that facilitate generating and comparing alternative solutions within a group environment.

CSDSS include the GVis and MCDM components discussed previously. However, the MCDM tools are now embedded with a group choice system that includes consensus-building technology and tools. As with SDSS, we can integrate off-the-shelf GIS software customized using a macro language or object-programming environment with MCMD software. Consensus building requires information technology for interactive geographic visualizations (e.g., a computer projector or a set of networked personal machines), information sharing, electronic voting, data and voice transmission, electronic whiteboards and computer conferencing (Jankowski et al. 1997; Nyerges et al. 1997).

The collaborative spatial decision making process extends the MCDM process described previously (figure 8-8). Geographic visualizations communicate solution or scenario properties to the group. Either openly or anonymously, participants can discuss and vote on the selection of criteria, criterion scores and preference weights or stepwise elimination based on attribute thresholds or rankings using structured group process techniques such as the Delphi or Nominal Group techniques. The system can display voting results either by person or group and rank selected alternatives based on these votes. Sensitivity analysis determines solution robustness and indicates which decision parameters should be considered and modified by the group (Jankowski et al. 1997).

Nyerges et al. (1997) provide extensive design guidelines for CSDSS in transportation decision making. They consider three representative transportation decision contexts, namely, an intergovernmental regional council, a local government and a public-private coalition. The conceptual model identifies the functional requirements for a *human-computer-human interaction* system to support these collaborative transportation decision processes. Nyerges et al. (1997) also provide practical experience with different information technologies and software for realizing this system.

## Conclusion

This chapter discussed the role of GISci and GIS in transportation planning. The objective of transportation planning is to guide development of a land-use/transportation system to achieve beneficial economic, social and environmental outcomes. Spatial analysis and GIS offer much to the transportation planning process, including spatial database management, geographic visualizations of possible scenarios and tools for processing geographic data into geographic information. GIS can also greatly enhance the quantity and quality of information flows among all components of transportation planning process, influencing decision making within this process.

This chapter mainly concerned support for tactical decision making and long-term, strategic planning of the entire transportation system. In the next chapter, we discuss geographic information technologies for operational or day-to-day transportation management within the context of *intelligent transportation systems* (ITS). ITS are integrated information technologies for monitoring and influencing a land-use/transportation system through direct control or indirect persuasion. ITS attempt to improve system efficiency through the use of advanced computing, real-time data sensor and communication technologies. The objective is to make transportation systems efficient, safer, and environmentally friendly. As we will see, ITS are very complex and require careful planning and configuration if they are to achieve these objectives.

# 9

# Intelligent Transportation Systems

In many parts of the world, mobility has vastly improved over the past century. Increasing mobility combined with urbanization and (in some cases) rapid population growth is creating undesirable negative impacts on economies and individual quality of life. In 1995, commuters in the United States spent more than 2 billion hours in traffic jams, generating US$100 billion in lost productivity (USDOT 1998a). In 1996, 41,907 people were killed and another 3.5 million were injured in the United States due to automobile accidents (USDOT 1999a). In Japan, over 10,000 people die in traffic accidents each year, about 5.6 billion person hours were spent in traffic congestion. Economic losses due to traffic congestion in urban areas amounted to 12 trillion yen per year (Japanese Ministry of Construction 1996). Traffic congestion also magnifies the impact of automobiles on ambient air quality since "stop-and-go" traffic generates more pollution than free-flow traffic.

Conventional approaches to tackling problems associated with transportation congestion attempt to increase transportation supply by building new highways and widening existing roads. Unfortunately, traffic congestion problems often occur again shortly after, if not before, a transportation project is completed. We cannot build our way out of congestion due to the transportation demand and changes in land use created by new transportation facilities. An alternative approach is *intelligent transportation systems* (ITS). ITS are integrated information technologies for monitoring and influencing a land-use/transportation system through direct control (e.g., traffic signals) or indirect persuasion (variable message signage, web pages). Rather than increase supply, ITS improves system efficiency through the use of advanced computing, real-time data sensor and communication technologies. The objective is to make transportation systems efficient, safer, and environmentally friendly.

This chapter provides an overview of intelligent transportation systems, particularly as they relate to geographic data, information, and GIS. We begin this chapter with a discussion of ITS development in three different national and pan-national settings (Japan, Europe, and the United States). We then identify and discuss the range of possible ITS user services. This sets the stage for a review of ITS architectures that logically configure the information and communication technologies and ITS services. We compare and contrast the three international settings mentioned above. Next we discuss the role of geographic information in ITS architectures, particularly with respect to location referencing and geographic data error. Finally, we provide examples of GIS-linked ITS services for in-vehicle navigation systems and Internet-based transportation information systems.

## ITS Development

Although integrated ITS are relatively recent, many ITS concepts could be traced back several decades (McQueen and McQueen 1999). Japan began its first period of ITS development in 1970s. The United States coined the term of *intelligent vehicle highway systems* (IVHS) in the 1980s. IVHS was renamed intelligent transportation systems in 1994 for inclusiveness with respect to multimodal transportation systems. In Europe, the theme was first known as *road transport informatics* (RTI) and was then renamed to *advanced transport telematics* (ATT).

ITS are large, complex, and often very idiosyncratic. To provide a context for discussions later in this chapter, the following sections describe ITS development efforts in Japan, Europe, and United States, respectively.

### ITS Development in Japan

Japan's research community started to explore advanced transportation technologies as far back as the 1960s (Nwagboso 1997). In 1973, the Japanese Ministry of International Trade and Industry initiated research on a comprehensive automobile traffic control system, including test deployment of a route guidance system (Japanese Ministry of Construction 1999b). During the 1980s, the Ministry of Construction began its work on *Road/Automobile Communication System* (RACS) and the National Police Agency initiated an *Advanced Mobile Traffic Information and Communication Systems* (AMTICS) project. These two projects were combined with communications standardization projects at the Ministry of Posts and Telecommunications to form the *Vehicle Information and Communication System* (VICS). In 1984, the *Highway Industry Development Organization* (HIDO) was established and supervised by the Ministry of Construction as a public service corporation to coordinate efforts of the government and private sector in improving roads. HIDO activities include research, development and marketing new technologies associated with roadways, including ITS.

Since the late 1980s, many ITS projects have been carried out by various ministries in Japan. These include *super smart vehicle systems* (SSVS) by Ministry of International Trade and Industry, *advanced safety vehicle* (ASV) by Ministry of Transport, *universal traffic management system* (UTMS) by National Police Agency, and contact-less card system for electronic toll collection system by Ministry of Posts and Telecommunications. In 1994, the *Vehicle, Road and Traffic Information Society* (VERTIS) was created as Japan's counterpart of the ITS America in United States and the *European Road Telematics Implementation Coordination Organization* (ERTICO) in Europe (see below). VERTIS involves the ministries and agencies above and is part of the Advanced Information and Telecommunication Society Promotion Headquarters headed by the Prime Minister of Japan to promote the use of information technology in society. VERTIS members include representatives from ITS-related organizations, industry, private business corporations, and academia. The main objectives of VERTIS are to further research and development of a broad range of ITS-related fields, the deployment of ITS and to exchange information with Europe and North America (Japanese Ministry of Construction 1999b).

In February 1995, Japanese government finalized the "Basic Guidelines on the Promotion of an Advanced Information and Telecommunications Society." In August 1995, the "Basic Government Guidelines of Advanced Information and Communications in the Fields of Roads, Traffic and Vehicles" was jointly prepared by the same five Japanese government agencies that sponsored the VERTIS to form a unified policy of ITS development. In July 1996, the five agencies compiled a "Comprehensive Plan for ITS in Japan" that represents a long-term vision of ITS development and implementation (Japanese Ministry of Construction 1996, 1999a). In the same year Japan allocated 59.6 billion yen for infrastructure improvements and for making ITS available for practical use, with additional 7.4 billion yen for ITS research and development.

Japan has an ambitious ITS deployment plan consisting of four phases (JHIDO 1999c). The first phase, called "Beginning of ITS," is expected to offer services of some leading ITS technologies such as in-vehicle navigation system and electronic toll collection by the year 2000. The second phase, "Traffic System Revolution," intends to achieve by 2005 the introduction of various ITS user services such as traffic management, assistance for safe driving, support for pedestrians, and support for public transport. The third phase, "Realization of A Dream," will start full-scale service of automated highway systems and establish a social and legal framework for ITS by 2010. After 2010, the "Maturity of ITS" phase will integrate ITS into a full-scale advanced information and telecommunication system through the deployment of a nationwide fiber optic network and innovative social systems to support these technologies.

## ITS Development in Europe

In mid 1970s, Germany took a lead role in the development of *Autofahrer Leit und Informations* (ALI) system. By the 1980s, automobile manufacturers began to envision the possibility of advanced transport systems that integrate electronics, computers, and telecommunications to reduce traffic congestion and to achieve more efficient and safer transportation (Nwagboso 1997). This resulted in an 8-year program known as the *Programme for a European Traffic with a Highest Efficiency and Unprecedented Safety* (PROMETHEUS), initiated in 1986. PROMETHEUS started with 15 European automobile manufacturers, but later on expanded to include academics and research institutes. Three subprograms were established under the PROMETHEUS. The PRO-CAR subprogram focused on in-vehicle systems such as collision avoidance and in-vehicle navigation systems. The PRO-ROAD subprogram researched vehicle-roadside communication systems and the PRO-NET subprogram concentrated on vehicle-vehicle communications. In 1987, PROMETHEUS became the European Union (EU)-backed EUREKA project. EUREKA is a pan-European cooperative research and development framework involving industry and research institutes to develop and exploit technologies crucial to improve quality of life and global competitiveness.

### DRIVE Program (1989-1991)

*Dedicated Road Infrastructure for Vehicle Safety in Europe* (DRIVE) was the first European Union (EU) research and development program for *road transport informatics* (RTI). RTI is the application of telematics for road transport, where "telematics" refers to the coupling of information and communication technologies (European Commission 1997a). DRIVE consisted of four working groups and 72 projects. Working Group 1 was responsible for selecting the most promising projects and deciding the best implementation approach. Working Group 2 focused on the behavioral aspects of telematics and traffic safety. Working Group 3 was charged with urban and interurban traffic control. Working Group 4 dealt with telecommunications and data flow for the *integrated road transport environment* (IRTE) and fleet operators.

DRIVE projects explored several telematics technologies. These included a test database for digital maps, automatic detection of traffic incidents by video image processing and prototypes for direct debiting of tolls and parking fees. Other projects were a pan-European trip planning system, an integrated fleet and freight management system, safety improvements for pedestrians at crossings and protocols for digital radio transmission (European Commission 1999c).

DRIVE also led to the establishment of *European Road Telematics Implementation Coordination Organization* (ERTICO, also known as ITS Europe) in 1991. ERTICO is a partnership of industry, users, transport and telecommunication operators, service providers and public authorities. Its mission is to support the creation of a successful pan-European market for ITS and to promote European ITS interests throughout the world. ERTICO therefore serves as the European counterpart of ITS-America in the United States and VERTIS in Japan.

### Advanced Transport Telematics (DRIVE II) Program (1992-1994)

The Advanced Transport Telematics (also known as DRIVE II) program was a follow-up to the earlier DRIVE program. A total of 57 projects under DRIVE

II contributed to seven areas of major operational interest, including demand management, travel and traffic information, integrated urban traffic management, integrated inter-urban management, driver assistance and cooperative driving, freight and fleet management, and public transport management (Nwagboso 1997).

*Transport Sector of the Telematics Applications Programme (1994–1998)*

The Telematics Applications Programme (TAP) is one of 19 specific programs that are supported under the European Union Fourth Framework Programme for *Research, Technological Development and Demonstration* (RTD&D). TAP focuses on societal applications of informatics and/or telecommunication technologies within different sectors of common interests. Of the twelve sectors under TAP, one is the Telematics Applications Programme-Transport Sector (TAP-T). The term *transport telematics* refers to telematics applications for all transportation modes and their interconnections (i.e., multimodal) (Keller 1999). A major difference between the TAP-T and the DRIVE and the DRIVE II programs is that TAP-T goes beyond road transport telematics to include other transport modes. TAP-T also puts a specific emphasis on intermodality of both passenger and freight transport.

The projects under TAP-T are categorized into six vertical areas of RTD&D and three horizontal areas that are related to common issues, common services, or EU policies (figure 9-1). The six vertical areas cover the transport modes of road, air, railway, and waterborne, plus traveler intermodality and freight intermodality. The three horizontal areas deal with the common issues (e.g., user needs, traffic safety, and standardization), telematics infrastructure and common services (e.g., system architecture, data exchange, and digital road map), and contribution to EU policies (e.g., demand management, integrated demonstrations, and deployment issues).

*European Community Fifth Framework Programme (1998–2002)*

The Fifth Framework Programme (FP5) sets up the priorities for the European Union's research, technological development and demonstration activities for the period of 1998–2002. FP5 is considerably different from its predecessors. It is a broadly conceived program intended to help solve problems and respond to the major socioeconomic challenges facing Europe.

FP5 consists of four thematic and three horizontal programs. The four thematic programs are "quality of life and management of living resources," "user-friendly information society," "competitive and sustainable growth" and "energy, environment, and sustainable development." The three cross-cutting, horizontal programs are "confirming the international role of community research," "promotion of innovation and encouragement of participation of small and medium-sized enterprises" and "improving human research potential and the socioeconomic knowledge base" (CORDIS 1999). There is no specific FP5

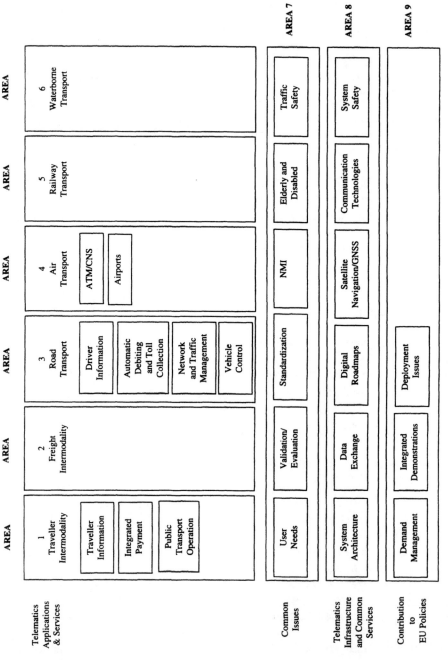

Figure 9-1: Concentration areas in the European Union's Telematics Applications Programme—Transport Sector (Keller 1999)

program that focuses on ITS although some projects are directly relevant to transport sector and ITS. For example, the Info-Mobility project investigates information-based transport that is aimed at compatible, interactive, and adaptive transport management systems within and across different transportation modes utilizing advanced control systems and communication facilities (European Commission 1999a).

### ITS Development in the United States

In 1986, the Institute of Transportation Studies at University of California, Berkeley, in collaboration with the California Department of Transportation (CalTrans) and other institutions and private industry, established the *Partners for Advanced Transit and Highway* (PATH) program. The mission of the PATH program is to apply advanced technology to improve highway capacity and safety, and to reduce air pollution, traffic congestion, and energy consumption. The PATH program was the first ITS research program in the United States.

In December 1991, the U.S. National ITS Program was authorized by the U.S. Congress as part of the *Intermodal Surface Transportation Efficiency Act* (ISTEA). The program was first known as the *Intelligent Vehicle-Highway Systems* (IVHS) program in the ISTEA legislation. The name was changed to *Intelligent Transportation Systems* (ITS) in 1994 to reflect its broader concerns beyond vehicles and highways. The U.S. Congress authorized the ITS program to explore "the use of advanced computer, communications, and sensor technologies to improve travel on highways and mass transit" (U.S. Congress 1995, p. vii). The goals of the ITS program in ISTEA include (U.S. Congress 1995, p. 3):

- enhance the capacity, efficiency, and safety of the federal-aid highway system and serve as an alternative to expanding the physical capacity of the highway systems
- enhance efforts to attain air quality goals established in the Clean Air Act
- improve safety on the highways
- develop and promote an IVHS industry in the United States
- reduce the societal, economic, and environmental costs of traffic congestion
- enhance U.S. competitiveness and productivity by improving the free flow of commerce and establish a significant U.S. presence in an emerging field of technology
- develop a technology base for IVHS, using the capabilities of national laboratories
- help transfer transportation technology from the national laboratories to the private sector

The U.S. Department of Transportation (USDOT) manages the National ITS program and has sponsored hundreds of ITS projects to research, develop, test, and deploy new technologies. In addition, the USDOT designated the *Intelligent Transportation Society of America* (ITS America, formerly known as IVHS America) as a federal advisory committee on the ITS program. ITS America includes members that cover state and local governments, public interest groups, universities and research institutes, motor vehicle manufacturers, commercial vehicle operators, and technology and consulting firms. It sponsors workshops and conferences for ITS participants to exchange ideas and experiences. It also produces reports and offers advice to the USDOT.

The ISTEA legislation of 1991 shifted the emphasis from highway construction to intermodal transportation management in the United States. The ITS program under ISTEA focused mainly on research and development projects. With the reauthorization of the National ITS program under the Transportation Equity Act for the 21st Century (TEA-21) passed by the U.S. Congress in 1998, the emphasis shifted towards ITS research and infrastructure deployment and integration (USDOT 1998b). TEA-21 also directs the U.S. Secretary of Transportation to establish a list of ITS standards that are critical to achieving national interoperability. In addition, TEA-21 requires ITS projects using Highway Trust Fund monies conform to applicable ITS standards/protocols and to the national ITS architecture.

According to the U.S. National ITS Program Plan, there are three stages of ITS deployments (Euler and Robertson 1995). The first stage, called the "Era of Travel Information and Fleet Management," covers the period of 1997–1999. This stage intended to achieve data sharing across all surface transportation modes between public agencies and private companies. With the development of shared travel information bases, states and metropolitan areas could integrate many ITS traffic, transit, safety, and commercial vehicle services. In addition, electronic clearance, automated vehicle identification, and weigh-in-motion systems would be operational on most major trucking corridors and international border crossings. Automobile manufacturers would also offer a variety of in-vehicle devices such as intelligent cruise control, autonomous route guidance system, and emergency mayday safety and security services. Electronic toll collection systems also would be deployed. Although many of the services envisioned in the National ITS Program Plan for the first deployment stage have been realized during the period of 1997–1999, it is evident that actual ITS deployment in the United States is slower and more limited than planned.

The second ITS deployment stage, "Era of Transportation Management," covers the period of 2000–2005. It is anticipated that the "smart traveler" will become a reality. More capable roadside-to-vehicle communications infrastructure will be implemented for real-time, adaptive traffic control and dynamic guidance systems. Universal electronic payment systems will be available for tolls, parking charges, transit fares, and other financial transactions. Smart vehicles equipped with advanced features such as lateral warning system, early collision avoidance system, and vehicle-to-vehicle communication system will be available on the market.

ITS deployment will move into its third stage, "Era of the Enhanced Vehicle," beyond the year of 2010. At this stage, vehicles will become even smarter with vision enhancement system, assisted braking and steering, and lateral and longitudinal space control. In other words, an *automated highway system* (AHS) will become a reality.

## ITS Applications

The three distinct ITS deployment strategies and experiences suggest a wide range of potential applications. These applications can be grouped into seven broad categories (U.S. Congress 1995):

- *Travel and transportation management* intends to keep highway traffic flowing smoothly and efficiently. These applications include technologies and tools for quick response to traffic accidents, controlling traffic signals, and providing information to travelers about transportation routes and services.
- *Travel demand management* attempts to reduce travel by single-occupancy vehicles. This include support for programs such as ride-matching and providing information about traffic conditions for pre-trip planning by travelers.
- *Public transportation operations* improve public transportation systems through technologies that monitor ridership demand and vehicle locations, provide en-route information to transit users, and enhance the safety of transit operations.
- *Electronic payment systems* allow travelers to pay for parking fees, tolls, and transit fares with "smart cards," wireless transmission, or other automated fee collection devices.
- *Commercial vehicle operations* streamline the commercial vehicle safety regulatory system and enhance its effectiveness in the trucking industry. The applications include systems such as electronic clearance, weigh-in-motion, automated roadside safety inspection, onboard safety monitoring systems, automated administrative processes, freight mobility systems, and hazardous materials incident response.
- *Emergency management* applications provide for quick notification of authorities and prompt response to emergencies through incident georeferencing, allocation, and routing of emergency response capabilities.
- *Advanced vehicle control and safety systems* support the development of automated highway systems through advanced control devices such as collision avoidance warning and automatic braking controls and to enhance safety through systems such as vision enhancement and pre-crash restraint deployment.

### Travel and transportation management

The fundamental goal of travel and transportation management is to accommodate more vehicles on existing highways through the use of advanced computing and communication technologies. This is more cost effective than expanding physical facilities: The U. S. Federal Highway Administration (FHWA) estimates $40 million to build but only $1 million to install electronic traffic surveillance system along one mile of freeway (U.S. Congress 1995).

Travel and transportation management applications can be further divided into two types, namely, *traffic management* and *traveler information systems*. Traffic management ITS applications normally involve a *traffic management center* (TMC). The TMC monitors the traffic flow through traffic sensors and video cameras installed on the roadways, identifies flow disruptions caused by traffic accidents or disabled vehicles and attempts to change the traffic pattern through traffic control signals, variable-message signage or broadcasting. The

*Advanced Driver and Vehicle Advisory Navigation Concept* (ADVANCE) Project in Chicago, for example, used vehicles as probes on highways to collect and transmit real-time traffic conditions to a TMC with traffic control devices.

Instead of direct control, traveler information systems provide timely information about routes, traffic congestion, and transit availability to travelers so they can make informed route choices. The TravTek Project, for example, was an early project involved 100 rental cars with in-vehicle navigation systems in Orlando, Florida. These rental cars can receive information about traffic congestion transmitted through digital data radio from a TMC.

### Travel demand management

Instead of enroute information, travel demand management ITS applications provide pre-trip information to support traveler's trip planning with respect to travel modes, destination choice and trip timing. The goal is to reduce traffic congestion and air pollution by making carpooling and transit more feasible and perhaps even attractive to commuters, especially during the peak hours. Methods for delivering information to travelers include phone, broadcasting, information kiosks and especially the Internet.

Several large U.S. cities have implemented traveler information systems. The SmarTraveler web site (http://www.smartraveler.com/), for example, includes real-time highway traffic and transit information for Boston, Chicago, Los Angeles, New York, San Francisco, Washington, DC, among others. The Boston web site, for example, provides both real-time closed circuit camera images as well as text description of traffic conditions. The web sites of Los Angeles and San Francisco, on the other hand, display the real-time traffic volumes at the locations of traffic monitoring devices installed along the major highways. Some web sites (e.g., San Francisco, Washington DC) also allow travelers to fill out information for ride-matching programs.

### Public transportation operations

ITS applications in public transportation operations help transit managers use resources more efficiently. The most common ITS application in public transportation operations is the *automatic vehicle location* (AVL) system. An AVL system allows the dispatch center to keep track of transit fleet locations, monitor their on-time performance, and reroute a vehicle if needed. Some public transportation ITS applications are more inclusive. For example, Orlando, Florida, USA operates a new high-tech bus known as Lymmo on an exclusive bus lane that is integrated with the traffic signal system. The Lymmo system provides passengers with up-to-the-minute information on bus arrival times through information boards and kiosk in Orlando. The Chicago Smart Intermodal System, sponsored by the Federal Transit Administration (FTA) and several departments of the City of Chicago, evaluates a Bus Service Management System that includes an AVL system, computer-aided dispatch and control, real-time passenger information signs, and adaptive signal timing.

### Electronic payment systems

Electronic fare collection can result in increased revenues for transit operations due to fewer evasions. Electronic toll collection can increase the highway capacity because vehicles do not stop at tollbooths and cause queues. An electronic payment system requires vehicles carry a tag that can be detected electronically. Early electronic payment devices were read-only; therefore, each user must set up an account with the toll collection authority. This raised concerns that the toll collection agency could monitor trips made by individuals and compromise privacy. Recent electronic payment systems have read-write capabilities that can update funds available on the user's tag; this does not require setting up an account in advance. These tags can be used for payments in multiple transportation services (e.g., tools, transit fares, and parking fees). The E-ZPass Interagency Group, for example, involves eight transportation agencies in New York, New Jersey, Pennsylvania, and Delaware. Together, they control 200 toll plazas, 1,500 miles of roadway, 4 tunnels, and 14 bridges, and collect 2/3 of total annual tolls collected in the United States (Gifford, Yermack, and Owens 1996). The Florida Turnpike has a read-write electronic toll collection system called SunPass.

### Commercial vehicle operations

This category of ITS applications is closely related to the private transportation industry. Carriers would like to improve the productivity of their fleets and drivers and provide better service to shippers. Many shipping companies already adopted technologies such as AVL systems to monitor their fleets and tracking systems to tell their customers the locations and the status of their shipments. ITS technologies have further improved the efficiency of commercial vehicle operations. For example, the *Advantage I-75 Corridor* project covers from Ontario, Canada to Florida, USA along the Interstate Highway 75 (I-75) corridor. Properly documented commercial vehicles equipped with transponders can travel along the I-75 corridor with minimal stopping at weight and inspection stations. This system uses electronic clearance, *weigh-in-motion* (WIM), electronic tag communications, and computerized credential checking technologies. The *Heavy vehicle Electronic License Plate* (HELP) Program extends from British Columbia, Canada along I-5 to California and then eastward along I-10 to Texas, USA. This program includes technologies such as *automatic vehicle identification* (AVI), *automatic vehicle classification* (AVC), and WIM.

The *Commercial Vehicle Information Systems and Networks* (CVISN) program is developing interoperable information systems between states in the United States. CVISN will support safety information exchange, credentials administration, and electronic screening among different jurisdictions. The safety information exchange will allow inspectors to automatically access current safety information on interstate and intrastate carriers, vehicles, and drivers. Credentials administration provides access to information such as carrier documentation, credential issuance, carrier registration, oversize/overweight, and hazardous materials. Roadside electronic screening will permit commercial vehicles to pass through inspection stations at highway speeds without stopping. California, for example, has the PrePass system that integrates AVI technologies, WIM sensors and transponders to electronically weigh and identify commercial vehicles and have their safety and State-required credentials checked as they approach weigh stations. In Section 5203(b)(6) of the Transportation Equity Act for the 21st Century (TEA-21), the U.S. Congress establishes a goal to complete deployment of CVISN in a majority of U.S. states by September 30, 2003.

### Emergency management

Emergency management ITS applications use advanced communications technologies to quickly notify authorities and deploy emergency services to a traffic incident site. One early test project was the *Puget Sound Help Me* (PUSHME) Mayday System implemented in northwest Washington state for emergency notification and response (U.S. Congress 1995). In recent years automobile manufacturers also started to offer drivers with emergency notification services. For example, the OnStar system installed in selected models of General Motors (GM) vehicles will automatically transmit a signal to the GM OnStar control center when the air bag is deployed or when the driver presses the OnStar button in the vehicle. The OnStar system combines the technologies of global positioning system (GPS), cellular phone, and digital street database to transmit an emergency signal along with the vehicle's location to the OnStar control center. Once an emergency signal is received by the control center, an OnStar advisor will call the cellular phone number to verify the situation and contact the appropriate agencies to dispatch emergency services to the vehicle's site. In addition to emergency services, the OnStar system also provides route guidance and other services such as roadside assistance, remote door unlock, and concierge services.

### Advanced vehicle control and safety systems

The main objective of *advanced vehicle control* (AVC) and safety systems is to prevent traffic accidents. AVC systems automatically detect impending accidents between a vehicle and surrounding traffic with sensing systems installed in the vehicle. A radar cruise control system, for example, can detect the speed of the vehicle ahead and automatically adjusts the speed to maintain a preselected time interval distance with the vehicle ahead. Once the vehicle ahead moves out of the traffic lane, the radar cruise control accelerates back up to the preset speed. Other sensing systems can detect another vehicle in the driver's blind spot and automatically sound an alarm to warn the driver when making a lane change.

The ultimate goal of advanced vehicle control and safety systems is to develop *automated highway systems* (AHS) that vehicles will be automatically guided in the traffic while drivers do not have to operate their vehicles. Operational tests of vehicle platoon have been successfully demonstrated on test tracks. Additional information of advanced vehicle control systems and AHS can be found on the web site of the PATH program at University of California, Berkeley.

## Marketing ITS: What services do travelers want?

With the wide possible configurations of applications within a given ITS implementation, an important but often overlooked question is: "Are we providing the types of ITS services that we want and need?" Los Angeles implemented a Smart Traveler *Automated Ridematching Service* (ARMS) as a touch-tone phone system to offer real-time ride-matching services for one time only ride-sharing and conventional carpool arrangements. There was no significant demand for this service (Giuliano, Hall, and Golob 1995). This type of mismatched ITS services highlights the importance of understanding the ITS market when we develop ITS strategic deployment plans.

Frayer and Kroot (1996) investigated the public perception and the likely acceptance of possible ITS initiatives in California. Among the nine ITS user services presented to the survey participants, only pre-trip traveler information systems seemed to meet users' needs. Services such as traveler information, ride-matching reservations, en-route driver information, and route guidance systems received mixed reactions. Hardly surprising in California, transit-related ITS services such as transit services information, en-route transit information, personalized public transit, public travel security, and intermodal electronic payment were poor in meeting user needs. Since this survey was based on the ITS technologies available then, there was no assessment of future ITS technologies. Nevertheless, it is a reminder that ITS also requires a supporting transportation system and therefore its applications must be tailored for the local setting.

## ITS architectures

In addition to the wide range of possible ITS application configurations, there are a large number of stakeholders involved in system development and deployment. These stakeholders include the public sector at all levels and often many private sector entities providing the technology required for the application bundle. Trying to put all of the pieces together can be challenging, particularly when integrating ITS across jurisdictions. For example, it can be difficult to reconcile different commercial vehicle electronic clearance systems across jurisdictions. Similarly, automobile manufacturers adopting incompatible vehicle-to-vehicle communication technologies can cause chaos in an automated highway system. In the case of ITS, system architectures are often national or pan-national enterprises to maximize compatibility across wide geographic areas. In the next few subsections, we will review emerging national and pan-national ITS architectures.

### National ITS Architecture in the United States

The United States started developing a national ITS architecture in September 1993. A 5,000-page document, *National ITS Architecture*, was completed in the summer of 1996. This document is available through the U.S. Federal Highway

Table 9-1: ITS User Services Under the U.S. National ITS Program

| User Services Bundle | User Services |
| --- | --- |
| Travel and transportation management | En-route driver information; route guidance; traveler services information; traffic control; incident management; emissions testing and mitigation; demand management and operations; pre-trip travel information; ride matching and reservation; highway rail intersection |
| Public transportation operations | Public transportation management; en-route transit information; personalized public transit; public travel security |
| Electronic payment | Electronic payment services |
| Commercial vehicle operations | Commercial vehicle electronic clearance; automated roadside safety inspection; on-board safety monitoring; commercial vehicle administration processes; hazardous materials incident response; freight mobility |
| Emergency management | Emergency notification and personal security; emergency vehicle management |
| Advanced vehicle control and safety systems | Longitudinal collision avoidance; lateral collision avoidance; intersection collision avoidance; vision enhancement for crash avoidance; safety readiness; pre-crash restraint deployment; automated highway system |
| Information management[a] | Data archiving |

Note: [a] This user service bundle and its user service were added to the U.S. National ITS Program in September 1998.
Source: http://www.odetics.com/itsarch.

Administration's ITS Electronic Document Library. The U.S. National ITS Architecture defines "the framework around which multiple design approaches can be developed, each one specifically tailored to meet the individual needs of the user, while maintaining the benefits of a common architecture" (Lockheed Martin Federal Systems 1997, p. 3).

The U.S. National ITS architecture consists of thirty *user services* grouped into six *user services bundles*. Table 9-1 describes these service bundles. This is an initial deployment set and is not exhaustive or final. In September 1998, a 31st user service, *archived data user service* (ADUS), was added to the list under a user services bundle called *Information Management*. The ADUS provides the Historical Data Archive Repositories and controls the archiving functionality for all ITS data (ITS America 1999).

The U.S. National ITS architecture includes logical architecture, physical architecture, and communication requirements. *Logical architecture* defines the processes (i.e., activities or functions) required to deliver ITS user services and the information exchanges between the processes. For example, "collecting traffic information" is a function that is required to deliver ITS user services such as traffic control or pre-trip travel information. The traffic information collected through this function is passed on to another function such as "processing traffic information" to support particular ITS user services. The logical architecture treats processes in a top-down fashion beginning with general processes

Intelligent Transportation Systems 309

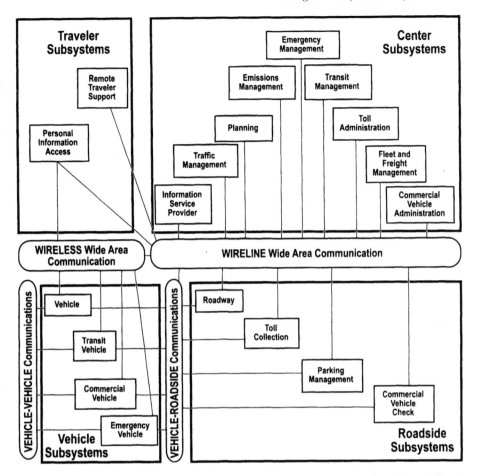

Figure 9-2: ITS architecture subsystems and communication elements (Lockheed Martin Federal Systems 1998)

(e.g., "Manage Traffic") that are then decomposed into more-detailed processes (e.g., "Provide Traffic Surveillance," "Monitor HOV Lane Use"). The National ITS Architecture also distinguishes the elements that are inside the ITS architecture or external to the ITS architecture. For example, vehicles are considered external to the ITS architecture, but the equipment used in a vehicle to support ITS user services is considered inside the ITS architecture. This distinction helps define the domain of ITS functions.

A deployed intelligent transportation system will consist of distinct but interoperating physical systems and subsystems. The *physical architecture* of the National ITS Architecture partitions the functions defined in the logical architecture into four systems and nineteen subsystems. See figure 9-2. The 19 subsystems are intended to capture the expected and likely subsystem boundaries for the present and the next 20 years (Lockheed Martin Federal Systems 1998).

They are grouped into four types of systems, namely, traveler system, center system, roadside system, and vehicle system.

The *traveler system* includes two subsystems for ITS functions related to travelers or carriers in support of multimodal traveling. These two subsystems are "remote traveler support" at a fixed location (e.g., kiosks for traveler information at public locations) and "personal information access" through home or portable computers for traveler information and emergency requests. The *center system* consists of nine subsystems that are not required to be on or adjacent to roadways. In other words, they can be located anywhere. This group of ITS subsystems is often implemented at TMCs and communicates with other subsystems through wide area network (WAN) wireline communications. The four subsystems comprising the *roadside system* require roadside locations for deployment of sensors, signals, programmable signs, or other interfaces with travelers and vehicles. These subsystems generally need wireline communications with center subsystems and short-range wireless communications with vehicles passing the roadside location where a roadside subsystem is deployed. *Vehicle system* covers the subsystems installed in a vehicle. Communication needs include one-way or two-way wide area network wireless communications to the center subsystems, vehicle-to-roadside communications for functions such as electronic toll collection and vehicle-to-vehicle communications in an automated highway system.

During the process of developing the U.S. National ITS Architecture, it appeared that some ITS user services were too broad in scope to be useful for planning ITS deployments. To support ITS analyses and deployment, ITS subsystems are further broken down into *equipment packages* with specific functional attributes (Lockheed Martin Federal Systems 1997). These equipment packages represent the smallest units of ITS that can be deployed. For example, transit vehicle tracking, transit fixed-route operations, demand response transit operations, transit passenger and fare management, transit security, transit maintenance, and multimodal coordination are separate equipment packages for transit management. In deployments, the character of a particular subsystem will be determined by the specific equipment packages chosen for the particular subsystem. For example, a bus transit authority may choose transit vehicle tracking, transit fixed-route operations, and demand response transit operations for its transit management system. A commuter rail authority may select transit vehicle tracking, transit fixed-route operations, and transit passenger and fare as the software packages in its transit management system deployment.

As mentioned above, various kinds of communication systems are needed between the subsystems defined in the U.S. National ITS Architecture. Wide area network (WAN) *wireline* communication elements can be implemented physically as fiber, coaxial, twisted-pair, or microwave networks between two fixed locations. WAN *wireless* communication elements are similar to the WAN wireline communication elements but do not require physical connections between two locations (e.g., cellular phone-based systems). These technologies are suitable for ITS services that disseminate information to users who require seamless coverage. *Short-range wireless* communications supports localized information transfer. The "vehicle-roadside communications" and the "vehicle-vehicle communications" shown in figure 9-2 represent two different types of short-range wireless communications between the ITS subsystems. Vehicle-to-vehicle short-range wireless communications are critical for ITS user services such as collision avoidance and Automated Highway Systems (AHS). Vehicle-vehicle communications differ from roadside-vehicle communications: The former handle communications between mobile units (i.e., moving vehicles) and the latter deal with communications between a fixed location (i.e., roadside unit) and mobile units.

*Dedicated short-range communications* (DSRC) are one-way or two-way communication channels that provide direct communication paths between vehicles (e.g., toll tags) and roadside equipment (e.g., beacons). Toll tags mounted in vehicles can be one-way (read only) or two-way (read-write) systems. A one-way tag system simply transmits the tag identification number to a roadside subsystem. A two-way tag system allows a roadside subsystem to communicate with the tag. For example, roadside subsystems at a toll both can transmit beacons to the passing vehicles and directly deduct tolls from a two-way tag mounted in the passing vehicle. A fundamental difference between DSRC and a WAN wireless network is that the DSRC-based ITS subsystem is out of touch with the ITS infrastructure most of the time. Beacon communication therefore is often used for ITS subsystems that require localized exchange of data such as toll and parking payment operations and roadside checking of commercial vehicle operation.

Another critical requirement of the U.S. National ITS is to ensure the ability of communicating and sharing information within and across geographic and jurisdictional boundaries. For example, an emergency mayday system installed in a vehicle should work across the entire nation. The U.S. National ITS Architecture identifies 12 key standards areas. These include:

   i. dedicated short range communications (DSRC)
   ii. personal and hazardous materials Mayday systems
   iii. intercenter data exchange for commercial vehicle operations
   iv. emergency management centers to other centers
   v. traffic management subsystems to other centers
   vi. traffic management subsystems to roadway devices
   vii. highway-rail interface
   viii. information service provider (ISP) to centers except emergency management
   ix. ISP wireless interfaces
   x. digital map data exchange and location referencing
   xi. signal priority for transit and emergency vehicles
   xii. transit management center to transit vehicle

These standards are needed to ensure national interoperability of ITS. The Transportation Equity Act for the 21st Century (TEA-21) of 1998 requires the Secretary of the USDOT to identify standards that are critical to ensuring national interoperability or critical to the development of other standards (USDOT 1999b).

TEA-21 identifies two types of critical ITS standards. Critical standards that are required to ensure national interoperability are *national standards*. Standards that are needed for the development of other critical standards are designated as *foundation standards*. For example, location-referencing standards are foundation since they are to develop national standards for various ITS user services. We will discuss location-referencing and related standards later in this chapter.

Shuman and Soloman (1996) discuss the standards and policy challenges of global interoperability for ITS. The International Standards Organization (ISO) *Technical Committee 204* (TC 204) has established many working groups that deal with standards ranging from electronic toll collection, traffic management, advanced vehicle control systems, to digital map databases. Of course, ITS standards alone will not guarantee a national or global interoperability of ITS. Other factors such as institutional issues and performance of information exchange could present problems to national interoperability (USDOT 1999c). For instance, an agency may insist on issuing its own electronic toll tags that are not accepted by other agencies. While methods are available for sharing this across administrative boundaries, these methods may not have the speed, accuracy, or reliability at a level to ensure interoperability.

## System architecture for ITS in Japan

Development of ITS system architecture in Japan is a joint effort of five government agencies, the National Police Agency, Ministry of International Trade and Industry, Ministry of Transport, Ministry of Posts and Telecommunications and Ministry of Construction. The "Comprehensive Plan for ITS in Japan" published in 1996 provided a long term vision of ITS development and implementation and identified a total of twenty ITS user services that were organized into nine ITS development areas (table 9-2). "Utilization of advanced information enabled in the advanced information and telecommunication society" was later on added to the list as the 21st user service (Japanese Ministry of Construction 1999a).

A comparison of table 9-2 with table 9-1 shows a significant overlap in the ITS user services between the two nations. Some differences exist due to the different transportation environments in the two nations. For example, Japan identifies an ITS development area of "support for pedestrians" that is not present among the U.S. ITS user service bundles due to the scarcity of American pedestrians.

As more ITS services were implemented, the Japanese government recognized the need to develop an architecture that could support an integrated and extensible ITS. This involves three major objectives (Japanese Ministry of Construction 1999a). The first objective is efficient construction of an integrated ITS that shares information and functions and interoperates among its constituent systems. The second objective is to maintain system expandability such that new user services and functions can be added in the future. The third objective is to promote domestic and international ITS standards through the identification of priority items to be standardized.

Development and deployment of ITS architecture in Japan involves four main stages (Japanese Ministry of Construction 1999a and 1999b). The first stage is *detailed definitions of user services*. This stage subdivides the 21 ITS *user services* into 56 *specific user services* and 172 *specific user sub-services*. Each sub-service has a detailed definition that describes its purpose and functions. For example, the "provision of route guidance traffic information" user service consists of four

**Table 9-2:** ITS User Services Under Japan's ITS program

| Development Areas | User Services |
|---|---|
| Advances in navigation systems | Provision of route guidance/traffic information; provision of destination-related information |
| Electronic toll collection systems | Electronic toll collection |
| Assistance for safe driving | Provision of driving and road conditions information; danger warning; assistance for driving; automated highway systems |
| Optimization of traffic management | Optimization of traffic flow; provision of traffic restriction information on incident management |
| Increasing efficiency in road management | Improvement of maintenance operations; management of special permitted commercial vehicles; provision of roadway hazard information |
| Support for public transport | Provision of public transport information; assistance for public transport operations and operations management |
| Increasing efficiency in commercial vehicle operations | Assistance for commercial vehicle operations management; automated platooning of commercial vehicles |
| Support for pedestrians | Pedestrian route guidance; vehicle-pedestrian accident avoidance |
| Support for emergency vehicle operations | Automatic emergency notification; route guidance for emergency vehicles and support for relief activities |

*Note:* "Utilization of advanced information enabled in the advanced information and telecommunication society" was later on added to the list as the 21st user service.
*Source:* Japanese Ministry of Construction 1996.

specific user services (i.e., "provision of route guidance information to drivers," "provision of information on other modes to drivers," "advanced provision of route guidance information" and "advanced provision of information on other modes"). The "provision of route guidance information to drivers" specific user service, in turn, is subdivided into five sub-services, namely, "provide optimum route information," "provide road traffic information," "provide required travel time when congested," "guide along the selected route," and "exchange information between running vehicles." A complete list of the specific user services and subservices is available at the web site of Highway Industry Development Organization (HIDO) of Japan.

The second stage, *development of the logical architecture*, creates an "information model" that clarifies the relationship between all information handled by ITS and a "control model" for each ITS sub-service that clarifies the relationship between information and functions needed to provide each ITS sub-service. The information model organizes information into a class hierarchy. Figure 9-3 provides an example. Moving body is a superclass that encompasses Human, Vehicle, and Goods subclasses. Vehicle in turn is a superclass of Basic vehicle, Transit vehicle, Commercial vehicle, and Emergency vehicle. An information model also represents the relationship between the generic items. For example, Moving body and Roadway have the relationship Move (i.e., a moving body moves along a roadway).

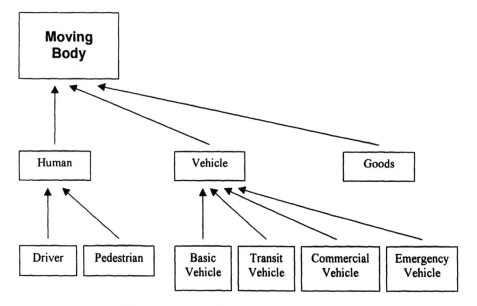

Figure 9-3: An example information model for Japan's ITS logical architecture (Japanese Ministry of Construction 1999a)

The *control model* specifies the information required to carry out functions that provide a particular ITS sub-service. For example, the ITS sub-service of "providing optimum route information" needs to collect information from driver (e.g., trip destination and route preference), vehicle (e.g., vehicle type), and other data sources (e.g., road and traffic information). These data become the input to a routing function that will find the optimum route and pass the result to the driver who requested the sub-service.

The third stage, *development of the physical architecture*, is to facilitate identification of functions and information that are shared among ITS user services. Four clusters of subsystems (i.e., human, center, vehicle, and roadside) are defined in the "subsystem interconnection diagram" in figure 9-4. These four clusters of subsystems are very similar to the U.S. National ITS Architecture (i.e., traveler, center, vehicle, and roadside in figure 9-2). The Japan's diagram also includes the "outside" elements that are external to the ITS. At the level of individual subsystems, there are differences between the U.S. ITS physical architecture and the Japanese ITS physical architecture. While the United States identifies 19 subsystems, Japan lists a total of 24 subsystems. Other important differences include, for example, the identification of "human interface" and "sensing" subsystems for human, vehicle, and roadside clusters in the Japanese version. This suggests that Japan's ITS architecture deals with sensing systems and human factors more explicitly than the U.S. National ITS architecture.

The last stage, *clarification of the candidate areas for standardization*, evaluates each subsystem and information transmission between subsystems. It then assigns scores based on the overall ITS system configuration clarified by the physical architecture. A subsystem or a communication cluster that is shared more frequently

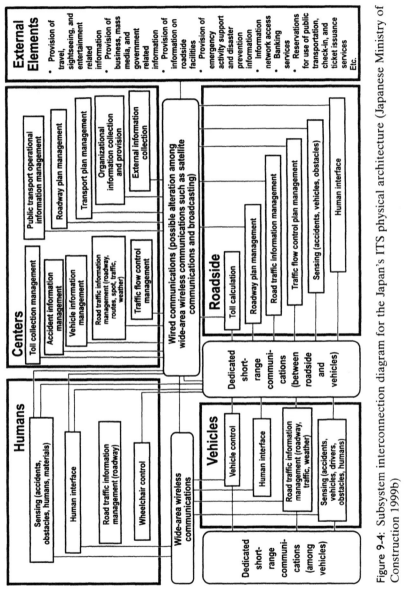

Figure 9.4: Subsystem interconnection diagram for the Japan's ITS physical architecture (Japanese Ministry of Construction 1999b)

among various ITS sub-services and is used frequently will get a higher score. In other words, they are good candidates for standardization.

## ITS architecture in Europe

Europe provides a contrasting case relative to the Japan and especially the United States, illustrating that ITS applications and architectures must be tailored for specific geographic and political contexts. For example, a report by ITS Focus evaluates the U.S. National ITS architecture and made recommendations for the United Kingdom (ITS Focus 1997). Suggested modifications of the U.S. National ITS Architecture include greater use of dedicated short-range communication (DSRC), a more attractive solution for ITS services due to the higher population densities and smaller geographic size of the United Kingdom. A more integrated set of wireless communications and a more complex deregulated transit environment creates both opportunities and challenges in the United Kingdom that are not present in the United States. A major criticism of the U.S. National ITS Architecture is that it is too encompassing and yet at the same time too permissive and does not provide specifications at a sufficient level of detail.

The comments by ITS Focus indicate contrasting ITS environments between the United States and Europe. Instead of a top-down approach where government and consortia develop an overarching (and arguably vague) design, Europe's experience is more of a "bottom up" design process where specific architectures developed independently for given user service bundles and domains that must be subsequently integrated. For example, the SOCRATES (*System of Cellular Radio for Traffic Efficiency and Safety*) consortium began developing an ITS architecture in 1989. By the mid-1990, SOCRATES was launching a set of ITS applications such as dynamic route guidance, driver information, and fleet management using GSM (*Global System for Mobile Communication*) digital cellular mobile telephone communications with support of trans-European data exchange. The GERDTEN project in the DRIVE program attempted to develop an architecture prototype for interurban applications of Advanced Transport Telematics (ATT) between 1991 and 1995 (Miles 1996).

Based on the results from the DRIVE I and II programs, the CONVERGE project provided the horizontal coordination of system architecture development across all projects in the Transport Telematics Applications Program (T-TAP). CONVERGE follows a decentralized approach due to the European context in which local requirements and legal and political frameworks require a flexible and modifiable architecture with central coordination to support system integration and serve as a stimulus for the European ATT market (Jesty et al. 1998).

The CONVERGE project guidelines identify three levels of ITS architecture (Jesty et al. 1998). Level 3 Architecture addresses the need for interoperability between autonomous authorities. Level 2 Architecture concerns the integration of functions and subfunctions within a single authority. Level 1 Architecture defines the overall structure of a system and the relationships among the subsystems.

Level 1 consists of four separate architectures. *Functional architecture* describes the ITS functions and subfunctions, data flows between the functions, and the main databases. For example, "data monitoring" could include pollution monitoring, weather monitoring, and speed monitoring as subfunctions. *Information architecture* describes the data needed by ITS and their relationships. *Physical architecture* deals with the grouping of the ITS functions into physical units or market packages as well as the communication paths between them. *Communication architecture* addresses the data flows between the physical units. This focuses on the physical transmission of information between spatially separated subsystems through the specifications such as the type of communication medium (wire, radio, infraraed, etc.), physical characteristics of data flow (volume, speed, encoding techniques, etc.), and logical characteristics of data flow (latency, information composition, etc.).

In 1997, the European Commission published a general strategy and framework for the deployment of *road transport telematics* (RTT) to avoid a fragmented system in Europe. Five priority areas were proposed for initial deployment during the period 1997–1999. These include *traffic information services, electronic fee collection, transport data exchange/information management, human machine interface*, and *system architecture*. The system architecture is expected to ensure interoperability between different RTT elements and applications as well as to identify priorities for standardization in the European Union. In addition to the five priority applications in first deployment phase, the EC also specified an "open list" of other priority RTT applications that will be revised as additional priorities become clear. This open list initially covers six other RTT applications, namely: (i) pre-trip and on-trip information and guidance; (ii) interurban and urban traffic management, operation, and control; (iii) other urban transport telematics services; (iv) collective transport; (v) advanced vehicle safety/control systems; and (vi) commercial vehicle operations (European Commission 1997a, 1997b).

Under the Telematics Application Programme of the EU Fourth Framework Programme, a two-year *Keystone Architecture Required for European Networks* (KAREN) project was launched in 1998 to develop a Framework Architecture for the deployment of ITS in Europe (European Commission 1999b). The KAREN project is designed to carry out the complete process that starts from the identification of requirements, through the development of a comprehensive European Transport Telematics Architecture Framework, to the consensus and endorsement of the Framework. This project is also expected to make recommendations for standardization and future actions of telematics in the transport sector of Europe. Participants in the KAREN project include public authorities and private companies from the member states of the European Union.

## ITS architectures and geographic information

### Overview

A critical aspect of the ITS architectures reviewed above is the need to communicate information among different subsystems. This communication requires transmitting geographic information based on an unambiguous location reference. For many ITS user services, these unambiguous location references must take place in real-time or near real-time (see Dane and Rizos 1998).

Several automobile manufacturers now offer an emergency notification system that automatically transmits the location of a vehicle to an emergency center when the driver presses an emergency button in the vehicle or when the vehicle is involved in an accident. The location reference transmitted to the emergency center is frequently in longitude and latitude coordinates derived from the GPS satellites. The coordinates must be matched against a street map database such that emergency vehicles can be dispatched to the location of the incident. The challenge here is to ensure that an accurate and unambiguous location is identified in the GIS street map database to pinpoint at the incident location.

Another scenario is a smart vehicle traveling from New York City to Los Angeles. To use ITS services, this smart vehicle must communicate with various ITS subsystems and understand the location references for traffic information and route guidance information in a multitude of jurisdictions. It is possible that different map databases and various location referencing methods could be used by different jurisdictions to support various ITS user services. We must ensure that location references can be exchanged and interpreted accurately and unambiguously among the various ITS services along the travel route.

### ITS requirements for geographic information

In 1993, the U.S. Oak Ridge National Laboratory (ORNL) at Oak Ridge, Tennessee, USA started a *Nationwide Map Database and Location Referencing System* project for the U.S. Federal Highway Administration (FHWA). The project identifies geographic data requirements at multiple levels for supporting ITS user services (Goodwin, Xiong, and Gordon 1994). Table 9-3 summarizes the geographic information requirements for these services. "Digital geographic data" refers to the need for one or more georeferenced thematic data including network data. As table 9-3 suggests, there is often a need for multiple location referencing systems within the same user service.

In addition to varying requirements for location referencing, ITS requires communication of these references using a heterogeneous communication system. Figure 9-2 provided communication linkages among technological components of the U.S. National ITS architecture. Figure 9-5 provides another view of the ITS communication system; this shows the information flows among different stakeholders involved an ITS architecture. Clearly, a major challenge of ITS is unambiguous capture and transmission of location referencing information.

The ITS geographic information requirements would suggest a central role for GIS. Unfortunately, these fields have evolved separately. ITS is often viewed as an engineering field that requires significant infrastructure built around information, sensing, and telecommunication technologies. GIS-T is viewed as an information technology that helps manage, analyze and visualize transportation data. A FHWA report from 1995 found "communication between the ITS and GIS communities to be practically non-existent." (UT-EERC

Table 9-3: Geographic Information Requirements for Some ITS User Services

| ITS User Service | Digital Geographic Data | Arc ID | Coordinates | Address | Linear Location Reference |
|---|---|---|---|---|---|
| Travel and transportation management | | | | | |
| En-route driver information | Y | Y | Y | Y | |
| Route guidance | Y | Y | Y | Y | |
| Traveler Information Services | Y | Y | Y | Y | |
| Traffic control | Y | | | | |
| Incident management | Y | | | | |
| Travel demand management | | | | | |
| Pre-trip travel information | Y | | | Y | |
| Ride-matching | Y | | | Y | |
| Demand management and operations | Y | Y | Y | | |
| Public transportation | | | | | |
| Public transportation management | Y | Y | Y | | Y |
| En-route transit information | Y | Y | Y | | Y |
| Personalized public transit | Y | Y | Y | Y | |
| Public transit security | Y | Y | Y | Y | |
| Commercial vehicle operations | | | | | |
| Commercial vehicle administration | Y | | Y | | |
| Hazardous material incidence response | Y | Y | Y | Y | |
| Commercial fleet management | Y | Y | Y | | Y |
| Emergency management | | | | | |
| Emergency notification | Y | | Y | | |
| Emergency vehicle management | Y | Y | Y | Y | Y |
| Advanced vehicle safety systems | | | | | |
| Longitudinal collision avoidance | Y | | | | |
| Lateral collision avoidance | Y | | | | |
| Intersection collision avoidance | Y | Y | Y | | |
| Safety readiness | Y | Y | Y | | Y |
| Automated highway systems | Y | Y | Y | | |

*Source*: After UT-EERC et al. 1995.

et al. 1995, p. 7). The report further recommends that designers of the U.S. National ITS architecture "recognize that ITS spatial data handling does not exist independently of the rest of the spatial data handling world; with respect to spatial data, ITS is a subset of GIS-T" (p. 47). Fortunately, this situation is changing. With the recently growing recognition of shared interests in spatially referenced transportation data, we are seeing increasing closer interaction between the GIS-T and ITS communities. The challenge of communicating location referencing messages accurately and unambiguously is not unique to ITS. In fact, this challenge is part of a broader challenge, known as *spatial data interoperability*. See chapters 2 and 3 for discussions.

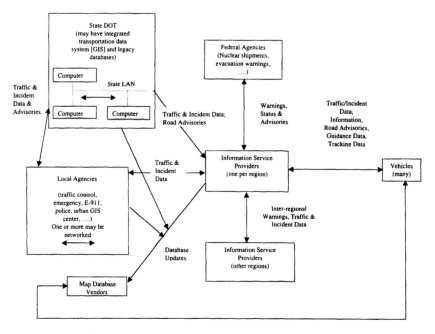

Figure 9-5: Information flows among ITS stakeholders (Goodwin 1996)

### Location referencing for ITS

When different location referencing methods are used in various ITS user services and across jurisdictions, how can we ensure that a location reference can be communicated accurately and unambiguously? First, the location referencing message being transmitted between different ITS components must be received and interpreted correctly. Secondly, the interpreted location referencing message must be translated successfully into a location in the geographic database at the receiving end. The accuracy of the identified location in the map database is a function of the quality of the map database.

Many different communication technologies could be employed to transmit location referencing messages between ITS subsystems (Lockheed Martin Federal Systems 1997). Two-way wide area wireless communication technologies include Global System for Mobile Communications (GSM), Special Mobile Radio (SMR), Personal Communications System (PCS), Cellular Digital Packet Data (CDPD), satellite communications, among others. One-way broadcast communications include AM subcarrier, FM subcarrier, and Highway Advisory Radio (HAR). Wireline communications could use private networks or public networks. Examples include Ethernet, Fiber Distributed Data Interface (FDDI), Asynchronous Transfer Mode (ATM), leased analog/digital lines, Integrated Services Digital Network (ISDN), and the Internet. For wireless short-range communications, there are two major types: vehicle-to-vehicle communication and Dedicated Short Range Communication (DSRC) (Polydoros and Panagiotou 1997). Some of the above communication channels have limited bandwidth. Communication channels therefore have a direct impact on the amount of data that can be effectively transmitted in location referencing messages. Standards for ITS communications have been evaluated at both the national and the international levels. Detailed discussion on this topic is beyond the scope of this book. See Polydoros and Panagiotou (1997) for a review of communication technologies and ITS. Additional information can be accessed at the U.S. National ITS Architecture home page and the ITS America Standards home page. We will focus our discussion on issues related to unambiguous and accurate location referencing between different digital geographic databases.

As part of the Nationwide Map Database and Location Referencing System project for the U.S. Federal Highway Administration, Oak Ridge National Laboratory evaluated the requirements of ITS applications for spatial data and location referencing. Gordon et al. (1994) review the status of spatial databases in both public and private sectors and conclude that no existing single spatial database is appropriate for national ITS deployment. In a follow-up study, Goodwin et al. (1994) evaluate the functional requirements of national map databases for ITS. They conclude that a common national location referencing system is neither necessary nor practical for most ITS applications. They suggest the development of application community-based standards within a national-level standard interoperability protocol framework. In other words, each application community (e.g., ATIS, CVO, or ATMS) defines the location referencing standards that meet the requirements of its functional and spatial domains. These community-based standards together provide inputs to a national interoperability protocol framework.

Location referencing interoperability in the U.S. National ITS architecture is further refined in another report that recommends the development of an interoperability framework for ITS spatial data (UT-EERC et al. 1995). Under the ITS interoperability framework, the study specifically recommended the development of an *ITS Datum* and the development of a national standard *Location Reference Message Protocol* (LRMP). The discussion below is based on various documents on LRMP and ITS Datum available at the ORNL Spatial Data Interoperability Project web site.

*ITS Datum*

In the United States, the ITS Datum system establishes a nationwide network of ground control points that will anchor spatial references across different map databases and serve as a skeleton map database for ITS product development (Siegel, Goodwin, and Gordon 1996a, 1996b). The system consists of two major components, namely, the ITS Geodetic Datum and the ITS Datum Network (UT-EERC et al. 1995).

*ITS Geodetic Datum* is a standard geodetic datum specification for ITS based on the World Geodetic System 1984 (WGS84) and an elevation datum. We discuss some valuable properties of the WGS 84 geoid in chapter 4. First, it supports elevation as well as latitude and longitude measurements. Second, it is geocentered and therefore can support the Global Positioning System (GPS). Third, it is a world standard and therefore supports integration of international data. Fourth, it is compatible with the common North American mapping datum, NAD-83.

The *ITS Datum Network* will include a node table and an arc table. The node table will include information on unique node ID, node latitude, node longitude, node elevation, and a list of the IDs of other nodes that are connected to the given node. The arc table will consist of information on unique arc ID and arc length. Arc IDs are a combination of "from node ID" and "to node ID." This network scheme is non-planar because it does not require a planar enforcement when a new datum point is inserted on an arc (e.g., a point on an overpass). An ITS Datum therefore is not a topological base map. ITS Datums only provide information about logical arcs (without geometric information) between datum nodes (Goodwin et al. 1998).

The density of the ITS Datum node network will be determined by the interoperability requirements. The ITS Datum prototype proposed by the ORNL is based on the roadway intersections (about 50,000 nodes) in the National Highway Planning Network (NHPN) database (see chapter 4). This prototype could provide an initial national level framework for location referencing. Denser node networks that include roadway intersections at regional or local levels can be appended to the national framework as needed.

An important issue is the geographic placement of ITS Datum nodes. These typically are placed at street and highway intersections. Figure 9-6 provides different road intersection layouts with possible placements for ITS Datum nodes. One possibility is to use the intersection of roadway centerlines (figure 9-6a). Alternatively, ITS Datum nodes can be placed at either *gore points* (i.e., small square symbol in figure 9-6b) or decision points (i.e., small circle symbol in figure 9-6b). The location of a gore point could change when a roadway is realigned or repaved. This can present problems for the ITS Datum. The decision points, where a turn decision must be made, sometimes are difficult to locate (e.g., decision point in figure 9-6b). The ITS Datum prototype proposed by ORNL uses intersections of roadway centerlines and decision points. Figure 9-6c illustrates that multiple ITS Datum nodes may be needed when detailed geometry of an interchange is represented in a GIS map database.

Another issue concerns accuracy requirements of the ITS datum network for ITS user services. Since the datum network is based on the U.S. National Highway Planning Network, accuracy is limited to the 1:100,000 scale of that network. This results in an approximately 80-meter radius of uncertainty around each node. This is not sufficient accuracy for many ITS services. This accuracy can be improved through the use of GPS (Siegel, Goodwin, and Gordon 1996a, 1996b).

The ITS Datum project is closely related to some other ongoing activities, especially the National Cooperative Highway Research Program's (NCHRP) linear referencing data model and the National Spatial Data Infrastructure (NSDI) Framework Road Data Model. We discussed the NCHRP linear referencing data model in chapter 3. ORNL suggests using the ITS Datum nodes as anchor points in the NCHRP generic linear referencing model (Goodwin et al. 1998). The NSDI framework develops a conceptual data model standard for identifying road segments as unique geo-spatial features that are independent of any cartographic or analytic network representation (FGDC 1999). These road segments will form the basis for maintaining the NSDI framework road data

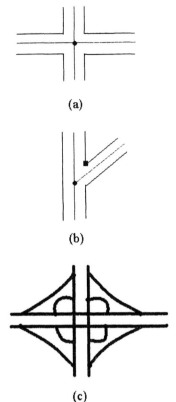

Figure 9-6: Placement of ITS Datum nodes at roadway intersections

and for establishing relationships between road segments and attribute data. The proposed framework is compatible with the NCHRP linear referencing data model; therefore, it is also related to the ITS Datum proposed by the ORNL.

These concurrent ongoing activities bring up an interesting question: What is the distinction between ITS Datum, GIS-T linear referencing data model, and NSDI Framework Road Data Model? Hickman (as quoted in Goodwin et al. 1998) commented at the ITS Datum Public Sector Requirements Workshop held in Knoxville, Tennessee in March, 1998.

> For ITS, the main role of the linear datum are to support database interoperability and real-time location message exchange. For GIS-T, the roles may be to support accurate dynamic application of road inventory attributes and linear event management. For NSDI transportation framework, a linear datum may support data improvements through conflation and transactions, and the representation of transportation features at multiple levels of detail. (p. 16)

Fletcher et al. (1998) suggest that transportation activities and data are divided into three functionally and institutionally separate domains—*transportation infrastructure management, civilian operations,* and *military operations.* Each domain collects and maintains separate (often redundant and sometimes inconsistent) information while little attention is given to the flow of information between the three domains. In a call for a unified linear reference system, Goodwin et al. (1998) argue that a unified linear datum will provide for the unambiguous transfer of location-based data both within and among the domains. They also argue that the overwhelming use of the unified linear datum will be for ITS applications. Therefore, a unified linear data model should be incorporated into the ITS Linear Reference Standard.

### Location reference message profile

ORNL also proposes a national standard Location Referencing Message Protocol that will provide a framework for developing standard location reference message formats. The *Location Referencing Message Specification (LRMS) Information Report* (SAE J2374) being developed by the Society of Automotive Engineers (SAE) is a result of the ORNL's proposal. The objective of the LRMS is to develop a set of standard interfaces for transmitting location references among different ITS components (USDOT 1999b, 1999c; Goodwin et al. 1996, 1999).

The LRMS is not a standard by itself because it does not define how the message transmissions are implemented. Instead, the LRMS is an agreement between applications to transfer location reference messages using well-documented public-domain formats (ORNL 1999b). Specifically, LRMS interfaces define the semantics (i.e., standard meaning) of the content of location reference messages and the syntax (i.e., public domain standard formats) for communicating location reference messages to application software. Since locations can be referenced in various ways, the LRMS Information Report consists of a family of interfaces (officially known as *profiles*). They include (USDOT 1999c):

- *Geometry Profile* for referencing locations defined by fundamental spatial objects such as points, nodes, links, and polygons.
- *Geographic Coordinate Profile* for referencing locations that are based on geographic coordinates of latitude, longitude, and altitude referenced to a geodetic datum.
- *Grid Profile* for referencing locations that are based on rectangular coordinate systems such as UTM or State Plane Coordinate Systems. This profile is intended mainly for applications with limited communication bandwidth.
- *Linear Referencing Profile* for linear referencing systems that reference locations as offsets from known locations on network links.
- *Cross Streets Profile* for referencing locations by intersecting street names.
- *Address Profile* for referencing locations by street addresses.
- *MDI (Model Deployment Initiatives) Profile* supports link and offset referencing and coordinate-based referencing using global coordinates and local offsets from reference nodes of various types. It also supports short references with respect to a locally defined grid.

Each of these profiles is based on a common location referencing method to meet the unique requirements of particular user communities. For example,

**Table 9-4:** Location Referencing Message Specification (LRMS)—Geographic Coordinate Profile, Globally Referenced Point Format Record

| Bit | Content | Values/Range |
|---|---|---|
| 0–3 | Start Code | 0010 |
| 4 | Unused | |
| 5–7 | Type | 000 = Globally Referenced Point |
| | | 001 = Globally Referenced Link |
| | | 010 = Locally Referenced Point |
| | | 011 = Locally Referenced Link |
| 8–10 | Horizontal Datum | 000 = WGS-84 EGM-96 |
| | | 001–111 = Other |
| 11–13 | Vertical Datum | 000 = NAVD88 |
| | | 001–111 = Other |
| 14 | Name Info Flag | 0 = No name info |
| | | 1 = Use name info |
| 15 | Street Name/Index Flag | 0 = Use street name |
| | | 1 = Use street index |
| 16–47 | Longitude | +/− 180,000,000 microdegrees |
| 48–79 | Latitude | +/− 90,000,000 microdegrees |
| 80–95 | Altitude | −8,191 ≥+57,344 decimeters |
| 96–103 | Street Name/Index Byte Count | Integer number of bytes of name data or street index (0–255) |
| 104–variable | Street Name | ASCII characters of street name |
| 104–variable | Street Index | Integer |

*Source:* From ORNL 1999b.

the geographic coordinate profile is more suitable for a user community with an extensive use of GPS, while many GIS-T applications need to employ the linear referencing profile. The profiles listed above are not exhaustive. It is expected that new profiles will be added to the LRMS list as needs arise. Since some user communities have more urgent needs than other communities on the development of location referencing standards, certain profiles have been developed by standards development organizations (e.g., SAE) further than other profiles.

Each LRMS profile consists of multiple *format records*. For example, the Geographic Coordinates Profile has five different format records: Globally Referenced Point, Globally Referenced Link, Locally Referenced Point, Locally Referenced Link, and Node Attribute (ORNL 1999a). Table 9-4 provides an example of bit-by-bit coding of the Globally Referenced Point Format Record under the Geographic Coordinates Profile.

*Location referencing error and ITS*

As discussed in chapter 4, several types of errors are common in digital geographic databases. Positional disagreements between different databases are inevitable due to errors associated with measurement methods and map generalization at different scales. Differences in map scale also introduce errors of inclusion/exclusion. For example, ramps at a freeway interchange may be represented on a large scale map while excluded on a small scale map. This implies different topological relationships in the street network between databases with different scales. Disagreements on street segment attributes (e.g., missing data, spellings of street names, alias street names, address ranges, and functional classification) also make data interoperability more difficult (Noronha 2000; Noronha and Goodchild 2000).

*Positional error*  Many ITS user services require coordinate location referencing, often in conjunction with other referencing systems (see table 9-3). Positional error is therefore a critical ITS concern. Goodchild and Hunter (1997) develop a *vector field* approach to evaluating positional distortion. Let the true location of a point be denoted by $(x, y)$ and its position recorded in a geographic database be denoted by $(x', y')$. We can represent the positional accuracy of this point through the vector $e(x, y)$. A vector is an ordered set of numbers (one for each spatial dimension) that together shows a direction and magnitude in space from a fixed location. We calculate the error vector by solving the following equations:

$$x' = x + e_x(x, y) \tag{9-1}$$

$$y' = y + e_y(x, y) \tag{9-2}$$

where:

$e_x$ = element of the vector indicating magnitude of error along the $x$ axis

$e_y$ = element of the vector indicating magnitude of error along the $y$ axis

Solving these equations for sample locations generates a vector field showing the distribution of positional error across space. We can use this information to adjust the position of each point in a database by a magnitude and direction defined by its neighborhood in the vector field. This is similar to the "rubber sheet" operation in GIS, where we adjust the underlying geometry by mathematically "stretching" and "shrinking" the space.

A vector field representation of positional error has direct implications for ITS (Church et al. 1998). For example, an ITS user in a vehicle using a digital map database developed by Vendor A receives a location reference message from a traffic management center (TMC) using a digital map database of Vendor B. Due to the likely positional disagreements between the two databases, we need a method to correctly match the position in one database with its corresponding position in another database. If each digital map database vendor includes the coordinates of ITS Datum control points in its product, it is then possible to establish a series of vector field such that the corresponding positions can be matched between the two databases.

The Vehicle Intelligence and Transportation Analysis Laboratory (VITAL) of the University of California, Santa Barbara tested a suite of proposed location reference messaging standards for the ITS industry using positional error measurement methods. Of the six street network databases for the County of Santa Barbara, California, one database was the TIGER/Line files developed by the U.S. Bureau of Census and another was from a local engineering firm with a claimed positional accuracy of 0.6 meters, 55% of the time. The other four databases are commercial products that are widely used in GIS-T and ITS applications in the United States.

Applying the vector field analysis to the six map databases indicates that the magnitude of displacement vector varies from less than 1 meter to more than 100 meters (Church et al. 1998). Positional differences of up to 50 meters were common (VITAL 1997). Road alignments in the test map databases differ as much as 100 meters in urban areas and 200 meters on winding mountain roads. In general, straight sections on major streets tend to have less error while sinuous roads in hilly suburban neighborhoods experience large positional displacements. These errors can cause serious problem for ITS user services that require positional and network location referencing (e.g., vehicle location and routing applications). A rubbersheeting algorithm can be used to adjust geometry of street network databases in real time using the ITS Datum (VITAL 1999a).

VITAL also tested the linear referencing profile (LRP) and the cross streets profile (XSP) among the seven LRMS profiles. The VITAL testing illustrates not only the performance of the United States ITS standards but also methods for evaluating ITS location referencing and spatial data error in general. The sections below will focus on the test findings of these two profiles.

*Cross-streets profile* A location message is a stream of digital data to describe a location. As mentioned earlier, the *cross-streets profile* (XSP) is a proposed U.S. national standard for communications between traffic management centers (TMCs), information service providers (ISPs) and vehicles (Noronha and Goodchild 2000). The original XSP specifies a point location in three components (VITAL 1998):

- the "on-street" on which the point is located, for example, Cumberland Ave. in figure 9-7
- the "from cross-street" and the "to cross-street," for example, NE 16th Street and NE 17th Street in figure 9-7, respectively
- the offset distance measured either as an absolute distance or as a proportion of the street segment between the "from cross-street" and the specific point

For example, the point in figure 9-7 can be described as "300 feet from NE 16th Street" or "60% between NE 16th Street and the NE 17th Street along Cumberland Avenue." To communicate a cross streets location referencing message from one ITS application to another with a different geographic database, the receiving system must unambiguously find the on-street, from-cross and to-cross street names in the receiving database and then must accurately locate the distance offset along the on-street.

VITAL performed separate tests of the two XSP requirements on four of the six digital map databases mentioned above (VITAL 1998). Approximately one third of the street records in the map databases have missing or blank names. Since XSP requires all three street names be identified in the receiving database, the missing street names in the test map databases will cause a high failure rate of XSP transmissions. Also, most freeways and ramps do not have associated names. Unfortunately, freeways and ramps are important elements in many

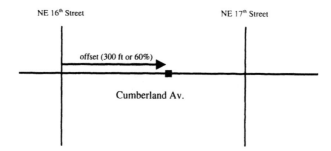

Figure 9-7: The LRMS cross-streets profile (after VITAL 1998)

ITS applications. For the database records with street names, the various ways of parsing and spelling street name, street prefix or suffix, and street type by different database vendors also reduce the success rate of finding matched names.

Complex topology among transportation features also complicates the street name-matching task. For example, figure 9-8 illustrates cases where a street can intersect with another street more than once (a, b, c) or can form a loop (d). In any of these cases, XSP could be ambiguous in identifying the crossing streets. Other possible topological complications exist. One situation could be database B consists of more streets than database A; therefore, a third crossing street could be present in database B between the two adjacent crossing streets identified in database A. Alternatively, wrong topology may be present in a database due to digitizing error. All of the above situations could cause a failure of using XSP to communicate a location reference between two map databases.

VITAL (1998) used data points collected from both a field survey and a laboratory test in the Phase I test of XSP. Their field survey used differential GPS (DGPS) to collect about 60 points within a 50-mile radius of the City of Santa Barbara. The laboratory test generated about 10,000 points at 1-km intervals along street centerlines in the test map databases. Test results indicate the hit rate for non-blank street names is in the range of 5–70%. Due to the high percentage of blank street names in the databases, the overall hit rate drops down to 1–40%. This success rate is lower than the requirement of most ITS applications, especially those mission-critical applications such as emergency services. This suggests that position coordinates should be included as part of the XSP to resolve ambiguities in some instances.

In Phase II, VITAL tested the revised cross streets profile (XSP2) that includes coordinates in addition to street names (SAE 1998). The XSP2 does not dictate whether the coordinates or street names should take precedence in the event of conflict. VITAL (1998) therefore conducted two sets of tests. One set of tests used coordinates as the data for locating a position and cross street names as tie-breakers. The second set of tests processed cross street names in increasingly sophisticated ways and reconciled ambiguous cases with assistance

Intelligent Transportation Systems    329

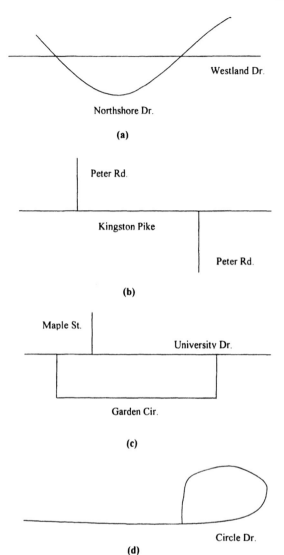

Figure 9-8: Ambiguous street intersections (after VITAL 1998)

from coordinates. These methods include (see VITAL 1998 and Noronha 2000 for details):

i. fuzzy name matching strategies
ii. allowing blank street names under certain conditions
iii. handling inconsistent topology due to the occurrence of intervening streets between the "from-cross street" and the "to-cross streets" in the destination database
iv. separating major streets from other streets to evaluate their respective hit rates
v. development of an intelligent algorithm to minimize the blank street names problem

Scores were assigned to each matching instance depending on the level of success of matching the triad of street names (i.e., on-street, from-cross street, and to-cross street) between the source database and the target database. Each outcome was classified into "likely," "possible," or "unlikely" categories. The first series of tests (i.e., use of coordinates) produce a 11% "likely" success rate and 61% "possible" success rate for County of Santa Barbara. Within the City of Santa Barbara, a 24% "likely" hit rate and a 26% "possible" hit rate were achieved. In the second series of test that use both street names and coordinates in the matching process, about 50–85% name matching cases were successful for all types of streets while major streets achieved 10–60% success rates. Most of the matches (40–96% on all streets and 50–97% on major streets) were located within 30 meters of the source point.

It is difficult to draw a definite conclusion from the XSP evaluation since the XSP does not specify how a message is processed at the receiving end; this plays a critical factor in the success of XSP (Noronha 2000). Nevertheless, the relatively low success rates of XSP transfers in the evaluation do not meet the requirement of the ITS industry, especially for applications such as emergency services. The low success rates of XSP transfers are mainly due to the quality of map databases rather than the XSP per se.

Three solutions can improve the performance of XSP (VITAL 1998; Noronha 2000). The ideal long-term solution is to resurvey the national street network to uniform quality standard. This approach has challenges that are both technical (e.g., finding a uniform quality standard that meets the needs of all stakeholders at a justifiable cost) and administrative (e.g., coordination between federal, state, local government agencies and the private sector). Two short-term approaches are database standardization and the ITS Datum. If all map database vendors would develop their products according to standards on street naming and attribute data, success rate of XSP transfers could improve significantly. Alternatively, the ITS Datum could provide an evolutionary framework for a high-quality national database. The ITS Datum would consist of a set of precisely surveyed control points that include street intersections. Disagreements between different map databases could be documented and error models could be developed (e.g., the vector field approach discussed above; also see Funk et al. 1998; Church et al. 1998). This would make corrections between map databases possible either statically or in real time for ITS applications.

A theoretically 100% successful protocol is the *location expression exchange* (LXX). This protocol, named LX-100, mimics the negotiation process that is often part of a verbal communication describing a location (Noronha and Goodchild 2000). In other words, the receiver processes the location referencing information passed by a sender and decides if it is sufficient to identify the location unambiguously and accurately. If the location referencing information is insufficient, the receiver will request additional information from the sender until the location can be unambiguously and accurately determined. This approach is similar to the concept of Intersection Location (ILOC) in the European standards development initiative of Extensive Validation of Identification Concepts in Europe (EVIDENCE). The success of LX-100 however will still depend on the quality of map databases.

*Linear referencing profile* Linear referencing is a widely used location referencing method in GIS-T and ITS applications (see chapter 3). The linear referencing profile (LRP) expresses a location in terms of offset distances measured from a fixed reference point along a route. A point event (e.g., railroad crossings) requires only a single offset while a linear event (e.g., median type) requires begin and end offsets. Offset distances can be expressed as either absolute distance or relative distance (i.e., percentage of the route length). The route for offset distance measurement is identified either by its name, an road index, or a logical arc reference defined by its begin and end node IDs. When different map databases are used, the geometry and length of a route are likely to vary between the databases due to differences in measurement, map scale and generalization. This can create errors in communicating linear referencing information.

In transportation surveying and engineering, a vehicle mounted *distance measurement instrument* (DMI) usually provides the offset distance measurement required for linear referencing. ITS applications, on the other hand, frequently use two-dimensional coordinate systems, often captured through GPS. As a consequence, conversion between 1-D linear referencing data and 2-D coordinates is often necessary when communicating location references among ITS applications. Table 9-5 provides five classes of error that may result in location exchanges using LRP (VITAL 1999b). Class I error (*DMI error*) results from the technology, calibration, and operation of distance measurement instrument. Class II error (*GPS error*) relates to the errors embedded in GPS data (see GPS section in chapter 4 for an overview). Class III error (*linear transformation error*) occurs when 2-D coordinates are transformed into a linear reference. Class IV error (*LRP transfer error*) occurs when a LRP message is transmitted from database A to database B and the location in database A cannot be accurately identified in database B. Class V error (*reverse transformation error*) takes place during a transformation of a linear reference to 2-D coordinates for purposes such as mapping in a GIS database.

VITAL developed three tests to evaluate the different error sources in LRP exchanges (see VITAL 1999b for detail). A sample of 15 stretches of road was selected from the six test map databases discussed above. The sample included straight urban road stretches, some minor suburban roads, and winding mountain roads. Field surveys of the roads used both Differential GPS (DGPS) and a vehicle-mounted DMI with a resolution level of 0.1 meter.

The first set of tests showed that DMI measurements were very close (i.e., within 0.1–0.5% or about 10 meters on the sampled road stretches) to the digital representations of road centerlines collected by field DGPS data and the test geographic database developed for engineering. Other test databases experienced larger discrepancies (i.e., about 5–15% or 100–150 meters error). Inability to defining the start and end points of a road stretch was found to be the greatest impedance to accurate measurement of route length. Poor representations of freeway ramps in the test databases also made length measurements based on the digital databases error prone. The test results suggested that differences of up to 10 meters in length measurements were inevitable in field surveys.

The second test set evaluated the likelihood that GPS coordinates could be matched to wrong streets and the magnitude of distance offset error when GPS

Table 9-5: Principal Classes of Error in Location Exchange Using LRP

|  | Determine Position | Express as LRP Relative to Database A | Transfer to Database B | Interpret Position as Required |
|---|---|---|---|---|
| GIS-T | DMI (m) Class I error | — | Class IV error | — |
| ITS | GPS (long, lat) Class II error | Class III error | Class IV error | Class V error |

Source: After VITAL 1999b.

coordinates are matched to the correct streets. Test results suggest accurate GPS readings would produce 90% success rate in identifying the correct streets and offset error would be in the 30–40-meter range. With GPS readings of +/–100 meters, the success rate dropped down to 58% and average offset error would be 135 meters.

The third test set examined the transfer of linear references between geographic databases. The sample points were transferred from each of the six map databases to each of the other five databases. The results suggest an average offset error of 50 meters.

The above discussions indicate the importance of GIS databases to ITS deployment. Since many ITS applications rely on digital GIS databases for locational referencing, the quality of GIS databases directly impact the success of correct and unambiguous identification of locations in ITS applications. Bespalko et al. (1996) point out that ITS require a wide variety information to deploy powerful and effective applications. Spatial data is an essential component of the information infrastructure. Unfortunately, development of spatial data in general falls outside the scope of ITS. Even though the ITS Datum and the LRMS are potential solutions to overcome the spatial interoperability issues, it is evident that these proposed solutions alone will not take care of all of the interoperability issues without additional actions to address the GIS database representation and quality issues. The next generation GIS need to be able to manage both historical as well as dynamic data at a level that meets the requirements of ITS applications (Bespalko et al. 1996).

## Integrating GIS and ITS: Some examples

Although there are still challenges to make GIS and ITS interoperate across the entire spectrum, there are some successful applications of GIS in several types of ITS user services. This section describes two ITS user services that employ GIS technology, namely, in-vehicle navigation systems and Internet-based transportation information systems.

### In-vehicle navigation systems

An in-vehicle navigation system provides driving guidance along the route from a trip origin to a trip destination. This service requires three critical components.

The first is determining the location of a vehicle and matching it to a location in a digital geographic database.

A conventional method for locating a vehicle or other moving objects is *dead reckoning*. Dead reckoning updates location over time based on a previous location, the distance traveled from the prior location, and the compass heading of the travel since the previous location. A problem with dead reckoning is that errors cumulate from one position to the next position and could eventually become too big to be useful in fixing a location. Map matching is one approach to taking care of the accumulated error from dead reckoning (White 1991). Through a comparison of travel path derived from dead reckoning with a digital map, the cumulative error in dead reckoning path can be identified and eliminated. With the availability and reduced cost of global positioning system, most in-vehicle navigation systems now include a GPS receiver to fix locations.

The second critical component of in-vehicle navigation systems is to provide route guidance to the user. This component involves the use of several GIS functions. First of all, the trip origin and the trip destination must be identified in the digital map database. Geocoding functions that find a map co-ordinate location through matching street addresses or cross street names, are frequently employed. Once the trip ends are identified, we need to apply a minimum cost path algorithm to find a route from the origin to the destination on the given street network. This step requires a vector GIS map database that includes topology (i.e., connectivity) of the street network and conjunction with a shortest path tree algorithm (see chapter 5). Multiple route attributes such as travel time, monetary expense and risk can be combined using methods discussed in the previous chapter to derive a pseudo cost for input to the routing algorithm.

The third critical component of in-vehicle navigation system is to effectively communicate the route guidance information to the user in real-time, often when the user is operating the vehicle. Since safety is a major concern, many vendors employ a map display with voice instructions to minimize the potential distraction of the driver. In the case that the driver fails to follow the route guidance offered by the system, the system must recompute the best route to the destination and provide guidance based on the updated route. This requires high-quality digital geographic data. Quality issues include, for example, positional accuracy, correct representations of physical restrictions (e.g., no "U-turns" on a highway with barriers along the median) and operational restrictions (e.g., one-way streets, prohibited left turn), and current and correct street address data (see Newcomb et al. 1993; Bennett 1994).

U.S. Department of Transportation sponsored a *TravTek* (Travel Technology) project in Orlando, Florida between 1990 and 1996. This project equipped 100 rental cars owned by Avis with in-vehicle navigation devices that can receive real-time traffic congestion information, via digital data radio broadcast, collected at a Traffic Management Center (TMC). A series of studies were conducted, including a rental user study of 4,354 drivers and a local user study of 53 local residents. Data collected from this project was used to conduct a series of evaluation of the technology (USDOT 1996).

Japan has been very active in the development of in-vehicle navigation and route guidance system. For example, Mercedes-Benz has developed an *Intelligent*

*Traffic Guidance System* (ITGS) that is available to travelers on main roads in several Japanese cities (JHIDO 1999a). ITGS is an interactive communication system that uses digital 800 MHz (9,600 bps) mobile or car phone to establish a link between the in-vehicle navigation system and traffic information center operated by the Advanced Traffic Information Service (ATIS) Corporation. The system offers dynamic route guidance using the latest traffic information. ITGS is available with monthly fees on selected models of Mercedes-Benz vehicles.

Toyota also started to offer its *Mobile Network* (MONET) service, a real-time travel information services to moving vehicles, in November 1997 (JHIDO 1999a). The MONET system provides a screen display plus voice guidance information to the driver. It includes not only traffic conditions en route to the destination, but also information services such as guidance to the facilities and amusements along the travel route, e-mail, news, weather forecast, etc. In case that the vehicle breaks down or the driver needs an emergency service, the MONET service can automatically dial a Toyota service center, hospital, or police station.

Japan has proposed a Smartway 2001 project that will create an Intelligent TRaffic System (ITRS) (JHIDO 1999b). The ITRS will consist of three advanced technologies, namely, *Smartway* (an intelligent road system), smart cars, and smart gateways (intelligent communications portals) that negotiate the flow of information between Smartway and smart cars. The Smartway component is envisioned as a road or highway that enables a wide range of information to be exchanged among all of its users (including drivers, cars, and pedestrians) and create a platform for a variety of ITS services. It contains the fiber optic cables that form the backbone of ITS and serves as the infrastructure that puts the "intelligence" in "Intelligent Traffic System."

### Internet-based GIS applications for ITS

The Internet started to gain its popularity in the early 1990s. One major impact of the Internet on GIS is the concept of *distributive geographic information* (Plewe 1997). With the use of the Internet, geographic information is no longer confined to standalone computers or computers on a local area network (LAN). It is now possible to share geographic information and GIS applications across all computers that are on the Internet.

For many ITS user services, one key task is to make the relevant information accessible to users in a timely fashion such that better decisions can be made (e.g., minimizing waiting time for transit services, and avoiding congested highway segments). As discussed above, many different communication technologies could be employed to deliver relevant traffic and transportation services information to users. The Internet is a convenient and low cost alternative of delivering some ITS user services, especially those in the "Travel and Transportation Management" bundle (see Table 9.1 for a list of ITS user services).

There are different strategies for distributing geographic information over the Internet (Plewe 1997). Methods range from simple *data download* and *static map display* to more advanced *dynamic map display* and *interactive GIS query and analysis*. Data download can be achieved by placing GIS data files on a server that can be downloaded to a client computer (i.e., computer at the user end) through protocols such as FTP (File Transfer Protocol). A protocol is a language accepted by the users to communicate between computers. Examples include HTTP (Hypertext Transfer Protocol) for the WWW, TCP/IP (Transfer Control Protocol/Internet Protocol) for communication across computer networks, and FTP for file transfers. The data download approach is not very useful for real-time ITS applications since the files must be processed by GIS software installed on the client computer before any meaningful information can be generated.

One critical component of any Internet-based GIS applications is the handling of maps over the Internet. The simplest approach is to display static maps on the Internet. This involves the following steps. First, the system converts cartographic output from GIS software into an Internet-compatible raster graphic format such as GIF (Graphics Interchange Format) or JPEG (Joint Photographic Experts Group). This raster graphic image then is incorporated into a web page for display. Users can view the map but cannot request different map displays. Only people who are responsible for the maintenance of the web page can make updates of the map display. The static map display approach has been used on some web pages to display, for example, maps of fixed transit routes or highway construction sites.

Static map display is of limited use for most ITS applications due to the dynamic nature of ITS data. ITS therefore need to provide users with the capability of requesting a map display of the current operational status of a transportation system. A dynamic map display allows users to assemble the map parameters on a client and send the request to a web server. The request then is passed to GIS software running on the server to generate a map according to the user-specified parameters. The map is sent back to the client for display. Web sites that allow users to zoom in/out and to pan around a map display are basic examples of dynamic map display.

Advanced Internet-based GIS applications include query and analysis capabilities such that a wide range of possible maps could be generated and displayed over the Internet. This approach will require both a user interface on the client for users to formulate their query and analysis requests as well as a GIS interface program on the server to execute the requests and pass the results (maps, tables, texts) back to the client.

The architecture underlying Internet-based GIS is the *client/server model* on which the WWW as well as most LANs are based. Different computers on a network may host different kinds of data and software. A computer (the client) can request data or that certain tasks be performed on another computer (the server) on the network. The server processes the request from a client and returns the requested information back to the client (see figure 9-9).

A *heavy server/thin client* implementation conducts most of the data processing on the server with the client only handles the function of a web browser (Plewe 1997). Advantages of this implementation strategy include centralized data and software servers that can be easier and cheaper to maintain than a distributed system. It also does not require much processing power on the client computers, meaning that it can be implemented more widely. There are also disadvantages. A heavy concentration of workload on the server could significantly

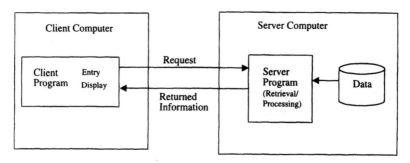

Figure 9-9: Client/server model on a network (after Plewe 1997)

slow down the performance when it is hit by many simultaneous requests. Some web sites overcome this problem by implementing multiple servers that can handle user requests concurrently. Heavy server/thin client implementation also means that clients must request information from the server for almost all data processing tasks. This can significantly increase the network traffic and the response time on the client computer.

An alternative is to distribute more workload on the client and minimize the load on the server. This strategy is known as *light server/thick client* implementation. For Internet-based GIS applications, this requires extending the capabilities of web clients for handling geographic information and map interaction operations (e.g., pan, zoom, turn on/off map layers), additional tools must be developed to work with web browsers. Common approaches to expanding the capabilities of a web browser include, for example, the use of Java applets or plug-ins between client and server computers for your specific Internet-based GIS application.

Java, developed by Sun Microsystems, is a platform-independent programming language that can run under different operating systems. A Java applet is a Java program that can be automatically downloaded to a client when the web page is hit by the client. Since Java is much more versatile than a web browser, an applet can be programmed to perform various map interaction operations (e.g., drag a rectangular box to zoom in on a map) on client computers. Java applets reside on the client for a single session only. The next time a client hits the same web site, Java applets must be downloaded again. This means it involves an overhead time to load a web page with large Java scripts. A plug-in is a program that is registered with a web browser for handling particular types of information. Unlike Java applet, a plug-in must be downloaded and installed on a client computer.

Both Java applet and browser plug-in can extend map interaction capabilities to handle vector GIS data. Internet-based GIS applications that use vector data rather than raster images such as GIF and JPEG not only result in smaller files to be transmitted over the Internet but also allow the manipulation of individual elements on a map display.

Given the brief review of Internet-based GIS concepts, we will now discuss some specific Internet GIS applications for ITS. The examples below are simply to illustrate how certain ITS user services could be delivered over the Internet using GIS technology. For additional examples, readers are encouraged to browse the WWW.

### ROMANSE project

ROad MANagement System for Europe (ROMANSE) project started as a pilot project in 1992 to make better use of existing infrastructure and resources as a flexible solution to increasing traffic congestion. The ROMANSE project is a consortium of partners from both the public and private sectors. The objective is to use advanced transport telematics to provide accurate, timely, and accessible information to travelers both before and during their journeys. With the information, people can make better informed decisions on time of journey, mode of travel, and route to travel to avoid congestion and maximize the efficiency of the existing transport systems.

Hampshire County and City of Southampton, United Kingdom were selected as the pilot project implementation site. The online ROMANSE provides the latest up-to-the-minute information over the Internet for the following ITS services.

- *Car parks selection.* This service shows the number of parking spaces available at different car parks. The information is updated every 5 minutes from the sensors installed at the car parks. A map on the web page displays color-coded car parks to show if a car park is full, nearly full, $1/2$ to $3/4$ full, or less than $1/2$ full. Users also can click a zone displayed on the map to retrieve a table of the actual number of parking spaces in that zone.
- *TRIPlanner.* This service provides details of journeys by public transport through the United Kingdom. Users will enter the start and end points of a journey (one of the trip ends must be in Southampton or Winchester) and the date and time of journey. TRIPlanner will calculate the best options available using public transport services and display the result on the web page.
- *Traffic congestion maps* of Southampton. This service displays maps of traffic congestion in Southampton. Again, the information is updated every 5 minutes according to the data collected by a network of sensors that monitor traffic volume on selected roads. The roads are color coded to reflect free flow traffic (green), moderate traffic flow (yellow), increasing congestion (orange), and heavy congestion (red).
- *Travel news.* This service displays text reports of the latest travel conditions on public transport services and the roads across the region. The web page refreshes automatically if any of the displayed information is changed.

Although the use of maps and GIS is rather limited in the online ROMANSE project, it demonstrates a good use of deploying certain ITS user services over the Internet.

### SmarTraveler

The SmarTraveler web site by SmartRoute Systems (Cambridge, MA) provides current traffic and other travel information for a number of cities across the United States. SmartRoute Systems maintains its proprietary databases in its regional operation centers. A network of electronic sensors, cameras, and links with multimodal public transportation agencies transmits current traffic and traveler information to the regional operation centers for processing. Real-time or near-real-time information is then displayed on the Internet for travelers to browse. Although the basic designs are very similar between different cities, differences also exist due to data availablity. For example, the web page for City of Boston includes:

- *Boston area traffic.* This service provides an area map that shows all freeways and selected major arterials. Users can click on a freeway or a major arterial to retrieve the latest traffic information.
- *Traffic cameras.* This service provides real-time photographs of traffic conditions taken by cameras installed at a number of locations in Greater Boston. The photo display on the web page refreshes automatically in every few seconds.
- *MBTA transit information.* This service provides current information of transit services in the metropolitan Boston area. It covers commuter rail, subways, bus, and water transit.

For San Francisco, the web page provides basically the same kind of information with a different map interface. The information services provided through the web page include:

- *Real-time maps.* This map displays real-time traffic conditions at selected points along freeways and selected major arterials. The point symbols are color coded for traffic flow of >50 mph (green), 31–50 mph (yellow), 15–30 mph (orange), and <15 mph (red). Users can click on a freeway/highway symbol on the map display to retrieve segment-by-segment traffic flow conditions along the selected highway.
- *Incidents.* This service provides a text report of traffic accidents and traffic hazards on roads in the San Francisco area.
- *Weather.* This service provides a link to a web site with the current weather and weather forecast.
- *Transit.* This service provides links to a number of local transportation and transit agencies for current transit information.

This site also plans to include video and road closure information in the near future. The SmarTraveler web site is developed specifically to support ITS. A unique feature is the display of real-time images of traffic flow taken by cameras.

### San Francisco bay area transportation information

This web site is an Internet version of the TR@KS (*Transportation Resources @ Kiosk Sites*). TRANSPAC (the Contra Costa Transportation Authority) and the Bay Area Air Quality Management District's Transportation Fund for Clean Air sponsor this project. This web site provides information about real-time road traffic condition, carpool, vanpool, bus/BART (Bay Area Rapid Transit), bikes, and other general information for the nine San Francisco Bay Area counties.

The real-time traffic map display incorporates some interactive functions. When a user places the cursor on a selected highway segment, travel speeds in both directions along the segment are automatically displayed on the web page.

If the user clicks on a highway, descriptions of segment-by-segment traffic condition are displayed. The carpool page allows users to fill in their home address, work address, and work schedule for ride matching. The Bus/BART page displays a map of the nine counties in the project area. Users can click any county on the map to access detailed transit information for the particular county. For example, the BART page gives the user a choice of accessing either a map-based or a form-based schedule and fare finder. The map-based schedule and fare finder allows a user to click the map to select an origin station and a destination station. The user then specify weekdays, Saturday, or Sunday, plus an optional time window on the specified day. This service then searches through the database and returns the schedule and fare information.

*Etak real-time traveler information*

This web site by Etak, Inc. (Menlo Park, CA) provides real-time traffic information for many major cities across the United States. Once a city is selected, the webpage displays a map with point symbols showing the locations of traffic incidents. This web page offers both a Java version and a JavaScript version. Under the Java version, if a user moves the mouse cursor over an incident symbol on the map, a pop-up message box describing the incident location is automatically displayed next to the incident symbol on the map. If the user clicks on the incident symbol, more detailed information about the incident is displayed in a tabular form. The JavaScript version performs the same functions. The only difference is that the JavaScript version displays the description of incident location in the status bar at the bottom of the web page when a user places the mouse cursor on top of an incident symbol shown on the map. This web site demonstrates the flexibility of using Java to develop an Internet-based GIS application.

*Seattle's BusView*

BusView is an application of the Seattle Smart Trek ITS Model Deployment Initiative sponsored by the U.S. Federal Highway Administration and the Washington Department of Transportation. The BusView application uses Java applets to display real-time transit vehicle locations (Dailey, Fisher, and Maclean 1999; Dailey, Maclean, and Pao 1999). This web page shows more advanced functions of interacting with maps on the Internet.

BusView's map-based user interface automatically refreshes itself to update vehicle locations. Users can drag the cursor to pan the displayed map or click on a vehicle location to show the bus progress along its route or route schedule. As the user places the cursor on top of street intersections on the map, a pop-up message box showing street names is displayed.

Several other interesting Internet-based ITS projects can be found at the web site of the ITS Research Program at University of Washington. They include, for example, the Seattle Smart Traveler project that develops an experimental WWW application of dynamic ridematching (Dailey, Loseff, and Meyers 1999) and the Traffic Data Acquisition and Distribution (TDAD) project that uses a web-based front end with a map display window.

## Conclusion

This chapter identified and highlighted issues surrounding the use of geographic information in intelligent transportation systems. ITS development in three national and pan-national settings (Japan, Europe, and the United States) indicate the range of possible ITS user service bundles and ways to configure and interconnect these services in a system architecture. Capturing, communicating, and integrating geographic information is central to ITS architectures. Positional error and ambiguous location referencing can seriously hamper the performance of these systems once implemented. Although GIS have not been linked to ITS in a comprehensive manner, in-vehicle navigation systems and Internet-based transportation information systems provide examples where integration for specific services can be very successful.

In the next chapter of this book, we discuss the use of geographic information in analyzing the relationships between transportation systems and the environment. This includes environment impacts of transportation facilities, the impacts of transportation systems on environmental quality and the transportation of hazardous materials.

# 10

# Transportation, Environment, and Hazards

Transportation facilities and systems are major sources of environmental degradation and risk in modern societies. Building transportation facilities requires major alterations to the physical environment, often creating negative impacts on sensitive environmental features and hydrological systems and visual blight in natural or historical areas. Automobile-oriented transportation systems are a major source of energy consumption and air pollution. Transportation systems also have an inherent degree of risk due to accidents, potentially causing harm to individuals and property. If hazardous materials (HazMats) are being transported when an accident occurs, the resulting material release can also cause substantial damage to people, property and the environment.

Transportation systems are also central to emergency response and management. Dispatching an emergency response team (such as police, fire, or HazMat teams) requires determining the closest (least travel time) response facility and routing the response vehicles through the network. In extreme circumstances, we may need to rapidly evacuate the population within an affected area, often using their personal transportation.

In this chapter, we will discuss principles and GIS tools for measuring the environmental impacts of transportation and mitigating the effects of transportation hazards. It is in this domain that transportation and geography are most tightly integrated. Assessing the environmental impacts of transportation systems and hazards require high resolution and accurate representations of the physical environment. Emergency routing and particularly evacuation requires detailed estimates of flow and congestion over geographic space and in real time. Although important in all GIS-T applications, the need for detailed and realistic geographic representations implies that geographic data error will be particularly critical issues in environmental and hazards assessment. Indeed, we will see data error issues rising repeatedly in this chapter.

In the next section of this chapter, we discuss the role of GIS in assessing the environmental impacts of transportation. This includes methods for assessing the environment impacts of transportation *facilities* and transportation *systems*. The former includes GIS methods for assessing impacts of built infrastructure on sensitive environmental features, hydrological systems and viewsheds. The latter includes methods for assessing the impact of operating transportation systems on air quality, a major environmental and political issue in many parts of the world. The next major section of this chapter examines transportation and hazards. This includes GIS methods for analyzing transportation accidents and assessing risk from HazMat transportation. We also discuss methods for real-time georeferencing of emergencies, routing emergency response teams to these incidences and locating emergency response service facilities. Finally, we will discuss GIS tools for formulating and assessing emergency evacuation plans.

## Transportation and the environment

### Transportation facilities and the environment

In this section of the chapter, we will discuss GIS tools for identifying and assessing the impacts of transportation infrastructure on the environment. We will generally restrict our attention to linear transportation facilities. Although we recognize the relevance of GIS tools to planning and design areal and even volumetric transportation facilities (e.g., airports, railyards, ports), these are complex topics that deserve fuller treatment in a text on GIS for site planning. We will consider three types of tools, namely, tools for *mapping sensitive environmental features*, calculating *hydrological impacts*, and *viewshed analysis*. In all three cases, we will see that spatial data error can have substantial impacts and therefore must be considered and controlled (if possible) in analysis and decision making.

### *Mapping sensitive environmental features*

In chapter 8, we examined tools for the right-of-way (ROW) planning problem. At a small cartographic scale this is a corridor location problem. We can solve the corridor location problem using tools for finding the least cost path through a surface. Location costs could be a composite reflecting a wide range of relevant factors including its environmental impacts and environmental constraints on its design. GIS-linked collaborative spatial decision support systems (SDSS) can help analysts, decision makers and stakeholders to assess tradeoffs and calculate decision weights that allow a composite cost to be calculated for each location. If tradeoffs among location-based factors are not possible, we can also use GIS and SDSS tools to identify a small set of good solutions and choose one through a process of elimination. At a larger cartographic scale the problem becomes a corridor design problem and requires closer interactions between a GIS for geographic data integration and spatial analysis and Computer-Aided Drafting and Design (CADD) software for engineering calculations.

Geographic data captured through technologies such as aerial photography, satellite-based remote sensing, radar and GPS-enabled field-based collection methods can be integrated with secondary or legacy geographic data using GIS (see, e.g., Ossinger, Schafer, and Cihon 1992). We have already discussed some of these issues earlier in this book (chapter 4 in particular). An important point to emphasize here is the effect of geographic data error on this decision process.

In chapter 8, we discussed geographic visualization (GVis) methods developed by Ehlschlaeger, Shortridge, and Goodchild (1997) for assessing the effects of uncertainty in regular square grid (RSG) terrain models on corridor location. This is a case where error in a single environmental variation (e.g., elevation) can influence the location of a transportation facility. A complexity of capturing environmental factors in corridor location and design is these often have multidimensional definitions. Interaction among measurement errors across multiple dimensions can have complex effects and should be factored into the decision process. Consider *slope stability*, a critical environmental constraint on linear transportation facilities. Stability is a function of factors such as slope, soil attributes, and land-cover characteristics. Sources of uncertainty and error include attribute errors associated with vegetation and soil polygons, continuous error in the elevation surface, and positional error associated with the polygon boundaries. These must be measured and integrated carefully into an overall soil stability measure. Davis and Keller (1997b) develop fuzzy set and simulation techniques for measuring uncertainty in multidimensional environmental measures such as soil stability.

Fuzzy set theory is a generalization of set theory that allows set membership to be characterized as a smooth transition rather than a crisp, binary relationship. For example, rather than saying "location $x$ is soil Type I" (implying certainty in this set membership belief), fuzzy set theory allows us to make statements such as "location $x$ is soil Type I with certainty $m$," where $m \in [0, 1]$. The fuzzy membership function $m$ is equal to unity when membership is certain, zero when not being a member is certain and $0 < m < 1$ for cases between these extremes. Note that this collapses to binary set theory at the extremes of the interval. The membership value can reflect uncertainty and error in the measurement process as well as inherent features of the phenomenon (examples of the latter include vegetation transition zones or wetlands). Membership functions can be estimated from expert panels (see Burrough 1989). It is important to keep in mind that membership functions are not probabilities and do not require accepting the assumptions of probability theory. See Fisher (1994, 1999, 2000) for discussions of fuzzy set theory in geographic representation and modeling data uncertainty.

Figure 10-1 compares traditional representations of a polygon boundary with the fuzzy set approach. Figure 10-1a illustrates the traditional crisp polygon representation along a linear transect that intersects the polygon boundary. Crisp representations imply that all locations inside the polygon are members

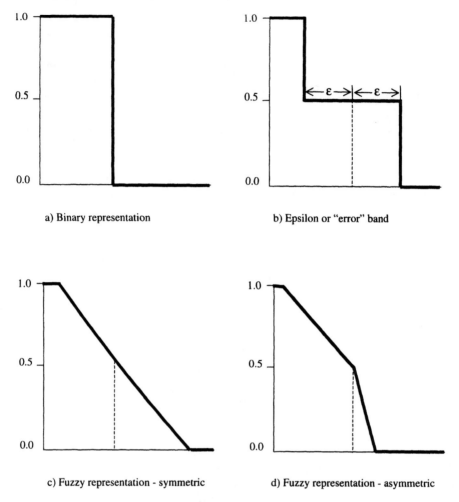

Figure 10-1: Polygon boundary representation using traditional (binary), error band, and fuzzy set methods

with absolute certainty and all locations outside are not members with absolute certainty. While these crisp boundaries may be available from secondary data sources, error in these data means that they may not be the true boundaries. The epsilon band approach in figure 10-1b recognizes possible error in the boundary by allowing an uncertainty band centered on the crisp polygon boundary (see chapter 4 for more detail). However, this approach somewhat unrealistically implies that polygon membership likelihood drastically changes to certainty outside this band. Figure 10-1c and 10-1d illustrates the fuzzy set approach that allows a smooth transition among polygon membership likelihood by location. Figure 10-1c shows the simple but unrealistic case that the fuzzy membership function is symmetric around the crisp boundary. The more general but complex case is the asymmetric function illustrated in figure 10-1d (in cases 10b–10d, a vertical dashed line shows the traditional crisp boundary).

Davis and Keller (1997b) convert crisp polygons to fuzzy polygons by deriving pairwise membership likelihood comparisons from an expert panel. An example is scores (between 0 and 1) showing the likelihood that a location with soil type I in the real world is classified as type I, type II, type III, and so on, based on soil intermixing characteristics in each location. Based on these scores, the crisp polygons are converted to fuzzy surfaces by assuming that the highest likelihood of correct classification occurs at the center of the polygon. Their procedure constrains the fuzzy membership function to $m = 0.5$ at the original crisp boundary and adjusts the membership function on either side of this boundary to reflect the uneven blend of these attributes on either side of a measured boundary in the real world. The result is a realistic asymmetric fuzzy membership function with respect to the boundary (see Davis and Keller 1997b).

A Monte Carlo simulation method captures uncertainty in the terrain data by generating random realizations of the terrain based on error bands for sampled elevations. Davis and Keller (1997a) integrate this with fuzzy polygon data on soil type and forest type in the soil stability calculation by generating a set of terrain realizations for each possible combination of soil type/forest type and evaluating the stability index. The result is a set of soil stability realizations that can be summarized with statistical measures of central tendency and dispersion in the aggregate as well as disaggregated by location. The enormous amount of information generated through this procedure can also be visualized using static and dynamic methods for exploration and communication to decision makers (see Davis and Keller 1997a).

*Hydrological impacts*

*Modeling hydrological systems: An overview* Assessing the hydrological impacts of transportation facilities requires methods for analyzing and predicting water flow and the transport of substances carried by water on the surface and through the subsurface of the Earth. We can use these methods to determine the hydrological conditions that will affect the facility design. We can also assess the hydrological system impacts of these facilities with respect to increased runoff, changes in flow and erosion patterns, and watershed pollution from oil and gasoline spills.

Hydrological modeling is a complex domain. Maidment (1993) is a classic overview of hydrological modeling and GIS tools. The remainder of this subsection follows Maidment (1993); also see Band (1999) and Moore et al. (1993).

Water flow over the Earth's surface and through its subsurface occurs when the amount of inflow at a location exceeds its storage capacity. Inflow can occur in the form of precipitation and infiltration from other locations. Any water that is not absorbed as storage (or lost to evaporation) must leave that location. The outflow travels to other locations based either on *momentum* or on movement toward a *minimum energy state* (a local minimum or "low point" in the terrain

**Table 10-1:** Taxonomy of Hydrological Models

|  |  | Time | |
| --- | --- | --- | --- |
| Space | Flow | Uniform Across Time | Varies Across Time |
| Uniform across space | Deterministic | Lumped and steady flow process | Lumped and unsteady flow process |
|  | Stochastic | Space-time independent | Time dependent |
| Varies across space | Deterministic | Distributed and steady flow processes | Distributed and unsteady flow processes |
|  | Stochastic | Space dependent | Space-time dependent |

*Source:* After Chow, Maidment, and Mays 1988; Maidment 1993.

with respect to some neighborhood). Momentum principles are more appropriate for modeling rapidly varying flow over time horizons of hours or days. Minimum energy principles are appropriate for modeling flow over long-term horizons such as months or years.

Material transported within the flow may move at different rates depending on advection, dispersion and transformation processes. *Advection* refers to passive motion of the material at the same rate as the flow; this is reasonable for some groundwater systems. *Dispersion* in the three spatial directions is more realistic; this occurs from mixing processes created by the flow. *Transformation* may occur as a result of degradation, absorption or chemical reactions involving the material.

The very skeletal theory outlined above suggests a central role for space and time in hydrological modeling. We can classify different types of hydrological models based on their treatment of flow and transport with respect to space and time. Table 10-1 provides a modeling taxonomy (Maidment 1993; also see Chow, Maidment, and Mays 1988). We can treat flow and transport process as *deterministic* or capable of being modeled without error (in principle). This suggests treating the process as a dynamical system and using differential or difference equations to describe its evolution over continuous or discrete time (respectively). *Lumped* models describe the flow process as constant across space for simplicity and analytical tractability. Complex but more realistic *distributed* models allow the flow process to vary with location. *Steady* flow models treat the dynamic flow or transport process as constant over time while *unsteady* flow models allow the process to change over time.

The equations describing the dynamical system can be difficult to solve using algebraic methods, particularly for distributed and unsteady flow processes. Numerical solution methods include finite difference and finite element methods. *Finite difference methods* (FDM) approximate a continuous dynamic process through difference equations that describe changes with respect to discrete and uniform "steps" over space and time. *Finite element methods* (FEM) are required for systems with nonuniform spatial or temporal intervals. FEM approximate the flow process using local flow equations that describe some discrete and bounded unit in space and time.

*Stochastic* models treat the flow process as partially random and use statistical methods to estimate the flow process. As in more general statistical methods, different methods are required depending on whether we assume that flow process is *independent* or *dependent* across space and time. The most realistic but complex stochastic hydrological models treat flow and transport processes as space-time dependent. Calibrating these models require measurement or calculation of real world hydrological system properties (e.g., terrain, hydrological network, watershed size and shape, flow rates, and water sample characteristics).

Hyrdrological systems that are particularly relevant to transportation facility design and impacts analysis include watersheds, stream channels and aquifers. *Watershed* models are especially sensitive to geographic representation. Properties such as vegetation, soil, slope and direction affect the amount and pattern of flow and transport through watershed systems (McCormack et al. 1993). The simplest and most common type of watershed model is a deterministic lumped system. This treats the watershed as a point location in space (although it may be associated with a polygon or areal object for display) and uses a single flow relationship representing the watershed's throughput over time. These models have high temporal resolution but low spatial resolution (Wegener 2000). "Linked-lumped" systems improve spatial resolution by disaggregating the watershed into interacting subwatersheds with independent lumped models. Distributed watershed models are fully spatially and temporally disaggregate and consequently quite complex (see Maidment 1993).

Distributed methods can also be used to model flow through linear features such as *stream channels* and through 3-D regions such as *aquifers*. Steady flow and transport through a linear feature can be modeled as a function of distance along the feature. More complex unsteady flow and transport processes require representing the linear feature as a set of discrete locations in linear space and time. This implies a 2-D lattice: one spatial dimension representing distance along the stream and a second dimension representing each location at regular time intervals. Flow between these locations in space-time can be solved using finite difference methods (see Maidment 1993). The geomorphology of aquifers can be approximated using a 2-D spatial tessellation such as triangulated irregular networks, raster cells or polygons. Some 3-D models and methods are also emerging (see Moore et al. 1993).

*Digital terrain models* (DTMs) maintained and processed within a GIS can provide high-resolution and detailed terrain information for input into hydrological models. RSG elevation models can be processed as a raster tessellation for input to finite difference hydrological modeling systems (see Gao, Sorooshian, and Goodrich 1993). It is also possible to model finite difference systems directly using raster GIS tools. Burrough (1998) describes a generic raster GIS toolbox for modeling a wide range of flow and transport processes. Irregular digital elevation models TINs can be used in conjunction with finite element hydrological modeling systems (see Richards, Jones, and Lin 1993).

*Extracting hydrological features from digital terrain models*  Critical information in a DTM for hydrological modeling is the drainage network for the study area. A *drainage network* is the collection of all watercourses (e.g., streams, rivers) in a given region. Logically, we can treat a drainage network as a collection of directed acyclic networks or *trees* (see chapter 5) that are rooted at and oriented toward the depressions or low points in the terrain. The geographic region that each network drains is the *drainage basin* of that system; this forms an irregular tessellation of the plane (Band 1986; van Kreveld 1997). The problem is to extract the drainage networks from terrain information implicit in a DTM.

Methods for extracting hydrologic features from a DTM include classification methods, pattern-matching algorithms and flow tracing algorithms (Band 1999). Classification and pattern-matching algorithms are usually applied to RSGs such as a raster tessellation. These methods classify each location into a type of topographic feature based on the pattern of elevations within a small neighborhood, typically 3 × 3 cell window centered on the location of interest. These perform poorly, particularly in the presence of spatial data error (Peucker and Douglas 1975).

Flow tracing procedures construct a drainage network by calculating paths for water movement over the terrain. Most methods examine the pattern of elevation differences among a given RSG or TIN node and its neighbors and trace single or multiple descent directions. Single direction methods are usually based on steepest descent. Multiple direction flow tracing methods include distributing the flow equally among all lower elevation neighbors, splitting flow using some combination of descent rate and direction (e.g., Quinn et al. 1991) or a stochastic simulation process (see Burrough and McDonnell 1998; Moore 1996). Single flow algorithms can only generate somewhat counterintuitive parallel and convergent flow networks while multiple direction algorithms also allow divergent networks to emerge (Desmet and Govers 1996).

The simplest single flow method is to examine the elevations within some neighborhood of a location and pick a flow direction, typically the steepest descent. An early approach by Peucker and Douglas (1975) finds the highest elevation within a roving 2 × 2 window; all other raster cells are considered part of the drainage network. The most common raster-based single-flow method in commercial GIS software is the *D8 algorithm* that finds the steepest descent direction in a three by three window. Although efficient, this procedure generates flow channels that are restricted to lateral or 45-degree angles (see Burrough and McDonnell 1998; Moore 1996). They can also generate a disconnected hydrological network, particularly if there are many flat spots. These methods can be improved by explicitly considering the accumulation and flow of water when tracing the drainage network. A method by O'Callaghan and Mark (1984) explicitly considers accumulation and flow, generating a more connected and realistic drainage network. Their method requires $O(n^2 \log n)$ time in the worse-case and $n^2$ storage for an $n \times n$ lattice (van Krevald 1997). Band (1986) develops a recursive version of the O'Callaghan and Mark (1984) algorithms that generates an efficient index-based storage scheme for the extracted drainage network.

Bennet and Armstrong (1996) develop an inductive approach for extracting a drainage network from a RSG terrain model. Their procedure examines elevation differences among the four transects that can be formed within a three by three window (a central point and its eight neighbors). This allows local classification based on the inferred topographic profile. The procedure then refines this preliminary network based on the terrain information gained in the previous step combined with principles based on geomorphological theory maintained in a knowledge base. Hydrological features are extracted from this refined network and stored in an efficient data model.

In chapter 5, we discussed the shortest path through terrain problem. We noted that RSG approximations of a continuous surface introduce elongation and deviation errors. These errors result from move permutations introduced by the RSG approximation. Elongation and deviation are independent of its density and cannot be improved through finer spatial sampling. Even a dense RSG with perfectly measured elevations can generate a drainage network that is distorted with respect to the true flow path through the continuous topographic surface. These errors can be alleviated somewhat through modification of the implicit neighborhood topology of the RSG; see chapter 5.

Yu, van Krevald, and Snoeyink (1996) develop a single direction flow method for TINs. Flow paths are defined based on the steepest descent direction from each TIN node. The steepest descent path from a TIN node is either a unique path across a triangular region or possibly multiple arcs. In the latter case, a single flow path must be selected. Since we must only trace a flow path until it meets an exiting path, the algorithm requires $O(n + k)$ time in the worse case, where $k$ is a measure of terrain complexity (van Krevald 1997).

As the discussion above suggests, extracting hydrological networks from DTMs is complex. The derived network can be different depending on the extraction algorithm. Desmet and Govers (1996) compare six different flow routing algorithms for a RSG terrain model. Derived networks are considerably different across algorithms, with a large difference between single flow and multiple flow algorithms. Terrain error has substantial impacts on the extracted drainage network. Lee, Snyder, and Fisher (1992) simulate spatially autocorrelated error within a RSG terrain model to assess its impact on floodplain extraction. Increasing positive spatial autocorrelation in the error decreases the rate of not including true floodplain cells. It also results in over-bounding from increased rates of including non-floodplain cells. However, the error magnitude mitigates the relationship between the degree of spatial autocorrelation and misclassification. Lower error magnitude decreases floodplain misclassification regardless of the degree of spatial autocorrelation among these errors (also see Lee 1991b).

Flow tracing procedures are particularly sensitive to local error in the RSG. Flow tracing procedures usually continue until flow reaches local minima or locations that are low points with respect to their neighbors. A problem with these procedures is spurious pits or depressions resulting from spatial data error. These artificial features can trap a flow tracing procedure and create topological errors in the hydrological network. A difficulty is that some of these depressions may be real rather than error.

One solution to spurious depressions is to smooth the RSG prior to routing (e.g., O'Callaghan and Mark 1984). This removes shallow depressions but cannot remove more drastic deviations without greatly distorting the natural terrain.

Vieux (1993) experiments with the effects of RSG aggregation and smoothing on hydrological runoff calculations. The effects of terrain generalization relate to the amount of rainfall, with low rainfall intensities creating greater error than high intensities. Vieux (1993) also develops a general measure that provides a global measure of runoff error at a given level of terrain generalization.

Another approach is to fill depressions by increasing the values of its constituent nodes in the RSG until they are equal to the bordering nodes (see Band 1986). A problem with these methods is that flow directions become undefined within filled regions and the resulting network may not be connected. A fill procedure by Jenson and Domingue (1988) assigns directions based on the mild condition that the outflow location does not point back to the inflow location. McCormack et al. (1993) use the stricter requirement that locations in the plateau drain to closest outlet (lower elevation) point along the plateau boundary.

*GIS-based hydrological analysis and transportation*  GIS software for hydrological analysis in transportation design include GISHYDRO and the *Hydrologic Data Development System* (HDDS) (Oliveria and Maidment 1998). The Maryland (U.S.) State Highway Administration developed GISHYDRO to integrate geographic data for input into the U.S. Soil Conservation Service TR-20 rainfall runoff model (Ragan 1991). GISHYDRO allows the user to run the TR-20 software based on land-use change scenarios to estimate impacts on watershed morphology as well as biochemical oxygen demand, nitrate, phosphate, and other transported material within the system. Smith (1995) developed HDDS as a prototype GIS that extracts drainage networks and measures summary parameters (such as maximum flow path length, average slope). HDDS also interfaces to a hydrological modeling software system developed by the Texas (U.S.) Department of Transportation.

Olivera and Maidment (1998) develop a GIS to support hydrological calculations when designing transportation facilities. The data preparation phase requires editing a digital terrain model derived from 15-minute and 30-minute USGS DEMs to match hydrological features derived from an independent linear hydrological feature database. A fill procedure eliminates spurious depressions. The *D8* steepest descent method extracts the drainage network. The system allows the user to select a location on the screen, delineate the upstream watershed and estimate water discharge parameters (such as the potential extreme discharge) for that location based on a peak discharge flow equation. This information can be used when designing bridges and highway drainage facilities such as culverts and storm drains.

*Viewsheds*

*Calculating viewsheds*  Assessing the aesthetic impacts of transportation facilities often requires determining the "viewshed" or the visible area from a given location. For example, we may wish to determine if a transportation facility is visible from sensitive locations such as residential areas, areas of outstanding natural beauty, wilderness areas and historic sites. We also may wish to reverse the problem and determine the visible areas from transportation facilities, e.g., for locating signage, observation points, scenic view parking areas and minimizing visual pollution.

Viewshed analysis involves solving *intervisibility problems* for a terrain model. Given a *viewpoint* on the terrain (possibly adjusted for different viewing heights such as platforms and towers), a *target* location is *intervisible* if we can form a line segment in three-dimensional Euclidean space from the viewpoint to the target that does not intersect the terrain surface. We can also consider obstructions caused by natural objects (e.g., trees) and built artifacts (e.g., buildings) located between the viewpoint and target location (Lee 1991a). Note that we are being careful about distinguishing between the viewpoint and the target location. The intervisibility relation is generally asymmetric: It is only symmetric when viewing heights at both locations are equal and the terrain surface is symmetric with respect to the midpoint between the two locations along the line-of-sight (Fisher 1996). This is highly unlikely in any real-world problem. A *viewshed* is the set of target locations that are inter-visible from a viewpoint given a digital terrain model and georeferenced objects. The *horizon*, a related construct, is the set of most distant points in the viewshed (De Floriani and Magillo 1994, 1999).

We can compute viewsheds based on RSG or TIN terrain models. Viewsheds defined on RSGs consist of discrete point locations that are inter-visible from the given viewpoint. We can solve for a viewshed on a RSG using a *ray-tracing* strategy. This algorithm first calculates the *sightline* between the viewpoint and a selected target location in the terrain model. The sightline is the ray that originates at the viewpoint and intersects the target. The algorithm then determines the set of intermediate terrain locations on the sightline between the viewpoint and target. We can determine this set by projecting the sightline to the 2-D plane and determining the intersection between this projected halfline and the RSG nodes or the calculated raster cells around each node. Finally, the algorithm calculates the elevations of the intermediate locations and checks if any intersect with the sightline in 3-D space. If we find an intersection between the sightline and the terrain, we conclude that the target is not inter-visible from the viewpoint (De Floriani and Magillo 1999; Fisher 1996; Wang, Robinson, and White 1996).

The basic ray tracing strategy repeats the above procedure for every target location (other than the viewpoint) in the RSG. This requires $O(n\sqrt{n})$ time in the worse-case for an RSG with $n$ nodes (equivalently, $O(m^2)$ for an $m \times m$ RSG). This is computationally burdensome for the large and dense RSGs required for sufficient geographic representation in real-world visibility problems. However, some of these sightline calculations are redundant since they overlap. Only computing sightlines between the viewpoint and the boundary of the region allows some computational savings at the expense of accuracy. Other heuristics include sampling a subset of the sightlines between a viewpoint and all target locations in the RSG (De Floriani and Magillo 1999). We also may only need to consider a subset of the targets if atmospheric conditions such as humidity limit visual acuity. Wang, Robinson, and White (1996) develop an efficient RSG viewshed algorithm that reduces the number of sightlines calculations based on slope and aspect in the neighborhood of the viewpoint and target location. The regular topology of a RSG also supports parallel processing implementations based on data parallelism (De Floriani and Magillo 1999).

Viewsheds defined on TINs are either a set of discrete point locations (a subset of the TIN nodes) or an unconnected region in continuous space constructed from the TIN triangles. The *front-to-back* algorithm solves the continuous space problem (see De Floriani and Magillo 1999). Front-to-back has two major phases. First, the TIN triangles are sorted in "front-to-back order": This is the order in which rays from the viewpoint would intersect the triangles (roughly in increasing Euclidean distance in the 2-D plane). We can sort a TIN in this order if it is based on the Delaunay triangulation (De Floriani et al. 1991). (A Delaunay triangulation consists of generally equilateral triangles; see Boots 1999). Second, a visibility phase processes the triangles in front-to-back order, incrementally constructing the viewshed by projecting the current horizon onto a candidate triangle being considered and determining which part of the triangle is visible. Asymptotic time complexity is roughly quadratic at $O[n^2\alpha(n)]$, where $\alpha(n)$ is a quantity that grows slowly with $n$ (the number of TIN nodes). Experiments suggest that run time in practice is nearly linear (De Floriani and Magillo 1994, 1999). De Floriani et al. (1994) develop a parallel implementation of the front-to-back algorithm. If we only require the viewshed at the resolution of the TIN triangles (i.e., the entire triangle is either visible or invisible) than we can modify this approach and sort the TIN edges instead of the triangles. We add a TIN triangle to the viewshed only if all three of its edges are visible. This allows some computational savings (see Lee 1991a).

There are several interesting problems related to basic intervisibility and viewshed analysis. *Visibility location problems* include finding a fixed number of locations so that the area of the composite viewshed is maximized or determining the minimum number of viewpoints so that an entire region is visible. An example application is fire observation towers in rugged terrain (see Goodchild and Lee 1989; Lee 1991a). Line of sight communication problems require solving for a visibility network such that adjacent nodes are mutually intervisible; example applications include locating microwave communication towers (De Floriani and Magillo 1999). A generalization of the viewshed problem is visualization of the terrain surface implied by a DTM. This is required for geographic visualizations based on natural terrain. De Berg (1997) provides a good overview of TIN visualization problems.

*Error and uncertainty in viewshed calculation*  Similar to other problems discussed in this chapter, intervisibility and viewshed calculations are sensitive to error and uncertainty. Intervisibility calculations are sensitive to resolution (terrain detail) proximal to the viewpoint. *Multiresolution* viewshed algorithms operate on terrain models that nest more detailed terrain information within more generalized representations in a hierarchical manner. These algorithms solve intervisibility problems with varying levels of spatial resolution based on distance from the viewpoint (see De Floriani and Magillo 1999).

An experimental analysis by Fisher (1991) found that elevation error in a RSG tends to bias the estimated viewshed downward, with greater error implying smaller calculated viewsheds. Elevation variability within the RSG and the height of the viewpoint relative to the surface also affect the estimated viewshed but in more complex ways. Fisher (1992) uses this error simulation method to calculate a *fuzzy viewshed* from the set of realized terrain surfaces.

The discrete approximation of continuous space through a RSG or TIN also introduces geometric error into the viewsheds. Approximating bounded regions in continuous space as a set of discrete points (cells) or through a finite set of triangular regions introduces errors that can be resolved through denser sampling. As we noted in chapter 5 with respect to the shortest path through a surface problem, a RSG can also introduce geometric error that is an artifact of the discrete spatial approximation. For example, unless the viewpoint and target are in the same row or column of the RSG, it is possible for intermediate cells along the sightlines to only partially block the target. Sorenson and Lanter (1993) develop two algorithms for calculating partial visibility cells. A quick heuristic uses vector geometry while a more expensive but accurate algorithm subdivides the target cell into visible and invisible subregions. Accuracy improvement is a function of RSG density.

As we also saw in chapter 5, although an algorithm such as the shortest path tree can appear straightforward and generic, implementation within a computational platform involves data handling decisions that can affect performance. Fisher (1993) examines different software implementations of the ray-tracing algorithm for calculating viewsheds on RSGs. Possible geometric strategies for representing the viewpoint and target include point to point, point to cell, cell to point and cell to cell. The different sightline calculations involved in each case can have dramatic effects on the calculated viewshed. Elevation interpolation strategies include interpolation to the four rook's case neighbor, interpolation to the eight queen's case neighbors, interpolation to a finer lattice and the simple stepped model where each cell has a constant elevation. This has subtler but complex effects on the viewshed. The results clearly suggest the need for more benchmarking tests and perhaps algorithm standards for GIS software. At the very least, it suggests the need to be an informed consumer of GIS software.

## Transportation systems and the environment

### Transportation and sustainable development

Automobile-oriented transportation systems are resource-intensive and have substantial environmental impacts. In addition to the resources required for construction, automobile-oriented transportation systems are major consumers of nonrenewable fossil fuels and major producers of air pollution. Consequently, transportation systems are an important component of *sustainable development* strategies (see Deen 1995).

There are several strategies for reducing the environmental impacts of transportation systems. Technological solutions include *vehicle design and manufacture*, *engine technologies*, *fuel technologies*, and *traffic control systems*. These intend to reduce the fuel consumption and waste output of the transportation system by improving efficiency. The first three are individual vehicle technologies while the last introduces technology at the system level. Gwilliams and Geerlings (1994) summarize the results of five European Commission studies on the potential environmental impacts of transportation technologies. They note that since transportation technologies and behavior are closely related, technological solutions can have unintended and complex effects that may negate and perhaps lead to increased environmental impacts. For example, efficiency gains from traffic control devices may be offset by the induced demand generated through the additional system capacity.

In addition to technological changes, we can reduce the environmental impacts of transportation systems through shifts to less-damaging travel behaviors or reducing the total amount of transportation demand (Gwilliams and Geerlings 1994). *Transportation control measures* (TCMs) are market-oriented incentives or disincentives that attempt to reduce travel demand or shift demands to more environmentally-friendly patterns. TCMs for *vehicular trip elimination* such as telecommuting centers and ridesharing programs attempt to reduce the number of commuting or shopping trips. *Trip length reduction* methods encourage shorter commutes through higher density zoning, mixed-use development and satellite work centers. *Mode switching* methods include transit subsidies, high occupancy vehicle (HOV) lanes and paratransit. *Schedule shifts* such as flex-time reduces peak period travel. Finally, *route change mechanisms* such as advanced traveler information systems can reduce congestion and auto-truck interaction (see Ferguson 1990). Of these TCMs, the most promising seem to be (i) more-efficient land-use arrangements, (ii) different types of travel pricing, (iii) information technology as a substitute and complement to the movement of people and goods (Southworth 1995).

As mentioned in the previous chapter, motives behind the development and deployment of intelligent transportation systems (ITS) include reducing environmental impacts through greater system efficiency and facilitating more environmentally friendly behaviors. ITS services such as non-stop electronic toll collection can reduce congestion bottlenecks at toll plazas and advanced traveler information systems can help drivers avoid congested areas. Advanced vehicle control systems can prevent accidents and therefore congestion. Commercial vehicle operation services can improve the operating efficiency of public and private vehicle fleets. These potential improvements depend on the travelers' responses to these ITS services (see Shladover 1993). Acceptance and use of ITS services is unlikely if they are not well tailored to the local transportation environment (see the discussion in the previous chapter). However, higher system efficiency and greater access to transportation information can be exploited through additional trip making, leading to an unintended and indirect increase in vehicle miles traveled (VMTs) and environmental impacts.

In chapter 8, we discussed methods for travel demand forecasting, land-use/transportation (LUT) modeling and spatial decision support tools for choosing among alternative future scenarios. Evaluating the effectiveness of TCMs in reducing environmental impacts introduces some stringent requirements for travel demand and LUT models. The subtlety of TCMs and the potential for indirect and unintended effects require models that are behaviorally realistic with tight interweaving of the land-use and transportation subsystems. This suggests the use of disaggregate, nonequilibrium dynamic approaches such as microsimulation and cellular automata (although much additional research is required). Also required are better methods for translating travel demands into environmental impacts (Guensler 1993; Southworth 1995).

*GIS and transportation-related air quality*

Contemporary transportation systems have major impacts on air quality at several geographic scales. Automobiles are a main source of unhealthy particulate matter such as $PM_{10}$ (airborne particulate matter with diameter less than 10 micrometers). Transportation systems are also major generators of precursors to ozone formulation. Transportation sources contribute about 31% of volatile organic compounds (VOC) and 43% of nitrogen oxides ($NO_x$) emissions in the United States, with the result that air quality standards are violated regularly in many U.S. cities (Roth 1990). Although not the largest contributor, greenhouse gas generated from transportation systems are a major contributor to global warming along with industry, agriculture, and building heating and cooling (see Greene 1993, Paswell 1990). Some estimates suggest that contribution from the transportation sector is close to one third of the annual carbon dioxide ($CO_2$) emissions in the United States (Hillsman and Southworth 1990). In addition to regional and global-scale effects such as acid rain and global warming, transportation-related air pollutants can have negative health effects on residents living in urban airsheds; see Horowitz (1982; chapter 4) for a review of the evidence.

Planning and designing sustainable transportation systems requires forecasting air quality impacts for different LUT or travel demand scenarios and monitoring airsheds to determine progress (or lack thereof). In addition to being a good idea, considering air quality in transportation planning is also mandatory in many locales due to increasing political pressure and mandates. Forecasting air quality impacts requires estimation of emission generation from mobile sources (automobiles and trucks) and modeling the dispersion of the emissions from these sources through the airshed. Monitoring air quality requires methods for locating the limited number of air sampling locations and interpolating the air quality data collected at these locations to the entire airshed.

*Modeling mobile emission sources* Three major factors affecting emission generation from an internal combustion vehicle are *vehicle characteristics*, *weather conditions*, and *driving behavior* (Anderson et al. 1996). Mobile emissions models can easily capture the first two factors, given available data. Vehicle characteristics include the type of engine, the energy efficiency of the vehicle (including power-to-weight ratio, vehicle age) and the presence of abatement devices such as catalytic converters. This requires data on vehicle mix at trip generation locations in the study area. Cold weather tends to reduce the efficiency of both vehicles and catalytic converters; this can be handled through simple "across-the-board" adjustments in a mobile emissions model.

Driving behavior is most difficult factor to capture in mobile source emissions modeling. Properties such as speed, acceleration and idling affect the emission generation rate from a vehicle. A single hard acceleration can create more emissions than the remainder of the trip. "Cold-starts" also result in higher emission rates early in the trip. Therefore, accurate mobile emissions modeling requires detailed spatial and temporal information on trip origin/destinations, vehicle mixes, and network flow conditions at a detailed level of spatial and temporal resolution (Bachman, Sarasua, and Guensler 1996).

Methods for estimating mobile emissions differ with respect to their treatment of driving behavior. Some models use average speed as a surrogate for the *driving cycle* or set of driving behaviors on a trip. These are typically applied to broad, urban and regional scale analysis. A popular average speed-based model in the United States is MOBILE. This is essentially a set of functions (in software form) that translate speeds for different vehicle types into $NO_x$, VOC and carbon monoxide (CO) per vehicle mile for different U.S. vehicle types (e.g., light-duty gasoline vehicles, light- and heavy-duty trucks, and motorcycles). The composite emission rate for the vehicle type includes factors such as the number of miles driven by the vehicle over its lifetime, average speed, air conditioning load, trailer towing and humidity correction factors (for $NO_x$). The average speed variable can be adjusted to account for congested versus free flow conditions, acceleration/deceleration events and idling in different types of streets and highways using supplementary network data (see Cottrell 1992).

As mentioned above, in transportation planning and policy analysis it is helpful to translate travel demand estimates to environmental impacts such as air quality. Since MOBILE is based on average speed, it is relatively straightforward to use flow and speed estimates from travel demand or urban models as input. Anderson et al. (1996) link MOBILE5.C (the Canadian version of the latest MOBILE software version 5) with the IMULATE urban model and a commercial GIS software package. IMULATE consists of three interlinked models representing movement across different time scales. A population and housing model estimates intraurban mobility with respect to residential location. A trip generation and distribution model estimates day by day changes in the origin-destination (O-D) flow matrix while a network assignment assigns the O-D flows to the network (using a static assignment procedure). This allows calculation of total emission of pollutant $p$ from vehicle type $k$ traveling within network arc $a$:

$$E_{pa}^k = e_p^k(s_a) x_a^k l_a \qquad (10\text{-}1)$$

$$s_a = \frac{l_a}{t_a} \qquad (10\text{-}2)$$

where $e_p^k(s_a)$ is the MOBILE emission factor for pollutant $p$ for vehicle type $k$ traveling at speed $s_a$, $x_a^k$ is the number of $k$ type vehicles flowing through arc $a$, $l_a$ is the length of arc $a$ and $t_a$ is the travel time on arc $a$ (calculated from the IMULATE network flow estimates).

As equations (10-1) and (10-2) suggest, network flow and travel time estimates are critical in linking travel demand and mobile emissions models. Stopher and Fu (1996) conduct a sensitivity analysis of the four-step travel demand model (TRANPLAN; see chapter 8) and emission estimates derived using MOBILE version 5. Their experimental design considers the effect of spatial and temporal aggregation on estimated emissions, including different levels of network aggregation and time-of-day, day-of-week and seasonal adjustments. Results suggest that estimating at the arc level (as opposed to a higher level of aggregation) make a small but consistent difference in the emissions estimates. However, results also suggest that arc capacity estimates have substantial impacts on estimated speeds, vehicle miles traveled, and, therefore, emission amounts. Adjusting the speeds estimated from the travel demand model to compensate for known biases in the speed estimates from Bureau of Public Roads (BPR) flow cost functions (see chapter 6) also had an impact on all three emission estimates ($NO_x$, VOC, and CO). This highlights a need for further research on these functions and speed adjustment. Emission estimates are also biased by the time period modeled (particularly the diurnal cycle and day-of-the-week), indicating the need to consider modeled time horizon carefully.

GIS software can provide critical functionality at the back-end and front-end of linked travel demand and mobile emissions modeling. Spatial data integration to facilitate this linked modeling include the traditional data collected for travel demand modeling (see chapter 8) but also supplementary data such as georeferenced vehicle registrations. This can allow more accurate estimation of fleet composition in the travel demand model as well as allow surrogate calculation of "off-network" emissions when using an aggregated or "sketch" transportation network. Road grades for network links can be calculated from DTMs to capture acceleration/deceleration events more accurately. Georeferenced data such as parking facilities, refueling locations and other emission generating locations can also be integrated into the analysis (see Bachman, Sarasua, and Guensler 1996). The resulting emission estimates can be visualized within the transportation network and overlain with land-use and socioeconomic data to assess uneven impacts of emissions generation on social and demographic groups (see Anderson et al. 1996).

*Airshed modeling*  After being emitted from mobile sources, pollutants disperse through the airshed based on wind speed, wind direction, and chemical reactions among these and other agents in the atmosphere. Horowitz (1982) provides a rigorous but clear review of airshed modeling strategies.

The *dynamical* approach to pollutant dispersion is similar to the deterministic hydrological flow and transport theory discussed previously in this chapter. The conservation of mass law says that change of a pollutant's mass within a small volume of the atmosphere is equal to the sum of (i) *net inflow* transported by wind, (ii) *net molecular diffusion* from neighboring locations, (iii) *net pollutant creation rate* from chemical reactions within the volume, and (iv) *pollution emissions rate* from sources within that volume. We use the term "net" with the first three components since in these cases the same process can lead to increases and decreases of the mass within the volume; for example, wind can transport pollutant out of as well as into the volume. We can state this law directly as a dynamical system using partial differential equations. This assumes perfect knowledge of difficult-to-measure quantities such as the emission rates and wind velocity at each location in space and time. Atmospheric turbulence makes these factors difficult to model completely. If we assume that these quantities are only partially measurable (similar to random utility choice models discussed in chapters 6 and 8), ignore random fluctuations in chemical reactions, and assume a simple "mixing" model to describe turbulence, we can derive a partial differential equation consisting of measurable quantities (Horowitz 1982).

Solving the dynamical system describing pollutant dispersion requires a numeric strategy similar to the finite element approach used to solve dynamic hydrological models. This imposes a discrete grid representation of 3-D space (and time). At each discrete step in time, we solve a finite approximation of the dynamical system based on observed or estimated emission rates, wind velocities, and chemical reaction rates in the neighborhood of each lattice point. Data requirements include (Horowitz 1982):

- *initial and boundary conditions* with respect to geographic space and real-world time
- *meteorological data*, including wind velocity and direction, intensity of solar radiation, air temperature, and pressure for each discrete location in space and for each time interval
- an *emissions inventory* or rates of emission generation for each discrete location in space (and over time if emission rates are not constant; this is the usual case for mobile sources)

Initial conditions include pollutant amounts at each grid location at the beginning of the modeled time period. Boundary conditions include the pollutant concentrations at the geographic border of the study airshed that can be transported into the airshed during the model's time horizon. These require careful measurement or estimation since airshed models are sensitive to these conditions. It is often not feasible to collect meteorological data at the dense level of spatial and temporal resolution required for modeling. Instead, these factors must be interpolated over space and time based on a limited sample. This is another major source of model error and therefore also requires careful consideration. We can reduce these intense data requirements by aggregating space into homogeneous columns or boxes within which pollutants disperse instantaneously (see Horowitz 1982).

GIS can provide the critical link between mobile source emission models and the spatiotemporal input data required for airshed models, particularly emission inventories or the results generated from mobile emissions models (Souleyrette et al. 1992). As discussed previously, GIS supports calculation of emission rates generated by network travel. Vehicle registration data at the U.S. census block level can be used to provide surrogates for "off-network" travel (i.e., on smaller, residential streets not included in the network database used for the flow modeling). These emission estimates can be transferred to a raster tessellation at the appropriate resolution for the dynamic model using spatial interpolation techniques. GIS software also supports geographic visualization and location-based querying of the model results (Bachman, Sarasua, and Guensler 1996).

The aggregate, dynamic flow approach is computationally complex but has the valuable feature of capturing chemical reactions. This is important for estimating ozone formulation and concentration over time. Another modeling strategy for airshed dispersion is to model the process at the disaggregate level of individual particles. Particle models are computationally simpler than dynamic models but cannot capture chemical reactions. They are more appropriate for measuring impacts of CO and $PM_{10}$ at larger geographic scales (e.g., roadways and intersections) (Horowitz 1982).

Since knowing the exact location in space and time of each pollutant particle is difficult, particle models tend to be probabilistic. A commonly used approach is the *Gaussian plume model*. This approach models pollutant dispersion from a point source as a 2-D (horizontal and vertical) normal distribution. The parameters of the distribution are calculated based on mean wind velocity and the source strength. The resulting plume follows the prevailing wind direction and diffuses with distance from the point source (i.e., the horizontal and vertical variances of the normal distribution increase with distance). We can use analytical calculations associated with normal distributions to calculate properties such as pollutant concentration at different locations in space and time. Adapting this approach to the linear case for modeling transportation networks requires geometric extensions that can handle curved roadways, intersections, elevated and depressed roadways, and turbulence generated by vehicle movement (Horowitz 1982).

The Gaussian plume model requires assuming that the atmosphere is stationary and homogeneous. This means that wind velocity is constant over space and time, the atmosphere is unbounded and the Earth is a perfectly reflective, mirror-like surface. It also requires assuming a "static" release (meaning that the plume is at a steady state and does not vary over time). A surrogate for capturing dynamic propagation is to assume a set of equal and successive discrete time intervals and derive a Gaussian plume for each time slice. Relaxing the restrictive assumptions to derive a nonstationary and inhomogeneous stochastic process usually requires Monte Carlo simulation methods for estimating the resulting model (Horowitz 1982).

Airshed dispersion models based on the Gaussian plume approach include the CALINE3 and CAL3QHC software for modeling CO concentration due to traffic. CALINE3 calculates a linear source-based Gaussian plume using average pollution emissions over the roadway. CAL3QHC extends CALINE3 to account for queuing at intersections. Both can use MOBILE-generated emission rates for transportation network arcs as input. Hallmark and O'Neill (1996) integrate commercial GIS software with CALINE3 and CAL3QHC. The output from these models is in the format of a RSG. This output grid can be processed, interpolated, analyzed and visualized using raster tools available in the commercial GIS. The raster tessellation is also used to integrate data on significant, off-network emission sources such as drive-through facilities. Although they note some data incompatibility between the GIS and airshed software, they demonstrate that a linked system can support impacts analysis and the identification of pollution hot spots.

## Transportation and hazards

### Accidents and safety analysis

Environmental degradation is not the only negative *externality* (side effect) of automobile-intensive transportation systems. The physics of large masses moving independently at high speeds within limited space means that collisions among automobiles and with structures are inevitable. The costs are staggering. In the United States, automobile accidents cause several hundred thousand injuries and several tens of thousands deaths per year, with an economic cost in the billions.

The incidence and severity of automobile accidents, injuries and costs are related to three major antecedent factors. *Driver characteristics* influence behaviors such as driver error, alcohol and drug use, speeding and seatbelt usage, although evidence on the relationship between age and sex characteristics and these behaviors is mixed. Driver behavior combined with *vehicle characteristics* and *environmental factors* influence the crash type (head-on, rollover, sideswipe). Crash type and seat-belt usage influences the severity of injury (Kim et al. 1994, 1995).

GIS software is an effective environment for integrating census, environmental, behavioral, travel demand and accident data, exploring accident patterns across time and space and analyzing and predicting accidents using mathematical and computational tools. In most parts of the world, accident location, time, and characteristics are recorded in some detail due to legal and liability concerns. Integrating accident data with other geographic data for spatial analysis and visualization requires recording or translating accident location data into a georeferencing system supported by the GIS.

### *Geocoding and georefencing accidents*

Accident locations are recorded in three ways depending on local practice. In some cases, particularly for highways, accident data are in a linear referencing system such as milepost. That allows the incidence to be maintained as a point event using a dynamic segmentation or related data model (see chapter 3). Faghri and Raman (1995) describe a prototype system that uses a linear referencing system with resolution to the hundredth mile. The dynamic segmentation model supports network arc-based summaries and an accident history database associated with each location within the linear referencing systems.

In other cases, accident data are recorded less precisely using the nearest street addresses or nearest intersection (although many accidents happen at intersections). We discussed address matching in chapter 4. Nearest intersection matching requires two data items, namely, the major street in which the accident occurs and the nearest intersecting street. Levine and Kim (1998) discuss the sources and effects of geocoding errors when referencing accidents using both systems. Common recording errors include neglecting street prefixes and numbers; using abbreviations, slang, and alternative names for streets; and substituting place names for streets. To achieve acceptable matching rates, GIS software often relaxes exact matching requirements for address-matching and intersection-matching algorithms. A street address consists of several different types of geographic references (street number, street prefix such as North or South, street name, and street type). Relaxing matching conditions will introduce different errors depending on the component relaxed. In the Honolulu case, Levine and Kim (1998) also note a spatial bias in matching error due to different types of street names used in the central city versus the suburbs.

Levine and Kim (1998) suggest some procedures for reducing the georeferencing errors in accident reporting. An ideal solution is for officers to use a standardized reporting system. Other strategies include pre-processing the accident database to translate typically used but nonstandard references into standardized references. Preprocessing the accident database to identify error patterns also may suggest systematic edits for the accident or network databases that can improve matching success rates. More intelligent intersection and address-matching algorithms are also required, including ones that can take into account ancillary information such as place names, nearby facilities, building, landmarks, recognized districts, and so on.

Continued improvements in global positioning system (GPS) receivers will make this georeferencing strategy more attractive in accident reporting. GPS receiver technology is becoming lighter, cheaper and user friendlier. The recent elimination of selective availability makes this alternative more attractive since the need for expensive differential GPS stations is eliminated. Unfortunately, as we noted in chapters 4 and 9, public and commercial street network databases may not be up to the task. Positional and topographic errors in these databases can make "snapping" GPS-derived locations to references with these networks ambiguous. However, the positional accuracy of street network databases should increase with continued improvements in GPS and high-resolution remote sensing technologies as applied to transportation network surveying (see chapter 4).

*Spatial analysis of automobile accidents*

Automobile accidents tend to have regular geographic patterns. These often reflect the geographic pattern of the demographic groups (and their travel demand patterns) that tend to be involved in accidents. Accidents may also be the result of environmental factors related to local geographic features (terrain, sun) combined with poor design of facilities. Unexpected flow patterns and interactions (particularly at intersections) can also generate accidents. Safety officials often wish to identify geographic *hotspots* with higher than expected accident rates for allocating additional enforcement and emergency response resources, investigating design or traffic control changes, and so on.

Kim and Levine (1996) describe a traffic accident GIS that supports spatial analysis for disaggregate point data as well as data summarized by network segment and zones. Although ridden with positional error, accidents can often be represented at a disaggregate level as *points* and analyzed using point-based spatial statistical methods. Mapping accidents with respect to driver characteristics (age, sex), driver behavior (driving under the influence, seat belt usage) and vehicle types can reveal surprising geographic patterns. Spatial summary measures such as mean center, spatial dispersion measures (standard deviation distance, standard deviation ellipse) and point-based spatial statistics (nearest neighbor, spatial autocorrelation) for the entire database or disaggregated by relevant attributes (such as age, sex, or seatbelt usage) can help identify geographic hotspots. These measures can also be calculated for different discrete time periods of the day or day of the week to reveal temporal changes in accident point patterns (see Levine, Kim, and Nintz 1995a). Arthur and Waters (1997) use a GIS to georeference accident point locations based on reported street addresses. They perform some basic statistics to relate accident rates to the type of location (intersection, mid-block) and estimated speeding rates and flow levels.

Accidents by theme can also be aggregated and analyzed at the network *segment*. Simple summary measures and rates such as accident per mile/kilometer can be mapped and visually explored. Other segment based data such as flow level, speed limits, actual speeds, and congestion levels can be related to these accident summary measures using network segments. Black (1992) develops a network autocorrelation statistic that extends spatial autocorrelation to the case where network topology, cost, or flow patterns influence spatial dependency. Black and Thomas (1998) apply the network autocorrelation method to accidents on motorways in Belgium. They decompose the autocorrelation statistic using the network structure to identify hotspot arcs with higher than expected accident rates.

Accident analysis at the *zonal* level tries to relate characteristics of trip origins and destinations to accident rates within each zone. Levine, Kim, and Nitz (1995b) use a linked GIS and spatial statistical software system to estimate a spatial autoregression model for accidents at a zonal level. As we discussed in chapter 8 with respect to trip generation model, a regression model that does not account for spatial autocorrelation can have serious problems with respect to parameter estimates and significance tests. The automobile accident model relates the number of crashes at the zonal level to travel demand factors (area, land use, employment, length of different street, and highway types). Levine, Kim, and Nitz (1995b) estimate 48 separate spatial lag models corresponding to different hours of the day for a weekday and weekend day. They note that accident rates tend to follow travel demand patterns in the sense that the factors that predict travel demand also predict accident rates. Temporal fluctuations in these relationships are substantial and qualitatively different for the weekday versus weekend. They note that adding the spatial lag variable substantially improved the model's performance with respect to prediction.

*Computational intelligence and knowledge-based approaches for accident analysis*

The error-ridden and qualitative nature of accident data has motivated some researchers to develop computational intelligence and knowledge-based approaches to accident analysis. Mussone, Rinelli, and Reitani (1996) train an *artificial neural network* (ANN) using a database of the number of accidents and a set of antecedent variables. They use the basic feedforward/backpropagation architecture (see chapter 8). The ANN relates the number of accidents on roadways in Italy to qualitative variables such as "day versus night," roadway type, within-roadway location (e.g., lane versus median), road "peculiarities" (e.g., rise, descent, left curve, and right curve), human factors, and meteorological conditions. They discuss the relative importance of these variables as indicated by the trained ANN weights but do not report details on model performance.

*Knowledge-based systems* (KBS) are computational methods that rely on a *knowledge base* (KB) or a set of "justifiable beliefs" about some domain. An example of a justifiable belief is "Land values within the central business district are at least four times greater than land values proximal to major intersections at the suburban fringe." This is not usually an explicit item in the database; rather, it must be deduced from atomic data items or *facts* in a database and a set of logical *rules* for deriving conclusions from these facts. An example of a rule is "a major intersection exists at every intersection of two or more major highways." Therefore, a justifiable belief means that it can be derived from the facts in the database and the set of logical rules. This deductive capability allows a KBS to capture complex information about a domain in a parsimonious manner. A common type of KBS is *expert systems*: These embody specialists' knowledge about a complex domain (e.g., medicine) usually in the form of "if-then" rules. Other types of KBS include inductive methods such as case-based reasoning that learn the logical rules implied by a database using a small set of primitive rules for deduction (see Nilsson 1998; Worboys 1995).

Since geographic data, properties and relationships can be complex, several researchers have developed *knowledge-based GIS* (KBGIS) for diverse spatial analysis and modeling domains, including air photo interpretation (Srinivasan and Richards 1993), wildfire representation (Yuan 1997), identifying ecosystems (Coughlan and Running 1996) and zoning system design (Chen et al. 1994). The terrain extraction procedure of Bennett and Armstrong (1996) discussed previously in this chapter is another example of a knowledge-based approach. KBGIS have also been applied to transportation applications such as pavement management (e.g., Sarasua and Jia 1995).

Figure 10-2 provides a high-level generic design for a KBGIS. Using a graphical user interface, a user can select some geographic location, feature or region for which some complex property must be estimated. The GIS component manages the geographic data and the spatial and non-spatial queries against the database. The KBS uses the extracted data from the GIS combined with knowledge stored in the KB and a computational inference engine (e.g., rule-based systems, deductive logic, and inductive reasoning) to infer the relevant properties of the selected feature. Results can be passed back to the GIS for visualization, exploration and assessment in combination with the geographic data.

KBGIS for accident analysis include Spring and Hummer (1995) and Panchanathan and Faghri (1995). Spring and Hummer (1995) develop their system to identify hazardous highway locations based on properties such as highway feature type (curve, bridge, intersections) and accident rates based on

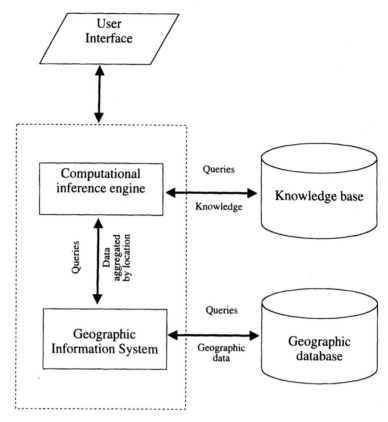

Figure 10-2: Conceptual design for a knowledge-based GIS

historic data and/or predictive models. The KB allows adjustment of these rates into a synthetic accident score and an associated level of confidence based on experts' judgements on the relative strength of these components. The KB also allows calculation of an overall hazard level based on the accident score and confidence.

Panchanathan and Faghri (1995) integrate a GIS with a KBS to identify the hazard levels and design remedies for highway-rail crossings. The user identifies a highway-rail crossing using GIS-based query tools. Data associated with each crossing include quantitative and qualitative factors such as track sight distance, road sight distance, illumination, type of land use, signage, proximity to other intersections, and accident history. The KBS integrates these site-specific factors with an index calculated from a U.S. Department of Transportation model using an inference mechanism. In addition to specifying a qualitative danger level, the KBS suggests remedies such as safety devices, design changes and even closure depending on cost and effectiveness measures derived from the KB.

## Transportation of hazardous materials

### Overview

An unfortunate consequence of technology intensive societies is the generation of hazardous material (HazMat) as a byproduct of industrial processes. The locations where these materials are generated are often not the same locations where they can be disposed. Therefore, HazMats must be transported from production to disposal sites, involving some risk due to accidents, material release and negative effects on the physical environment, humans, and property. Since many of the factors that comprise risk are geographic, GIS can play a central role in analyzing and engineering HazMat transportation systems.

HazMat transport involves finding a route through a (possibly multimodal) transportation network from a generation site to a disposal site (or possibly a tour of several generation sites beginning and ending at a disposal depot). Each possible route (tour) can be characterized by its risk. The problem is to determine the routes (tours) that minimize risk. This problem can be treated in a straightforward manner through some surrogate cost measure and standard network shortest path routing and scheduling algorithms available in commercial GIS software (see chapter 5). However, "risk" is a complex concept that requires careful measurement and treatment.

### Estimating HazMat transportation risk

HazMat transportation risk consists of three major components, namely, *accident likelihood*, material *release probability* given an accident, the *consequence* given a release event. Another possible component is *risk preference* to reflect the perceptions and attitudes of the public and other stakeholders. For example, one accident killing 100 people is usually perceived as much worse than 100 accidents that kill one person each (Lepofsky and Abkowitz 1993). Emergency response capacity and other mitigation strategies can also be factored into risk assessment (Pijawka, Foote, and Soesilo 1985).

A formal description of the HazMat transport risk at any location $x$, with "location" broadly construed to be a point, line or region (see Lepofsky and Abkowitz 1993):

$$r_x = P(a_x|k)P(r_x|a_x,k)E(c_x|r_x, g_x, v_{g(x)})^\alpha \qquad (10\text{-}3)$$

where $r_x$ is the risk associated with location $x$, $P(a_x|k)$ is the probability of an accident at location $x$ given transport by mode $k$, $P(r_x|a_x, k)$ is the conditional probability of a release given an accident and transport by mode $k$, $E(c_x|r_x, g_x, v_{g(x)})$ is the expected consequence of a release at location $x$ given a release, the geographic properties of that location that affect the material dispersion ($g_x$) and the vulnerable human and physical elements within the affected area ($v_{g(x)}$). (We treat the HazMat properties as implicit for notational simplicity.) The parameter $\alpha \in [0,1]$ reflects risk preference ($\alpha = 1$ reflects risk neutrality, $\alpha > 1$ reflects risk aversion and $\alpha < 1$ reflects risk inclination). Expected consequence is usually measured in terms of monetary cost, vulnerable impacts or area affected. Preparedness and response capabilities can be factored in as reducing the geographic dispersion ($g_x$) or reducing the vulnerability within the affected area ($v_{g(x)}$; e.g., through quick evacuations or land-use restrictions).

Equation (10-3) specifies a probability-based risk model. In practice, the required measurements to estimate probabilities and expectations can be difficult to obtain, particularly at a disaggregate level of spatial (and temporal) resolution. Consequently, there are several modeling strategies for approximating risk depending on the particular situation and (especially) the available data. Common risk estimation strategies include statistical inference, fault and event trees, analytical and simulation modeling, subjective assessment, Bayesian analysis and hybrid approaches based on some combination of the above (Abkowitz and Cheng 1989).

*Statistical inference* is the most common risk assessment strategy: this involves estimating a statistical model (e.g., a regression model) corresponding to equation (10-3). This requires an adequate record of previous incidences; this is not always available. Temporal stationarity may also be a problem, e.g., a policy or technical change after a serious accident can change the risk process and render the statistical estimation ineffectual. *Fault trees and event trees* are logical trees that describe the basic events that must occur to cause a consequence. Fault trees are based on binary events (something occurring or not occurring) and the logical operators AND and OR. A particular combination of events may lead to some qualitative output (e.g., a fatal release). An event tree also consists of basic events but these are treated as random variables with probability distributions. This allows probability statements to be made. *Analytical and simulation modeling* involves mathematical and/or computational modeling of the system under different "what-if?" scenarios while manipulating variables that can affect the transport risk. *Subjective assessment* involves experts' opinions, perhaps maintained as a KBS. *Bayesian analysis* involves optimal combinations of multiple evidence in a statistically optimal manner (see Harwood et al. 1989; Saccomanno et al. 1989).

Accident probabilities, material dispersion processes and vulnerable human and physical entities all vary by geographic location and therefore the relevant data can be managed, manipulated and visualized within a GIS environment. Data requirements for HazMat transportation risk assessment include transportation network data, social and demographic variables, and physical factors such as meteorology, terrain, and surface and subsurface hydrological features. Other useful data include previous or estimated accident rates by geographic location, traffic flows within the network (for both accident probability and consequence estimation), the location of special populations, special facilities (e.g., schools) and emergency response capabilities (see Abkowitz, Cheng, and Lepofsky 1990).

*HazMat routing using GIS*

As noted in chapter 8 with respect to transportation planning, GIS provide a particularly effective environment for "what-if?" scenario modeling and assessment.

GIS tools can support calculation and comparison of the attributes associated with alternative HazMat routes. If we can estimate a maximum affected distance from a release at each location within the transportation network, we can use GIS buffering, overlay, and interpolation operations to assign geographic attributes to a candidate HazMat route (Abkowitz, Cheng, and Lepofsky 1990). This allows querying of attributes such as population and natural features at risk. A HazMat dispersion model to capture the material transport process through the airshed or the hydrological system over time can also replace the simple buffer around a route. We can estimate affected areas by applying hydrological transport or airshed plume models at high accident probability or other sample locations within the route (see Chakraborty and Armstrong 1996; Lepofsky and Abkowitz 1993). The calculated route attributes can also be used as input into risk models (Souleyrette and Sathisan 1994).

A GIS environment can also allow visualization and comparison of alternative routes' attributes in conjunction with their geographic setting. Coutinho-Rodriques et al. (1996) develop a spatial decision support system for HazMat routing that includes some simple but clever and powerful visualization tools. The Best AGAinst Least (BAGAL) plot allows the alternative routes to be visualized with respect to their scores against several attributes all normalized to a common scale. The BAGAL plot resembles a bagel (or doughnut), with each attribute scale sharing the same origin, an inner boundary plotted against these scales showing the minimum acceptable score for each attribute, and an outer boundary showing the maximum acceptable score. Each route is plotted as a closed polyline within the BAGAL (showing its performance along the selected attribute dimensions) and as a geographic route within a cartographic display.

GIS tools for calculating and comparing route attributes are particularly powerful when combined with methods for network routing and location. Possible criteria for routing HazMat shipments through a transportation network include minimizing cost, accident probability, expected consequence or total population or subpopulations such as children or elderly at risk (see Saccomanno and Chan 1985; Souleyrette and Sathisan 1994). Brainard, Lovett, and Parfitt (1996) use the network analysis capabilities of a commercial GIS software to generate a set of optimal HazMat routes based on minimizing travel time, accident probability, consequence, population at risk and encouraging particular routes. Their method is predictive rather than prescriptive: It estimates the amount of HazMat flows within a transportation network based on limited data on the actual shipments.

Facility location and routing tools from the decision science, management science and operations research communities also have tremendous potential for enhancing GIS-based routing and decision support for HazMat transportation. List et al. (1991) provide a comprehensive review of the basic formulations. Many HazMat routing models are single objective and find an optimal or good set of routes that minimize some function of either cost or risk. However, single-objective models cannot capture tradeoffs among cost and risk. The cost versus risk tradeoff is important to decision makers since the most appropriate solution is often a compromise between these criteria (particularly if there are decreasing amounts of risk reduction with added cost). Zografos and Davis (1989) use *goal programming* (a multiobjective optimization technique that penalizes deviations from optimality for each criterion) to route HazMats based on the general population at risk, special populations at risk, potential property damage, and travel time (as a surrogate for cost). Also available are models that simultaneously locate HazMat disposal sites and route shipments to these sites in order to minimize single or multiple criteria; e.g., see Helander and Melachrinoudis (1997). There are stochastic models that account for random events such as weather (List et al. 1991).

Many of the risk measures used in HazMat routing and location are vague, particularly when they are subjective assessments from experts and/or stakeholders. Klein (1991) uses fuzzy sets to incorporate imprecise information into the transport risk analysis. A routing algorithm considers fuzzy network attributes such as the safety level on each segment, with membership values based on expert panel ratings. *Dynamic programming* techniques that optimize decisions over multiple stages solve for the optimal paths. A facility location model determines the location of the disposal facilities in addition to the HazMat flows.

## Emergency management and evacuation routing

### Emergency management

*Emergency management* refers to using scientific, technological and management tools to deal with extreme events that injure human beings, damage property, and disrupt community life. These extreme events can be natural, technological or human-caused (the latter sometimes purposeful as with terrorist activities). The emergency management process can be divided into four distinct phases, namely, mitigation, preparedness, response after a disaster and recovery. The recovery phase transitions into a new round of mitigation, preparedness and so forth as humans (hopefully) learn from the disaster experience, adapt to the hazardous environment and prepare for future extreme events (Cova 1999).

Cova (1999) provides a good review of the literature on GIS in emergency management. *Mitigation* activities supported by GIS include risk mapping and assessment. This involves answering geographic questions such as "What are the geographic variations in hazards and vulnerability?" and "What location and transportation strategies can be used to reduce the effects of a hazard and human vulnerability to that hazard?" HazMat transportation analysis is an example of attempted mitigation within a specific domain. Mitigation occurs when tools are used to reduce risk through changes in the land-use and transportation system, improving emergency response planning and increasing response capabilities in high hazard or vulnerability areas.

From the perspective of GIS, the *preparedness* and *response* phases are highly integrated: GIS tools used for plan formulation in the preparedness phase are often used for plan execution in the response phase. Geographic questions answered in this phase include determining the spatial extent and expected losses of an affected area, finding the best routes for emergency response vehicles and determining the populations to be evacuated and the evacuation plans. Other requirements include the need for real-time georeferencing of emergency locations and emergency callers (as in enhanced or E911 emergency calling systems). We will discuss the role of GIS in emergency vehicle response and evacuation planning below.

GIS can support the coordination of logistic and planning activities during the *recovery* phase after a disaster. This includes detailed spatial accounting of the losses, tracking recovery and rebuilding efforts, developing an historical spatial database of hazard events and their consequences and education of the public on risks, mitigation and preparedness/response plans. Logistics tools (see chapters 5 and 11) can also aid in transporting food, medicine, building supplies, personnel and other resources into the affected area.

*Emergency vehicle routing and facility location*

The role of GIS in emergency vehicle routing is straightforward but involves some difficult technical issues. For tactical operations, the software system must be able to locationally reference a reported emergency with respect to the transportation network, identify the *emergency response service* (ERS) facility that should respond to the emergency and determine the shortest path between facility and the emergency. All of this must occur very quickly, ideally fast enough to support decision making in real-time. Also desirable is the ability to track the vehicle using embedded GPS receiver technologies and perhaps alter transportation system control signals (such as traffic lights) and variable message signage to improve response time and decrease risk associated with the response. For strategic planning, we can use a GIS in conjunction with facility location and routing tools to determine the optimal or near-optimal locations of ERS facilities based on the historical or estimated spatial demands for emergency response.

*Location referencing in emergency 911 services*  In the United States, "911" was introduced in the 1960s as a universal public access phone number for reporting emergencies via telephony to centralized *Public Safety Answering Points* (PSAPs). *Enhanced 911* (E911) services, developed in the 1970s, automatically provide the phone number and address of the person reporting the emergency. If the caller is using traditional wireline telephony, the telephone number is fixed to a single location. Obtaining locational information is a relatively simple matter of matching the caller's phone number with address data stored in a relational database indexed by phone numbers. (Note that, in general, there is a one-to-many relationship between addresses and phone numbers since a location may have more than one phone line).

GIS address matching can locate a caller within the transportation network given a sufficiently accurate network database with address ranges. We discussed address matching and related data issues in chapter 4. A GIS can allow the caller's address to be linked to other address-indexed data files, such as site and architectural layouts, building materials used in the structure, the presence and type of hazardous materials, available utilities and the presence of special populations (children, elderly, or disabled persons at the address). This can be combined with routing information (see below) to greatly enhance the preparedness of the emergency response team being dispatched to the emergency location.

The increasing use of wireless mobile telephony enhances the ability of the public to report emergencies, particularly in locations without wireline telephony. In 1996, the U.S. Federal Communications Commission (FCC) mandated the development of wireless E911 services. By definition, mobile telephones are not linked to fixed geographic locations. This complicates the E911 location-referencing problem.

There are two basic technologies for location referencing a wireless mobile phone (Rappaport, Reed, and Woerner 1996). One possibility is to use embedded GPS receivers in the mobile phone to collect and report locational information. The recent elimination of selective availability by the U.S. Department of Defense and continued improvements in GPS receiver technology may make this option more attractive in the future. To date, however, this option has not been widely adopted by mobile phone manufacturers and wireless service providers due to the added costs, complexity and power requirements. Also, GPS receivers require line-of-sight communication with at least three (preferably four) of the satellites in the GPS constellation. In urban environments, tall buildings may block reception (Reed et al. 1998; Zagami et al. 1998). In the previous chapter on Intelligent Transportation Systems, we also noted problems with reconciling GPS location references with available transportation network databases due to topological and positional errors in these databases. Location messaging standards such as the U.S. cross-streets profile can help resolve these problems, although empirical tests of these message standards are less than encouraging (see chapter 9).

Another strategy for location referencing mobile phone callers is to exploit the cellular network used to support communications in wireless mobile systems using *radiolocation* methods. Cellular communications systems operate by defining connected regions (*cells*), each associated with a transmission facility or *base station*. The size and shape of each region depends on the transmission quality from that location. A mobile phone located within a cell uses its base station to relay signals. The phone must be "handed-off" to the next base station when crossing cell boundaries. In practice, transmission cells have fuzzy boundaries and overlap. Therefore only coarse georeferencing is required. Wireless E911 service creates demand for higher resolution locational referencing and matching these references to digital geographic data. In the United States, FCC requirements mandate that these systems generate location references with positional error less than 125 meters 67% of the time (Zagami et al. 1998).

Radiolocation strategies for deriving locational information from wireless transmission signals exploit the fact that radio waves move at a constant rate (the speed of light) along straight lines in 3-D Euclidean space. One strategy is to measure the *angle of arrival* (AOA) of signals from multiple base stations at the mobile unit. Simple triangulation provides the estimated mobile unit location based on the known locations of the base stations. However, this system requires installation and precise calibration of a custom antenna array; this can be expensive. *Time-based strategies* can exploit the existing cellular antenna hardware. One time-based strategy is to measure the *time of arrival* (TOA) for a signal at multiple base stations. Similar to the strategy behind GPS, this defines spheres in Euclidean space centered on each base station that ideally (but rarely if ever in practice) intersect at a point corresponding to the mobile unit's location. This requires the mobile unit to report time accurately. Embedded GPS receivers in the mobile unit can improve time referencing accuracy, but as mentioned above coverage can be poor in urban environments. The *time difference of arrival* (TDOA) eliminates the need for accurate time measurement at the mobile unit. In this strategy, we measure the differences between arrival times at multiple pairs of base stations. This defines hyperbolae with foci on base stations indicating a constant time difference isocurve between a base station and some other base station. TDOA requires precise time synchronization among base stations; this can be obtained through GPS receivers at each base station (which are almost certainly visible to a sufficient number of satellites due to transmission requirements). However, this system requires higher bandwidth for transmitting time references among mobile and base stations, possibly restricting the ability of the network to provide other information services (Caffery and Stüber 1998; Reed et al. 1998; Zagami et al. 1998).

Sources of error caused by the communication technology include multipath propagation, refraction/deflection and cross-channel interference. Multiple signals arriving at a receiver from the same source can obviously confound the AOA strategy since the signal that followed the direct path must be discerned. In time-based strategies, multiple signals arriving within the same discrete clock cycle of the system cannot be distinguished. Signals can also be refracted or deflected by the atmosphere and other media and hence arrive at the receiver without taking the direct path. Interference from other channels can also affect timing references. Benchmarking with simulated error suggests that TOA outperforms AOA by a substantial amount, although this advantage decreases with larger numbers of base stations (Caffery and Stüber 1998). Hybrid AOA/TOA systems may provide greater accuracy, particularly in challenging environments for signal propagation (Reed et al. 1998).

*Routing algorithms and information services for emergency vehicles*   The emergency vehicle routing problem requires solving the shortest (least cost) path problem from the nearest available ERS facility to the georeferenced location of the emergency. The overwhelming criteria is minimum travel time, although other criteria such as using major routes and minimizing the number of left turns may also be factors. Ideally, up-to-the-minute network travel time data is available through ITS data collection devices or fast estimation methods. An optimal route is required very quickly; this may dictate parallel processing implementations or heuristics such as A* or restricting the search to certain levels of the network hierarchy (except possibly in the neighborhood of the emergency). See chapter 5 for discussion of shortest path heuristics and parallel implementations.

GIS can greatly enhance emergency vehicle routing by providing ancillary geographic information and supporting data collection for on-the-fly routing and providing routing information for subsequent response teams. The *Integrated Network Fire Operations* (INFO) system developed by the city of Winston-Salem, North Carolina (U.S.) provides a good case in point. INFO integrates a commercial GIS to provide a comprehensive suite of emergency routing support services. INFO address matches the emergency location obtained through E911 services and determines an effective route from the ERS facility. In-vehicle displays provide route information along with other critical information linked to the emergency address (such as registered HazMats, building plans, number and locations of physically challenged occupants) drawn from a database stored in the mobile unit. The displays are touch sensitive and response teams can identify any blocked or congested streets for on-the-fly updating of the emergency vehicle routes (ESRI 1999a).

*Locating emergency response service facilities*  A geographic factor contributing to emergency vehicle response times is the location of the ERS facilities. As we noted in chapter 6, since maximum response times from ERS facilities is an important locational criterion, we can treat the ERS facility location problem as *set covering* and *maximal covering location problems* (SCLP and MCLP, respectively). The SCLP requires finding the number and locations of facilities such that all demand points are "covered" within a prespecified travel time (or cost). The MCLP requires finding the locations for a fixed number of facilities such that the maximum amount of demand is covered; this is more realistic since available budget often restricts the number of facilities to be opened. We can measure spatial demand for these facilities by using surrogate measures such as population, vulnerable subpopulations, historical incidence patterns or estimates of future incidence patterns. We also may wish to consider vulnerability measures (e.g., property values). Although the MCLP and SCLP are *NP*-hard and difficult to solve optimally, very effective heuristics are available (see chapter 6).

While the SCLP and MCLP are good approximations of the ERS facility location problem, these basic formulations ignore some complex factors inherent in emergency response systems (Marianov and ReVelle 1995). One factor is different levels of service. For example, EMS teams are sometimes configured as either "basic life support" or "advanced life support" teams with different capabilities, costs, expected utilization rates and covering requirements (see Fotheringham and O'Kelly 1989). The problem is not only to decide which facilities to open but also determine the level of service at each open facility by assigning different types of response units to the open facilities. The *Facility Location Equipment Emplacement Technique* (FLEET) model is a customized MCLP that locate prespecified numbers of the two types of fire-fighting units with different covering requirements (Marianov and ReVelle 1995; Schilling et al. 1979). The FLEET solution method solves a linear approximation and uses branch and bound to resolve nonsensical fractional allocations of equipment. (See chapters 5 and 6 for brief discussions of these techniques and references to the literature.)

Another complicating factor that should be considered is the reliability of the emergency response system. If a response unit is busy at an emergency it is not available to respond if other emergencies occur. We may wish to build "back-up" coverage in the system to minimize this possibility. A deterministic strategy is to measure and maximizes the amount of redundancy in the system, where "redundancy" is the difference between the number of facilities covering a demand location and the minimum number required (usually one). *Backup covering models* maximize redundancy while still requiring or maximizing coverage of demand points by the first facility. The trick is to distribute the redundant coverage evenly since there is a tendency for highly weighted demand locations to have multiple redundant coverage with most other demand locations only being singly-covered (Marianov and ReVelle 1995). Hogan and ReVelle (1986) solve this problem by maximizing first and second coverage only; multiple coverage beyond the second facility is ignored. See Daskin, Hogan, and ReVelle (1988) for a review of these and related models.

Another approach is to calculate the actual redundancy in the system by estimating the likelihood that a response unit will be busy based on the expected demand for emergency services. The *maximum expected covering location model* (MECLM) maximizes the expected population covered given a fixed number of facilities (Daskin 1983). The expected coverage of a demand location is calculated through the probability that at least one of its covering facilities will be available at a given moment in time. We estimate this probability by assuming a systemwide probability that any single facility is occupied with an emergency (a *busy fraction*) and a binomial distribution for calculating the probability that multiple covering facilities will be occupied. In a different approach, the *Probabilistic Set Covering Location Problem* (PSCLP) minimizes the number of facilities to be sited such that the probability that each demand location has an available facility within the covering distance exceeds a minimum threshold (ReVelle and Hogan 1988, 1989). The model allows a facility's busy fraction to vary based on the demand in its coverage area. ReVelle and Hogan (1989) formulate a maximal covering version, the *Maximum Availability Location Problem* (MALP), that sites a fixed number of facilities to maximize the number of demand points covered with a probability greater than a minimum threshold (Marianov and ReVelle 1995). Ball and Lin (1993) formulate a maximal covering model with system reliability measures that consider service call rates from demand nodes and service availability at facilities over time using Poisson stochastic processes.

Variations on the $p$-median or "location-allocation" model (see chapter 6) have also been used to solve ERS facility location problems. Mirchandani and Reilly (1987) formulate a $p$-median location model that considers the probability of emergencies in demand zones and the expected response times (and their variances) for primary and backup units based on availability and demand zonal geometry. Fotheringhamn and O'Kelly (1989) incorporate a spatial interaction model describing demand for emergency services into a $p$-median model for ERS facility location with two levels of services.

ERS facility location also depends on the dispatching policy and the routes used between the facility and the emergency location. Dispatching policy refers to which facility should respond if the emergency requires more than one response team. Most of the facility location models discussed above use simple Euclidean distance or average travel times (with the exception of Mirchandani and Reilly 1987); this does not capture the expected response times conditioned by the transportation network. Daskin (1987) formulates a covering-like model that simultaneously solves the ERS facility location, dispatching and routing problems for a single-service level. The multiobjective model treats travel times within the network as stochastic and considers the mean and variance of travel times within the network for single responses and covariances of travel times for vehicles in double responses. The model uses these estimates to calculate the probability that a demand location can be covered by two response teams within a specified covering time.

As this quick review should suggest, the basic covering and $p$-median location problems discussed in chapter 6 are at the foundation of some sophisticated ERS facility location models that can capture the complexity of real-world ERS systems. This review should also suggest that the state-of-the-art is well beyond these basic covering location problems. However, at the time of this writing commercial GIS software that incorporate facility location modeling tools do not go much beyond the basic covering formulations. This is somewhat understandable since sophisticated and realistic ERS location models are special purpose and probably will not have wide, mass-market appeal. The increasing availability of customizable and extensible GIS software compliant with standard programming environments will allow these and other specialized tools to be effectively developed and implemented, even for single-application deployments. These custom ERS facility tools will necessarily be more complex than the basic facility loaction models, but proper design of user interfaces, decision support and geographic visualization tools as well as on-line documentation and knowledge-based systems developed within the GIS environment can help users apply these sophisticated models appropriately.

### Evacuation planning

*Evacuation flow routing* Evacuation planning involves finding optimal routes for moving people within an affected area to safe locations outside that area. The usual criterion is minimizing travel time, although risk may also be an issue (e.g., increased accidents or if the evacuation paths can be affected by the emergency). Since time is often critical, many evacuation plans require individuals to use personal transportation to leave the affected area. Typically, the origins for this travel tend to be clustered geographically while destinations tend to be dispersed. Consequently, we could view this problem as a special case of the network flow problem discussed in chapter 6. Dunn (1992) discusses the use of an "out-of-kilter" flow algorithm (see Ford and Fulkerson 1962; Fulkerson 1961) within a GIS environment for calculating evacuation routes from one evacuation origin to a single destination and to multiple destinations.

Another property of evacuation events is the large amount of flow that must traverse the subnetwork in a relatively short amount of time. Therefore, evacuation models should capture the dynamic flow, congestion and queuing effects that will occur under these extreme transportation flows. Similar to the network flow simulation models discussed in chapter 6, there are three levels at which we can model evacuation flows within transportation networks (Pidd, Eglese, and De Silva 1997). *Macroscopic* models treat the evacuation flows at the aggregate level, typically as a problem in fluid dynamics. These methods are fast and scalable so they are useful for quick, even real-time, but rough answers. Examples include NETVAC1 (Sheffi, Mahmassani, and Powell 1982) and MASSVAC (Hobeika and Jamie 1985).

Southworth and Chin (1987) provide an example of macro-level evacuation modeling. They adopt the MASSVAC model for dam failure events, an application that requires very quick answers. MASSVAC combines a temporal flow loading relationship to describe evacuation "travel demand" over time. The model loads flow onto the network according to a logistic curve:

$$P(t) = \{1 + \exp[-\alpha(t-\beta)]\}^{-1} \qquad (10\text{-}4)$$

where $P(t)$ is the cumulative percentage of flow loaded on the network at (discrete) time $t$ and $\alpha, \beta$ are parameters controlling the steepness and length of the curve (respectively). Calibration requires finding the parameter values that minimizes the total evacuation time; this can be viewed (crudely) as a problem in finding the optimal evacuation schedule. A spatial interaction model (see chapter 8) provides the "trip distribution" of evacuees to shelters. A stochastic user equilibrium model (SUO; see chapter 6) determines the likely routes from the origin-destination flows between evacuation origins to the shelters. The Dial (1971) mulitpath algorithm determines the routes from the origins to destinations for each time period based on SUO principles. The SUO flow pattern captures variations in perceptions and route decision-making; this is appropriate given the highly unusual travel situation being modeled. However, a SUO flow pattern evaluated across successive time intervals cannot capture real-world dynamic flow propagation and is therefore inappropriate unless the time intervals are long.

At the other end of the continuum are *microscopic* evacuation models that represent individual-level vehicles (Pidd, Eglese, and De Silva 1997). These can capture more realistic behaviors such as car-following, platooning, queuing and even irrational responses to random events such as vehicle breakdowns or infrastructure failures. Due to their flexibility, the microsimulation models mentioned in chapter 6 can be adapted for these purposes.

Figure 10-3 shows a conceptual design for a microscopic evacuation simulation model that incorporates communication, traffic monitoring/control, and related geographic information technologies for evaluating the use and performance of information system under different evacuation plans. This design is a generalized version of a model developed by Stern and Sinuany-Stern (1989). An *evacuation information system* provides evacuation plans to households; this could be a component of an intelligent transportation system (ITS; see chapter 9). Household properties such as size and car availability are simulated based on the aggregate statistical data for each evacuation location. Expected preparation time and information lag components capture the expected time for evacuees to react to evacuation information. A network flow simulator estimates the likely routes and network performance (e.g., clearance time) given the flows from evacuation origins to destinations. A traffic monitoring and control system determines traffic control strategies and updates evacuation plans to be disseminated by the evacuation information system.

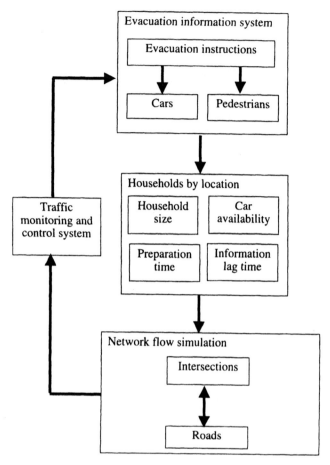

Figure 10-3: Conceptual design for a microscopic evacuation model with geographic information technology

In the middle ground between the macro and micro are *mesoscopic* evacuation models. Again, similar to the flow simulation models discussed in chapter 6, these models are "halfway" between macro- and microscopic simulation models. Some models describe the distributions of flow conditions within the network while others aggregate vehicles to packets or convoys so that the system is event driven (e.g., a vehicle's current condition only changes when it moves or a similar event occurs) Examples of mesoscopic evacuation models include the Integrated Emergency Management System developed by the U.S. Federal Emergency Management Administration (FEMA) (see Pidd, Eglese, and De Silva 1997).

*Decision support for evacuation flow routing*  As we have stressed throughout this book, the network algorithms for solving evacuation flow problems do not require a GIS environment. GIS can support evacuation modeling and plan formulation through geographic data integration and data management capabilities. Data preparation to support evacuation modeling is particularly burdensome: High accuracy and spatial resolution are required to capture the dynamic congestion and queuing conditions that affect travel times and total clearance times during these unusual and complex network events. GIS can make data preparation activities much less onerous (Pidd, Eglese, and De Silva 1997).

GIS-based visualization and decision support tools are very useful in evacuation planning. An emergency situation requiring evacuation is stressful and time pressured decision environment. Evacuation planners must be able to understand evacuation plans and communicate instructions clearly and quickly. An early example of the use of visualization and decision support is the evacuation software system developed by Southworth and Chin (1987). Their system integrates the evacuation modeling system with a state-of-the-art (at the time) graphics program (MVOPL) for geographic visualization of evacuation origin-to-shelter flows, evacuation routes and link flow volumes. The *Transportation Evacuation Decision Support System* (TEDSS) (Hobeika, Kim, and Beckwith 1994) continues this development trajectory. TEDSS integrates the MASSVAC evacuation system with decision support and knowledge-based tools for evaluating evacuation plans. In addition to evacuation routes and link flows, information generated by MASSVAC can be used to identify highly congested links and network clearance times. TEDSS allows the user to specify "what-if" evacuation scenarios and compare simulated results using these performance measures. TEDSS also maintains a knowledge base derived from actual evacuations, including maximum evacuation times, public response behavior, location of emergency response vehicles, capabilities for shoulder use on highways by emergency vehicles and network closures. The knowledge base guides the development of evacuation management strategies. An induction engine updates the knowledge base based on real and simulated evacuations, supporting additional search for effective evacuation plans.

The *Configurable Emergency Management and Planning System* (CEMPS) is an evacuation decision support system that integrates commercial GIS software with an evacuation microsimulation model (Pidd, Eglese, and De Silva 1997). In addition to the geographic data management and geographic visualization tools, CEMPS uses network routing and allocation tools in the GIS to support a customized object-oriented evacuation simulation module. A menu driven graphical user interface allows the user to query and visualize solution information from the microsimulation as well as ancillary geographic data.

*Identifying evacuation bottlenecks using GIS* In a very innovative approach, Cova and Church (1997) sidestep the complex issues involved in evacuation flow modeling and instead use network analytical tools within a GIS environment to identify probable evacuation "bottlenecks." Bottlenecks are likely congestion locations in the network during an evacuation event; these are network locations where a simple measure of evacuation difficulty (the ratio of maximum flow to network capacity) is high. This worse-case measure ignores the temporal spread of evacuation flows. Therefore, their system is a vulnerability analysis tool rather than an operational support tool. However, it can clearly identify potential problem locations within the network, suggesting strategic design changes or where to concentrate emergency response capabilities for tactical responses.

Central to the Cova and Church (1997) evacuation vulnerability model is a network construct known as *critical cluster*. Solving the critical cluster problem for a particular root node identifies the potentially worse evacuation scenario faced by that node with respect to evacuation difficulty. This is a special case of a more general graphic theoretic problem known as *partitioning problems* that attempt to divide a graph into sub-graphs no greater than given maximum size such that the connections among subgraphs are minimized. The critical cluster problem requires finding the subnetwork no greater than a specified maximum size connected to a root node that maximizes its flow to capacity ratio, assuming that all flows originating within that sub-network travel through that node. This problem can be stated as an integer programming (IP) problem: this is a constrained optimization problem where solutions are required to have integer values (typically, binary; see chapters 5 and 6). It is difficult to solve the IP to optimality, so Cova and Church (1997) develop a heuristic solution algorithm that grows a cluster from the root node based on expected gains in the flow to capacity ratio.

Cova and Church (1997) loosely couple the critical cluster software with a commercial GIS using file import and export tools. The critical information required by the cluster software is the *forward star structure* (FSS) for a node in the transportation network. Recall from chapter 5 that a FSS is an efficient data storage strategy that organizes network data by nodes and the arcs leaving each node. The FSS is central to many network algorithms including the shortest path problem. Tighter integration using the GIS software macro language proved infeasible. The relatively glacial execution speed of the interpreted macro language meant that the FSS could not be accessed at a fast enough rate for reasonable execution times with respect to large real-world networks. Dynamic link libraries within the Windows environment that allow compiled software to make FSS queries from the GIS can allow tighter and more efficient integration. See chapter 7 for a discussion of GIS integration and customization strategies.

## Conclusion

In this chapter, we discussed principles and GIS tools for assessing the environmental impacts and mitigating the effects of hazards related to transportation facilities and systems. The tight interlinking between transportation and geographic environment in this problem domain requires detailed and sometimes complex measurements of environmental characteristics and transportation system performance. We noted that geographic data error can have substantial and often ill-understood effects on these geographic information products generated from these input data. We also discussed the role of GIS and transportation in environmental hazard and risk analysis, either in the mitigation phase

(HazMat routing) or in the response phase (emergency vehicle routing and evacuation management).

In the next and final chapter of this book, we consider the role of GIS in solving logistics problems. Logistics involves planning and managing flows of material, services, information or capital within an organization. Since many organizations are geographically dispersed, logistics increasingly involves sophisticated geographic information and communication technologies to monitor, coordinate, and optimize these intricate systems.

# 11

# Logistics

*Logistics* refers to the planning and management of material, service, information or capital flows within an organization (Cox 1999). The organization is typically a private sector enterprise, although military and government organizations also require logistic services. Since many organizations are geographically dispersed, sometimes at global scales, logistics involves planning and managing transportation networks and/or services. Logistics increasingly involves sophisticated information and communication technologies to coordinate these dispersed and intricate systems.

This chapter, written by Bruce Ralston, considers the interface between GIS and logistics, particularly within business environments. It may seem obvious that GIS has something to offer logistics. Logistics is mainly concerned with issues surrounding the transportation or transmission of materials, services and information across geographic space. There are some impressive GIS tools designed for analyzing and managing transportation systems. These include network routing (particularly traveling salesman, vehicle routing, and related procedures) and facility location algorithms (see chapters 5 and 6).

An increasing trend in many private sector enterprises is the deployment of *just-in-time* (JIT) logistic systems. These attempt to reduce costs and risks associated with storage of materials by timing the system such that the materials arrive at a facility just as they are needed in a process. This requires intricate and efficient timing of shipments. Many of the intelligent transportation systems user services discussed in chapter 9 are relevant to JIT systems. These include commercial vehicle operations such as electronic clearance, weigh-in-motion, automated roadside safety inspection, onboard safety monitoring systems, automated administrative processes, freight mobility systems, and hazardous materials incident response. Travel and transportation management ITS services, particularly as implemented within proprietary networks, support vehicle tracking and en-route information provision, sometimes in response to unplanned disturbances (e.g., accidents, construction, and bad weather).

In this chapter, we are particularly concerned with the mismatch between narrow spectrum of commercial GIS software and the full range and complexity of most logistics problems. Commercial GIS applications to logistics usually address narrowly defined problems and often simplify the relationship between *shippers* (those who wish to move goods) and *carriers* (those who actually move the goods). This simplification is due to many factors, including the complexity of many logistics problems, the biases of GIS software vendors, and the limits of the current state-of-the-art in transportation theory and methods.

There is ample opportunity for transportation analysts in general, and the GIS-T community in particular, to play an important role in the evolving world of logistics. Several trends are combining to provide new and exciting applications of GIS tools. Many companies are investing in new information systems that allow for the capture and real time analysis of business transactions. Leading industries are beginning to rethink their corporate culture, trying to have their various units (for example, procurement, manufacturing, marketing, logistics, and sales) act in harmony. Further, different firms that together form a *chain* from raw materials through manufacturing to delivered goods are finding they all can benefit if they harmonize their operations. To do this, these organizations require better tools for handling and analyzing geographic information. They also need the proper platforms and models for solving optimization problems that maximize the space and time utility of their products. GIS has the potential to address many of these needs.

Logistics is being transformed by the revolution in how businesses capture, share, use, and evaluate information. Computer hardware and software have advanced to the point where we can handle large quantities of geographic data and begin to evaluate the contribution of a firm's activities to its costs or profitability. Deregulation has led to an environment in which decisions on route choice, mode choice, warehouse location and customer allocation more closely reflect those we would expect based on classical spatial economic theories and models—some of which have been integrated into existing GIS packages. In North America, the emergence of *third-party logistics* (3PLs) firms has given rise to an industry whose focus, in many respects, is to act as *spatial brokers* who try to efficiently organize movements in space and time. It is an interesting time with much opportunity. However, it is fair to say that many transportation analysts and GIS-T practitioners largely have been left out of the loop. This is at best curious. At the worst, it represents a failure on our part, and one that is more profound than providing the wrong answers to logistics questions. We, the GIS-T community (both academic and vendors), in many cases are asking the wrong questions.

This chapter begins with a discussion of logistics analysis. As with many aspects of modern commerce, logistics is undergoing a revolution driven by deregulation, and advances in information technology and electronic commerce. In many respects there are striking parallels between the revolution in business information systems and GIS. This is followed by a discussion of areas in which GIS can aid logistics analysis, with reference to some specific examples. We also review some of the currently available logistics solutions available from the major commercial GIS software vendors. The chapter ends with a discussion of the weaknesses in the current state-of-the-art in GIS for logistics and presents some areas for future research.

## Supply chains and logistics

Logistics is fluid, vague and difficult to define in concrete and detailed terms. In part, this is due to the rapidly changing way in which business is conducted. Logistics is defined by the Council of Logistics Management as:

> part of the supply chain process that plans, implements, and controls the efficient, effective flow and storage of goods, services and related information from the point of origin to the point of consumption in order to meet customers' requirements. (CLM 1998)

Note the reference to the *supply chain process*. This is a relatively new term, and it reflects a holistic view of business, with the emphasis not on the firm (as in classical economics) but on the relationships between the various actors involved in delivering goods and services.

*Supply chains* constitute "the integration of business processes from end user through original suppliers that provides products, services and information that add value for customers" (Monczka et al. 1997). Thus, the supply chain approach, of which logistics is an important part, looks at the relationships between actors, and the movements of goods, value and information. The goal is to manage these relationships and movements so that some measure of economic attainment, usually measured in profits, costs, or customer service, is optimized. This requires synthesizing large amounts of information, sharing information within and between companies in ways that are understandable to all, making decisions that affect the space and time utility of products and services, and understanding the distribution of customers, suppliers, and inventory. These are all tasks at which GIS can excel.

Logistics was not always viewed as an activity that could contribute to the smooth, efficient functioning of firms. In a famous paper in *Fortune Magazine* in 1962, Peter Drucker referred to logistics as "Economy's Dark Continent." That is, logistics was a cost sink to firms and what went on there was poorly understood. One was forced to ask, "How much of the distribution cost is really 'value added' and how much is merely 'waste added?'" (Corbin 1970). Much has changed since that paper. The last two to three decades have seen many economies move to deregulation in transportation. We also have witnessed the growth of global markets. These trends, along with advances in information gathering techniques and computational power, now allow firms to harmonize their purchasing, production, shipping, and storage activities, to track the source of their costs and profits, and to consider the array of transportation and storage options available to them.

The results have been dramatic. Logistics costs as a percentage of Gross Domestic Product (GDP) in the United States declined from approximately 17% in 1980 to just over 10% in 1993 (Delany 1998). This mirrored a decline in inflation over the same period, although it is not clear which is cause and which is effect. In a similar time frame (1983–1997), the value of inventory as a percentage of U.S. GDP has fallen from 24% to less than 17%. To squeeze even more savings out of the logistics arena, firms are gathering and analyzing information on logistics activities. As a result, logistics is no longer the "Dark Continent" of business. It is an activity about which much data are gathered and analyzed. In the "9th Annual State of Logistics Report," Delany (1998, p. 6) states:

> Since 1982, the improvements in inventory efficiency have been huge as we learned how to replace inventory with more nimble and reliable transportation service. The challenge for logistics managers now is to replace inventory and transportation with advanced planning systems and communications.

A key to enhancing logistics value is to make sense out of all the information being captured. Indeed, a major contribution of GIS to logistics is simply to allow firms to understanding the huge volume of data they are now collecting. As DeWitt, Langley, and Ralston (1997, p. 35) state:

> Perhaps most important to logistics is the ability of GIS to integrate all components in the logistics chain and to display their relationships in an intuitive and understandable manner. Before companies can re-think their logistics operations . . . they must first understand their current situation . . . GIS gives logistics an effective decision support capability.

GIS professionals interested in logistics must look at two related questions. Given the current functions in GIS, what can GIS-T contribute to the study of logistics? What are the current opportunities for improving GIS for logistics?

### GIS and logistics management

There are many potential applications of GIS technology to the logistics problems facing businesses and other organizations. We will discuss GIS applications in three major domains. First is GIS for *descriptive analysis*; in other words, for summarizing and exploring logistics data. Second, we will discuss the role of GIS in *logistics information management*, that is, monitoring and supervising a logistics operation. Third, we will discuss the role of GIS in *prescriptive analysis* to support strategic planning and decision making when configuring the system.

### Descriptive analysis: Visualizing and querying logistics data

GIS-based visualization and database querying are the most obvious areas of contribution for GIS in logistics. Geographic visualization can illuminate hidden information and relationships in logistics data, yielding new insights for effective planning and management. Most companies keep information on the locations of inputs, production, storage, and markets as well as on flows between these locations. By geocoding these locations, integrating the information in different geographic databases and building the proper flow relationships, it is possible to track logistics costs and productivity at a very detailed level.

In the spring of 1997, a large U.S.-based retail chain purchased an even larger competitor. The chain went from having three *distribution centers* (DCs) and 1,200 stores to having nine distribution centers and over 4,000 stores. Before management could even consider optimizing their supply chain, they needed to see where their stores were located, the relationships between stores and DCs, and how the movements of goods through their system affected costs. Stores acquired their stock from distribution centers, and each store was assigned to a DC. When a store's DC carried a needed commodity, the DC would pick the good form the warehouse and ship it to the store. This resulted in what were called *pick costs* and *direct shipment costs* (respectively). When a DC did not carry the good required by one of its stores, the good was picked at another DC and then shipped to the store's DC. The good was then off-loaded and transferred onto a truck that would go to the store. The costs involved in this case are pick costs at the origin DC, a transshipment cost (DC to DC), a cross-docking cost at the receiving DC, and a direct shipment cost from the DC to the store.

The levels of organization and types of movements in the scenario described above generated many different types of data. An additional complexity is that each DC had its own handling and cross-docking costs that vary by product. Transportation costs also varied by product. A simple question facing the corporation was "Can we see the direct logistics costs for each unit of product, by type, at each retail outlet? Can we also show the flows involved in getting those products from DCs to stores?"

An advantage of logistics shared by many transportation systems is their network foundation. Since networks are discrete structures, we can store these data using a relational format (see chapter 2). Although limited with respect to managing transportation facility data, logistics data is often flow based and therefore more easily represented using the relational and perhaps matrix data models (see chapter 3). To answer the questions posed in the previous paragraph, the analyst constructed several relations, including:

```
Store Demand (Store ID, Demand by Product)

Direct Shipment (DC ID, Store ID, Transport Cost per unit
by product)

DC (DC ID, pick cost per unit by product, supply by product,
cross-dock cost per unit by product)

Transshipment (From DC ID, To DC ID, Transport cost per unit
by product)
```

Underlines indicate primary keys. Functional dependencies maintained through foreign keys (see chapter 2) include:

```
Store Demand.Store ID → Direct Shipment.Store ID,
DC.DC ID → Direct Shipment.DC ID
```

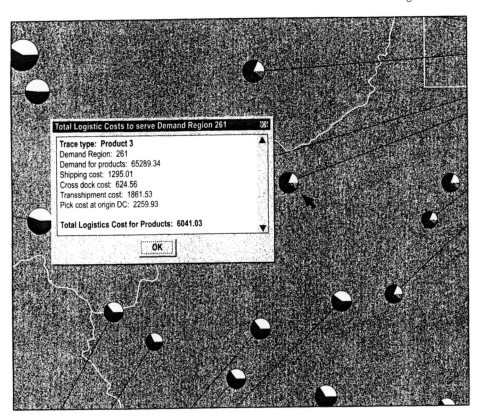

Figure 11-1: Example query of product movements within a logistics system

```
DC.DC ID → Transshipment.From DC ID
DC.DC ID → Transshipment.To DC ID
```

After geocoding the stores and DCs, it is possible to map all the direct (out-of-pocket) costs associated with the goods on the shelf in each store using the linkages described above. Figure 11-1 shows an example. In this figure we show the movements for a particular type of product. The pie slices represent the direct logistics costs (transportation from DC to store, pick costs, cross-docking costs, and transshipment costs) at each store for this category.

The previous example, explained in more detail in Bennett (1998), illustrates how relationships between data tables can be exploited, so that some of the costs of logistics activities can be analyzed. There are other logistics costs that were not considered. In particular, inventory carrying costs and other time-based measures of logistics costs are missing. Nonetheless, the management of this company found maps of this type quite useful.

GIS spatial analysis tools such as buffering, overlay and interpolation cannot be found in other information management tools. These capabilities allow on-the-fly generation of new spatial entities and spatial queries. In the retail chain

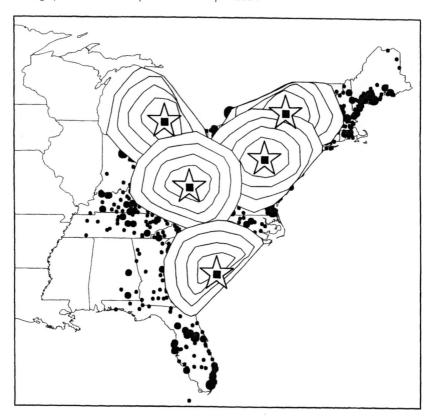

Figure 11-2: Buffers corresponding to national-level distribution centers in a logistics system

example discussed above, buffering, overlay, and routing combined to provide management with insights regarding how its distribution system compared with those of its competitors. In this case, it was possible to geocode the location of retail outlets and distribution centers for each of the firm's major competitors. With these two GIS layers, and a highway database, the analyst generated buffers around each retail chain's distribution centers in increasing highway mileage increments. See figure 11-2 for an illustration.

The highway mileage buffers allowed the analyst to calculate the percentage of retail outlets within given distance zone of the closest DC in the competing operation. Using spatial query tools, the analyst constructed histograms indicating the percentage of stores within each zone. See figure 11-3. Management then used this information to see how their location strategies match up against their competitions' strategies. Interestingly, no two chains appeared to have the same spatial strategy.

### GIS as a platform for logistics information management

Many GIS are interoperable (at some level) with other database management systems (DBMS), including systems that can handle ill-structured data (e.g., imagery) and display of real-time tracking data (e.g., captured through vehicle-

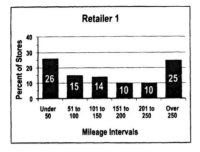

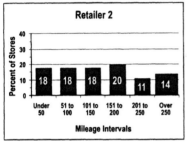

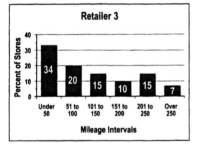

Figure 11-3: Query results: retail outlets by geographic proximity to distribution centers

mounted GPS receivers). Increasing numbers of GIS applications support database access and display over the Internet. All these capabilities, along with the ability to display large amounts of information visually, combine to make GIS a powerful information integration tool for managing logistics operations.

The ability to integrate information is particularly important when we consider supply chain management. Efficient supply chains are those where information flows in both directions. Information is the facilitator that makes the flow of product and money between the links of the chain possible. Supply chains that do not work well are those with information "firewalls" between the various components that impede the flow of goods and money. Gathering, analyzing, and sharing information as key to the success of logistics operations is not lost on the logistics industry. The self-proclaimed strategy at one logistics firm is "to leverage three leading technologies, namely electronic data interchange (EDI), imaging, and the Internet" (Delany, 1999).

*Enterprise resource planning software*

The emphasis on information gathering and analysis has pushed many firms to embrace *enterprise resource planning* (ERP) software. ERP software are information systems for comprehensive management and coordination of financial, manufacturing, sales, distribution and human resources across an enterprise (see Jacobs and Whybark 2000; Shtub 1999). ERP software is often quite expensive; a single company's implementation can exceed US$100 million. Although these systems have proponents and detractors, an undisputed fact is the enormous amount of data collected by these systems.

The similarities between the debates surrounding ERP and GIS are quite striking. First, much like the GIS marketplace, there are a few large firms that dominate the ERP software market (e.g., SAP, Oracle, and PeopleSoft), and their proponents and critics are quite vocal. Second, successful ERP systems are seen as more than just software packages. They are viewed as tools that allow firms to look at their operations in new ways, that is, to change the way firms think about problems. This is an argument that has been made about GIS—that it helps us to think about problems in new ways. Third, the success of an ERP adoption depends on expert management (see Ploszaj 1999). Anyone experienced in implementing GIS in an organization will recognize the uncanny parallels between these two technologies. Finally, as in the GIS user community, there are increasing calls in the ERP user community for more open systems.

Given the similarities between these two GIS and ERP technologies, it is not surprising that there should be ties between them. ERP software allows firms to capture more data than ever before, but making it understandable and allowing access to it are major hurdles. GIS is ideally suited to these tasks, particularly for logistics where the subject matter deals with locations of supplies and demands and the flows between them. SAP and ESRI have developed interfaces between their products, resulting in an ArcView extension for SAP Release 3 (ESRI 1999c). Similarly, Oracle and MapInfo have formed a partnership to market the "Spatial Internet Solution on Oracle8$i$" (MapInfo 1999). Both of these initiatives illustrate the potential for using GIS as a data integration and access tool in a large database environment.

Like much of the IT industry, GIS has moved from the mainframe to the personal computer and now to the client-server model, particularly over the web. The same is true for ERP software. Web based GIS is still in its early stages, but several tools now exist for developing web applications (see chapter 9 for a discussion of distributing geographic information). In the client-sever arena the major GIS vendors are most actively pushing technologies, such as Active X components, that support mapping capabilities on the server in response to requests from the client.

*Event tracking data*

One task for which GIS is particularly well suited is *event tracking*. The ability to locate and track assets is of particular interest to carriers. For example, Schneider National, Inc., North America's largest trucking firm, has GPS units in each of its tractors. (There are approximately 14,000 tractors.) Recently, they have put transponders on all their trailers (over 43,000) that utilize low earth orbit satellite data and messaging communication services (Schneider National 1999). The ability to track all their assets makes fleet management and shipment tracking possible throughout North America, even when trailers are not connected to tractors, as is the case when they piggyback on rail.

If shipments have a method for reporting their location, via GPS or other location tracking systems, these data can be displayed and updated in real time. The ability to display such information is built into several commercial GIS packages. Also see chapter 8 for a related discussion in location tracking in activity-based travel demand analysis and chapter 9 for an extensive discussion of communicating and integrating real-time coordinate referencing data with digital geographic data in intelligent transportation systems.

### GIS for prescriptive analysis

"Prescriptive analysis" in logistics involves finding an optimal or near-optimal system configuration. This includes flow of materials, services or information among components such as factories, warehouses, distribution centers, retail outlets, and customer locations. This can also involve the location of these components.

Flow and location are two sides of the same coin in a logistics system. The location of factories, warehouses, DCs, and retail outlets determines the optimal flows of materials among these system components. At the same time, system flows determine the optimal locations of the facilities. Although solved more appropriately as a simultaneous system, we often solve these separately, that is, flows given a set of facility locations or facility locations given a set of flows. This is for tractability but also since the entire system may not be under the control of the logistics operations. For example, some locations such as facilities (under short-term planning horizons) or customer residences may be fixed and exogenous to the problem.

Combining logistics and GIS tools can provide a highly effective suite of techniques for solving logistics problems. Similar to the transportation planning process (see chapter 8), GIS can change the flow of information and the decision processes involved in logistics analysis and planning. The recent experience of a U.S. company, Procter and Gamble provides an excellent case study.

In 1993, Procter and Gamble (P&G) began a "fundamental reexamination of P&G's North American product supply chain" that combined GIS and optimization models of P&G's operations (Camm et al. 1997, p. 129). The reexamination placed particular emphasis on plant location and consolidation and on the assignment of markets and distribution centers to production plants. The P&G analysts first considered building a large, comprehensive operations research model of their activities. However, building such models takes a long time and requires developing mathematical models that take into account all salient relationships and constraints. Large mathematical models are often "back room" operations that do not give members of management a sense of involvement. P&G had existing software for logistics optimization. However, since it was run on a mainframe, management could not easily change assumptions or input new scenarios. Nor did the existing system allow for interactive "drill-down" on data to examine particular cases and determine why the model made a given decision (Camm et al. 1997).

P&G built a modeling system that combined a facility location model for distribution centers (a *p*-median problem), a transportation problem for product sourcing and a commercial GIS package (MapInfo). The system produced cost savings of over US$250 million annually. There were other more subtle but perhaps more important impacts.

GIS helped sell the entire modeling process to the product management teams at P&G. The use of GIS "greatly facilitated user ... acceptance of the analytical techniques." It helped the decision makers at P&G better understand the relationships between constraints (capacities), costs, and possible solutions (distribution system design). Combining the mathematical models with the GIS "provided a laboratory in which product-strategy teams could test ideas and develop insights." A further benefit of the use of GIS for solution visualization was that it highlighted counterintuitive assignments of customers to DCs. With further study, it became clear that such assignments were the result of database errors. The leaders of the P&G concluded that GIS led to a renaissance of logistics analysis at P&G (Camm et al. 1997).

Because of increasing demands for transportation logistics tools, many commercial GIS software vendors have incorporated prescriptive logistics tools in their off-the-shelf and toolkit systems. Other systems involve integration between GIS and special-purpose logistics software. The experience of P&G and other companies clearly indicates that these systems can produce substantial returns on investment.

Although the integration of GIS and prescriptive logistics tools is welcome, we must be careful not to sacrifice sophisticated routing, scheduling and location tools for better geographic information processing. GIS vendors have traditionally implemented these tools in limited and inflexible ways. While this is sufficient for many users, it prevents the advanced analyst who has a deeper understanding of these problems from tailoring the heuristic for the particular situation at hand.

Another problem with many combined GIS and logistic software tools is inflexibility when representing the logistics problems within the GIS. Since the data models in a GIS are designed for representing physical artifacts, they are relatively poor at representing many of the logical decisions in logistics problems. Many real-world problems involve multiple modes and time periods, constraints that are specific to a company, or other special characteristics that make using standard, built in approaches problematic.

A third problem is that logistics reorganization and supply chain optimization are often not well-defined problems. As Camm et al. (1997) state, there is a need for "a simple interactive tool that would allow product-strategy teams to quickly evaluate options (choices of plant locations and capacities), make revisions, evaluate the new options, and so on ... a system that would guide users to better options in an evolutionary fashion." GIS software tools are becoming better in this respect over time, although fully developed spatial decision support systems (see chapter 8) are still not evident.

*Routing and scheduling tools*

Routing and scheduling tools combined with GIS data integration and management capabilities can greatly improve the efficiency of logistics systems. A well-known example is SEARS (Mitchell 1997). SEARS has an extensive home delivery operation. The traditional routing and scheduling method involved manual assignment of shipments to trucks and constructing tours for these trucks that began and terminated at a central depot. The result was 4-hour delivery windows that were met about 80% of the time.

A new home delivery scheduling GIS adopted by SEARS automates much of the process described above. First, the system uses address-matching tools to locate customers. The software assigns the georeferenced customers to truckloads using a (proprietary) algorithm that takes into account the volume and weight restrictions of trucks. The assignment procedure also considers crew capabilities; for example, only certified technicians can hook up a gas appliance. The spatial distribution and delivery time windows of stops are considered. The routing algorithm determines the order, based on the time windows, in which the stops should be visited. It then routes the trucks to honor the time windows while minimizing the cost of transportation. The result has been that SEARS' time windows have been cut from 4 hours to 2 hours, and the proportion of on-time deliveries has increased from 80% to as much as 95%. SEARS now uses the system to route forklifts in warehouses to optimize the picking of products.

The SEARS example is a particularly successful case of applying routing, scheduling and GIS tools. As discussed in chapter 5, there are a large number of techniques available for solving routing and scheduling problems. In particular, we discussed two very important routing problems, namely, the *traveling salesman problem* (TSP) and the *vehicle routing problem* (VRP).

The TSP involves finding a network tour (i.e., a path that starts and ends at the same node) that covers a given set of nodes (known as *stops*). The VRP is essentially a fleet version of the TSP that routes a fixed number of delivery or pickup vehicles through a number of demand locations such that the total cost is minimized and vehicle capacity constraints are not violated. Typically, there is a designated location known as the *depot* that is not a member of the demand locations. The depot is where all vehicles must start and end their tours.

The TSP and VRP are enormously difficult problems. There are a large number of methods for solving the TSP and VRP. Powerful but complex optimization strategies can solve enormous TSPs (over 7,000 stops). Common heuristic strategies for the TSP include *construction procedures* and *improvement procedures*. Construction procedures start with an arbitrary node and add nodes to the tour based on some selection rule. Criteria for selecting the next node on the tour include *arbitrary insertion, nearest neighbor, nearest insertion, farthest insertion,* and *cheapest insertion*. Improvement procedures start with an arbitrary solution and attempts to improve it through some perturbation. For example, the *r-opt strategy* perturbs a current solution by changing $r$ components, with larger $r$ meaning better but more computationally intensive search.

Powerful optimization methods are also available for the VRP. Heuristics include *cluster first-route second, route first-cluster second, construction procedures,* and *improvement procedures*. "Cluster first-route second" first groups demand nodes and then designs optimal routes independently for each cluster using a TSP method. "Route first-cluster second" is the reverse: We first construct a large route for all stops then partition it into smaller, feasible routes. Construction and improvement procedures for the VRP are similar to the TSP.

Solution methods for the TSP and VRP involve trading-off time complexity and performance. This trade-off depends on the properties of the particular problem being solved. For example, some solution heuristics are more effective if the travel costs are metric (obey symmetry and triangular inequality properties) versus non-metric. Network travel costs are usually nonmetric at large map scales (e.g., urban scales) but become metric at small map scales (e.g., regional scales and above). It is also possible to combine smaller heuristics into effective composite procedures. An example is the Golden et al. (1980) composite procedure for the Euclidean TSP (discussed in chapter 5) that uses a fast procedure to obtain a good initial solution and more computationally demanding procedures to fine-tune the solution. Finally, side constraints such as time windows for delivery can help narrow the search for an optimal or near-optimal solution by eliminating many solutions.

GIS software that includes TSP and VRP tools tend to implement a single or very few solution procedures. While this is sufficient for many users and for many problems, it prevents the advanced user from tailoring solution methods when solving complex and large problems. It also prevents the analyst from developing composite solution procedures from simpler techniques. Using high-level macro languages to link separate software tools is not always a viable option since software commands usually involve considerable computational overhead. The rise of component-based software that allows the analyst to generate custom software by linking small footprint GIS objects using OLE/COM-compliant languages is more promising. However, very large and complex problems that require high efficiency may dictate the use of independent software written in a lower level language (e.g., C) with direct access to the GIS data files. Results can be ported from and to a GIS for geographic data management and visualization/querying.

*Flow modeling tools*

In chapter 6, we mentioned the *minimum cost flow problem* (MCFP) as a fundamental flow problem for uncongested networks. We can consider a logistics systems as an abstract network without congestion, although when these systems are embedded within a transportation system congestion can become a factor (see Miller, Wu, and Hung 1999).

The MCFP determines the minimum cost routes through a network for a set of fixed origin-destination flows. Special cases of this problem include the *transportation problem* where source nodes (e.g., factories) are directly linked to demand nodes (e.g., retail outlets) and the problem is to find the minimal cost flows such that all supplies and demands are satisfied. The *transshipment problem* allows a set of intermediate nodes (e.g., warehouses). The maximum flow problem determines the throughput capacity of a network. All of these problems can be solved using the simplex method for linear programming (LP) problems or specially tailored simplex-like methods.

Some commercial GIS software incorporate mathematical programming algorithms, albeit on a limited scale. TransCAD, for example, includes tools for solving the transportation and related problems. Powerful, special-purpose LP software can allow for more flexibility in solving flow cost problems, particularly with respect to sensitivity analysis. Many LP software allow the analyst to compute the *dual* of the given LP. The dual is the inverse of the original problem. Solving the dual generates solution values known as *shadow prices*. These show the value (i.e., the improvement in the objective function) that can be obtained by relaxing the corresponding constraint (see Hillier and Lieberman 1990). This can guide rational planning of logistics systems by indicating the components or linkages that should be improved and the rational limit of these investments. This full spectrum of LP tools is usually not available within GIS toolkits for solving the transportation and related problems.

An example of combining GIS with powerful LP tools is the TRAILMAN system developed for the United States Agency for International Development (Ralston, Liu, and Zhang 1990). TRAILMAN solves food distribution problems for multimodal logistics networks over a multi-period time horizon. When food aid is needed in emergency situations, logistics experts must determine the best ways to move food from donor countries to distribution centers in deficit regions. This usually takes place over several months and can involve several modes of transportation. Decisions about where to put food into storage and when to distribute it (typical logistics inventory control issues) are also critical. Key questions addressed by TRAILMAN include the following:

- What is the least expensive way to move food aid from supply points to demand points so that all demands are met?
- What is the fastest way to move food aid from supply points to demand points so that all demands are met?
- What is the best combination of time and cost minimization?
- What is the maximum amount of food aid that can be moved from supply points to demand points in a fixed amount of time?
- What is the shortest distance, cost, time, or cost and time path between any two nodes by a given set of modes?
- What is the best strategy for moving food aid from major distribution centers to outlying demand nodes?

Geographic data required for the food aid distribution problem include the supplies and demands in each period, the storage capacities at different locations, and the network costs and capacities. The multi-modal logistics network can include roads, rails, and water transport with modal transfer nodes. Figure 11-4 illustrates the geometric configuration of an intermodal transfer node. This view is sufficient for cartographic display. A greatly expanded view of a transfer node is required for the LP to capture all of the logistical decisions involved at these locations. Critical properties are that intermodal transfers need not be symmetric in either cost or capacity. Similar to the intersection turn direction problem discussed in chapter 3, we have expanded from one node/six arcs to five

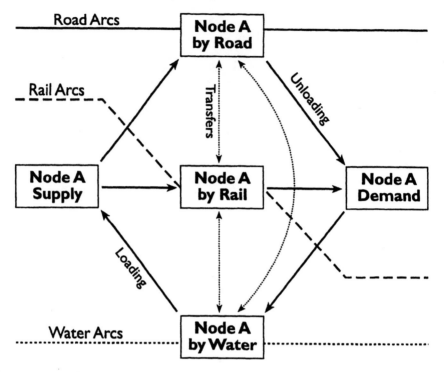

Figure 11-4: Logical configuration of an intermodal transfer point in a logistics system

nodes/eighteen arcs (the six original arcs, six intermodal transfer arcs, three loading, and three unloading arcs).

Clearly, the physical transportation network is not sufficient to represent the range of intermodal transfer decisions required for the multimodal logistics problem. If these logical relationships are not present in the LP database, we will not be able to capture some important movements in logistics problems, such as movements in to and out of storage. One strategy is to build *logical nodes* and *logical arcs* out of functional dependencies among the tables in the georelational database describing the modal networks and other facility data. This structure becomes more complex when we add time to this problem, since we need to link network entities for one period to those from the preceding and following periods.

Figure 11-5 illustrates output from the TRAILMAN system. This illustrates visualization of optimal flows and system components with nonzero shadow prices. Depending on the system component, the shadow prices tell us the marginal value of more supply at a given location, the marginal savings of increasing capacity along a link or the marginal value of increasing storage capacity.

*Facility location tools*

The P&G case study discussed above illustrates the power of integrating facility location tools with GIS. The facility location model identified inefficient distrib-

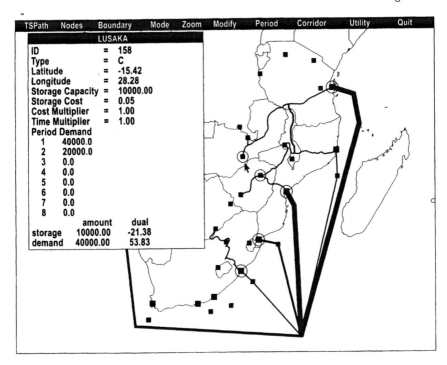

Figure 11-5: Visualizing optimal flows and shadow prices using TRAILMAN

ution centers (DC) that should be closed. This was a large part of the US$250 million cost savings. The visualization capabilities allowed decision-makers to identify and modify undesirable assignments of customers to DCs, enhancing the usefulness of the model results for implementation in the real world.

In chapter 6, we discussed facility location within networks. These problems involve siting $p$ facilities among a set of $m$ discrete candidate sites to serve $n$ demand sites to maximize some objective. *Median problems* require finding locations that minimize some function of total or average cost. Variations on this problem include maximum distance constraints, maximizing attendance, using power function of distances to improve equity. *Center* problems attempt to minimize the maximum travel cost incurred based on travel to/from the facility. *Requirements problems* locate facilities according to some prespecified performance such as being within a set distance of a facility. Variations on these problems exist depending on whether the facilities and demand are located at nodes, at vertices, or anywhere.

A wide range of median problems as well as "demand covering" requirements problem can be derived as a special case of the *p-median* or "location-allocation" problem. Heuristics for the *p*-median problem include drop, add, interchange and global/regional interchange. The drop strategy starts with all sites occupied and drops facilities one at a time until $p$ remain. The add strategy starts with no sites occupied and adds one at a time until $p$ are occupied. Interchange locates $p$

random facilities and exchanges occupied sites with unoccupied sites until no improvement can be gained. The global/regional interchange algorithm (GRIA) using drop at a "global" level then fine-tunes solutions through interchange within each facility's service region.

Solution algorithms for the $p$-median perform very well and can tackle very large problems. Interchange heuristics are particularly powerful. Nevertheless, like many heuristics, the performance of these methods degrades as the problem size increases. In chapter 7, we mentioned that few off-the-shelf GIS support a very simple yet powerful method for improving performance of interchange for very large problems. Since interchange heuristics require an arbitrary (random) starting configuration, we can improve solution quality simply by *restarting* the heuristic several times with different random starting solutions and then choosing the best of the results across all restarts. Experimental results from Church and Sorenson (1996) clearly suggest that restarting interchange heuristics is a powerful strategy that does not require a major increase in computational time.

## Some complex aspects of logistics problems

### Inventory-transport cost tradeoffs

Given the orientation of GIS to geographic information management and analysis, it is not surprising that the early applications of GIS in logistics focused on routing and scheduling. Since routes are often measured in terms of time or distance, and these are often directly related to distance measures, the relevance of GIS to such problems appears obvious. But the almost singular focus of GIS on routing ignores other measures that are important to logistics decisions.

There are aspects of logistics that are not explicitly related to flows and routing. Strategic planning decisions in logistics must also consider inventory-related costs such as product picking from storage and cross-dock transfers in warehouses. There are cost functions that explicitly account for the tradeoff between inventory costs and transportation costs, but they have yet to be integrated into commercial GIS software (Langley 1981).

Even more critical is the lack of time-based measures and analytical tools in commercial GIS logistics software. In many of these toolkits, time is considered only as it relates to time constraints on routing. However, time is as important as transportation costs in some logistics problems. Time is obviously important in highly valued and perishable products. The rise of express package delivery systems and the trend to JIT systems over the past two decades have placed new emphasis on temporal representation and tools in logistics software toolkits. JIT systems in particular attempt to reduce inventory costs through intricate timing of logistics systems so that materials or services arrive just as they are needed in a process. To capture the properties of the logistics systems, GIS software must include time sensitive measures such as inventory carrying costs and in-transit inventory carrying costs.

Ignoring inventory costs can result in inferior and even nonsensible answers to logistical problems. Figure 11-6 illustrates the results from a real case study

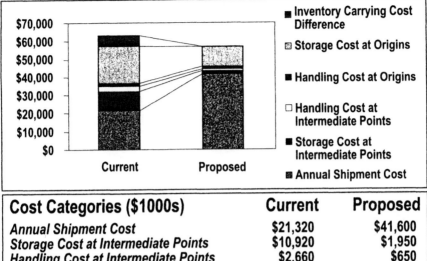

| Cost Categories ($1000s) | Current | Proposed |
|---|---|---|
| Annual Shipment Cost | $21,320 | $41,600 |
| Storage Cost at Intermediate Points | $10,920 | $1,950 |
| Handling Cost at Intermediate Points | $2,660 | $650 |
| Handling Cost at Origins | $1,300 | $1,430 |
| Storage Cost at Origins | $20,930 | $11,700 |
| Inventory Carrying Cost Difference | $6,500 | $0 |
| Average Cycle Time | 14 days | 2 days |
| Total ($1000s) | $63,630 | $57,330 |

Figure 11-6: Example of inventory costs in a supply chain analysis

from a logistics analysis in the semiconductor industry. Note that the proposed strategy is to nearly double transportation costs. This is a result that most commercial GIS software would not generate. In other words, many commercially available GIS software for logistics are not giving us wrong answers but instead are leading us to ask the wrong questions!

### The logistics marketplace

The GIS view of logistics is usually quite simplified. When we consider solutions available from vendors, we see that the shipper is usually the carrier, or at most there is one shipper and one carrier. While this may work for modeling home deliveries by an appliance store, it does not hold true for large operations. Current commercial GIS-based logistics tools have no capability for handling the interaction between shippers and carriers. The role of *third-party logistics* (3PL) firms is completely missing from commercially available GIS-based logistics tools. Yet GIS can aid all players in the logistics marketplace (shippers, carriers, and 3PLs) in understanding their markets.

Third-party logistics have the opportunity to become the geographic brokers between shippers and carriers. Consider, for example, the *continuous move problem*. When shippers negotiate shipping costs with carriers, a key issue is rolling stock utilization. For example, if a shipper can chain its shipments so that a carrier's rolling stock is kept busy, then that carrier should give the shipper a discount on successive legs since the carrier's rolling stock is generating continuous revenue.

How would a shipper know if such cost saving opportunities exist? Usually, there is a *deadhead zone* around deliveries. These are areas for which there is no charge for traveling from the last delivery to the next pickup. There also are time constraints, for example, the next pickup must not only be geographically proximal to the previous drop off, but it must be within a certain number of hours. GIS tools can be used to solve this problem. The analyst can build network-based buffers around deliveries, overlaying these with a GIS layer showing pickup locations and selecting all pickups that lie within the spatial buffer and within the time window between each delivery and the next pickup. This produces a GIS layer of potential continuous move opportunities such as illustrated in figure 11-7. By having this information, the shipper is in a better position to negotiate discounts from the carrier.

The simple GIS methodology outlined above is tractable for situations involving small numbers of shipments and perhaps a few carriers. However, large corporations often have too many shipments and deal with too many potential shippers to explore such problems using simple buffer and overlay operations. This is also true for 3PL service providers who deal with multiple shippers and carriers. In such instances, GIS can be used to generate inputs into a mathematical programming technique and to study the results in a manner similar to those discussed earlier in this chapter. The shipper or 3PL can use the GIS to generate legs or movements. These legs would be space and time specific (e.g., a leg from Chicago to Pittsburgh on Tuesday). The shipper or 3PL could then enter an electronic bidding process where carriers could bid on single legs or bundles of legs.

The result of the above process could be a matrix like figure 11-8. Each bid would be a column in the matrix. At the end of the bidding process, the shipper would need to find that combination of bids that minimized transportation costs while covering all the legs. (This problem is a type of set covering facility location problem that can also be solved using the techniques discussed in chapter 6.)

## Conclusion

The field of logistics, like that of geography, has undergone a revolution due to changes in information technology. The amount and timeliness of information on logistics has exploded. GIS are well-suited to make sense of this avalanche of data. Perhaps even more profound has been the change in the political economy in many countries. Deregulation, combined with intermodalism, has led to decision making that more closely reflects what we would expect from economic and location theories.

Logistics 399

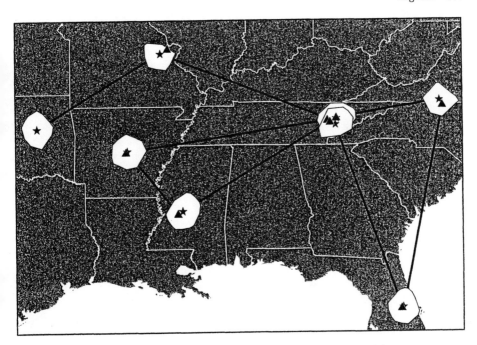

Figure 11-7: Buffering to identify potential continuous move opportunities

We are no longer facing a world where decisions are made solely by firms. We are now in a world where increasingly decisions are made by alliances which form a supply chain. This requires that actors in the chain (shippers, carriers, manufactures, marketers, and logistics managers) use information to harmonize their activities. Again, GIS can help manage this new environment, both in terms of model formulation and visualization, and in terms of each actor's understanding of the relationships between the various parts of the supply chain. There are many technical issues that need to be resolved, and the GIS-T community faces some

|  |  | Carrier Bids | | | | | | | | |
|---|---|---|---|---|---|---|---|---|---|---|
|  |  | Bid 1 | Bid 2 | Bid 3 | Bid 4 | Bid 5 | Bid 6 | Bid 7 | Bid 8 | Bid 9 | Bid 10 |
|  | Leg 1 | 1 |  | 1 |  | 1 | 1 |  | 1 |  |
| L | Leg 2 |  | 1 |  |  | 1 | 1 | 1 |  | 1 |
| E | Leg 3 |  |  | 1 |  |  | 1 |  | 1 |  |
| G | Leg 4 |  |  | 1 |  | 1 |  |  |  |  |
| S | Leg 5 |  |  |  | 1 |  |  | 1 |  | 1 |
|  | Leg 6 |  |  |  |  | 1 |  | 1 |  |  |
|  | Leg 7 |  |  | 1 |  | 1 |  |  |  | 1 |

Figure 11-8: Matrix of potential continuous move opportunities

Table 11-1: Business Literature Citations in GIS, Transportation, and Logistics, 1986–1995

| Topic | 1994–1995 | 1992–1993 | 1986–1991 |
|---|---|---|---|
| GIS | 280 | 167 | 138 |
| Transportation | 3,980 | 3,466 | 4,726 |
| Logistics | 1,531 | 966 | 1,547 |
| GIS and transport | 12 | 2 | 9 |
| GIS and logistics | 2 | 0 | 0 |

*Source*: From DeWitt and Ralston 1996.

difficult modeling tasks. However, the potential benefits, whether measured in profits, customer satisfaction, or better quality of life, are enormous.

There is a natural relationship between logistics, geography and GIS. Coyle, Bardi, and Langley (1996), a major business logistics text on the revolution in supply chain optimization, discusses topics such as node-arc data models, network design problems and facility location. Also discussed are major location theories that relate location to transportation costs such as von Thünen's theory of land use and productivity, Alfred Weber theory of industrial location and modification to this theory by Edgar M. Hoover and Melvin Greenhut. In other words, topics that are central to location theory and GIS are now becoming major themes in business logistics.

However, if we look at location theory, transportation geography and GIS texts, we see little if any discussion of how information technology has changed logistics and supply chain optimization. A review of literatures and dissertations on GIS and logistics revealed only a few references that combined both GIS and logistics (see tables 11-1 and 11-2). Similarly, the logistics software being offered by GIS vendors tend to embody narrow definitions of what constitutes logistics (usually only transportation costs) and an extremely simplistic view of the logistics industry (i.e., the shipper is also the carrier). It is perhaps not surprising that GIS vendors would first address logistics problems that are tractable (e.g., least cost routing and scheduling). As the software industry matures, more realistic and complex views of logistics may be addressed.

Table 11-2: Theses and Dissertations in GIS, Transportation, and Logistics, 1986–1995

| Category | 1993–1995 | 1988–1992 | 1982–1987 |
|---|---|---|---|
| GIS | 493 | 117 | 24 |
| Transportation | 1,169 | 1,273 | 1,151 |
| Logistics | 1,098 | 946 | 548 |
| GIS and transport | 24 | 4 | 1 |
| GIS and logistics | 9 | 0 | 0 |

*Source*: From DeWitt and Ralston 1996.

# References

Aashtiani, H. Z. and Magnanti, T. L. (1981) "Equilibria on a congested transportation network," *SIAM Journal of Algebraic and Discrete Methods*, 2, 213–226.

Abkowitz, M. and Cheng, P. D. M. (1989) "Hazardous waste materials transport risk estimation under conditions of limited data availability," *Transportation Research Record*, 1245, 14–22.

Abkowitz, M., Cheng, P. D.-M., and Lepofsky, M. (1990) "Use of geographic information systems in managing hazardous materials shipments," *Transportation Research Record*, 1261, 35–43.

Abler, R. F. (1987) "The National Science Foundation National Center for Geographic Information and Analysis," *International Journal of Geographical Information Systems*, 1, 303–326.

Abrahamsson, T. and Lundqvist, L. (1999) "Formulation and estimation of combined network equilibrium models with applications to Stockholm," *Transportation Science*, 33, 80–100.

Adams, T. M., Vonderohe, A. P., Butler, J. A., and Koncz, N. (1999) "Model considerations for a multimodal, multidimensional location reference system," *GIS-T 1999 Proceedings*, American Association of State Highway and Transportation Officials.

Adrabinski, A. and Syslo, M. M. (1983) "Computational experiments with some approximation algorithms for the traveling salesman problem," *Zastosowania Matematyki*, 18, 91–95.

Adriaans, P. and Zantinge, D. (1996) *Data Mining*, Harlow, U.K.: Addison-Wesley Longman.

Aho, A. V., Hopcraft, J. E., and Ullman, J. D. (1974) *The Design and Analysis of Computer Algorithms*, Reading, MA: Addison-Wesley.

Ahuja, R. K., Magnanti, T. L., and Orlin, J. B. (1993) *Network Flows: Theory, Algorithms and Applications*, Englewood Cliffs, NJ: Prentice Hall.

Ahuja, R. K., Mehlhorn, K., Orlin, J. B., and Tarjan, R. E. (1990) "Faster algorithms for the shortest path problem," *Journal of the Association for Computing Machinery*, 37, 213–223.

Alfelor, R. M. (1995) "GIS and the integrated highway information system," in H. J. Onsrud and G. Rushton (eds.) *Sharing Geographic Information*, New Brunswick, NJ: Center for Urban Policy Research.

Alonso, W. (1964) *Location and Land Use*, Cambridge: Harvard University Press.

Anas, A. (1983) "Discrete choice theory, information theory and the miultinomial logit and gravity models," *Transportation Research B*, 17B, 13–23.

Anderson, M. D. and Souleyrette, R. R. (1996) "Geographic information system-based transportation forecast model for small urbanized areas," *Transportation Research Record*, 1551, 95–104.

Anderson, W. P., Kanaroglou, P. S., Miller, E. J., and Buliung, R. N. (1996) "Simulating automobile emissions in an integrated urban model," *Transportation Research Record*, 1520, 71–80.

Angel, S. and Hyman, G. M. (1976) *Urban Fields: A Geometry of Movement for Regional Science*, London: Pion.

Ansari, N. and Hou, E. (1997) *Computational Intelligence for Optimization*, Boston: Kluwer Academic.

Anselin, L. (1988a) "Lagrangian multiplier test diagnostics for spatial dependency and spatial heterogeneity," *Geographical Analysis*, 20, 1–17.

Anselin, L. (1988b) *Spatial Econometrics: Methods and Models*, Dordrecht, The Netherlands: Kluwer Academic.

Anselin, L. (1989) "What is special about spatial data? Alternative perspectives on spatial data analysis," Technical paper 89-4, National Center for Geographic Information and Analysis, Santa Barbara, CA.

Anselin, L. (1992) *SpaceStat: A Program for the Analysis of Spatial Data*, National Center for Geographic Information and Analysis, University of California, Santa Barbara, CA.

Anselin, L. (1993) "Discrete space autoregressive models," in M. F. Goodchild, B. O. Parks, and L. T. Steyaert (eds.) *Environmental Modeling with GIS*, New York: Oxford University Press, 454–469.

Anselin, L. (1996) "The Moran scatterplot as an ESDA tool to assess local instability in spatial association," in M. Fischer, H. J. Scholten, and D. Unwin (eds.) *Spatial Analytical Perspectives on GIS*, London: Taylor and Francis, 111–125.

Anselin, L., Dodson, R. F., and Hudak, S. (1993) "Linking GIS and spatial data analysis in practice," *Geographical Systems*, 1, 3–23.

Anselin, L. and Getis, A. (1992) "Spatial statistical analysis and geographic information systems," *Annals of Regional Science*, 26, 19–33.

Anselin, L. and Getis, A. (1993) "Spatial statistical analysis and geographic information systems," in M. Fischer and P. Nijkamp (eds.) *Geographic Information Systems, Spatial Modeling, and Policy Evaluation*, Berlin: Springer-Verlag, 35–49.

Anselin, L. and Griffith, D. A. (1988) "Do spatial effects really matter in regression analysis?" *Papers of the Regional Science Association*, 65, 11–34.

Anselin, L. and Rey, S. (1991) "Properties of tests for spatial dependence in linear regression models," *Geographical Analysis*, 23, 112–131.

Armstrong, M. P., Densham, P. J., Lolonis, P., and Rushton, G. (1992) "Cartographic displays to support locational decision making," *Cartography and Geographic Information Systems*, 19, 154–164.

Armstrong, M. P., Densham, P. J., and Rushton, G. (1986) "Architecture for a microcomputer-based spatial decision support system," *Proceedings of the Second*

*International Symposium on Spatial Data Handling*, Columbus: OH: International Geographical Union, 120–131.

Aronoff, S. (1989) *Geographic Information Systems: A Management Perspective*, Ottawa, Canada: WDL Publications.

Arthur, R. M. and Waters, N. M. (1997) "Formal scientific research of traffic collision data utilizing GIS," *Transportation Planning and Technology*, 21, 121–137.

Assad, A. A. (1988) "Modeling and implementation issues in vehicle routing," in B. L. Golden and A. A. Assad (eds.) *Vehicle Routing: Methods and Studies*, Amsterdam: North-Holland, 7–45.

Azar, K. T. and Ferriera, J. (1992) "Visualizing transit demand for current and proposed transit routes," *GIS-T 1992 Proceedings*, 133–146.

Azar, K. T. and Ferreira, J. (1995) "Integrating geographic information systems into transit ridership forecast models," *Journal of Advanced Transportation*, 29, 263–279.

Baaj, M. H. and Mahmassani, H. S. (1991) "An AI-based approach for transit route system planning and design," *Journal of Advanced Transportation*, 25, 187–210.

Bachman, W., Sarasua, W., and Guensler, R. (1996) "Geographic information system framework for modeling mobile-source emissions," *Transportation Research Record*, 1551, 123–132.

Bailey, T. C. and Gatrell, A. C. (1995) *Interactive Spatial Data Analysis*, Essex, U.K.: Longman Group Ltd.

Balakrishnan, V. K. (1995) *Network Optimization*, London: Chapman and Hall.

Balas, E. and Christofides, N. (1981) "A restricted lagrangean approach to the traveling salesman problem," *Mathematical Programming*, 21, 19–46.

Ball, M. O. and Lin, F. L. (1993) "A reliability model applied to emergency service vehicle location," *Operations Research*, 41, 18–36.

Band, L. E. (1986) "Topographic partition of watershed with digital elevation models," *Water Resources Research*, 22, 15–24.

Band, L. E. (1999) "Spatial hydrography and landforms," in P. A. Longley, M. F. Goodchild, D. J. Maguire, and D. W. Rhind (eds.) *Geographic Information Systems, Vol. 1: Principles and Technical Issues*, 2d ed., New York: Wiley, 527–542.

Barcelo, J., Ferrer, J. L., Garcia, D., Florian, M., and Le Saux, E. (1998) "Parallelization of microscopic traffic simulation for ATT systems analysis," in P. Marcotte and S. Nguyen (eds.) *Equilibrium and Advanced Transportation Modeling*, Boston: Kluwer Academic, 1–26.

Batty, M. (1976) "Entropy in spatial aggregation," *Geographical Analysis*, 8, 1–31.

Batty, M. (1996) "Visualizing urban dynamics," in P. Longley and M. Batty (eds.) *Spatial Analysis: Modelling in a GIS Environment*, Cambridge, U.K.: GeoInformation International, 297–320.

Batty, M. and Sikdar, P. K. (1982a) "Spatial aggregation in spatial interaction models. 1. An information theoretic framework," *Environment and Planning A*, 14, 377–405.

Batty, M. and Sikdar, P. K. (1982b) "Spatial aggregation in spatial interaction models. 2. One dimensional population density models," *Environment and Planning A*, 14, 525–553.

Batty, M. and Sikdar, P. K. (1982c) "Spatial aggregation in spatial interaction models. 3. Two dimensional trip distribution and location models," *Environment and Planning A*, 14, 629–658.

Batty, M. and Sikdar, P. K. (1982d) "Spatial aggregation in spatial interaction models. 4. Generalisations and large-scale applications," *Environment and Planning A*, 14, 795–822.

Batty, M. and Xie, Y. (1994a) "Modelling inside GIS: Part 1. Model structures, exploratory data analysis and aggregation," *International Journal of Geographical Information Systems*, 8, 291–307.

Batty, M. and Xie, Y. (1994b) "Modelling inside GIS: Part 2. Selecting and calibrating urban models using ARC-INFO," *International Journal of Geographical Information Systems*, 8, 451–470.

Batty, M. and Xie, Y. (1997) "Possible urban automata," *Environment and Planning B: Planning and Design*, 24, 175–192.

Beale, R. and Jackson, T. (1990) *Neural Computing: An Introduction*, Bristol, U.K.: Adam Hilger.

Beardwood, J. E. and Kirby, H. R. (1975) "Zone definition and the gravity model: The separability, excludability and compressibility properties," *Transportation Research*, 9, 363–369.

Beckmann, M., McGuire, G. B., and Winsten, C. B. (1956) *Studies in the Economics of Transportation*, New Haven, CT: Yale University Press.

Bedard, Y., Merrett, T., and Han, J. (2001) "Fundamentals of geospatial data warehousing for geographic knowledge discovery," in H. J. Miller and J. Han (eds.) *Geographic Data Mining and Knowledge Discovery*, London: Taylor and Francis, in press.

Bell, M. G. H. (1983) "The estimation of an origin-destination matrix from traffic counts," *Transportation Science*, 17, 198–217.

Bell, M. G. H. and Iida, Y. (1997) *Transportation Network Analysis*, New York: Wiley.

Bell, M. G. H., Lam, W. H., Ploss, G., and Inaudi, D. (1993) "Stochastic user equilibrium assignment and iterative balancing," in C. F. Daganzo (ed.) *Transportation and Traffic Theory*, New York: Elsevier Science, 427–439.

Bellman, R. E. (1958) "On a routing problem," *Quarterly Journal of Applied Mathematics*, 16, 87–90.

Ben-Akiva, M. E. and Bowman, M. E. (1998) "Activity based travel demand model systems," in P. Marcotte and S. Nguyen (eds.) *Equilibrium and Advanced Transportation Modeling*, Boston: Kluwer Academic, 27–46.

Ben-Akiva, M., Koutsopoulos, H. N., Mishalani, R. G., and Yang, Q. (1997) "Simulation laboratory for evaluating dynamic trafic management systems," *Journal of Transportation Engineering*, 123, 283–289.

Ben-Akiva, M. and Lerman, S. R. (1979) "Disaggregate travel and mobility-choice models and measures of accessibility," in D. A. Hensher and P. R. Stopher (eds.) *Behavioural Travel Modelling*, London: Croom-Helm, 654–679.

Ben-Akiva, M. and Lerman, S. (1985) *Discrete Choice Analysis: Theory and Application to Travel Demand*, Cambridge: MIT Press.

Bennett, D. A. (1997) "A framework for the integration of geographical information systems and modelbase management," *International Journal of Geographical Information Science*, 11, 337–357.

Bennett, D. A. and Armstrong, M. P. (1996) "An inductive knowledge-based approach to terrain feature extraction," *Cartography and Geographic Information Systems*, 23, 3–19.

Bennett, K. (1998) *Using Geographic Information Systems for Business Logistics Analysis*, Master's thesis, Department of Geography, University of Tennessee Knoxville, TN, U.S.A.

Bennett, R. (1994) "Validating a route guidance database," *GIS-T 1994 Proceedings*, 151–163.

Berman, O., Hodgson, M. J., and Krass, D. (1995) "Flow-interception problems," in

Z. Drezner (ed.) *Facility Location: A Survey of Applications and Methods*, Berlin: Springer, 389–426.
Bernard, L., Schmidt, B., Streit, U., and Uhlenkuken, C. (1998) "Managing, modeling and visualizing high-dimensional spatio-temporal data in an integrated system," *GeoInformatica*, 2, 59–77.
Bernstein, D. and Eberlein, X. J. (1992) "A relational GIS-T," *GIS-T 1992 Proceedings*, Portland, OR: American Association of State Highway and Transportation Officials, 79–96.
Bespalko, S. J., Ganter, J. H., and Van Meter, M. D. (1996) "Geospatial data for ITS," in L. M. Branscomb and J. H. Keller (eds.) *Converging Infrastructures: Intelligent Transportation and the National Information Infrastructure*, Cambridge: MIT Press, 209–226.
Bespalko, S. J., Sutton, J. C., Wyman, M., Van Der Veer, J. A., and Sindt, A. D. (1998) "Linear referencing systems and three dimensional GIS," Paper presented at the 1998 annual meeting of the Transportation Research Board, TRB paper No. 981404.
Bhat, C. R. (2000) "A multi-level cross-classified model for discrete response variables," *Transportation Research B*, 34B, 567–582.
Bhat, C., Govindarajan, A., and Pulugurta, V. (1998) "Disaggregate attraction-end choice modeling: Formulation and empirical analysis," *Transportation Research Record*, 1645, 60–68.
Bhat, C. and Koppelman, F. S. (2000) "Activity-based travel demand analysis: History, results and future directions," *Proceedings of the 79th annual meeting of the Transportation Research Board*, Washington, DC, Jan. 9–13, CD-ROM.
Bielle, M., Ambrosino, G., and Boero, M. (eds.) (1994) *Artificial Intelligence Applications to Traffic Engineering*, Utrecht, The Netherlands: VSP.
Biggs, N. L. (1985) *Discrete Mathematics*, Oxford, U.K.: Oxford University Press.
Bishop, I. (1994) "The role of visual realism in communicating and understanding spatial change and process," in H. M. Hearnshaw and D. J. Unwin (eds.) *Visualization in Geographical Information Systems*, Chichester, U.K.: Wiley, 60–64.
Bishr, Y. (1998) "Overcoming the semantic and other barriers to GIS interoperability," *International Journal of Geographical Information Science*, 12, 299–314.
Bitenc, V. (1994) "Geographical data description directory," *Proceedings of the European GIS Conference*, 1358–1362.
Bivand, R. S. (1984) "Regression modeling with spatial dependence: An application of some class selection and estimation techniques," *Geographical Analysis*, 16, 25–37.
Black, W. R. (1992) "Network autocorrelation in transportation network and flow systems," *Geographical Analysis*, 24, 207–222.
Black, W. R. (1995) "Spatial interaction modeling using artificial neural networks," *Journal of Transport Geography*, 3, 159–166.
Black, W. R. and Thomas, I. (1998) "Accidents on Belgium's motorways: A network autocorrelation analysis," *Journal of Transport Geography*, 6, 23–31.
Bodin, L. and Berman, L. (1979) "Routing and scheduling of school buses by computer," *Transportation Science*, 13, 113–129.
Bodin, L. and Golden, B. (1981) "Classification in vehicle routing and scheduling," *Networks*, 11, 97–108.
Bolduc, D. (1999) "A practical technique to estimate multinomial probit models in transportation," *Transportation Research B*, 33, 63–79.
Booch, G. (1994) *Object-Oriented Analysis and Design with Applications*, 2d ed., Redwood City, CA: Benjamin/Cummings.
Boots, B. N. (1999) "Spatial tessellations," in P. A. Longley, M. F. Goodchild, G. J. Maguire,

and D. W. Rhind (eds.) *Geographical Information Systems Vol. 1: Principles and Technical Issues*, 2d ed., New York: Wiley, 503–526.

Boyce, D. E. (1984) "Urban transportation network-equilibrium and design models: Recent achievements and future prospects," *Environment and Planning A*, 16, 1445–1474.

Boyce, D. E. (1990) "Network equilibrium models of urban location and travel choices: A research agenda," in M. Chatterji and R. E. Kuenne (eds.) *New Frontiers in Regional Science: Essays in Honor of Walter Isard, Vol. 1*, New York: New York University Press, 238–256.

Boyce, D. (1998) "Long-term advances in the state of the art of travel forecasting methods," in P. Marcotte and S. Nguyen (eds.) *Equilibrium and Advanced Transportation Modeling*, Boston: Kluwer Academic, 73–86.

Boyce, D. E., LeBlanc, L. J., and Chon, K. S. (1988) "Network equilibrium models of urban location and travel choices: A retrospective survey," *Journal of Regional Science*, 28, 159–183.

Boyce, D. E., Lee, D.-H., Janson, B. N., and Berka, S. (1997) "Dynamic route choice model of large-scale traffic network," *Journal of Transportation Engineering*, 123(4), 276–282.

Boyce, D. E., Zhang, Y.-F., and Lupa, M. R. (1994) "Introducing feedback into four-step travel forecasting procedure versus equilibrium solution of combined model," *Transportation Research Record*, 1443, 65–74.

Brainard, J., Lovett, A., and Parfitt, J. (1996) "Assessing hazardous waste transport risks using a GIS," *International Journal of Geographical Information Systems*, 10, 831–849.

Branscomb, L. M. and Keller, J. H. (1996) "Introduction: Converging infrastructures," in L. M. Branscomb and J. H. Keller (eds.) *Converging Infrastructures: Intelligent Transportation and the National Information Infrastructure*, Cambridge: MIT Press, 1–19.

Branston, D. (1976) "Link capacity functions: A review," *Transportation Research*, 10, 223–236.

Brunsdon, C., Fotheringham, A. S., and Charlton, M. E. (1996) "Geographically weighted regression: A method for exploring spatial nonstationarity," *Geographical Analysis*, 28, 281–298.

Buntine, W. (1996) "Graphical models for discovering knowledge," in U. M. Fayyad, G. Piatetsky-Shapiro, P. Smyth, and R. Ulthurusamy (eds.) *Advances in Knowledge Discovery and Data Mining*, Cambridge, MA: MIT Press, 59–82.

Burdea, G. and Coiffet, R. (1994) *Virtual Reality Technology*, New York: Wiley.

Bureau of Transportation Statistics (1998) *Resource Guide on the Implementation of Linear Referencing Systems in Geographic Information Systems*, Bureau of Transportation Statistics, U.S. Department of Transportation, CD-ROM no. BTS-CD-22.

Burkard, R. E., Deineko, V. G., Van Dal, R., Van Der Veen, J. A. A., and Woeginger, G. J. (1998) "Well-solvable special cases of the traveling salesman problem: A survey," *SIAM Review*, 40, 495–546.

Burns, L. D. (1979) *Transportation, Temporal and Spatial Components of Accessibility*, Lexington, MA: Lexington Books.

Burridge, P. (1980) "On the Cliff-Ord test for spatial autocorrelation," *Journal of the Royal Statistical Society B*, 42, 107–108.

Burrough, P. A. (1986) *Principles of Geographic Information Systems for Land Resource Assessment*, Oxford, U.K.: Clarendon Press.

Burrough, P. A. (1989) "Fuzzy mathematical methods for soil survey and land evaluation," *Journal of Soil Science*, 40, 477–492.

Burrough, P. A. (1998) "Dynamic modeling and geocomputation," in P. A. Longley, S. M. Brooks, R. McDonnell, and B. Macmillan (eds.) *Geocomputation: A Primer*, New York: Wiley, 165–191.

Burrough, P. A. and McDonnell, R. A. (1998) *Principles of Geographical Information Systems*, Oxford, U.K.: Oxford University Press.

Burrough, P. A. and Frank, A. U. (eds.) (1996) *Geographic Objects with Indeterminate Boundaries*, London: Taylor and Francis.

Burrough, P. A., van Deursen, W., and Heuvelink, G. (1988) "Linking spatial process models and GIS: A marriage of convenience of a blossoming partnership?" *GIS/LIS '88 Proceedings*, Vol. 2, 598–607.

Butler, J. A. (1998) *A Comprehensive Plan to Establish and Operate a Regional Geographic Data Network (Geonet)*, Technical report, Department of Engineering, Hamilton County General Government, Chattanooga, TN.

Button, K. J. (1993) *Transport Economics*, 2d ed., Aldershot, U.K.: Edward Elgar.

Caffery, J. J. and Stüber, G. L. (1998) "Overview of radiolocation in CDMA cellular systems," *IEEE Communications Magazine*, 36(4), 38–45.

Câmara, A. S., Ferreira, F., and Castro, P. (1996) "Spatial simulation modeling," in M. Fischer, H. J. Scholten, and D. Unwin (eds.) *Spatial Analytical Perspectives on GIS*, GISDATA IV, London: Taylor and Francis, 201–212.

Camm, J. D., Chorman, T. E., Dill, F. A., Evans, J. R., Sweeney, D. J., and Wegryn, G. W. (1997) "Blending OR/MS, judgment, and GIS: Restructuring P&G's supply chain," *Interfaces*, 27, 128–142.

Campbell, J. (1997) *Map Use and Analysis*, New York: McGraw-Hill.

Cantarella, G. E. and Cascetta, E. (1995) "Dynamic processes and equilibrium in transportation networks: Towards an unifying theory," *Transportation Science*, 29, 305–329.

Carey, M. (1992) "Nonconvexity of the dynamic traffic assignment problem," *Transportation Research B*, 26B, 127–133.

Carey, R., Bell, G., and Marrin, C. (1997) *ISO/IEC 14772-1:1997 Virtual Reality Modeling Language (VRML97)*, The VRML Consortium Inc., available at http://www.web3d.org/fs_technicalinfo.htm.

Carpaneto, G. and Toth, P. (1980) "Some new branching and bounding criteria for the asymmetric traveling salesman problem," *Management Science*, 26, 736–743.

Carpenter, S. and Jones, P. (1983) *Recent Advances in Travel Demand Analysis*, Aldershot, U.K.: Gower.

Carver, S. J. (1991) "Integrating multi-criteria evaluation with geographical information systems," *International Journal of Geographical Information Systems*, 5, 321–339.

Carver, S. J. and Brunsdon, C. F. (1994) "Vector to raster conversion error and feature complexity: An empirical study using simulated data," *International Journal of Geographical Information Systems*, 8, 261–270.

Cascetta, E. and Nguyen, S. (1988) "A unified framework for estimating or updating origin/destination matrices from traffic counts," *Transportation Research B*, 22B, 437–455.

Casillas, P. A. (1987) "Data aggregation and the p-median problem in continuous space," in A. Ghosh and G. Rushton (eds.) *Spatial Analysis and Location-Allocation Models*, New York: Van Nostrand Reinhold, 327–344.

Center for Transportation Research (1994) *Development and Testing of Dynamic Traffic Assignment and Simulation Procedures for ATIS/ATMS Applications*, Technical report DTFH6 1-90-R-00074-FG, Center for Transportation Research, University of Texas at Austin.

Cervero, R. (1986) *Suburban Gridlock*, New Brunswick, NJ: Center for Urban Policy Research.

Chakraborty, J. and Armstrong, M. P. (1996) "Using geographic plume analysis to assess community vulnerability to hazardous accidents," *Computers, Environment and Urban Systems*, 19, 341–356.

Chang, G.-L. and Su, C.-C. (1995) "Predicting intersection queue with neural network models," *Transportation Research C: Emerging Technologies*, 3, 175–191.

Chard, B. M. and Lines, C. J. (1987) "TRANSYT: The latest developments," *Traffic Engineering and Control*, 28 (July/Aug.), 387–390.

Chen, C., Ran, B., and Vonderohe, A. (1997) "Spatial/temporal data modeling and integration problems in dynamic network routing," *GIS/LIS '97 Proceedings*, 849–857.

Chen, H.-K. (1999) *Dynamic Travel Choice Models: A Variational Inequality Approach*, Berlin: Springer-Verlag.

Chen, J., Newkirk, R. T., Davidson, G., and Gong, P. (1994) "The development of a knowledge-based geographical information system for the zoning of rural areas," *Environment and Planning B: Planning and Design*, 21, 179–190.

Chen, M. and Alfa, A. S. (1991) "Algorithms for solving Fisk's stochastic traffic assignment model," *Transportation Research B*, 25B, 405–412.

Cherkassky, B. V., Goldberg, A. V., and Radzik, T. (1993) "Shortest path algorithms: Theory and experimental evaluation," Discussion paper, Department of Computer Science, Stanford University, Stanford, CA.

Chou, Y.-H. (1991) "Map resolution and spatial autocorrelation," *Geographical Analysis*, 23, 228–246.

Chow, V. T., Maidment, D. R., and Mays, L. W. (1988) *Applied Hydrology*, New York: McGraw-Hill.

Chrisman, N. (1997) *Exploring Geographic Information Systems*, New York: Wiley.

Christofides, N. (1976) *Worse Case Analysis of a New Heuristic for the Traveling Salesman Problem*, Report 388, Graduate School of Industrial Administration, Carnegie-Mellon University, Pittsburgh, PA.

Christofides, N. (1985) "Vehicle routing," in E. L. Lawler, J. K. Lenstra, A. H. G. Rinnooy Kan, and D. B. Shmoys (eds.) *The Traveling Salesman Problem: A Guided Tour of Combinatorial Optimization*, New York: Wiley, 431–448.

Christofides, N. and Eilon, S. (1969) "An algorithm for the vehicle dispatching problem," *Operations Research Quarterly*, 20, 309–318.

Christofides, N., Mingozzi, A., and Toth, P. (1980) "Exact algorithms for the vehicle routing problem, based on spanning tree and shortest path relaxations," *Mathematical Programming*, 20, 255–282.

Christofides, N., Mingozzi, A., and Toth, P. (1981) "State-space relaxation procedures for the computation of bounds to routing problems," *Networks*, 11, 145–164.

Christofides, N. and Viola, P. (1971) "The optimum location of multicenters on a graph," *Operational Research Quarterly*, 22, 145–154.

Chung, J.-H. and Goulias, K. G. (1996) "Access management using geographic information systems and traffic management tools in Pennsylvania," *Transportation Research Record*, 1551, 114–122.

Church, R. L., Curtin, K., Fohl, P., Goodchild, M., Kyriakidis, P., and Noronha, V. (1998) "Positional distortion in geographic data sets as a barrier to interoperation," Paper presented at the annual meeting of the American Congress on Surveying and Mapping, Baltimore MD, available at http://www.ncgia.ucsb.edu/vital.

Church, R. L. and ReVelle, C. S. (1976) "Theoretical and computational linkages between the p-median, location set-covering, and the maximal covering location problem," *Geographical Analysis*, 8, 406–415.

Church, R. L. and Sorenson, P. (1996) "Integrating normative models into GIS: Problems and prospects with the p-median model," in P. Longley and M. Batty (eds.) *Spatial*

*Analysis: Modelling in a GIS Environment*, Cambridge, U.K.: GeoInformational International, 167–183.

Clarke, G. and Wright, J. W. (1964) "Scheduling vehicles from a central depot to a number of delivery points," *Operations Research*, 12, 568–581.

Clarke, K. C. and Gaydos, L. J. (1998) "Loose-coupling a cellular auotmaton model and GIS: Long-term urban growth prediction for San Francisco and Washington/Baltimore," *International Journal of Geographical Information Science*, 12, 699–714.

Clarke, K. C., Hoppen, S., and Gaydos, L. (1997) "A self-modifying cellular automaton model of historical urbanization in the San Francisco Bay area," *Environment and Planning B: Planning and Design*, 24, 247–261.

Cliff, A. D. and Ord, J. K. (1972) "Testing for spatial autocorrelation among regression residuals," *Geographical Analysis*, 4, 267–284.

Clifford, J. L., Yermack, L., and Owens, C. A. (1996) "E-ZPass: Case study of institutional and organizational issues in technology standards development," *Transportation Research Record*, 1537, 10–22.

Coelho, J. D. and Wilson, A. G. (1976) "The optimum location and size of shopping centres," *Regional Studies*, 10, 413–421.

Community Research and Development Information Service (CORDIS) (1999) *FP5: Programme Structure and Content*, available at http://www.cordis.lu/fp5/src/struct.htm.

COMSIS (1996) *Incorporating Feedback in Travel Forecasting: Methods, Pitfalls and Common Concerns*, Final report to Federal Highway Administration, U.S. Department of Transportation, Contract no. DTFH61-93-C-00216, COMSIS Corp.

Cook, D., Symanzik, J., Majure, J., and Cressie, N. (1997) "Dynamic graphics in a GIS: More examples using linked software," *Computers and Geosciences*, 23, 371–385.

Cook, W. J., Cunningham, W. H., Pulleyblank, W. R., and Schrijver, A. (1998) *Combinatorial Optimization*, New York: Wiley.

Corbin, A. (1970) "The impact of Drucker on marketing," in T. H. Bonaparte and J. E. Flaherty (eds.) *Peter Drucker, Contributions to the Business Enterprise*, New York: NYU Press, 147–165.

Cormon, T. H., Leiserson, C. E., and Rivest, R. L. (1998) *Introduction to Algorithms*, New York: McGraw-Hill.

Cottrell, W. D. (1992) "Comparison of vehicular emissions in free-flow and congestion using MOBILE4 and highway performance monitoring system," *Transportation Research Record*, 1366, 75–82.

Couclelis, H. (1997) "From cellular automata to urban models: New principles for model development and implementation," *Environment and Planning B: Planning and Design*, 24, 165–174.

Coughlan, J. C. and Running, S. W. (1996) "Biophysical aggregations of a forested landscape using an ecological diagnostic system," *Transactions in GIS*, 1, 25–39.

Council of Logistics Management (1998) *CLM's Definition of Logistics*, available at http://www.clm1.org/Mission/Logistics.asp.

Coutinho-Rodriques, J., Current, J., Climaco, J., and Ratick, S. (1996) "An interactive spatial decision support system for multiobjective hazmat location-routing problems," Working Paper WPS 95-39, Max M. Fisher College of Business, The Ohio State University, Columbus.

Cova, T. J. (1999) "GIS in emergency management," in P. A. Longley, M. F. Goodchild, G. J. Maguire, and D. W. Rhind (eds.) *Geographical Information Systems, Vol. 2: Applications and Management Issues*, 2d ed., New York: Wiley, 845–858.

Cova, T. J. and Church, R. L. (1997) "Modelling community evacuation vulnerability using GIS," *International Journal of Geographical Information Systems*, 11, 763–784.

Cova, T. J. and Goodchild, M. F. (1994) "Spatially distributed navigable databases for intelligent vehicle highway systems," *GIS/LIS '94 Proceedings*, 191–200.

Cowen, D. J. (1988) "GIS versus CAD versus DBMS: What are the differences?" *Photogrammetric Engineering and Remote Sensing*, 54, 1551–1554.

Cox, M. D. (1999) *WWW Virtual Library of Logistics Version 4* (Sept. 1999), Logistics World, Inc.; available at http://www.logisticsworld.com.

Coyle, J. J., Bardi, E. J., and Langley, C. J. (1996) *The Management of Business Logistics*, St. Paul, MN: West Publishing.

Cremer, M. and Keller, H. (1987) "A new class of dynamic methods for the identification of origin-destination flows," *Transportation Research B*, 21B, 117–132.

Cullen, F. H., Jarvis, J. J., and Ratliff, H. D. (1981) "Set partitioning-based heuristics for interactive routing," *Networks*, 11, 125–143.

Current, J. and Min, H. (1986) "Multiobjective design of transportation networks: Taxonomy and annotation," *European Journal of Operational Research*, 26, 187–201.

Current, J. R. and Schilling, D. A. (1987) "Elimination of source a and b errors in p-median location problems," *Geographical Analysis*, 19, 95–110.

Current, J. R. and Schilling, D. A. (1990) "Analysis of errors due to demand data aggregation in the set covering and maximal covering location problem," *Geographical Analysis*, 22, 116–126.

Dafermos, S. (1980) "Traffic equilibrium and variational inequalities," *Transportation Science*, 14, 42–54.

Dafermos, S. (1982) "The general multimodal network equilibrium problem with elastic demand," *Networks*, 12, 57–72.

Daganzo, C. F. (1994) "The cell transmission model: A dynamic representation of highway traffic consistent with the hydrodynamic theory," *Transportation Research B*, 28B, 269–287.

Daganzo, C. F. (1980a) "An equilibrium algorithm for the spatial aggregation problem of traffic assignment," *Transportation Research B*, 14B, 221–228.

Daganzo, C. F. (1980b) "Network representation, continuum approximations and a solution to the spatial aggregation problem of traffic assignment," *Transportation Research B*, 14B, 229–239.

Daganzo, C. F. (1984a) "The distance traveled to visit $N$ points with a maximum of $C$ stops per vehicle: An analytical model and an application," *Transportation Science*, 18, 331–350.

Daganzo, C. F. (1984b) "The lengths of tours in zones of different shapes," *Transportation Research B*, 18B, 135–145.

Daganzo, C. F. (1997) *Fundamentals of Transportation and Traffic Operation*, New York: Pergamon Press.

Daganzo, C. F. and Sheffi, Y. (1977) "On stochastic models of traffic assignment," *Transportation Science*, 11, 253–274.

Dailey, D. J., Fisher, G., and Maclean, S. (1999) "Busview and Transit Watch: An update on two products from the Seattle Smart Trek model deployment initiative," Paper presented at the Sixth Annual World Congress on Intelligent Transport Systems, Toronto, Omtaino, Canada.

Dailey, D. J., Loseff, D., and Meyers, D. (1999) "Seattle Smart Traveler: Dynamic ridematching on the World Wide Web," *Transportation Research C: Emerging Technologies*, 7, 17–32.

Dailey, D. J., Maclean, S., and Pao, I. (1999) *Busview: An APTS Precursor and a Deployed Applet*, Research report, ITS Research Program, University of Washington, Seattle.

Daly, A. (1987) "Estimating tree logit models," *Transportation Research B*, 21B, 251–267.
Damberg, O., Lundgren, J. T., and Patriksson, M. (1996) "An algorithm for the stochastic user equilibrium problem," *Transportation Research B*, 30B, 115–131.
Dana, P. H. (2000a) "Geodetic datum overview," *The Geographer's Craft Project*, Department of Geography, The University of Colorado at Boulder, available at http://www.colorado.edu/geography/gcraft/notes/datum/datum_f.html.
Dana, P. H. (2000b) "Global Positioning System overview," *The Geographer's Craft Project*, Department of Geography, The University of Colorado at Boulder, available at http://www.colorado.edu/geography/gcraft/notes/gps/gps.html.
Dane, C. and Rizos, C. (1998) *Positioning Systems in Intelligent Transportation Systems*, Boston: Artech House.
Dantzig, G. B., Fulkerson, D. R., and Johnson, S. M. (1954) "Solution of a large-scale traveling salesman problem," *Operations Research*, 7, 58–66.
Dargay, J. M. and Goodwin, P. B. (1995) "Evaluation of consumer surplus with dynamic demand," *Journal of Transport and Policy*, 29, 179–193.
Daskin, M. S. (1983) "A maximum expected covering location model: Formulation, properties and heuristic solution," *Transportation Science*, 17, 48–70.
Daskin, M. S. (1987) "Location, dispatching and routing models for emergency services with stochastic travel times," in A. Ghosh and G. Rushton (eds.) *Spatial Analysis and Location-Allocation Models*, New York: Van Nostrand Reinhold, 224–265.
Daskin, M. S., Hogan, K., and ReVelle, C. (1988) "Integration of multiple, excess, backup, and expected covering models," *Environment and Planning B: Planning and Design*, 15, 15–35.
Daskin, M. S. and Stern, E. (1981) "A multi-objective set covering model for EMS vehicle deployment," *Transportation Science*, 15, 137–152.
Date, C. J. (1995) *An Introduction to Database Systems*, 6th ed. Reading, MA: Addison-Wesley.
Davis, T. J. and Keller, C. P. (1997a) "Modelling and visualizing multiple spatial data uncertainties," *Computers and Geosciences*, 23, 397–408.
Davis, T. J. and Keller, C. P. (1997b) "Modelling uncertainty in natural resouce analysis using fuzzy sets and Monte Carlo simulation: Slope stability prediction," *International Journal of Geographical Information Science*, 11, 409–434.
de Berg, M. (1997) "Visualization of TINs," in M. van Kreveld, J. Nievergelt, T. Roos, and P. Widmayer (eds.) *Algorithmic Foundations of Geographic Information Systems*, Lecture Notes in Computer Science 1340, Berlin: Springer-Verlag, 79–97.
de Berg, M. and van Kreveld, M. (1997) "Trekking in the Alps without freezing or getting tired," *Algorithmica*, 18, 306–323.
De Floriani, L., Falcidieno, B., Nagy, G., and Pienovi, C. (1991) "On sorting triangles in a Delaunay tessellation," *Algorithmica*, 6, 522–532.
De Floriani, L. and Magillo, P. (1994) "Visibility algorithms on triangulated digital terrain models," *International Journal of Geographical Information Systems*, 8, 13–41.
De Floriani, L. and Magillo, P. (1999) "Intervisibility on terrains," in P. A. Longley, M. F. Goodchild, D. J. Maguire, and D. W. Rhind (eds.) *Geographical Information Systems, Vol. 1: Principles and Technical Issues*, 2d ed., New York: Wiley, 543–556.
De Floriani, L., Montani, C., and Scopigno, R. (1994) "Parallelizing visibility computations on triangulated terrains," *International Journal of Geographical Information Systems*, 8, 515–532.
de la Barra, T. (1989) *Integrated Land Use and Transport Modelling*, Cambridge, U.K.: Cambridge University Press.

Deen, T. B. (1995) "Theodore M. Matson award—Transportation and a sustainable environment: An opportunity for transportation engineers," *ITE Journal*, 65(10), 18–25.

Defense Mapping Agency (1984) *Geodesy for the Layman*, Report no. DMA TR 80-003, Washington, DC.

Delany, R. (1998) "We, the people, demand logistics productivity," *Ninth Annual State of Logistics Report*, Chicago: Cass Information Systems.

Delany, R. (1999) "A look back in anger at logistics productivity," *Tenth Annual State of Logistics Report*, Chicago: Cass Information Systems.

DeMers, M. N. (1997) *Fundamentals of Geographic Information Systems*, 2d ed., New York: Wiley.

Denardo, E. V. and Fox, B. L. (1979) "Shortest-route methods: 1. Reaching, pruning and buckets," *Operations Research*, 27, 161–186.

Denning, P. J. and Metcalfe, R. M. (eds.) (1997) *Beyond Calculation: The Next Fifty Years of Computing*, New York: Copernicus.

Densham, P. J. (1996) "Visual interactive locational analysis," in P. Longley and M. Batty (eds.) *Spatial Analysis: Modeling in a GIS Environment*, Cambridge, UK: GeoInformation International, 185–205.

Densham, P. J. and Armstrong, M. P. (1998) "Spatial analysis," in R. Healy, S. Dowers, B. Gittings, and M. Moneter (eds.) *Parallel Processing Algorithms for GIS*, London: Taylor and Francis, 387–413.

Densham, P. J. and Rushton, G. (1992a) "A more efficient heuristic for solving large p-median problems," *Papers in Regional Science*, 71, 307–329.

Densham, P. J. and Rushton, G. (1992b) "Strategies for solving large location-allocation problems by heuristic methods," *Environment and Planning A*, 24, 289–304.

Dent, B. D. (1998) *Cartography: Thematic Map Design*, Dubuque, IA: Wm. C. Brown.

Desmet, P. J. J. and Govers, G. (1996) "Comparison of routing algorithms for digital elevation models and their implications for predicting ephemeral gullies," *International Journal of Geographical Information Science*, 10, 311–331.

Desrochers, M., Desrosiers, J., and Solomon, M. (1992) "A new optimization algorithm for the vehicle routing problem with time windows," *Operations Research*, 40, 342–354.

DeWitt, W. J., Langley, C. J., Jr., and Ralston, B. A. (1997) "The impact of GIS on supply chains," *Logistics International 1997*, 34–37.

DeWitt, W. J. and Ralston, B. A. (1996) "GIS: Existing and potential applications for logistics and transportation," *Proceedings, 1996 Business Geographics Meeting*.

Dial, R. B. (1969) "Algorithm 360: Shortest path forest with topological ordering," *Communications of the ACM*, 12, 632–633.

Dial, R. B. (1971) "A probabilistic multipath traffic assignment model which obviates path enumeration," *Transportation Research*, 5, 83–111.

Dial, R. B. (1996) "Bicriterion traffic assignment: Basic theory and elementary algorithms," *Transportation Science*, 30, 93–111.

DiBiase, D. (1990) "Visualization in the earth and mineral sciences," *Bulletin of the College of Earth and Mineral Sciences*, Pennsylvania State University, 59(2), 13–18.

DiBiase, D., MacEachren, A. M., Krygier, J. B., and Reeves, C. (1992) "Animation and the role of map design in scientific visualization," *Cartography and Geographic Information Systems*, 19, 201–214.

Dijkstra, E. W. (1959) "A note on two problems in connection with graphs," *Numeriche Mathematik*, 1, 269–271.

Ding, C. (1994) "Impact analysis of spatial data aggregation on transportation forecasted demand: A GIS approach," *Proceedings of the Urban and Regional Information Systems Association (URISA) Conference*, 362–375.

Ding, C. (1998) "The GIS-based human-interactive TAZ design algorithm: Examining the impacts of data aggregation on transportation-planning analysis," *Environment and Planning B: Planning and Design*, 25, 601–616.

Diplock, G. and Openshaw, S. (1996) "Using simple genetic algorithms to calibrate spatial interaction models," *Geographical Analysis*, 28, 262–279.

Donnelly, J. D. (1993) *Design Considerations of a Spatiotemporal GIS Database for Data Exploration*, Master's thesis, Department of Geography, Florida Atlantic University, Boca Raton.

Dougherty, M. (1995) "A review of neural networks applied to transport," *Transportation Research C: Emerging Technologies*, 3, 247–260.

Dougherty, M. S. and Cobbett, M. R. (1997) "Short-term inter-urban traffic forecasts using neural networks," *International Journal of Forecasting*, 13, 21–31.

Dougherty, M. S. Kirby, H. R., and Boyle, R. D. (1994) "Using neural networks to recognise, predict and model traffic," in M. Bielle, G. Ambrosino, and M. Boero (eds.) *Artificial Intelligence Applications to Traffic Engineering*, Utrecht, The Netherlands: VSP, 233–250.

Dougherty, M. S. and Schintler, L. A. (1997) "Forecasting carbon monoxide concentrations near a sheltered intersection using video surveillance and neural networks: A comment," *Transportation Research D*, 2, 221–222.

Dowsland, K. A. (1993) "Simulated annealing," in C. R. Reeves (ed.) *Modern Heuristic Techniques for Combinatorial Problems*, New York: Wiley, 20–69.

Dueker, K. J. and Butler, J. A. (1997) "GIS-T enterprise data model with suggested implementation choices," Discussion paper, Center for Urban Studies, Portland State University, Portland, OR.

Dueker, K. J. and Butler, J. A. (1999) "A framework for GIS-T data sharing," Discussion paper 99-04, Center for Urban Studies, Portland State University, Portland, OR.

Dueker, K. J. and Vrana, R. (1992) "Dynamic segmentation revisited: A milepoint linear data model," *URISA Journal*, 4, 94–105.

Dunn, C. E. (1992) "Optimal routes in GIS and emergency planning appliactions," *Area*, 24, 259–267.

Earnshaw, R. A., Gigante, M. A., and Hones, J. (1993) *Virtual Reality Systems*, London: Academic Press.

Eastman, R. J., Jin, W., Kyem, P. A. K., and Toledano, J. (1995) "Raster procedures for multi-criteria/multi-objective decisions," *Photogrammetric Engineering and Remote Sensing*, 61, 539–547.

Egenhofer, M. J. (1991) "Extending SQL for graphical display," *Cartography and Geographic Information Systems*, 18, 230–245.

Egenhofer, M. J. (1992) "Why not SQL!" *International Journal of Geographical Information Systems*, 6, 71–85.

Egenhofer, M. J., Glasgow, J., Gunther, O., Herring, J. R., and Peuquet, D. J. (1999) "Progress in computational methods for representing geographical concepts," *International Journal of Geographical Information Science*, 13, 775–796.

Ehlschlaeger, C. R., Shortridge, A. M., and Goodchild, M. F. (1997) "Visualizing spatial data uncertainty using animation," *Computers and Geosciences*, 23, 387–395.

Elmasri, R. and Navathe, S. B. (1994) *Fundamentals of Database Systems*, 2d ed., Redwood City, CA: Benjamin/Cummings.

Emerson, S. L. Darnovsky, M., and Bowman, J. (1989) *The Practical SQL Handbook: Using Structured Query Language*, Reading, MA: Addison-Wesley.

Environmental Systems Research Institute (ESRI) (1998) *Spatial Database Engine*, ESRI White Paper, available at http://www.esri.com.

Environmental Systems Research Institute (ESRI) (1999a) "City of Winston-Salem's In-Vehicle Fire Response Mapping System Wins Global Bangemann Challenge," press release, Environmental Systems Research Institute, Inc., Aug. 13, available at http://www.esri.com.

Environmental Systems Research Institute (ESRI) (1999b) *Customizing ArcInfo 8*, ESRI White Paper, available at http://www.esri.com.

Environmental Systems Research Institute (ESRI) (1999c) *ESRI and SAP: Spatially Enabling SAP R/3*, press release available at http://www.esri.com/.

Erlenkotter, D. (1978) "A dual-based procedure for uncapacitated facility location," *Operations Research*, 26, 992–1009.

Essinger, R. and Lanter, D. (1992) "User-centered software design in GIS: Designing an icon-based flowchart that reveals the structure of Arc/Info data graphically," *Proceedings of 12th Annual ESRI User Conference*, 245–256.

Euler, G. W. and Robertson, H. D. (eds.) (1995) *National ITS Program Plan—Synopsis*, Washington, DC: ITS America.

European Commission (EC) (1997a) *Community Strategy and Framework for the Deployment of Road Transport Telematics in Europe and Proposals for Initial Actions Document*, available at http://www.ertico.com.

European Commission (EC) (1997b) *Telematics Applications Programme—Telematics for Transport European Commission*, Telematics Applicatons Programme of DG XIII-C/E, available at http://www.cordis.lu.

European Commission (EC) (1999a) *Consultation Meeting on Info-Mobility*, available at http://www.cordis.lu.

European Commission (EC) (1999b) *Guide to the 1997–1998 Telematics Applications Projects*, available at http://www.cordis.lu.

European Commission (EC) (1999c) *Telematics Applications for Transport European Union*, Telematics Applications Programme of DG XIII-C, available at http://www.cordis.lu.

Evans, J. R. and Minieka, E. (1992) *Optimization Algorithms for Networks and Graphs*, 2d ed., New York: Marcel Dekker.

Evans, S. P. (1976) "Derivation and analysis of some models for combining trip distribution and assignment," *Transportation Research*, 10, 37–57.

Faghri, A. and Raman, N. (1995) "A GIS-based traffic accident information system," *Journal of Advanced Transportation*, 29, 321–334.

Fairbairn, D. and Parsley, S. (1997) "The use of VRML for cartographic presentation," *Computers and Geosciences*, 23, 475–481.

Fayyad, U. M., Piatetsky-Shapiro, G., Smyth, P., and Ulthurusamy, R. (eds.) (1996) *Advances in Knowledge Discovery and Data Mining*, Cambridge: MIT Press.

Federal Geographic Data Committee (FGDC) (1994) *Content Standards for Digital Geospatial Metadata*, available at http://www.fgdc.gov/.

Federal Geographic Data Committee (FGDC) (1999) "NSDI Framework Transportation Identification Standard," working draft prepared by Ground Transportation SubCommittee, FGDC, available at http://www.bts.gov/.

Ferguson, E. (1990) "Transportation demand management: Planning, development and implementation," *Journal of the American Institute of Planners*, 56, 442–456.

Ferguson, M. R. and Kanaroglou, P. S. (1998) "Representing the shape and orientation of destinations in spatial choice models," *Geographical Analysis*, 30, 119–137.

Fernández, J. E. and Friesz, T. L. (1983) "Equilibrium predictions in transportation markets: The state of the art," *Transportation Research B*, 17B, 155–172.

Fischer, M. M. (1997) "From conventional to CI-based spatial analysis," Paper presented at the 36th Annual Meeting of the Western Regional Science Association, Hawai'i, February 23–27.

Fischer, M. M. and Gopal, S. (1994) "Artificial neural networks: A new approach to modeling interregional telecommunication flows," *Journal of Regional Science*, 34, 503–527.

Fischer, M. M. and Leung, Y. (1998) "A genetic algorithms based evolutionary computational neural network for modeling spatial interaction data," *Annals of Regional Science*, 32, 437–458.

Fischer, M. M. and Nijkamp, P. (eds.) (1993) *Geographic Information Systems, Spatial Modeling, and Policy Evaluation* Berlin: Springer-Verlag.

Fischer, M. M., Scholten, H. J., and Unwin, D. (1996) "Geographic information systems, spatial data analysis and spatial modeling: An introduction," in M. Fischer, H. J. Scholten, and D. Unwin (eds.) *Spatial Analytical Perspectives on GIS*, London: Taylor and Francis, 3–19.

Fisher, M. L. (1990) "Optimal solution of vehicle routing problems using minimum K-trees," Working paper, Department of Decision Sciences, University of Pennsylvania, Philadelphia.

Fisher, M. L. (1995) "Vehicle routing," in M. O. Ball, T. L. Magnanti, C. L. Monma, and G. L. Nemhauser (eds.) *Network Routing, Handbooks in Operations Research and Management Science Vol. 8*, Amsterdam: Elsevier, 1–33.

Fisher, M. L. and Jaikumar, R. (1981) "A generalized assignment heuristic for vehicle routing," *Networks*, 11, 109–124.

Fisher, P. F. (1991) "First experiments in viewshed uncertainty: The accuracy of the viewshed area," *Photogrammetric Engineering and Remote Sensing*, 57, 1321–1327.

Fisher, P. F. (1992) "First experiments in viewshed uncertainty: Simulating fuzzy viewsheds," *Photogrammetric Engineering and Remote Sensing*, 58, 345–352.

Fisher, P. F. (1993) "Algorithm and implementation uncertainty in viewshed analysis," *International Journal of Geographical Information Systems*, 7, 331–347.

Fisher, P. F. (1994) "Probable and fuzzy models of the viewshed operation," in M. F. Worboys (ed.) *Innovations in GIS, Vol. 1*, London: Taylor and Francis, 161–176.

Fisher, P. F. (1996) "Reconsideration of the viewshed function in terrain modelling," *Geographical Systems*, 3, 33–58.

Fisher, P. F. (1999) "Models of uncertainty in spatial data," in P. A. Longley, M. F. Goodchild, D. J. Maguire, and D. W. Rhind (eds.) *Geographic Information Systems, Vol. 1: Principles and Technical Issues*, New York: Wiley, 191–205.

Fisher, P. (2000) "Fuzzy modeling," in S. Openshaw and R. J. Abrahart (eds.) *GeoComputation*, London: Taylor and Francis, 161–186.

Fisk, C. (1980) "Some developments in equilibrium traffic assignment," *Transportation Research B*, 14B, 243–255.

Fisk, C. and Boyce, D. E. (1983) "Alternative variational inequality formulations of network equilibrium travel choice problem," *Transportation Science*, 17, 454–463.

Fleischmann, B. (1985) "A cutting plane procedure for the travelling salesman problem on road networks," *European Journal of Operational Research*, 21, 307–317.

Fletcher, D. (1987) "Modelling GIS transportation networks," *Proceedings of the 25th Annual Meeting of the Urban and Regional Information Systems Association*, 84–92.

Fletcher, D. (2000) "Geographic information systems for transportation: A look forward," in *Transportation in the New Millenium: State of the Art and Future Directions*, Washington, DC: Transportation Research Board, CD-ROM.

Fletcher, D., Espinoza, J., Mackoy, R. D., Gordon, S., Spear, B., and Vonderohe, A. (1998) "The case for a unified linear reference system," *URISA Journal*, 10(1), electronic journal archive available at http://www.urisa.org/Journal/.

Florian, M. and Nguyen, S. (1976) "An application and validation of equilibrium trip assignment methods," *Transportation Science*, 10, 374–390.

Florian, M., Nguyen, S., and Ferland, J. (1975) "On the combined distribution-assignment of traffic," *Transportation Science*, 9, 43–53.

Florian, M., Wu, J. H., and He, S. (1999) "A multi-class multi-mode variable demand network equilibrium model with hierarchical logit structures," unpublished paper, Center for Research on Transportation, Université de Montréal, Montréal, Quebec, Canada.

Florida Department of Transportation (FDOT) (1996) *GIS-T Standards Manual*, FDOT Planning, Analysis and Implementation Project, Nov. 19, 1996, Tallahassee, FL: Florida Department of Transportation.

Flowerdew, R., Green, M., and Kehris, E. (1991) "Using areal interpolation methods in geographic information systems," *Papers in Regional Science*, 70, 303–315.

Fohl, P., Curtin, K. M., Goodchild, M. F., and Church, R. L. (1996) "A non-planar, lane-based navigable data model for ITS," *Proceedings of the 7th International Symposium on Spatial Data Handling*, Delft, The Netherlands, 7B.17–7B.29.

Ford, L. D. and Fulkerson, D. R. (1962) *Flows in Networks*, Princeton, NJ: Princeton University Press.

Forer, P. (1998) "Geometric approaches to the nexus of time, space and microprocess: Implementing a practical model for mundane socio-spatial systems," in M. J. Egenhofer and R. G. Golledge (eds.) *Spatial and Temporal Reasoning in Geographic Information Systems*, Oxford, U.K.: Oxford University Press, 171–190.

Fotheringham, A. S. (1981) "Spatial structure and distance-decay parameters," *Annals of the Association of American Geographers*, 71, 425–436.

Fotheringham, A. S. (1983) "A new set of spatial interaction models: The theory of competing destinations," *Environment and Planning A*, 15, 15–36.

Fotheringham, A. S. (1984) "Spatial flows and spatial patterns," *Environment and Planning A*, 16, 529–543.

Fotheringham, A. S. (2000) "GIS-based spatial modelling: A step forwards or a step backwards?" in A. S. Fotheringham and M. Wegener (eds.) *Spatial Models and GIS: New Potential and New Models*, London: Taylor and Francis, 21–30.

Fotheringham, A. S., Charlton, M., and Brunsdon, C. (1996) "The geography of parameter space: An investigation of spatial non-stationarity," *International Journal of Geographical Information Systems*, 10, 605–627.

Fotheringham, A. S., Charlton, M., and Brunsdon, C. (1997) "Two techniques for exploring non-stationarity in geographical data," *Geographical Systems*, 4, 59–82.

Fotheringham, A. S. and O'Kelly, M. E. (1989) *Spatial Interaction Models: Formulations and Applications*, Dordrecht, The Netherlands: Kluwer Academic.

Fotheringham, A. S. and Rogerson, P. A. (1993) "GIS and spatial analytical problems," *International Journal of Geographical Information Systems*, 7, 3–19.

Fotheringham, S. and Rogerson, P. (eds.) (1994) *Spatial Analysis and GIS*, London: Taylor and Francis.

Fotheringham, A. S. and Wong, D. W. S. (1991) "The modifiable areal unit problem in multivariate statistical analysis," *Environment and Planning A*, 23, 1025–1044.

Francis, R. L. and Lowe, T. J. (1992) "On worst-case aggregation analysis for network location problems," *Annals of Operations Research*, 40, 229–246.

Francis, R. L., Lowe, T. J., and Rayco, M. B. (1996) "Row-column aggregation for rectilinear distance p-median problems," *Transportation Science*, 30, 160–174.
Francis, R. L., Lowe, T. J., Rushton, G., and Rayco, M. B. (1999) "A synthesis of aggregation methods for multifacility location problems: Strategies for containing error," *Geographical Analysis*, 31, 67–87.
Frank, A. (1982) "MAPQUERY: Database query language for retrieval of geometric data and their graphical representation," *Computer Graphics*, 16(3), 199–207.
Frayer, C. D. and Kroot, L. (1996) "California consumer perceptions of potential intelligent transportation system innovations," *Transportation Research Record*, 1537, 30–37.
Fredman, M. L. and Tarjan, R. E. (1987) "Fibonacci heaps and their uses in improved network optimization problems," *Journal of the ACM*, 34, 596–615.
French, G. T. (1996) *Understanding the GPS: An Introduction to the Global Positioning System*, Bethesda, MD: GeoResearch, Inc.
Friesz, T. L., Bernstein, D., Smith, T. E., Tobin, R. L., and Wie, B.-W. (1993) "A variational inequality formualtion of the dynamic network user equilibrium problem," *Operations Research*, 41, 179–191.
Friesz, T. L., Luque, F. J., Tobin, R. L., and Wie, B.-W. (1989) "Dynamic network traffic assignment considered as a continuous time optimal control problem," *Operations Research*, 37, 893–901.
Fulkerson, D. R. (1961) "An out-of-kilter method for minimal cost flow problems," *Journal of the Society for the Industrial Applications of Mathematics*, 9, 18–27.
Funk, C., Curtin, K., Goodchild, M. F., Montello, D., and Noronha, V. (1998) "Formulation and test of a model of positional distortion fields," paper presented at the Third International Symposium on Spatial Accuracy Assessment in Natural Resources and Environmental Sciences, Quebec City, available at http://www.ncgia.ucsb.edu/vital.
Gahegan, M. (1999) "Four barriers to the development of effective exploratory visualisation tools for the geosciences," *International Journal of Geographical Information Science*, 13, 289–309.
Gahli, M. O. and Smith, M. J. (1995) "A model for the dynamic system optimum traffic assignment problem," *Transportation Research B*, 29B, 155–170.
Gallo, G. and Pallottino, S. (1988) "Shortest path algorithms," *Annals of Operations Research*, 13, 3–79.
Galperin, D. (1977) "On the optimality of A*," *Artificial Intelligence*, 8, 69–76.
Gan, C.-T. (1994) "A GIS-aided procedure for converting census data for transportation planning," *ITE Journal*, 64 (Nov.), 34–40.
Ganter, J. H. (1993) "Metadata management in an environmental GIS for multidisciplinary users," *GIS/LIS'93 Proceedings*, 233–245.
Ganter, J. H., Goodwin, C. W. H., and Xiong, D. (1995) "The open geodata interoperability specifications (OGIS) as a technology for geospatial transportation computing," *GIS-T 1995 Proceedings*, 535–553.
Gao, X., Sorooshian, S., and Goodrich, D. C. (1993) "Linkage of a GIS to a distributed rainfall-runoff model," in M. F. Goodchild, B. O. Parks, and L. T. Steyaert (eds.) *Environmental Modeling with GIS*, New York: Oxford University Press, 182–187.
Gardels, K. (1996) "The Open GIS approach to distributed geodata and geoprocessing," *Proceedings of the Third International Conference/Workshop on Integrating GIS and Environmental Modeling*, Santa Fe, NM, Jan. 21–26.
Garey, M. R. and Johnson, D. S. (1979) *Computers and Intractability: A Guide to the Theory of NP-Completeness*, New York: W. H. Freeman.
Garling, T., Kwan, M.-P., and Golledge, R. G. (1994) "Computational-process modeling of household activity scheduling," *Transportation Research B*, 26B, 355–364.
Geertman, S. C. M. and Van Eck, J. R. R. (1995) "GIS and models of accessibility

potential: An application in planning," *International Journal of Geographical Information Systems*, 9, 67–80.

Geoffrion, A. (1983) "Can OR/MS evolve fast enough?" *Interfaces*, 13, 10–25.

German, D. and Jedynak, B. (1996) "An active testing model for tracking roads in satellite images," *IEEE Transactions on Pattern Matching and Machine Intelligence*, 18, 1–14.

Ghali, M. and Smith, M. J. (1993) "Traffic assignment, traffic control and road pricing," *Proceedings of the 12th International Symposium on Transportation and Traffic Theory*, Amsterdam: Elsevier Science, 147–170.

Ghosh, A. and McLafferty, S. L. (1987) *Location Strategies for Retail and Service Firms*, Lexington, MA: Heath.

Gill, P. E., Murray, W., and Wright, M. H. (1981) *Practical Optimization*, London: Academic Press.

Gillett, B. E. and Miller, L. R. (1974) "A heuristic algorithm for the vehicle dispatch problem," *Operations Research*, 22, 340–349.

Giuliano, G., Hall, R. W., and Golob, J. M. (1995) *Los Angeles Smart Traveler—Field Operational Test Evaluation*, California PATH Research Report UCB-ITS-PRR-95-41, University of California, Berkeley.

Glover, F., Glover R., and Klingman, D. (1984) "Computational study of an improved shortest path algorithm," *Networks*, 14, 25–37.

Glover, F., Klingman, D., and Phillips, N. (1985) "A new polynomial bounded shortest path algorithm," *Operations Research*, 33, 65–73.

Glover, F. and Laguna, M. (1993) "Tabu search," in C. R. Reeves (ed.) *Modern Heuristic Techniques for Combinatorial Problems*, New York: Wiley, 70–150.

Goldberg, A. V. and Radzik, T. (1993) "A heuristic improvement of the Bellman-Ford algorithm," *Applied Mathematics Letters*, 6, 3–6.

Golden, B. L., Bodin, L. D., Doyle, T., and Stewart, W. (1980) "Approximate traveling salesman algorithms," *Operations Research*, 28, 694–711.

Golden, B. L. and Stewart, W. R. (1985) "Empirical analysis of heuristics," in E. L. Lawler, J. K. Lenstra, A. H. G. Rinnooy Kan, and D. B. Shmoys (eds.) *The Traveling Salesman Problem: A Guided Tour of Combinatorial Optimization*, Chichester, U.K.: Wiley, 207–249.

Golledge, R. G. (1998) "The relationship between geographic information systems and disaggregate behavioral travel modeling," *Geographical Systems*, 5, 9–17.

Golledge, R. G., Kwan, M.-P., and Garling, T. (1994) "Computational process modeling of household travel decisions using a geographical information system," *Papers in Regional Science*, 73, 99–117.

Golledge, R. G. and Stimson, R. J. (1997) *Spatial Behavior: A Geographic Perspective*, New York: Guilford Press.

Golob, J. M. and Giuliano, G. (1996) "Smart traveler automated ridematching service lessons learned for future ATIS initiatives," *Transportation Research Record*, 1537, 23–29.

Goodchild, M. F. (1977) "An evaluation of lattice solutions to the problem of corridor location," *Environment and Planning A*, 9, 727–738.

Goodchild, M. F. (1979) "The aggregation problem in location-allocation," *Geographical Analysis*, 11, 240–255.

Goodchild, M. F. (1987) "A spatial analytical perspective on geographical information systems," *International Journal of Geographical Information Systems*, 1, 327–334.

Goodchild, M. F. (1992) "Geographical information science," *International Journal of Geographical Information Systems*, 6, 31–45.

Goodchild, M. F. (1998) "Geographic information systems and disaggregate transportation modeling," *Geographical Systems*, 5, 19–44.

Goodchild, M. F. (2000) "GIS and transportation: status and challenges," *GeoInformatica*, 4, 127–139.
Goodchild, M. F., Anselin, L., and Deichman, U. (1993) "A framework for the areal interpolation of socioeconomic data," *Environment and Planning A*, 25, 383–397.
Goodchild, M. F. and Gopal, S. (eds.) (1989) *Accuracy of Spatial Databases*, London: Taylor and Francis.
Goodchild, M. F., Haining, R., and Wise, S. (1992) "Integrating GIS and spatial data analysis: Problems and possibilities," *International Journal of Geographical Information Systems*, 6, 407–423.
Goodchild, M. F. and Hunter, G. J. (1997) "A simple positional accuracy measure for linear features," *International Journal of Geographical Information Science*, 11, 299–306.
Goodchild, M. F., Klinkenberg, B., and Janelle, D. G. (1993) "A factorial model of aggregate spatio-temporal behavior," *Geographical Analysis*, 4, 277–294.
Goodchild, M. F. and Lam, N. S.-N. (1980) "Areal interpolation: A variant of the traditional spatial problem," *Geo_Processing*, 1, 297–312.
Goodchild, M. F. and Lee, J. (1989) "Coverage problems and visibility regions on topographic surfaces," *Annals of Operations Research*, 18, 175–186.
Goodchild, M. F. and Noronha, V. (1983) *Location-Allocation for Small Computers*, Research Monograph #8, Department of Geography, The University of Iowa, Iowa City.
Goodchild, M. F. and Noronha, V. T. (1987) "Location-allocation and impulse shopping: The case of gasoline retailing," in A. Ghosh and G. Rushton (eds.) *Spatial Analysis and Location-Allocation Models*, New York: Van Nostrand Reinhold, 121–136.
Goodchild, M. F., Parks, B. O., and Steyaert, L. T. (eds.) (1993) *Environmental Modeling with GIS*, New York: Oxford University Press.
Goodwin, C. (1996) "Location referencing for ITS," Paper prepared for Oak Ridge National Laboratory by Energy, Environment, and Resources Center, The University of Tennessee, Knoxville.
Goodwin, C. W. H. (1994) "Towards a common language for IVHS databases," *GIS/LIS '94 Proceedings*, 351–360.
Goodwin, C., Gordon, S., Haas, R., Lau, J., and Siegel, D. (1999) *Location Referencing Message Specification Revision C, Spatial Data Interoperability Protocol For ITS*, Project report prepared for Federal Highway Administration, Office of Safety and Traffic Operations by Oak Ridge National Laboratory, Oak Ridge, TN.
Goodwin, C., Latham, F. E., Siegel, D., and Gordon, S. (1998) *Intelligent Transportation Systems Datum Public Sector Requirements*, Workshop report prepared for Federal Highway Administration by Oak Ridge National Laboratory and Viggen Corporation.
Goodwin, C., Siegel, D., and Gordon, S. (1996) *Location Reference Message Specification: Final Design, Task B: Spatial Data Interoperability Protocol for ITS*, Project report prepared for Federal Highway Administration, Office of Safety and Traffic Operations by Oak Ridge National Laboratory, Oak Ridge, TN.
Goodwin, C., Xiong, D., and Gordon, S. (1994) *Functional Requirements for National Map Databases for IVHS, Task A3: National Map Database and Location Referencing System*, Project report prepared for Federal Highway Administration, Office of Safety and Traffic Operations by Oak Ridge National Laboratory, Oak Ridge, TN.
Goodwin, P. (1998) "The end of equilibrium," in T. Garling, T. Laitila, and K. Westin (eds.) *Theoretical Foundations of Travel Choice Modeling*, New York: Elsevier Science, 103–132.

Gopal, S. and Fischer, M. M. (1996) "Learning in single hidden-layer feedforward network models: Backpropagation in a spatial interaction modeling context," *Geographical Analysis*, 28, 38–55.

Gordon, S., Goodwin, C., and Xiong, D. (1994) *Final Report on Status of Spatial/Map Databases, Task A2: Assessment of Current Transportation Spatial/Map Databases,* Report prepared for Federal Highway Administration, Office of Safety and Traffic Operations by Oak Ridge National Laboratory, Oak Ridge, TN.

Gorry, G. and Morton, M. (1971) "A framework for management information systems," *Sloan Management Review*, 13, 56–70.

Gottsegen, J., Goodchild, M., and Church, R. (1994) "A conceptual navigable database model for intelligent vehicle highway systems," *GIS/LIS '94 Proceedings*, 371–380.

Graham, I. (1993) *Object Oriented Methods*, 2d ed., Wokingham, U.K.: Addison-Wesley.

Greaves, S. and Stopher, P. (1998) "A synthesis of GIS and activity-based travel-forecasting," *Geographical Systems*, 5, 59–89.

Green, M. and Flowerdew, R. (1996) "New evidence on the modifiable areal unit problem," in P. Longley and M. Batty (eds.) *Spatial Analysis: Modelling in a GIS Environment*, Cambridge, U.K.: GeoInformation International, 41–54.

Greene, D. L. (1993) "Transportation and energy: The global environmental challenge," *Transportation Research A*, 27A, 163–166.

Griffith, D. A. (1982) "Geometry and spatial interaction," *Annals of the Association of American Geographers*, 72, 332–346.

Griffith, D. A. (1983) "The boundary value problem in spatial statistical analysis," *Journal of Regional Science*, 23, 377–387.

Griffith, D. A. (1985) "An evaluation of correction techniques for boundary effects in spatial statistical analysis: Contemporary methods," *Geographical Analysis*, 17, 81–88.

Grötschel, M. (1980) "On the symmetric traveling salesman problem: Solution of a 120-city problem," *Mathematical Programming Study*, 12, 61–77.

Grötschel, M. and Padberg, M. W. (1985) "Polyhedral theory," in E. L. Lawler, J. K. Lenstra, A. H. G. Rinnooy Kan, and D. B. Shmoys (eds.) *The Traveling Salesman Problem: A Guided Tour of Combinatorial Optimization*, Chichester, U.K.: Wiley, 251–305.

Gruen, A. and Li, H. (1997) "Semi-automatic linear feature extraction by dynamic programming and LSB-snakes," *Photogrammetric Engineering and Remote Sensing*, 63, 985–995.

Guensler, R. (1993) "Critical research needs in land use, transportation and air quality," *TR News*, 167 (July–Aug.), 21–23.

Guimaraes Pereira, A. (1996) "Generating alternative routes by multicriteria evaluation and a genetic algorithm," *Environment and Planning B: Planning and Design*, 23, 711–720.

Gunnink, J. L. and Burrough, P. A. (1996) "Interactive spatial analysis of soil attribute patterns using exploratory data analysis (EDA) and GIS," in M. Fischer, H. J. Scholten, and D. Unwin (eds.) *Spatial Analytical Perspectives on GIS*. London: Taylor and Francis, 87–99.

Guo, B. and Poling, A. D. (1995) "Geographic information systems/global positioning systems design for network travel time study," *Transportation Research Record*, 1497, 135–144.

Gurney, C. M. (1980) "Threshold selection for line detection algorithms," *IEEE Transactions on Geoscience and Remote Sensing*, GRS-18, 204–211.

Gwilliam, K. M. and Geerlings, H. (1994) "New technologies and their potential to reduce

the environmental impact of transportation," *Transportation Research A*, 28A, 307–319.
Hagerstrand, T. (1970) "What about people in regional science?" *Papers of the Regional Science Association*, 24, 7–21.
Hakimi, S. L. (1964) "Optimum locations of switching centers and the absolute centers and medians of a graph," *Operations Research*, 12, 450–459.
Hakimi, S. L. (1965) "Optimum distribution of switching centers in a communication network and some related graph theoretic problems," *Operations Research*, 13, 462–475.
Hall, R.W. and Chatterjee, I. (1996) "Application of computer-integrated transportation to commercial vehicle operations," *Transportation Research Record*, 1537, 1–9.
Hallmark, S. and O'Neill, W. A. (1996) "Integrating geographic information systems for transportation and air quality models for microscale analysis," *Transportation Research Record*, 1551, 133–140.
Halpern, J. and Maimon, O. (1982) "Algorithms for the m-center problems: A survey," *European Journal of Operational Research*, 10, 90–99.
Han, J., Stefanovic, N., and Koperski, K. (1998) "Selective materialization: An efficient method for spatial data cube construction," *Proceedings of the 1998 Pacific-Asia Conference on Knowledge Discovery and Data Mining*, Berlin: Springer-Verlag Lecture Notes in Computer Science 1394, 144–158.
Hansan, M. K. and Al-Gadhi, S. A. H. (1998) "Application of simultaneous and sequential transportation network equilibrium models to Riyadh, Saudi Arabia," *Transportation Research Record*, 1645, 127–132.
Hansen, P., Labbe, M., Peeters, D., and Thisse, J.-F. (1987) "Facility location analysis," in P. Hansen, M. Labbe, D. Peeters, J.-F. Thisse, and J. V. Henderson (eds.) *Systems of Cities and Facility Location*, Chur, Switzerland: Harwood Academic, 1–70.
Hansen, W. G. (1959) "How accessibility shapes land use," *Journal of the American Institute of Planners*, 25, 73–76.
Hanson, S. (1980) "Spatial diversification and multipurpose travel: Implications for choice theory," *Geographical Analysis*, 12, 245–257.
Hanson, S. (1995) "Getting there: Urban transportation in context," in S. Hanson (ed.) *The Geography of Urban Transportation*, 2d ed., New York: Guilford Press, 3–25.
Harinarayan, V., Rajaraman, A., and Ullman, J. D. (1996) "Implementing data cubes efficiently," *SIGMOD Record*, 25, 205–216.
Hartgen, D. T. and Li, Y. (1994) "Geographic information systems applications to transportation corridor planning," *Transportation Research Record*, 1429, 57–66.
Harwood, D. W., Russell, E. R., and Viner, J. G. (1989) "Characteristics of accidents and incidents in highway transportation of hazardous materials," *Transportation Research Record*, 1245, 23–33.
Haslett, J., Wills, G., and Unwin, A. (1990) "SPIDER: An interactive statistical tool for the analysis of spatially distributed data," *International Journal of Geographical Information Systems*, 3, 285–296.
Hayes-Roth, B. and Hayes-Roth, F. (1979) "A cognitive model of planning," *Cognitive Science*, 3, 275–310.
Healy, R. G. (1991) "Database management systems," in D. J. Maguire, M. F. Goodchild, and D. W. Rhind (eds.) *Geographical Information Systems: Principles and Applications*, New York: Wiley, 251–267.
Hearnshaw, H. M. and Unwin, D. J. (eds.) (1994) *Visualization in Geographical Information* Systems, Chichester, UK: Wiley.
Helander, M. E. and Melachrinoudis, E. (1997) "Facility location and reliable route

planning in hazardous material transportation," *Transportation Science*, 31, 216–226.
Hepner, G. F. (1984) "Use of value functions as a possible suitability scaling procedure in automated composite mapping," *Professional Geographer*, 36, 468–472.
Herring, J. R., Larsen, R. C., and Shivakumar, J. (1988) "Extensions to the SQL query language to support spatial analysis in a topological database," *GIS/LIS '88 Proceedings*, 741–750.
Hillier, F. S. and Lieberman, G. J. (1990) *Introduction to Operations Research*, 5th ed., New York: McGraw-Hill.
Hillsman, E. L. (1980) *Heuristic Solutions to Location-Allocation Problems: A User's Guide to ALLOC IV*, Research Monograph #7, Department of Geography, The University of Iowa, Iowa City.
Hillsman, E. L. (1984) "The p-median structure as a unified linear model for location-allocation analysis," *Environment and Planning A*, 16, 305–318.
Hillsman, E. L. and Rhoda, R. (1978) "Errors in measuring distances from populations to service centers," *Annals of Regional Science*, 12, 74–88.
Hillsman, E. L. and Southworth, F. (1990) "Factors that may influence responses of the U.S. transportation sector to policies for reducing greenhouse gas emissions," *Transportation Research Record*, 1267, 1–11.
Hobeika, A. G. and Jamie, B. (1985) "MASSVAC: A model for calculating evacuation times under natural disasters," *Proceedings of the Conference on Computer Simulation in Emergency Planning*, 5, 23–28.
Hobeika, A. G., Kim, S., and Beckwith, R. E. (1994) "A decision support system for developing evacuation plans around nuclear power stations," *Interfaces*, 24, 22–35.
Hodgson, M. E. and Gaile, G. L. (1996) "Characteristic mean and dispersion in surface orientations for a zone," *International Journal of Geographical Information Systems*, 10, 817–830.
Hodgson, M. J. (1990) "A flow-capturing location-allocation model," *Geographical Analysis*, 22, 270–279.
Hodgson, M. J., Rosing, K. E., and Zhang, J. (1996) "Locating vehicle inspection stations to protect a transportation network," *Geographical Analysis*, 28, 299–314.
Hogan, K. and ReVelle, C. (1986) "Concepts and applications of backup coverage," *Management Science*, 34, 1434–1444.
Holmes, J., Williams, F. B., and Brown, L. A. (1972) "Facility location under a maximum travel restriction: An example using day care facilities," *Geographical Analysis*, 7, 199–204.
Horn, M. (1996) "Analysis and computational schemes for p-median heuristics," *Environment and Planning A*, 28, 1699–1708.
Horn, M. E. T. (1995) "Solution techniques for large regional partitioning problems," *Geographical Analysis*, 27, 230–248.
Horowitz, E. and Sahni, S. (1976) *Fundamentals of Data Structures*, Woodland Hills, CA: Computer Science Press.
Horowitz, J. L. (1982) *Air Quality Analysis for Urban Transportation Planning*, Cambridge: MIT Press.
Huang, H.-J. (1995) "A combined algorithm for solving and calibrating the stochastic traffic assignment problem," *Journal of the Operational Research Society*, 46, 977–987.
Huber, D. L. and Church, R. L. (1985) "Transmission corridor location modeling," *Journal of Transportation Engineering*, 111, 114–130.
Huisman, O. and Forer, P. (1998) "Computational agents and urban life spaces:

A preliminary realisation of the time-geography of student lifestyles," Paper presented at GeoComputation 98: 3rd International Conference on GeoComputation, University of Bristol, U.K.

Hunt, J. D. (1990) "A logit model of public transport route choice," *ITE Journal*, 60(12), 26–31.

Hunt, J. D. and Simmonds, D. C. (1993) "Theory and application of an integrated land-use and transport modelling framework," *Environment and Planning B: Planning and Design*, 20, 221–244.

Hunter, G. J. and Goodchild, M. F. (1997) "Modeling the uncertainty of slope and aspect estimates derived from spatial databases," *Geographical Analysis*, 29, 35–49.

Hush, D. R. and Horne, B. G. (1993) "Progress in supervised neural networks: What's new since Lippmann," *IEEE Signal Processing Magazine*, 10(1), 8–39.

Illingworth, V. and Pyle, I. (eds.) (1996) *Dictionary of Computing*, New York: Oxford University Press.

ITS America (1999) *Archived Data User Service Resource Page*, available at http://www.itsa.org/.

ITS Focus (1997) *ITS Focus Report on System Architecture—Evaluation of the US National Architecture and Recommendations for the U.K.*, Berkshire, U.K.: ITS Focus.

Jacobs, F. R. and Whybark, D. C. (2000) *Why ERP? A Primer on SAP Implementation*, New York: McGraw-Hill.

Jahne, B. (1991) *Digital Image Processing*, Berlin: Springer-Verlag.

Janelle, D. and Hodge, D. (1998) *Measuring and Representing Accessibility in the Information Age*, Report of a Specialist Meeting of Project Varenius' Geographies of the Information Society, National Center for Geographic Information and Analysis.

Janelle, D. G., Goodchild, M. F., and Klinkenberg, B. (1988) "Space-time diaries and travel characteristics for different levels of respondent aggregation," *Environment and Planning A*, 20, 891–906.

Janelle, D. G., Klinkenberg, B., and Goodchild, M. F. (1998) "The temporal ordering of urban space and daily activity patterns for population role groups," *Geographical Systems*, 5, 117–137.

Jankowski, P. (1989) "Mixed data multicriteria evaluation for regional planning: A systematic approach to the decisionmaking process," *Environment and Planning A*, 21, 349–362.

Jankowski, P. (1995) "Integrating geographical information systems and multiple criteria decision-making methods," *International Journal of Geographical Information Systems*, 9, 251–273.

Jankowski, P., Nyerges, T. L., Smith, A., Moore, T. J., and Horvath, E. (1997) "Spatial group choice: A SDSS tool for collaborative spatial decision making," *International Journal of Geographical Information Science*, 11, 577–602.

Jankowski, P. and Richard, L. (1994) "Integration of GIS-based suitability analysis and multicriteria evaluation in a spatial decision support system for route selection," *Environment and Planning B: Planning and Design*, 21, 323–340.

Janson, B. N. (1991a) "Convergent algorithm for dynamic traffic assignment," *Transportation Research Record*, 1328, 69–80.

Janson, B. N. (1991b) "Dynamic traffic assignment for urban road networks," *Transportation Research B*, 25B, 143–161.

Janson, B. N. and Robles, J. (1995) "Quasi-dynamic traffic assignment model," *Transportation Research Record*, 1493, 199–206.

Janson, B. N. and Southworth, F. (1992) "Estimating departure times from traffic counts using dynamic assignment," *Transportation Research B*, 26B, 3–16.

Japanese Highway Industry Development Organization (JHIDO) (1999a) *ITS in Industry*, available at http://www.hido.or.jp/ITSHP_e/IiI/cont_IinI.htm.
Japanese Highway Industry Development Organization (JHIDO) (1999b) *Making Smartway A Reality*, available at http://www.nihon.net/ITS/index.html.
Japanese Highway Industry Development Organization (JHIDO) (1999c) *What's ITS?*, available at http://www.hido.or.jp/ITSHP_e/wi/wi.htm.
Japanese Ministry of Construction (1996) *Comprehensive Plan for ITS in Japan*, available at http://www.its.go.jp.
Japanese Ministry of Construction (1999a) *ITS Handbook 1999-2000*, available at http://www.its.go.jp.
Japanese Ministry of Construction (1999b) *System Architecture for ITS in Japan—Summary*, available at http://www.vertis.or.jp.
Jara-Diaz, S. R. (1990) "Consumer's surplus and the value of travel time savings," *Transportation Research B*, 24B, 73-77.
Jara-Diaz, S. R. and Friesz, T. L. (1982) "Measuring the benefits derived from a transportation investment," *Transportation Research B*, 16B, 57-77.
Jarke, M., Lenserini, M., Vassiliou, Y., and Vassiliadis, P. (2000) *Fundamentals of Data Warehouses*, Berlin: Springer.
Jayakrishnan, R., Tsai, W. K., and Chen, A. (1995) "A dynamic traffic assignment model with traffic-flow relationships," *Transportation Research C: Emerging Technologies*, 3, 51-72.
Jayakrishnan, R., Tsai, W. K., Prashker, J. N., and Rajadhyaksha, S. (1994) "Faster path-based algorithm for traffic assignment," *Transportation Research Record*, 1443, 75-83.
Jensen, J. R. (1996) *Introductory Digital Image Processing: A Remote Sensing Perspective*, Upper Saddle River, NJ: Prentice Hall.
Jensen, J. R. and Cowen, D. C. (1999) "Remote sensing of urban/suburban infrastructure and socio-economic attributes," *Photogrammetric Engineering and Remote Sensing*, 65, 611-622.
Jensen, J. R., Narumalani, S., Weatherbee, O., and Mackey, H. E. (1993) "Measurement of seasonal and yearly cattail and waterlily distribution using remote sensing and GIS techniques," *Photogrammetric Engineering and Remote Sensing*, 59, 519-525.
Jenson, S. K. and Domingue, J. O. (1988) "Extracting topographic structure from digital elevation data for geographic information system analysis," *Photogrammetric Engineering and Remote Sensing*, 54, 1593-1600.
Jesty, P. H., Gaillet, J.-F., Giezen, J., Franco, G., Leighon, I., and Schultz, H.-J. (1998) *Guidelines for the Development and Assessment of Intelligent Transport System Architectures*, European Road Telematics Implementation Coordination Organization (ERTICO), Brussels, Luxemburg.
Jha, M. K. and Schonfeld, P. (2000) "GIS-based analysis of right-of-way cost for highway optimization," *Proceedings of the 79th Annual Meeting of the Transportation Research Board*, Washington, DC, Jan. 9-13, CD-ROM.
Jia, W. and Ford, B. (2000) "Transit GIS applications in Fairfax County, Virginia," *Proceedings of the 79th Annual Meeting of the Transportation Research Board*, Washington, DC, Jan. 9-13, CD-ROM.
Johnston, R. A. and de la Barra, T. (2000) "Comprehensive regional modeling for long-range planning: Linking integrated urban models and geographic information systems," *Transportation Research A*, 34, 125-136.
Jones, K. and Duncan, C. (1996) "People and places: The multilevel model as a general framework for the quantitative analysis of geographical data," in P. Longley and M. Batty (eds.) *Spatial Analysis: Modelling in a GIS Environment*, Cambridge, U.K.: GeoInformation International, 79-104.

Jones, P. (1990) *Developments in Dynamic and Activity-Based Approaches to Travel Analysis*, Aldershot, U.K.: Gower.

Josephson, M. C., Downey, T., and Ott, E. K. (1995) "Developments in the GIS-T marketplace," *GIS-T 1995 Proceedings*, American Association of State Highway and Transportation Officials.

Kalawsky, R. S. (1994) *The Science of Virtual Reality and Virtual Environments*, Cambridge, U.K.: Cambridge University Press.

Kanaroglou, P. S. and Ferguson, M. R. (1996) "Discrete spatial choice models for aggregate destinations," *Journal of Regional Science*, 36, 271–290.

Karimi, H. A., Hummer, J. E., and Khattak, A. J. (2000) "Collection and Presentation of Roadway Inventory Data," *National Cooperative Highway Research Program Report 437*, Washington, DC: National Academy Press.

Karimi, H. A. and Hwang, D. (1997a) "Overcoming the time complexity barrier of routing for real-time GIS-T applications: A parallel routing algorithm," *Journal of Geographic Information and Decision Analysis*, 1, 144–150.

Karimi, H. A. and Hwang, D. (1997b) "A parallel algorithm for routing: Best solutions at low computational costs," *Geomatica*, 51, 45–51.

Kariv, O. and Hakimi, S. L. (1979a) "An algorithmic approach to network location problems. I: The p-centers," *SIAM Journal on Applied Mathematics*, 37, 513–538.

Kariv, O. and Hakimi, S. L. (1979b) "An algorithmic approach to network location problems. II: The p-medians," *SIAM Journal on Applied Mathematics*, 37, 539–560.

Keller H. (ed.) (1999) *Telematics Applications Programme: Concertation and Achievements*, Report of the Transport Sector (CARTS) Co-ordinated Disseminationin Europe (CODE) Deliverable 2.2, available at http://www.cordis.lu.

Kemp, Z. and Groom, J. (1994) "Incorporating generic temporal capabilities in a geographical information system," *Proceedings of European GIS (EGIS)*, 86–95.

Kessell, S. R. (1990) "An Australian geographical information and modeling system for natural area management," *International Journal of Geographical Information Systems*, 4, 333–362.

Khoshafian, S. and Abnous, R. (1995) *Object Orientation: Concepts, Analysis and Design, Languages, Databases, Graphical User Interfaces, Standards*, 2d ed., New York: Wiley.

Khumawala, B. M. (1972) "An efficient branch and bound algorithm for the warehouse location problem," *Management Science*, 18, B718–B731.

Khumawala, B. M. (1973) "An efficient algorithm for the p-median with maximum distance," *Geographical Analysis*, 5, 309–321.

Kikuchi, S. and Tanaka, M. (2000) "Estimating an origin-destination table under repeated counts of in-out volumes at highway ramps: Use of artificial neural networks," *Transportation Research Record*, 1739, 59–66.

Kim, K. and Levine, N. (1996) "Using GIS to improve highway safety," *Computers, Environment and Urban Systems*, 20, 289–302.

Kim, K., Nitz, L., Richardson, J., and Li, L. (1994) "Analyzing the relationship between crash types and injuries in motor vehicle collisons in Hawaii," *Transportation Research Record*, 1467, 9–13.

Kim, K., Nitz, L., Richardson, J., and Li, L. (1995) "Personal and behavioral predictors of automobile crash and injury severity," *Accident Analysis and Prevention*, 27, 469–481.

Kim, T. J. (1989) *Integrated Urban Systems Modeling: Theory and Practice*, Norwell, MA: Martinus Nijhoff.

Kirby, H. R. (1997) "Buffon's needle and the probability of intercepting short-distance trips by multiple screen-line surveys," *Geographical Analysis*, 29, 64–71.

Kitamura, R. (1984) "Incorporating trip chaining into analysis of destination choice," *Transportation Research B*, 18B, 67–81.

Kitamura, R., Chen, C., and Pendyala, R. M. (1997) *Generation of Synthetic Daily Activity-Travel Patterns*, Technical Report #69, National Institute of Statistical Sciences, Research Triangle Park, NC.

Klein, C. M. (1991) "A model for the transportation of hazardous waste," *Decision Sciences*, 22, 1091–1108.

Kolata, G. (1991) "Math problem, long baffling, slowly yields," *New York Times*, March 12, 1991, p. B-5.

Köninger, A. and Bartel, S. (1998) "3D-GIS for urban purposes," *GeoInformatica*, 2, 79–103.

Kraak, M. J., Smets, G., and Sidianin, P. (1998) "Virtual reality, the new 3D interface for geographical information systems," in A. Camara and J. Raper (eds.) *Spatial Multimedia and Virtual Reality*, London: Taylor and Francis.

Kulkarni, R. G., Stough, R. R., and Haynes, K. E. (1996) "Spin glass and the interactions of congestion and emissions: An exploratory step," *Transportation Research C: Emerging Technologies*, 4C, 407–424.

Kwan, M.-P. (1997) "GISICAS: An activity-based travel decision support system using a GIS-interfaced computational-process model," in D. F. Ettema and H. J. P. Timmermans (eds.) *Activity-Based Approaches to Travel Analysis*, Oxford, U.K.: Elsevier Science, 263–282.

Kwan, M.-P. (1998) "Space-time and integral measures of accessibility: A comparative analysis using a point-based framework," *Geographical Analysis*, 30, 191–216.

Kwan, M.-P. and Golledge, R. G. (1995) "Integration of GIS with activity-based models in ATIS," Paper presented at the 74th annual meeting of the Transportation Research Board, Washington, DC.

Kwan, M.-P., Golledge, R. G., and Speigle, J. M. (1996) "A review of object-oriented approaches in geographic information systems for transportation modeling," Draft, Department of Geography, Ohio State University, Columbus.

Kwan, M.-P. and Hong, X.-D. (1998) "Network-based constraints-oriented choice set formation using GIS," *Geographical Systems*, 5, 139–162.

Landis, J. D. (1994) "The California urban futures model: A new generation of metropolitan simulation models," *Environment and Planning B: Planning and Design*, 21, 399–420.

Landis, J. and Zhang, M. (1998a) "The second generation of the California urban futures model. Part 1: Model logic and theory," *Environment and Planning B: Planning and Design*, 25, 657–666.

Landis, J. and Zhang, M. (1998b) "The second generation of the California urban futures model. Part 2: Specification and calibration results of the land-use change submodel," *Environment and Planning B: Planning and Design*, 25, 795–824.

Langley, C. J., Jr. (1981) "The inclusion of transportation costs in inventory models: Some considerations," *Journal of Business Logistics*, 2, 106–125.

Langran, G. (1992) *Time in Geographic Information Systems*, London: Taylor and Francis.

Langran, G. and Chrisman, N. (1988) "A framework for temporal geographic information," *Cartographica*, 25(3), 1–14.

Laporte, G. (1992) "The traveling salesman problem: An overview of exact and approximate algorithms," *European Journal of Operational Research*, 59, 231–247.

Laporte, G., Nobert, Y., and Desrochers, M. (1985) "Optimal routing under capacity and distance constraints," *Operations Research*, 33, 1050–1073.

Larson, R. C. and Odoni, A. R. (1981) *Urban Operations Research*, Englewood Cliffs, NJ: Prentice Hall.

LeBlanc, L. J., Morlock, E. K., and Pierskalla, W. P. (1975) "An efficient approach to solving the road network equilibrium traffic assignment problem," *Transportation Research*, 9, 309–318.

Lee, H. and Clover, P. (1995) "GIS-based highway design review system to improve constructability of design," *Journal of Advanced Transportation*, 29, 375–388.

Lee, H., Wong, M., Clover, P., and Anderson, K. (1991) "Development of a hypertext-linked highway constructability improvement systems," *Transportation Research Record*, 1310, 9–16.

Lee, J. (1991a) "Analyses of visibility sites on topographic surfaces," *International Journal of Geographical Information Systems*, 5, 413–429.

Lee, J. (1991b) "Comparison of existing methods for building triangular irregular network models of terrain from grid digital elevation models," *International Journal of Geographical Information Systems*, 5, 267–285.

Lee, J., Snyder, P. K., and Fisher, P. F. (1992) "Modeling the effect on data errors on feature extraction from digital elevation models," *Photogrammetric Engineering and Remote Sensing*, 58, 1461–1467.

Lenntorp, B. (1976) *Paths in Space-Time Environments: A Time-Geographic Study of the Movement Possibilities of Individuals*, Lund Studies in Geography, Series B, Number 44.

Lenntorp, B. (1978) "A time geographic simulation model of individual activity programmes," in T. Carlstein, D. Parkes, and N. Thrift (eds.) *Timing Space and Spacing Time, Vol. 2: Human Activity and Time Geography*, London: Edward Arnold, 162–180.

Lenstra, J. K. and Rinnooy Kan, A. H. G. (1981) "Complexity of vehicle routing and scheduling problems," *Networks*, 11, 221–227.

Leontief, W. (1967) *Input-Output Economics*, New York: Oxford University Press.

Lepofsky, M. and Abkowitz, M. (1993) "Transportation hazards analysis in integrated GIS environment," *Journal of Transportation Engineering*, 119, 239–254.

Leung, Y. and Yan, J. (1998) "A locational error model for spatial features," *International Journal of Geographical Information Science*, 12, 607–620.

Leurent, F. M. (1995) "Contributions to logit assignment model," *Transportation Research Record*, 1493, 207–212.

Levine, N. and Kim, K. E. (1998) "The location of motor vehicle crashes in Honolulu: A methodology for geocoding intersections," *Computers, Environment and Urban Systems*, 22, 557–576.

Levine, N., Kim, K. E., and Nitz, L. H. (1995a) "Spatial analysis of Honolulu motor vehicle crashes: I. Spatial patterns," *Accident Analysis and Prevention*, 27, 663–674.

Levine, N., Kim, K. E., and Nitz, L. H. (1995b) "Spatial analysis of Honolulu motor vehicle crashes: II. Zonal generators," *Accident Analysis and Prevention*, 27, 675–685.

Lewis, S. (1990) "Use of geographical information systems in transportation modeling," *ITE Journal*, 60(March), 34–38.

Li, X. and Yeh, A. G.-O. (2000) "Modelling sustainable urban development by the integration of constrained cellular automata and GIS," *International Journal of Geographical Information Science*, 14, 131–152.

Lillesand, T. M. and Kiefer, R. W. (2000) *Remote Sensing and Image Interpretation*, 4th ed. New York: Wiley.

Lin, S. (1965) "Computer solutions of the traveling salesman problem," *Bell Systems Technical Journal*, 44, 2245–2269.

Lin, S. and Kernighan, B. W. (1973) "An effective heuristic for the traveling salesman problem," *Operations Research*, 21, 498–516.

List, G. F., Mirchandani, P. B., Turnquist, M. A., and Zografos, K. G. (1991) "Modeling and analysis for hazardous materials transportation: Risk analysis, routing/scheduling and facility location," *Transportation Science*, 25, 100–114.

Little, J. D. C., Murty, K. G., Sweeney, D. W., and Karel, C. (1963) "An algorithm for the traveling salesman problem," *Operations Research*, 11, 972–989.

Liu, R. and Van Vilet, D. (1996) "DRACULA: A dynamic microscopic model of road traffic," *Proceedings of the International Transport Symposium*, Beijing, China, 160–170.

Lockheed Martin Federal Systems (1997) *Intelligent Transportation Systems Architecture*, Executive Summary Report prepared for U.S. Department of Transportation (USDOT), Federal Highway Administration, Washington, DC.

Lockheed Martin Federal Systems (1998) *National ITS Architecture—Theory of Operations*, Report prepared for U.S. Department of Transportation (USDOT), Federal Highway Administration, Washington, DC.

Lolonis, P. and Armstrong, M. P. (1993) "Location-allocation as decision aids in delineating administrative regions," *Computers, Environment and Urban Systems*, 17, 153–174.

Lombard, K. and Church, R. L. (1993) "The gateway shortest path problem: Generating alternative routes for a corridor location problem," *Geographical Systems*, 1, 25–45.

Longley, P. A. (2000) "The academic success of GIS in geography: Problems and prospects," *Journal of Geographical Systems*, 2, 37–42.

Longley, P. and Batty, M. (eds.) (1996) *Spatial Analysis: Modeling in a GIS Environment*, Cambridge, U.K.: GeoInformation International.

Longley, P. A., Goodchild, M. F., Maguire, D. J., and Rhind, D. W. (1999) "Introduction," in P. A. Longley, M. F. Goodchild, D. J. Maguire, and D. W. Rhind (eds.) *Geographical Information Systems, Vol. 1: Principles and Technical Issues*, 2d ed., New York: Wiley, 1–20.

Lopez, C. (1997) "Locating some types of random errors in digital terrain models," *International Journal of Geographical Information Science*, 11, 677–698.

Love, R. F., Morris, J. G., and Wesolowsky, G. O. (1988) *Facilities Location: Models and Methods*, Amsterdam: North-Holland.

Lowry, I. S. (1964) *A Model of Metropolis*, Report RM-4035-RC, RAND Corp., Santa Monica, CA.

Lowry, J. H., Miller, H. J., and Hepner, G. F. (1995) "A GIS-based sensitivity analysis of community vulnerability to hazardous contaminants on the Mexico/U.S. border," *Photogrammetric Engineering and Remote Sensing*, 61, 1347–1359.

MacDougall, E. B. (1992) "Exploratory analysis, dynamic statistical visualization, and geographical information systems," *Cartography and Geographic Information Systems*, 19, 237–246.

MacEachren, A. M. (1994a) "Time as a cartographic variable," in H. M. Hearnshaw and D. J. Unwin (eds.) *Visualization in Geographical Information Systems*, Chichester, U.K.: Wiley, 115–130.

MacEachren, A. M. (1994b) "Visualization in modern cartography: Setting the agenda," in A. M. MacEachren and D. R. F. Taylor (eds.) *Visualization in Modern Cartography*, London: Pergamon Press, 1–12.

MacEachren, A., Bishop, I., Dykes, J., Doring, D., and Gatrell, A. (1994) "Introduction to advances in visualizing spatial data" in H. M. Hearnshaw and D. J. Unwin (eds.) *Visualization in Geographical Information Systems*, Chichester, U.K.: Wiley, 51–59.

MacEachren, A. M. and Kraak, M.-J. (1997) "Exploratory cartographic visualization: Advancing the agenda," *Computers and Geosciences*, 23, 335–343.

MacEachren, A. M. and Monmonier, M. (1992) "Introduction to a special issue on geographic visualization," *Cartography and Geographic Information Systems*, 19, 197–200.

MacEachren, A. M., Wachowicz, M., Edsall, R., and Haug, D. (1999) "Constructing knowledge from multivariate spatiotemporal data: Integrating geographical visualization with knowledge discovery in database methods," *International Journal of Geographical Information Science*, 13, 311–334.

Magnanti, T. L. (1981) "Combinatorial optimization and vehicle fleet planning: Perspectives and prospects," *Networks*, 11, 179–213.

Magnanti, T. L. and Wong, R. T. (1984) "Network design and transportation planning: Models and algorithms," *Transportation Science*, 18, 1–55.

Maguire, D. J., Smith, R., and Jones, S. (1993) "GIS on the move: Some transportation applications of GIS," *Proceedings of the Thirteenth Annual ESRI User Conference*, vol. 3, 39–46.

Maher, M. J. and Hughes, P. C. (1997) "A probit-based stochastic user equilibrium assignment model," *Transportation Research B*, 31B, 341–355.

Mahmassani, H. S., Hu, T., and Jayakrishnan, R. (1992) "Dynamic traffic assignment and simulation for advanced network informatics (DYNASMART)," Paper presented at the Second International Capri Seminar on Urban Traffic Networks, Capri, Italy.

Mahmassani, H. S. and Peeta, S. (1995) "System optimal dynamic assignment for electronic route guidance in a congested traffic network," in N. H. Gartner and G. Improta (eds.) *Urban Traffic Networks: Dynamic Flow Modeling and Control*, Berlin: Springer-Verlag, 3–36.

Maidment, D. R. (1993) "GIS and hydrological modeling," in M. F. Goodchild, B. O. Parks, and L. T. Steyaert (eds.) *Environmental Modeling with GIS*, New York: Oxford University Press, 147–167.

Manar, A. and Baass, K. G. (1996) "Traffic platoon dispersion modeling on arterial streets," *Transportation Research Record*, 1566, 49–53.

MapInfo (1999) *Oracle and MapInfo Combine Products and Services for Internet Business Applications*, press release available at http://www.mapinfo.com/.

Maranzana, F. E. (1964) "On the location of supply points to minimize transport costs," *Operations Research Quarterly*, 15, 261–270.

Marble, D. F. (1991) "The extended data dictionary: A critical element in building viable spatial databases," *Proceedings of the 11th Annual ESRI User Conference*, 169–176.

Marble, D. F., Gou, Z., Liu, L., and Saunders, J. (1997) "Recent advances in the exploratory analysis of interregional flows in space and time," in Z. Kemp (ed.) *Innovations in GIS 4*, London: Taylor and Francis 75–88.

Marcotte, P. and Nguyen, S. (1998) "Hyperpath formulation of traffic assignment problems," in P. Marcotte and S. Nguyen (eds.) *Equilibrium and Advanced Transportation Modeling*, Boston: Kluwer Academic, 175–200.

Marianov, V. and ReVelle, C. (1995) "Siting emergency services," in Z. Drezner (ed.) *Facility Location: A Survey of Applications and Methods*, New York: Springer, 199–223.

Mark, D. M. (1999) *Geographic Information Science: Critical Issues in an Emerging Cross-Disciplinary Research Domain*, Report from a workshop on Geographic Information Science and Geospatial Activities at the NSF, Jan. 14–15, 1999, available at http://www.geog.buffalo.edu/.

Mark, D. M. and Csillag, F. (1989) "The nature of boundaries on area-class maps," *Cartographica*, 26, 65–77.

Martin, R. J. (1987) "Some comments on correction techniques for boundary effects and missing value techniques," *Geographical Analysis*, 19, 273–282.

Mayberry, J. P. (1973) "Structural requirements for abstract-mode models of passenger transportation," in R. E. Quandt (ed.) *The Demand for Travel: Theory and Measurement*, Lexington, MA: Heath, 103–125.

McCormack, E. and Nyerges, T. (1997) "What transportation modeling needs from a GIS: A conceptual framework," *Transportation Planning and Technology*, 21, 5–23.

McCormack, J. E., Gahegan, M. N., Roberts, S. A., Hogg, J., and Hoyle, B. S. (1993) "Feature-based derivation of drainage networks," *International Journal of Geographical Information Systems*, 7, 263–279.

McCormick, B. H., DeFanti, T. A., and Brown, M. D. (1987) *Visualization in Scientific Computing*, Report to the National Science Foundation, Panel on Graphics, Image Processing and Workstations, Baltimore, MD: ACM SIGGRAPH.

McHarg, I. L. (1971) *Design with Nature*, New York: Wiley.

McNally, M. G. (1998) "Activity-based forecasting models integrating GIS," *Geographical Systems*, 5, 163–187.

McQueen, B. and McQueen, J. (1999) *Intelligent Transportation Systems Architecture*, Norwood, MA: Artech House.

Medyckyj-Scott, D., Newman, I., Ruggles, C., and Walker, D. (eds.) (1991) *Metadata in the Geosciences*, Leicestershire, U.K.: Group D Publications.

Mesev, V., Longley, P., and Batty, M. (1996) "RS-GIS: Spatial distributions from remote imagery," in P. Longley and M. Batty (eds.) *Spatial Analysis: Modelling in a GIS Environment*, Cambridge, U.K.: GeoInformation International, 123–148.

Mikula, B., Mathian, H., Pumain, D., and Sanders, L. (1996) "Integrating dynamic spatial models with GIS," in P. Longley and M. Batty (eds.) *Spatial Analysis: Modelling in a GIS Environment*, Cambridge, U.K.: GeoInformation International, 282–295.

Miles, J. C. (ed.) (1996) *Intelligent Transport System (ITS) Architecture: Proceedings of A TRL International Workshop*, Berkshire, U.K.: Transportation Research Laboratory (TRL).

Miller, D. L. and Pekny, J. F. (1991) "Exact solution of large asymmetric traveling salesman problems," *Science*, 251, 754–761.

Miller, H. J. (1991) "Modeling accessibility using space-time prism concepts within geographical information systems," *International Journal of Geographical Information Systems*, 5, 287–301.

Miller, H. J. (1993) "Consumer search and retail analysis," *Journal of Retailing*, 69, 160–192.

Miller, H. J. (1996) "GIS and geometric representation in facility location problems," *International Journal of Geographical Information Systems*, 10, 791–816.

Miller, H. J. (1997) *Towards Consistent Travel Demand Estimation in Transportation Planning: A Guide to the Theory and Practice of Equilibrium Travel Demand Modeling*, Research report, Bureau of Transportation Statistics, U.S. Department of Transportation.

Miller, H. J. (1999a) "Measuring space-time accessibility benefits within transportation networks: Basic theory and computational methods," *Geographical Analysis*, 31, 187–212.

Miller, H. J. (1999b) "Potential contributions of spatial analysis to geographic information systems for transportation (GIS-T)," *Geographical Analysis*, 31, 373–399.

Miller, H. J. (2000) "Geographic representation in spatial analysis," *Journal of Geographical Systems*, 2, 55–60.

Miller, H. J. and Han, J. (2000) "Discovering geographic knowledge in data-rich environments: Report from a specialist meeting," *Explorations: The Newletter of the ACM Special Interest Group on KDD*, 2, electronic newsletter available at http://www.acm.org/sigkdd/explorations.

Miller, H. J. and Han, J. (eds.) (2001) *Geographic Data Mining and Knowledge Discovery*, London: Taylor and Francis, in press.

Miller, H. J. and Storm, J. D. (1996) "Geographic information system design for network equilibrium-based travel demand models," *Transportation Research C: Emerging Technologies*, 4C, 373–389.
Miller, H. J., Storm, J. D., and Bowen, M. (1995) "GIS design for multimodal network analysis," *GIS/LIS'95 Proceedings*, 750–759.
Miller, H. J. and Wu, Y.-H. (2000) "GIS software for measuring space-time accessibility in transportation planning and analysis," *GeoInformatica*, 4, 141–159.
Miller, H. J., Wu, Y.-H., and Hung, M.-C. (1999) "GIS-based dynamic traffic congestion modeling to support time-critical logistics," *Proceedings of the Hawaii International Conference on Systems Science (HICSS-32)*, CD-ROM.
Miller, R. E. and Blair, P. D. (1985) *Input-Output Analysis: Foundations and Extensions*, Englewood Cliffs, NJ: Prentice-Hall.
Mirchandani, P. B. (1990) "The p-median problem and its generalizations, " in P. B. Mirchandani and R. L. Francis (eds.) *Discrete Location Theory*, New York: Wiley, 55–117.
Mirchandani, P. B. and Reilly, J. M. (1987) "Spatial distribution design for fire fighting units," in A. Ghosh and G. Rushton (eds.) *Spatial Analysis and Location-Allocation Models*, New York: Van Nostrand Reinhold, 186–223.
Mitchell, A. (1997) *Zeroing In: Geographic Information Systems at Work in the Community*, Redlands, CA: ESRI Press.
Mitchell, J. S. B., Mount, D. M., and Papadimitriou, C. H. (1987) "The discrete geodesic problem," *SIAM Journal of Computing*, 16, 647–668.
Mitchell, J. S. B. and Papadimitriou, C. H. (1991) "The weighted region problem: Finding shortest paths through a weighted planar subdivision," *Journal of the Association for Computing Machinery*, 38, 18–73.
Moellering, H. (1976) "The potential uses of a computer animated film in the analysis of geographical patterns of traffic crashes," *Accident Analysis & Prevention*, 8, 215–227.
Moellering, H. and Tobler, W. (1972) "Geographical variances," *Geographical Analysis*, 4, 34–50.
Monczka, R., Ragatz, G., Hanfield, R., Trent, R., and Frayer, D. (1997) "Supplier integration into new product development: A strategy for competitive advantage," *Global Procurement and Supply Chain Benchmarking Initiative*, Lansing, MI: The Eli Broad Graduate School of Management, Michigan State University.
Monmonier, M. (1990) "Strategies for the interactive exploration of geographic correlation," *Proceedings of the Fourth International Symposium on Spatial Data Handling*, 512–521.
Moore, E. F. (1959) "The shortest path through a maze," in *Proceedings of the International Symposium on the Theory of Switching*, Cambridge: Harvard University Press, 285–292.
Moore, I. D. (1996) "Hydrological modeling and GIS," in M. F. Goodchild, L. T. Steyaert, B. O. Parks, C. Johnston, D. Maidment, and S. Glendinning (eds.) *GIS and Environmental Modeling: Progress and Research Issues*, Ft. Collins, CO: GIS World Books, 143–148.
Moore, I. D., Turner, A. K., Wilson, J. P., Jenson, S. K., and Band, L. E. (1993) "GIS and land-surface-subsurface process modeling," in M. F. Goodchild, B. O. Parks, and L. T. Steyaert (eds.) *Environmental Modeling with GIS*, New York: Oxford University Press, 196–230.
Morehouse, S. (1985) "ARC/INFO: A geo-relational model for spatial information," *Proceedings of AUTO CARTO*, 7, 388–397.
Morrill, R. L. and Symons, J. (1977) "Efficiency and equity aspects of optimal location," *Geographical Analysis*, 9, 215–225.

Morris, J. M., Dumble, P. L., and Wigan, M. R. (1979) "Accessibility indicators for transport planning," *Transportation Research A*, 13A, 91–109.

Mozolin, M., Thill, J.-C., and Usery, E. L. (2000) "Trip distribution forecasting with multilayer perceptron neural networks: A critical evaluation," *Transportation Research B*, 34B, 53–73.

Murakami, E. and Wagner, D. P. (1999) "Can using global positioning system (GPS) improve trip reporting?" *Transportation Research C: Emerging Technologies*, 7, 149–165.

Murray, A. T. and Estivill-Castro, V. (1998) "Cluster discovery techniques for exploratory data analysis," *International Journal of Geographical Information Science*, 12, 431–443.

Murray, A. T. and Gottsegen, J. M. (1997) "The influence of data aggregation on the stability of p-median location model solutions," *Geographical Analysis*, 29, 200–213.

Mussone, L., Rinelli, S., and Reitani, G. (1996) "Estimating the accident probability of a vehicular flow by means of an artificial neural network," *Environment and Planning B: Planning and Design*, 23, 667–675.

Nagel, K. (1998) "From particle hopping models to traffic flow theory," *Transportation Research Record*, 1644, 1–9.

Nagurney, A. (1993) *Network Economics: A Variational Inequality Approach*, Dordrecht, The Netherlands: Kluwer Academic.

Narula, S. C., Ogbu, U. I., and Samuelsson, H. M. (1977) "An algorithm for the p-median problem," *Operations Research*, 25, 709–712.

National Center for Geographic Information and Analysis (NCGIA) (1992) *GEOLINUS: Data Management and Flowcharting for Arc/Info User Guide, Version 1.0*, National Center for Geographic Information and Analysis, technical software series S-92-2, Santa Barbara, CA.

National Cooperative Highway Research Program (NCHRP) (1987) *Integrated Highway Information Systems*, National Cooperative Highway Research Program, Synthesis of Highway Practice Report No. 133, Transportation Research Board, National Research Council, Washington, DC.

National Cooperative Highway Research Program (NCHRP) (1997) *A Generic Data Model for Linear Referencing Systems*, National Cooperative Highway Research Program, Research Results #218, Transportation Research Board, National Research Council, Washington, DC.

Nemhauser, G. L. and Wolsey, L. A. (1988) *Integer and Combinatorial Optimization*, New York: Wiley.

Neves, N., Silva, J. P., Goncalves, P., Muchaxo, J., Silva, J., and Camara, A. (1997) "Cognitive space and metaphors: A solution for interacting with spatial data," *Computers and Geosciences*, 23, 483–488.

Nevita, R. and Babu, K. R. (1980) "Linear feature extraction and description," *Computer Graphics and Image Processing*, 13, 257–269.

Newcomb, M., Medan, J., and Smartt, B. (1993) "Data requirements for route guidance," *GIS-T 1993 Proceedings*, 209–220.

Newell, G. F. (1986) "Design of multiple vehicle delivery tours: III," *Transportation Research B*, 20B, 377–390.

Newell, G. F. and Daganzo, C. F. (1986a) "Design of multiple vehicle delivery tours: I," *Transportation Research B*, 20B, 345–364.

Newell, G. F. and Daganzo, C. F. (1986b) "Design of multiple vehicle delivery tours: II," *Transportation Research B*, 20B, 365–376.

Newton, R. and Thomas, W. (1974) "Bus routing in a multi-school system," *Computers and Operations Research*, 1, 213–222.

Nicholson, A. J., Elms, D. G., and Williman, A. (1976) "A variational approach to optimal route location," *Journal of the Institute of Highway Engineers*, 23, 22–25.

Nielsen, O. A. (1995) "Using GIS in Denmark for traffic planning and decision support," *Journal of Advanced Transportation*, 29, 335–354.

Nievergelt, J. and Widmayer, P. (1997) "Spatial data structures: Concepts and design choices," in M. van Kreveld, J. Nievergelt, T. Roos, and P. Widmayer (eds.) *Algorithmic Foundations of Geographic Information Systems*, Berlin: Springer-Verlag, 153–197.

Nijkamp, P., Reggiani, A., and Tritapepe, T. (1996) "Modelling inter-urban transport flows in Italy: A comparison between neural network analysis and logit analysis," *Transportation Research C: Emerging Technologies*, 4C, 323–338.

Nilsson, N. J. (1998) *Artificial Intelligence: A New Synthesis*, San Francisco: Morgan Kaufmann.

Niskanen, E. (1987) "Congestion tolls and consumer welfare," *Transportation Research B*, 21B, 171–174.

Noronha, V. (2000) "Towards ITS map database interoperability: Database error and rectification," *GeoInformatica*, 4, 201–213.

Noronha, V. and Goodchild, M. (2000) "Map accuracy and location expression in Transportation: Reality and prospects," *Transportation Research C: Emerging Technologies*, 8C, 53–69.

Noronha, V., Goodchild, M., Church, R., and Fohl, P. (1999) "Location expression standards for ITS: Testing the LRMS cross street profile," *Annals of Regional Science*, 33, 197–212.

Novak, K. and Nimz, J. (1997) "Transportation infrastructure management," *GIS-T 1997 Proceedings*, American Association of State Highway and Transportation Officials.

Nwagboso, C. (1997) "Introduction to intelligent transportation systems," in C. O. Nwagboso (ed.) *Advanced Vehicle and Infrastructure Systems: Computer Application, Control and Automation*, Chichester, U.K.: Wiley, 3–32.

Nyerges, T. L. (1990) "Locational referencing and highway segmentation in a geographic information systems," *ITE Journal*, 60(3), 27–31.

Nyerges, T. L. (1991a) "Analytical map use," *Cartography and Geographic Information Systems*, 18, 11–22.

Nyerges, T. L. (1991b) "Geographic information abstractions: Conceptual clarity for geographic modeling," *Environment and Planning A*, 23, 1483–1499.

Nyerges, T. L. (1992) "Coupling GIS and spatial analytical models," *Proceedings of the 5th International Symposium on Spatial Data Handling*, Vol. 2, 534–543.

Nyerges, T. L. (1993) "Understanding the scope of GIS: Its relationship to environmental modeling," in M. Goodchild, B. Parks, and L. Steyaert, *Environmental Modeling Within GIS*, Oxford, U.K.: Oxford University Press, 75–93.

Nyerges, T. L., Montejano, R., Oshiro, C., and Dadswell, M. (1997) "Group-based geographic information systems for transportation improvement site selection," *Transportation Research C: Emerging Technologies*, 5C, 349–369.

O'Banion, K. (1980) "Use of value functions in environmental decisions," *Environmental Management*, 4, 3–6.

O'Callaghan, J. F. and Mark, D. M. (1984) "The extraction of drainage networks from digital elevation data," *Computer Vision Graphics and Image Proceedings*, 28, 323–344.

O'Kelly, M. E. and Miller, E. J. (1984) "Characteristics of multistop multipurpose travel: An empirical study of trip length," *Transportation Research Record*, 976, 33–39.

O'Neill, W. A. (1991) "Developing optimal transportation analysis zones using GIS," *ITE Journal*, 61(Dec.), 33–36.

O'Neill, W. A. (1995) "A comparison of accuracy and cost of attribute data in analysis of transit service areas uisng GIS," *Journal of Advanced Transportation*, 29, 299–320.

O'Neill, W. A., Ramsey, R. D., and Chou, J. (1992) "Analysis of transit service areas using geographic information systems," *Transportation Research Record*, 1364, 131–138.

O'Sullivan, D., Morrison, A., and Shearer, J. (2000) "Using desktop GIS for the investigation of accessibility by public transport: An isochrone approach," *International Journal of Geographical Information Science*, 14, 85–104.

Oak Ridge National Laboratory (ORNL) (1999a) *The Geographic Coordinate Profile*, available at http://itsdeployment.ed.ornl.gov/spatial/datum/geograp.htm.

Oak Ridge National Laboratory (ORNL) (1999b) *Overview of the Location Reference Message Specification*, available at http://itsdeployment.ed.ornl.gov/spatial/files/lrmsovr.htm.

Okabe, A. and Kitamura, M. (1996) "A computational method for market area analysis on a network," *Geographical Analysis*, 28, 330–349.

Okabe, A. and Miller, H. J. (1996) "Exact computational methods for calculating distance between objects in a cartographic database," *Cartography and Geographic Information Systems*, 23, 180–195.

Olivera, F. and Maidment, D. (1998) "Geographic information system use for hydrologic data development for design of highway drainage facilities," *Transportation Research Record*, 1625, 131–138.

Open GIS Consortium (OGC) Technical Committee (1998) *Introduction to Interoperable Geoprocessing and the OpenGIS Specification*, 3d ed., Open GIS Consortium, Inc., Wayland, MA, available at http://www.opengis.org/.

Openshaw, S. (1977a) "A geographical solution to scale and aggregation problems in region building, partitioning and spatial modeling," *Transactions of the Institute for British Geographers*, 2(new series), 459–472.

Openshaw, S. (1977b) "Optimal zoning systems for spatial interaction models," *Environment and Planning A*, 9, 169–184.

Openshaw, S. (1978) "An empirical study of some zone-design criteria," *Environment and Planning A*, 10, 781–794.

Openshaw, S. (1996) "Developing GIS-relevant zone-based spatial analysis methods," in P. Longley and M. Batty (eds.) *Spatial Analysis: Modeling in a GIS Environment*, Cambridge, UK: GeoInformation International, 55–73.

Openshaw, S., Charlton, M., and Wymer, C. (1987) "A Mark I geographic analysis machine for the automated analysis of point pattern data," *International Journal of Geographical Information Systems*, 1, 335–350.

Openshaw, S. and Clarke, G. (1996) "Developing spatial analysis functions relevant to GIS environments," in M. Fischer, H. J. Scholten, and D. Unwin (eds.) *Spatial Analytical Perspectives on GIS*, London: Taylor and Francis, 21–37.

Openshaw, S., Cross, A., and Charlton, M. (1990) "Building a prototype geographical correlates exploration machine," *International Journal of Geographical Information Systems*, 4, 297–311.

Openshaw, S. and Schmidt, J. (1996) "Parallel simulated annealing and genetic algorithms for re-engineering zoning systems," *Geographical Systems*, 3, 201–220.

Openshaw, S. and Taylor, P. J. (1979) "A million or so correlation coefficients: Three experiments on the modifiable areal unit problem," in N. Wrigley (ed.) *Statistical Applications in the Spatial Sciences*, London: Pion, 127–144.

Openshaw, S., Waugh, D., and Cross, A. (1994) "Some ideas about the use of map animation as a spatial analysis tool," in H. M. Hearnshaw and D. J. Unwin (eds.) *Visualization in Geographical Information Systems*, Chichester, U.K.: Wiley, 131–138.

Oppenheim, N. (1995) *Urban Travel Demand Modeling: From Individual Choices to General Equilibrium*, New York: Wiley.

Oracle Corporation (1999a) *Oracle8i Objects and Extensibility Option—Feature Overview*, Feb. 1999, Redwood Shores, CA: Oracle Corp., available at http://www.oracle.com/.

Oracle Corporation (1999b) *Oracle Spatial—Data Sheet*, March 1999, Redwood Shores, CA: Oracle Corp., available at http://www.oracle.com/.

Oracle Corporation (1999c) *Spatially Enabling the Enterprise—An Oracle Business White Paper*, Feb. 1999, Redwood Shores, CA: Oracle Corp., available at http://www.oracle.com/.

Ord, J. K. and Cliff, A. D. (1976) "The analysis of commuting patterns," *Environment and Planning A*, 9, 941–946.

Ortúzar, J. D. (1983) "Nested logit models for mixed-mode travel in urban corridors," *Transportation Research B*, 17B, 283–299.

Ortúzar, J. and Willumsen, L. G. (1994) *Modelling Transport*, 2d ed., New York: Wiley.

Osborn, S. L. and Heaven, T. E. (1986) "The design of a relational database system with abstract data types for domains," *ACM Transactions on Database Systems*, 11, 357–373.

Ossinger, M. C., Schafer, J. A., and Cihon, R. F. (1992) "Method to identify, inventory and map wetlands using aerial photography and geographic information systems," *Transportation Research Record*, 1366, 35–40.

Padberg, M. W. and Grötschel, M. (1985) "Polyhedral computations," in E. L. Lawler, J. K. Lenstra, A. H. G. Rinnooy Kan, and D. B. Shmoys (eds.) *The Traveling Salesman Problem: A Guided Tour of Combinatorial Optimization*, Chichester, U.K.: Wiley, 307–360.

Padberg, M. W. and Hong, S. (1980) "On the symmetric traveling salesman problem: A computational study," *Mathematical Programming Study*, 12, 78–107.

Pallottino, S. (1984) "Shortest-path methods: Complexity, interrelationships and new propositions," *Networks*, 14, 257–267.

Pallottino, S. and Scutellá, M. G. (1998) "Shortest path algorithms in transportation models: Classical and innovative aspects," in P. Marcotte and S. Nguyen (eds.) *Equilibrium and Advanced Transportation Modeling*, howell, MA: Kluwer Academic, 245–281.

Panchanathan, S. and Faghri, A. (1995) "Knowledge-based geographic information system for safety analysis at rail-highway grade crossings," *Transportation Research Record*, 1497, 91–100.

Pape, U. (1974) "Implementation and efficiency of Moore's algorithm for the shortest root problem," *Mathematical Programming*, 7, 212–222.

Paswell, R. E. (1990) "Air quality and the transportation community: Executive committee of the Transportation Research Board addresses this issue in a special session," *TR News*, 148(May–June), 5–10.

Pavlidis, T. (1982) *Algorithms for Graphics and Image Processing*, Berlin: Springer-Verlag.

Peterson, C. and Söderberg, B. (1993) "Artificial neural networks," in C. R. Reeves (ed.) *Modern Heuristic Techniques for Combinatorial Problems*, New York: Wiley, 197–242.

Peuker, T. K. and Douglas, D. H. (1975) "Detection of surface specific points by local parallel processing of discrete terrain elevation data," *Computer Graphics and Image Processing*, 4, 375–387.

Peuquet, D. J. (1994) "It's about time: A conceptual framework for the representation of temporal dynamics in geographic information systems," *Annals of the Association of American Geographers*, 84, 441–461.

Peuquet, D. J. and Duan, N. (1995) "An event-based spatiotemporal data model (ESTDM) for temporal analysis of geographical data," *International Journal of Geographical Information Systems*, 9, 359–384.

Peuquet, D. J. and Wentz E. A. (1994) "An approach for time-based spatial analysis of spatio-temporal data," *Proceedings of Advances in GIS Research*, 1, 489–504.

Philbrick, A. K. (1953) "Toward a unity of cartographical forms and geographical content," *Professional Geographer*, 5(5), 11–15.

Phillips, D. T. and Garcia-Diaz, A. (1981) *Fundamentals of Network Analysis*, Englewood Cliffs, NJ: Prentice Hall.

Pidd, M., Eglese, R., and De Silva, F. N. (1997) "CEMPS: A prototype spatial decision support system to aid in planning emergency evacuations," *Transactions in GIS*, 1, 321–334.

Pijawka, K. D., Foote, S., and Soesilo, A. (1985) "Risk assessment of transporting hazardous material: Route analysis and hazard management," *Transportation Research Record*, 1020, 1–6.

Plane, D. A. and Rogerson, P. A. (1994) *The Geographical Analysis of Population with Applications to Planning and Business*, New York: Wiley.

Plane, D. R. and Hendrick, T. E. (1977) "Mathematical programming and the location of fire companies for the Denver fire department," *Operations Research*, 25, 563–578.

Plewe, B. (1997) *GIS Online: Information, Retrieval, Mapping, and the Internet*, Santa Fe, NM: OnWord Press.

Ploszaj, T. (1999) *ERP Implementation: Win, Lose, or Crawl*, available at http://www.geocities.com/~erppro/winlosecrawl.html.

Polydoros, A. and Panagiotou, P. (1997) "Communication technologies for AHS," in P. A. Ioannou (ed.) *Automated Highway Systems*, New York: Plenum Press, 173–193.

Poundstone, W. (1988) *Labyrinths of Reason: Paradox, Puzzles and the Fraility of Knowledge*, New York: Doubleday.

Prastacos, P. (1986a) "An integrated land-use-transportation model for the San Francisco region: 1. Design and mathematical structure," *Environment and Planning A*, 18, 307–322.

Prastacos, P. (1986b) "An integrated land-use-transportation model for the San Francisco region: 2. Empirical estimation and results," *Environment and Planning A*, 18, 511–528.

Prastacos, P. (1992) "Integrating GIS technology in urban transportation planning and modeling," *Transportation Research Record*, 1305, 123–130.

Preparata, F. P. and Shamos, M. I. (1985) *Computational Geometry: An Introduction*, Berlin: Springer-Verlag.

Putnam, S. H. (1991) *Integrated Urban Models*, 2, London: Pion Ltd.

Putnam, S. H. and Chung, S.-H. (1989) "Effects of spatial system design on spatial interaction models. 1: The spatial system definition problem," *Environment and Planning A*, 21, 27–46.

Pyle, I. C. and Illingworth, V. (eds.) (1997) *Dictionary of Computing*, 4th ed., Oxford U.K.: Oxford University Press.

Quinn, P. F., Beven, K. J., Chevallier, P., and Planchon, O. (1991) "The prediction of hillslope flow paths for distributed hydrological modelling using digital terrain models," *Hydrological Processes*, 5, 59–79.

Ragan, R. M. (1991) *A Geographic Information System to Support Statewide Hydrologic and Nonpoint Pollutant Modeling*, Research report, Department of Civil Engineering, University of Maryland, College Park.

Raju, K. A., Sikdar, P. K., and Dhingra, S. L. (1996) "Modelling mode choice by means of an artificial neural network," *Environment and Planning B: Planning and Design*, 23, 677–683.
Rakha, H. A. and van Aerde, M. W. (1996) "Comparison of simulation modules of TRANSYT and INTEGRATION models," *Transportation Research Record*, 1566, 1–7.
Rakha, H., van Aerde, M., Bloomberg, L., and Huang, X. (1998) "Construction and calibration of a large-scale microsimulation model of the Salt Lake area," *Transportation Research Record*, 1644, 93–102.
Ralston, B. A. (1999) *Visual Basic Programming with MapObjects*, available at http://www.gistools.com.
Ralston, B. A., Liu, C., and Zhang, M. (1990) *The Transportation and Inland Logistics Manager*, Washington, DC: USAID.
Ramirez, A. I. and Seneviratne, P. N. (1996) "Transit route design applications using geographic information systems," *Transportation Research Record*, 1557, 10–14.
Ran, B. and Boyce, D. (1996) *Modeling Dynamic Transportation Networks: An Intelligent Transportation System Approach*, 2d ed., Berlin: Springer.
Ran, B., Boyce, D. E., and LeBlanc, L. J. (1993) "A new class of instantaneous dynamic user-optimal traffic assignment models," *Operations Research*, 41, 192–202.
Raper, J. and Livingston, D. (1993) Interfacing GIS with virtual reality technology," *Proceedings of the Association for Geographic Information Conference*, 3, 1–4.
Raper, J., McCarthy, T., and Williams, N. (1997) Intergrating ArcView and geographically referenced VRML models in real time," *Proceedings of the ESRI User Conference*, available at http://www.esri.com.
Rappaport, T. S., Reed, J. H., and Woerner, B. D. (1996) "Position location using wireless communication on highways of the future," *IEEE Communications Magazine*, 34(10), 33–41.
Rathi, A. K. and Santiago, A. J. (1990) "New NETSIM simulation program," *Traffic Engineering and Control*, 31, 317–320.
Rayco, M. B., Francis, R. L., and Lowe, T. J. (1996) "Error-bound driven demand point aggregation for the rectlinear distances $p$-center model," *Location Science*, 4, 213–235.
Recker, W. W., McNally, M. G., and Root, G. S. (1986a) "A model of complex travel behavior: Part I—theoretical development," *Transportation Research A*, 20A, 307–318.
Recker, W. W., McNally, M. G., and Root, G. S. (1986b) "A model of complex travel behavior: Part II—an operational model," *Transportation Research A*, 20A, 319–330.
Reed, J. H., Krizman, K. J., Woerner, B. D., and Rappaport, T. S. (1998) "An overview of the challenges and progress in meeting E-911 requirement for location service," *IEEE Communications Magazine*, 36(4), 30–37.
Reeves, C. R. (1993) "Genetic algorithms," in C. R. Reeves (ed.) *Modern Heuristic Techniques for Combinatorial Problems*, New York: Wiley, 151–196.
ReVelle, C. (1987) "Urban public facility location," in E. S. Mills (ed.) *Handbook of Regional and Urban Economics, Vol. 11*, 1053–1096.
ReVelle, C. and Hogan, K. (1988) "A reliability-constrained siting model with local estimates of busy fractions," *Environment and Planning B: Planning and Design*, 15, 143–152.
ReVelle, C. and Hogan, K. (1989) "The maximum availability location problem," *Transportation Science*, 23, 192–200.
ReVelle, C. and Swain, R. W. (1970) "Central facilities location," *Geographical Analysis*, 2, 30–42.

Rhind, D. (1998) "The incubation of GIS in Europe," in T. W. Foresman (ed.) *The History of GIS: Perspectives from Pioneers*, Upper Saddle River, NJ: Prentice Hall, 293–306.

Rhyne, T. M. (1997) "Going virtual with geographic information and scientific visualization," *Computers and Geosciences*, 23, 489–491.

Richards, D. R., Jones, N. L., and Lin, H. C. (1993) "Graphical innovations in surface water flow analysis." in M. Goodchild, B. Parks, and L. Steyaert (eds.) *Environmental Modeling with GIS*, Oxford, U.K.: Oxford University Press, 188–195.

Richards, J. A. and Jia, X. (1999) *Remote Sensing Digital Image Analysis: An Introduction*, 3d ed., Berlin: Springer-Verlag.

Robinson, W. (1950) "Ecological correlation and the behavior of individuals," *American Sociological Review*, 15, 351–357.

Robusté, F., Daganzo, C. F., and Souleyrette, R. R., II (1990) "Implementing vehicle routing models," *Transportation Research B*, 24B, 263–286.

Roddick, J. F. and Lees, B. (2001) "Paradigms for spatial and spatio-temporal data mining," in H. J. Miller and J. Han (eds.) *Geographic Data Mining and Knowledge Discovery*, London: Taylor and Francis, in press.

Rogerson, P. A. (1990) "Buffon's needle and the estimation of migration distances," *Mathematical Population Studies*, 2, 229–238.

Rosenkrantz, D. J., Stearns, R. E., and Lewis, P. M., II (1977) "An analysis of several heuristics for the traveling salesman problem," *SIAM Journal of Computing*, 6, 563–581.

Rosing, K. E., Hillsman, E. L., and Rosing-Vogelaar, H. (1979) "A note comparing optimal and heuristic solutions to the p-median problem," *Geographical Analysis*, 11, 86–89.

Roth, P. M. (1990) "Effects of transportation on ozone in cities: Appraisal of the air quality planning process," *TR News*, 148(May–June), 11–16.

Rumbaugh, J., Blaha, M., Premerlani, W., Eddy, F., and Lorensen, W. (1991) *Object-Oriented Modeling and Design*, Englewood Cliffs, NJ: Prentice Hall.

Rushton, G. and Kohler, J. A. (1973) *ALLOC: Heuristic Solutions in Multi-Facility Location Problems on a Graph*, Research Monograph #6, Department of Geography, The University of Iowa, Iowa City.

Saccomanno, F. F. and Chan, A. Y. W. (1985) "An economic evaluation of routing strategies for hazardous road shipments," *Transportation Research Record*, 1020, 12–18.

Saccomanno, F. F., Shortreed, J. H., van Aerde, M., and Higgs, J. (1989) "Comparison of risk measures for the transport of dangerous commodities by truck and rail," *Transportation Research Record*, 1245, 1–13.

Safwat, K. N. A. (1987) "Computational experience with a convergent algorithm for the simultaneous prediction of transportation equilibrium," *Transportation Research Record*, 1120, 60–67.

Safwat, K. N. A. and Magnanti, T. L. (1988) "A combined trip generation, trip distribution, modal split, and trip assignment model," *Transportation Science*, 18, 14–30.

Safwat, K. N. A. and Walton, C. M. (1988) "Computational experience with an application of a simultaneous transportation equilibrium model to urban travel in Austin, Texas: Computational results," *Transportation Research B*, 22B, 457–467.

Sanders, L. (1996) "Dynamic modelling of urban systems," in M. Fischer, H. J. Scholten, and D. Unwin (eds.) *Spatial Analytical Perspectives on GIS*, GISDATA IV, London: Taylor and Francis, 229–244.

Sapeta, K. and Barnell, B. (1999) "3-D models of lower Manhattan produce myriad benefits," *GeoWorld*, March, 39.

Sarasua, W. and Jia, X. (1995) "A framework for integrating GIS-T with KBES: A pavement management system example," *Transportation Research Record*, 1497, 153–163.

Scarponcini, P. (1997) "Relational databases now ready for spatial GIS-T data," Paper presented at International Highway Engineering Exchange Program Conference, Portland, ME.

Schilling, D. A., Elzinga, D., Cohon, J., Church, R., and ReVelle, C. (1979) "The TEAM/FLEET models for simultaneous facility and equipment siting," *Transportation Science*, 13, 163–175.

Schilling, D. A., Jayaraman, V., and Barkhi, R. (1993) "A review of covering problems in facility location," *Location Science*, 1, 25–55.

Schneider National (1999) *Schneider National Selects ORBCOMM Trailer Tracking Solution*, press release available at http://www.schneider.com/news/orbcommnews.html.

Schulze, J. and Fahle, T. (1999) "A parallel algorithm for the vehicle routing problem with time window constraints," *Annals of Operations Research*, 86, 585–607.

Sedgewick, R. (1992) *Algorithms in C++*, Reading, MA: Addison-Wesley.

Shaw, S.-L. (1989) "Design considerations for a GIS-based transportation network analysis system," *GIS/LIS '89 Proceedings*, 1, 20–29.

Shaw, S.-L. (1993) "GIS for urban travel demand analysis: Requirements and alternatives," *Computers, Environment and Urban Systems*, 17, 15–29.

Shaw, S.-L. (2000) "Moving toward spatiotemporal GIS for transportation applications," *Proceedings of the ESRI Users Conference*, available at http://www.esri.com.

Shaw, S.-L. and Donnelly, J. D. (1994) "Spatiotemporal evolution of airline networks with visualization considerations," Paper presented at the annual meeting of the Association of American Geographers, San Francisco, CA.

Shaw, S.-L. and Wang, D. (2000) "Handling disaggregate spatiotemporal travel data in GIS," *GeoInformatica*, 4, 161–178.

Sheffi, Y. (1985) *Urban Transportation Networks: Equilibrium Analysis with Mathematical Programming Methods*, Englewood Cliffs, NJ: Prentice Hall.

Sheffi, Y. and Daganzo, C. F. (1980) "Computation of equilibrium over transportation networks: the case of disaggregate demand models," *Transportation Science*, 14, 155–173.

Sheffi, Y., Mahmassani, H., and Powell, W. B. (1982) "A transportation network evacuation model," *Transportation Research A*, 16A, 209–218.

Sheffi, Y. and Powell, W. B. (1981) "A comparison of stochastic and deterministic traffic assignment over congested networks," *Transportation Research B*, 15B, 53–64.

Sheffi, Y. and Powell, W. B. (1982) "An algorithm for the equilibrium assignment problem with random link times," *Networks*, 12, 191–207.

Shekhar, S. and Fetterer, A. (1996) "Path computation in advanced traveler information systems," available at http://www.cs.umn.edu/Research/shashi-group/.

Shekhar, S., Kohli, A., and Coyle, M. (1993) "Path computation algorithms for advanced traveller information system (ATIS)," *Proceedings of the 9th International Conference on Data Engineering*, Los Alamitos, CA: IEEE Computer Society Press, 31–39.

Shekhar, S., Lu, C., Tan, X., Chawla, S., and Vatsavai, R. R. (2001) "Map cube: A visualization tool for spatial data warehouses," in H. J. Miller and J. Han (eds.) *Geographic Data Mining and Knowledge Discovery*, London: Taylor and Francis, in press.

Sherali, H. D., Arora, N., and Hobeika, A. G. (1997) "Parameter optimization methods for estimating dynamic origin-destination trip tables," *Transportation Research B*, 31B, 141–157.

Shi, W. and Liu, W. (2000) "A stochastic process-based model for the positional error of line segments in GIS," *International Journal of Geographical Information Science*, 14, 51–66.

Shladover, S. E. (1993) "Potential contributions of intelligent vehicle/highway systems (IVHS) to reducing transportation's greenhouse gas production," *Transportation Research A*, 27A, 207–216.

Shtub, A. (1999) *Enterprise Resource Planning (ERP): The Dynamics of Operations Management*, Dordrecht, The Netherlands: Kluwer Academic.

Shuman, V. and Soloman, R. J. (1996) "Global interoperability for the NII and ITS: Standards and Policy Challenges," in L. M. Branscomb and J. H. Keller (eds.) *Coverging Infrastructure—Intelligent Transportation and the National Information Infrastructure*, Cambridge: The MIT Press, 191–208.

Siegel, D., Goodwin, C., and Gordon, S. (1996a) *ITS Datum Final Design Report*, prepared for Federal Highway Administration, Office of Safety and Traffic Operations by Oak Ridge National Laboratory, Oak Ridge, TN.

Siegel, D., Goodwin, C., and Gordon, S. (1996b) *ITS Datum Prototype Design Report Task C: Spatial Data Interoperability Protocol for ITS Project*, Federal Highway Administration, IVHS Research Division, McLean, VA.

Sinton, D. F. (1978) "The inherent structure of information as a constraint to analysis: Mapped thematic data as a case study," in G. Dutton (ed.) *Harvard Papers on Geographic Information Systems, Vol. 6*, Reading: MA: Addison-Wesley, 1–17.

Sipser, M. (1997) *Introduction to the Theory of Computation*, Boston: PWS Publishing.

Slavin, H. (1995) "An integrated, dynamic approach to travel demand forecasting," *Transportation*, 22, 1–40.

Smith, B. L. and Demetsky, M. J. (1997) "Traffic flow forecasting: Comparison of modeling approaches," *Journal of Transportation Engineering*, 123, 261–266.

Smith, J. R. (1997) *Introduction to Geodesy: The History and Concepts of Modern Geodesy*, New York: Wiley.

Smith, M. J. (1979) "The existence, uniqueness and stability of traffic equilibria," *Transportation Research B*, 13B, 295–304.

Smith, P. (1995) *Hydrological Data Development System*, unpublished M.Sc. thesis, Department of Civil Engineering, University of Texas, Austin.

Smith, T. R., Peng, G., and Gahinet, P. (1989) "Asynchronous, iterative, and parallel procedures for solving the weighted-region least cost path problem," *Geographical Analysis*, 21, 147–166.

Smyth, C. S. (2001) "Mining mobile trajectories," in H. J. Miller and J. Han (eds.) *Geographic Data Mining and Knowledge Discovery*, London: Taylor and Francis, in press.

Snodgrass, R. T. (1992) "Temporal databases," in A. U. Frank, I. Campari, and U. Formentini (eds.) *Theories and Methods of Spatio-Temporal Reasoning in Geographic Space*, Berlin: Springer-Verlag Lecture Notes in Computer Science #639, 22–64.

Snyder, J. P. (1987) *Map Projections—A Working Manual*, U.S. Geological Survey Professional Paper 1395. Washington, DC: United States Government Printing Office.

Society of Automotive Engineers (SAE) (1998) *Surface Vehicle Information Report—Location Referencing Message Specification Information*, Report J2374, Warrendale, PA: Society of Automotive Engineers.

Solomon, M. M. and Desrosiers, J. (1988) "Time window constrained routing and scheduling problems," *Transportation Science*, 22, 1–13.

Sorensen, P. A. and Church, R. L. (1995) *A Comparison of Strategies for Data Storage Reduction in Location-Allocation Problems*, Santa Barbara, CA: National Center for Geographical Information and Analysis Technical Report 95-4.

Sorenson, P. A. and Church, R. L. (1996) "A comparison of strategies for data storage reduction in location-allocation problems," *Geographical Systems*, 3, 221–242.

Sorensen, P. A. and Lanter, D. P. (1993) "Two algorithms for determining partial visibility and reducing data structure induced error in viewshed analysis," *Photogrammetric Engineering and Remote Sensing*, 59, 1149–1160.

Souleyrette, R. R. and Sathisan, S. K. (1994) "GIS for radioactive materials transportation," *Microcomputers in Civil Engineering*, 9, 295–303.

Souleyrette, R. R., Sathisan, S. K., James, D. E., and Lim, S.-T. (1992) "GIS for transportation and air quality analysis," in R. L. Wayson (ed.) *Transportation Planning and Air Quality*, New York: American Society of Civil Engineers, 182–194.

Souleyrette, R. R. and Strauss, T. R. (1999) "Transportation," in S. Easa and Y. Chan (eds.) *Planning and Development Applications of GIS*, Reston, VA: American Society of Civil Engineers, 117–133.

Southworth, F. (1995) *A Technical Review of Urban Land Use-Transportation Models as Tools for Evaluating Vehicle Reduction Strategies*, Research report, Oak Ridge National Laboratory, ORNL-6881.

Southworth, F. and Chin, S.-M. (1987) "Network evacuation modelling for flooding as a result of dam failure," *Environment and Planning A*, 19, 1543–1558.

Spear, B. D. and Lakshmanan, T. R. (1998) "The role of GIS in transportation planning analysis," *Geographical Systems*, 5, 45–58.

Spiekermann, K. and Wegener, M. (2000) "Freedom from the tyranny of zones: Towards new GIS-based spatial models," in A. S. Fotheringham and M. Wegener (eds.) *Spatial Models and GIS: New Potential and New Models*, GISDATA 7, London: Taylor and Francis, 45–61.

Spiess, H. and Florian, M. (1989) "Optimal strategies: A new assignment model for transit networks," *Transportation Research B*, 23B, 83–102.

Spring, G. S. and Hummer, J. (1995) "Identification of hazardous highway locations using knowledge-based GIS: A case study," *Transportation Research Record*, 1497, 83–90.

Srinivasan, A. and Richards, J. A. (1993) "Analysis of GIS spatial data using knowledge-based methods," *International Journal of Geographical Information Systems*, 7, 479–500.

Stern, E. and Sinuany-Stern, Z. (1989) "A behavioural-based simulation model for urban evacuation," *Papers of the Regional Science Association*, 66, 87–103.

Stopher, P. R. and Fu, H. (1996) "Travel demand analysis impacts on estimation of mobile emissions," *Transportation Research Record*, 1520, 104–113.

Stopher, P. R., Hartgen, D., and Li, Y. J. (1996) "SMART: Simulation model for activities, resources and travel," Paper presented at the 73rd annual meeting of the Transportation Research Board, Washington, DC.

Stopher, P. R. and Wilmot, C. G. (2000) "Some new approaches to designing household travel surveys: Time-use diaries and GPS," *Proceedings of the 79th Annual Meeting of the Transportation Research Board*, Washington, DC, Jan. 9–13, CD-ROM.

Streit, U. and Wisemann, K. (1996) "Problems of integrating GIS and hydrological models," in M. Fischer, H. J. Scholten, and D. Unwin (eds.) *Spatial Analytical Perspectives on GIS*, London: Taylor and Francis, 161–173.

Summers, M. and Southworth, F. (1998) "Design of a testbed to assess alternative traveler behavior models within an intelligent transportation system architecture," *Geographical Systems*, 5, 91–115.

Sutton, J. (1997) "Data attribution and network representation issues in GIS and transportation," *Transportation Planning and Technology*, 21, 25–44.

Sutton, J. and Bespalko, S. (1995) *Network Pathologies Phase 1 Report*, Sandia National

Laboratories, Transportation Systems Analysis GIS Project, Document No. AH-2266.

Sutton, J. and Gillingwater, D. (1997) "Geographic information systems and transportation: Overview," *Transportation Planning and Technology*, 21, 1–4.

Swain, R. (1974) "A parametric decomposition algorithm algorithm for the solution of uncapacitated location problems," *Management Science*, 21, 189–198.

Swann, R., Hawkins, D., Westwell-Roper, A., and Johnstone, W. (1988) "The potential for automated mapping from geocoded digital image data," *Photogrammetric Engineering and Remote Sensing*, 54, 187–193.

Taaffe, E. J., Gauthier, H. L., and O'Kelly, M. E. (1996) *Geography of Transportation*, 2d ed., Upper Saddle River, NJ: Prentice Hall.

Talvitie, A. (1992) "How to code the generalized cost of accessing a transit system," *Transportation Research Record*, 1357, 1–7.

Tansel, B. C., Francis, R. L., and Lowe, T. J. (1983a) "Location on networks: A survey. Part I: The p-center and p-median problems," *Management Science*, 29, 482–497.

Tansel, B. C., Francis, R. L., and Lowe, T. J. (1983b) "Location on networks: A survey. Part II: Exploiting tree structure," *Management Science*, 29, 498–511.

Taori, S. and Rathi, A. K. (1996) "Comparison of NETSIM, NETFLO I and NETFLO II traffic simulation models for fixed signal control," *Transportation Research Record*, 1566, 20–30.

Tarjan, R. E. (1983) *Data Structures and Network Algorithms*, Philadelphia, PA: Society for Industrial and Applied Mathematics.

Taylor, D. R. F. (1991) "Geographic information systems: The microcomputer and modern cartography," in D. R. F. Taylor (ed.) *Geographic Information Systems: The Microcomputer and Modern Cartography*, Oxford, U.K.: Pergamon, 1–20.

Teitz, M. B. and Bart, P. (1968) "Heuristic methods for estimating the generalized vertex median of a weighted graph," *Operations Research*, 16, 955–961.

Thill, J.-C. and Thomas, I. (1987) "Toward conceptualizing trip-chaining behavior: A review," *Geographical Analysis*, 19, 1–17.

Tobler, W. R. (1970) "A computer movie simulating urban growth in the Detroit region," *Economic Geography*, 46, 234–240.

Tobler, W. R. (1979) "Cellular geography," in S. Gale and G. Olsson (eds.) *Philosophy in Geography*, Dordrecht, The Netherlands: D. Reidel, 379–386.

Toregas, C., Swain, R., ReVelle, C., and Bergman, L. (1971) "The location of emergency service facilities," *Operations Research*, 19, 1363–1373.

Trietsch, D. (1987a) "Comprehensive design of highway networks," *Transportation Science*, 21, 26–35.

Trietsch, D. (1987b) "A family of methods for preliminary highway alignment," *Transportation Science*, 21, 17–25.

Turban, E. and Meredith, J. R. (1985) *Fundamentals of Management Science*, 3d ed., Plano, TX: Business Publications.

Turk, A. (1994) "Cogent GIS visualization," in H. M. Hearnshaw and D. J. Unwin (eds.) *Visualization in Geographical Information Systems*, Chichester, U.K.: Wiley, 26–33.

Turner, A. K. and Hansen, J. H. (2000) *Advances in Remote Sensing and Data Capture Technologies for Transportation Applications*, Workshop materials, 79th annual meeting of the Transportation Research Board, Washington, DC.

U.S. Congress (1995) *High-Tech Highways: Intelligent Transportation Systems and Policy*, Washington, DC: Congressional Budget Office.

U.S. Department of Transportation (USDOT) (1996) *Technical Summary—TravTek Operational Test Evaluation Final Report*, Federal Highway Administration, available at http://www.its.dot.gov/.

U.S. Department of Transportation (USDOT) (1998a) *Developing Intelligent Transportation Systems Using the National ITS Architecture—An Executive Edition for Senior Transportation Managers*, Washington, DC: Federal Highway Administration, Intelligent Transportation Systems Joint Program Office.

U.S. Department of Transportation (USDOT) (1998b) *Guidance for Congressionally-designated ITS Projects*, Washington, DC: Federal Highway Administration, available at http://www.its.dot.gov/.

U.S. Department of Transportation (USDOT) (1999a) *Fatality Analysis Reporting Systems* (FARS), *Traffic Safety Facts 1996*, Washington, DC: National Highway Traffic Safety Administration (NHTSA), available at http://www.nhtsa.dot.gov/.

U.S. Department of Transportation (USDOT) (1999b) *Intelligent Transportation Systems: Critical Standards*, Washington, DC, available at http://www.itsdocs.fhwa.dot.gov/jpodocs/repts_pr/49701!.pdf.

U.S. Department of Transportation (USDOT) (1999c) *Intelligent Transportation Systems Standards Fact Sheet—SAE J2374*, Draft Location Referencing Message Specification (LRMS) Information Report, available at http://www.its.dot.gov/.

U.S. Government (1994) "Coordinating geographic data acquisition and access: The National Spatial Data Infrastructure," *Federal Register*, 59(71), 17671–17674.

U.S. National Institute of Standards and Technology (1992) *Federal Information Processing Standards Publication 173: Spatial Data Transfer Standard*, Washington, DC: U.S. Department of Commerce.

University Consortium for Geographic Information Science (UCGIS) (1998) "Interoperability of geographic information," *UCGIS Research Policy White Paper*, available at http://www.ucgis.org.

University of Tennessee Energy, Environment, and Resources Center (UT-EERC), ALK Associates Inc., Oak Ridge National Laboratory, and Viggen Corporation (1995) *Meeting National ITS Spatial Data Needs, Task C2: Nationwide Map Database and Location Referencing System*, Report prepared for Federal Highway Administration, Office of Safety and Traffic Operations by Oak Ridge National Laboratory, Oak Ridge, TN.

Unwin, A. (1996) "Exploring spatio-temporal data," in M. Fischer, H. J. Scholten, and D. Unwin (eds.) *Spatial Analytical Perspectives on GIS*, London: Taylor and Francis, 101–110.

Usami, Y. and Kitaoka, M. (1997) "Traveling salesman problem and statistical physics," *International Journal of Modern Physics B*, 11, 1519–1544.

van Aerde, M. (1992) "INTEGRATION: A dynamic simulation/assignment model," Presented at the IVHS Dynamic Traffic Assignment and Simulation Workshop, Federal Highway Administration, McLean, VA.

van Bemmelen, J., Quak, W., van Hekken, M., and van Oosterom, P. (1993) "Vector vs. raster algorithms for cross-country movement planning," *Proceedings of AutoCarto 11*, 304–317.

van der Knaap, W. G. M. (1997) "Analysis of time-space activity patterns in tourist recreation complexes: A GIS-oriented methodology," in D. F. Ettema and H. J. P. Timmermans (eds.) *Activity-Based Approaches to Travel Analysis*, Oxford, U.K.: Elsevier Science, 283–311.

van Krevald, M. (1997) "Digital elevation models and TIN algorithms," in M. van Krevald, J. Nievergelt, T. Roos, and P. Widmayer (eds.) *Algorithmic Foundations of Geographic Information Systems*, Berlin: Springer Lecture Notes in Computer Science 1340, 37–78.

van Oosterom, P. (1999) "Spatial access methods," in P. A. Longley, M. F. Goodchild, D. J. Maguire, and D. W. Rhind (eds.) *Geographical Information Systems, Vol. 1: Principles and Technical Issues*, 2d ed., New York: Wiley, 385–400.

van Vliet, D. (1978) "Improved shortest path algorithms for transport networks," *Transportation Research*, 12, 7–20.

Varian, H. R. (1992) *Microeconomic Analysis*, 3d ed., New York: W. W. Norton.

Vause, M. (1997) "A rule-based model of activity scheduling behavior," in D. F. Ettema and H. J. P. Timmermans (eds.) *Activity-Based Approaches to Travel Analysis*, Oxford, U.K.: Elsevier Science, 73–88.

Vehicle Intelligence & Transportation Analysis Laboratory (VITAL) (1997) *Interoperability of Map Databases—Development of Experimental Infrastructure*, Report prepared for California Department of Transportation, Test Center for Interoperability.

Vehicle Intelligence & Transportation Analysis Laboratory (VITAL) (1998) *The Cross Streets Profile with Coordinates—Techinical Evaluation*, Report prepared for U.S. Department of Transportation, Federal Highway Administration.

Vehicle Intelligence & Transportation Analysis Laboratory (VITAL) (1999a) *ITS Datum—Rubbersheeting*, available at http://www.ncgia.ucsb.edu/vital/.

Vehicle Intelligence & Transportation Analysis Laboratory (VITAL) (1999b) *The LRMS Linear Referencing Profile—Technical Evaluation*, Report prepared for U.S. Department of Transportation, Federal Highway Administration.

Verbree, E., Van Maren, G., Gerns, R., Jansen, F., and Kraak, M-J. (1999) "Interaction in Virtual World View—Linking 3D GIS with VR," *International Journal of Geographical Information Science*, 13, 385–396.

Veregin, H. (1995) "Developing and testing of an error propagation model for GIS overlay operations," *International Journal of Geographical Information Systems*, 9, 595–619.

Veregin, H. (1999) "Data quality parameters," in P. A. Longley, M. F. Goodchild, D. J. Maguire, and D. W. Rhind (eds.) *Geographical Information Systems, Vol. 1: Principles and Technical Issues*, 2d ed., New York: Wiley, 177–189.

Versenyi, J. H., Holdstock, D. A., and Fischer, T. (1994) "Considering the alternatives: GIS identifies nine possibilities for highway connection," *Geographic Information Systems*, 9(4), 43–45.

Vieux, B. E. (1993) "DEM aggregation and smoothing effects on surface runoff modeling," *Journal of Computing in Civil Engineering*, 7, 310–337.

Visvalingam, M. (1994) "Visualization in GIS, cartography and ViSC," in H. M. Hearnshaw and D. J. Unwin (eds.), *Visualization in Geographical Information Systems*, Chichester, U.K.: Wiley, 18–25.

Vonderohe, A. and Hepworth, T. (1996) *A Methodology for Design of a Linear Referencing System for Surface Transportation*, Research report, Sandia National Laboratory, Project AT-4567.

Vonderohe, A. and Hepworth, T. (1998a) "A methodology for the design of measurement systems for linear referencing," *URISA Journal*, 10(1), electronic journal acrhive available at http://www.urisa.org/Journal/.

Vonderohe, A. and Hepworth, T. (1998b) "Analysis and adjustment of measurement systems for linear referencing," *URISA Journal*, 10(1), electronic journal archive available at http://www.urisa.org/Journal/.

Vonderohe, A. P., Travis, L., Smith, R. L., and Tsai, V. (1993) *Adaptation of Geographic Information Systems for Transportation*, Washington, DC: Report 359, National Cooperative Highway Research Program, National Academy Press.

Vonderohe, A. and others (1995) "On the results of a workshop on generic data model for linear referencing systems," *GIS-T '95 Proceedings*, American Association of State Highway and Transportation Officials.

Vovsha, P. and Bekhor, S. (1998) "Link-nested logit model of route choice: Overcoming route overlapping problem," *Transportation Research Record*, 1645, 133–142.

Walker, P. A. and Moore, D. M. (1988) "SIMPLE—An inductive modeling and mapping tool for spatially-oriented data," *International Journal of Geographical Information Systems*, 2, 347–363.
Wang, J., Robinson, G. J., and White, K. (1996) "A fast solution to local viewshed computation using grid-based digital elevation models," *Photogrammetric Engineering and Remote Sensing*, 62, 1157–1164.
Wang, J., Treitz, P. M., and Howarth, P. J. (1992) "Road network detection from SPOT imagery for updating geographical information systems in the rural-urban fringe," *International Journal of Geographical Information Systems*, 6, 141–157.
Wardrop, J. G. (1952) "Some theoretical aspects of road traffic research," *Proceedings of the Institution of Civil Engineers, Part II*, 1, 325–362.
Waters, N. (1999) "Transportation GIS: GIS-T," in P. A. Longley, M. F. Goodchild, D. J. Maguire, and D. W. Rhind (eds.) *Geographical Information Systems, Vol. 2: Management Issues and Applications*, 2d ed., New York: Wiley, 827–844.
Watterson, W. T. (1990) "Adapting and applying existing urban models: DRAM and EMPAL in the Seattle region," *URISA Journal*, 2(2), 35–46.
Watterson, W. T. (1993) "Linked simulation of land use and transportation systems: Developments and experience in the Puget Sound region," *Transportation Research A*, 27A, 193–206.
Waugh, T. C. and Healey, R. G. (1987) "The GEOVIEW design: A relational data base approach to geographical data handling," *International Journal of Geographical Information Systems*, 1, 101–118.
Webber, M. J. (1977) "Pedagogy again: What is entropy?" *Annals of the Association of American Geographers*, 67, 254–266.
Webber, M. J. (1980) "A theoretical analysis of aggregation in spatial interaction models," *Geographical Analysis*, 12, 129–141.
Wegener, M. (1994) "Operational urban models: State of the art," *Journal of the American Planning Association*, 60(Winter), 17–29.
Wegener, M. (2000) "Spatial models and GIS," in A. S. Fotheringham and M. Wegener (eds.) *Spatial Models and GIS: New Potential and New Models*, GISDATA 7, London: Taylor and Francis, 3–20.
Weibull, J. (1976) "An axiomatic approach to the measurement of accessibility," *Regional Science and Urban Economics*, 6, 357–379.
Weibull, J. W. (1980) "On the numerical measurement of accessibility," *Environment and Planning A*, 12, 53–67.
Weiss, P. (1999) "Stop-and-go science," *Science News*, 156, 8–10.
Werner, C. (1968) "The law of refraction in transportation geography: Its multivariate extension," *Canadian Geographer*, 12, 28–40.
Werner, C. (1985) *Spatial Transportation Modeling*, Beverely Hills, CA: Sage.
White, M. (1991) "Car navigation systems," in D. Maguire, M. Goodchild, and D. Rhind (eds.) *Geographical Information Systems: Principles and Applications*, Essex, U.K.: Longman Group, 115–125.
White, R. and Engelen, G. (1997) "Cellular automata as the basis of integrated dynamic regional modelling," *Environment and Planning B: Planning and Design*, 24, 235–246.
Wie, B.-W., Friesz, T. L., and Tobin, R. L. (1990) "Dynamic user optimal traffic assignment on congested multidestination networks," *Transportation Research B*, 24B, 431–442.
Wiggins, L., Deuker, K., Ferreira, J., Merry, C., Peng, Z., and Spear, B. (2000) "Application challenges for geographic information science: Implications for research, education and policy for transportation planning and management," *URISA Journal*, 12, 52–59.

Wilkinson, G. G. (1996) "A review of current issues in the integration of GIS and remote sensing data," *International Journal of Geographical Information Systems*, 10, 85–101.

Willer, D. (1990) *A Spatial Decision Support System for Bank Location: A Case Study*, NCGIA Technical Report 90–9, National Center for Geographic Information and Analysis, Santa Barbara, CA.

Williams, N. A. (1999) "Four-Dimensional Virtual Reality GIS (4D VRGIS): Research guidelines," in B. Getting (ed.) *Innovations in GIS 6: Integrating Information Infrastructures with GIS Technology*, London: Taylor and Francis, 201–214.

Wills, G., Bradley, R., Haslett, J., and Unwin, A. (1990) "Statistical exploration of spatial data," *Proceedings of the Fourth International Symposium on Spatial Data Handling*, 491–500.

Wilson, A. G., Coelho, J. D., Macgill, S. M., and Williams, H. C. W. L. (1981) *Optimization in Locational and Transport Analysis*. New York: Wiley.

Wilson, A. G. (1967) "A statistical theory of spatial distribution models," *Transportation Research*, 1, 253–269.

Wilson, A. G. (1976) "Retailers' profits and consumers' welfare in a spatial interaction shopping model," in I. Masser (ed.) *Theory and Practice in Regional Science*, London: Pion, London Papers in Regional Science 6, 42–57.

Wilson, J. D. (1998) "ModelCity Philadelphia elevates GIS to the next level," *Professional Surveyor*, 17(2), 20–24.

Wolfgang, T. (1991) *Geodesy*, New York: Walter De Gruyter.

Wong, D. and Amrhein, C. (1996) "Research on the MAUP: Old wine in a new bottle or real breakthrough?" *Geographical Systems*, 3, 73–76.

Worboys, M. F. (1992) "A model for spatio-temporal information," *Proceedings of the 5th International Symposium on Spatial Data Handling*, 602–611.

Worboys, M. F. (1995) *GIS: A Computing Perspective*, London: Taylor and Francis.

Worrall, L. (ed.) (1991) *Spatial Analysis and Spatial Policy Using Geographic Information Systems*, London: Belhaven Press.

Wrigley, N. (1982) "Quantitative methods: Developments in discrete choice modeling," *Progress in Human Geography*, 6, 547–562.

Wrigley, N. (1985) *Categorical Data Analysis for Geographers and Environmental Scientists*, London: Longman.

Wu, F. L. (1998) "SimLand: A prototype to simulate land conversion through the integrated GIS and CA with AHP-derived transition rules," *International Journal of Geographical Information Science*, 12, 63–82.

Wu, Y-H., Miller, H. J., and Hung, M-C. (2001) "A GIS-based decision support system for analysis of route choice in congested urban road networks," *Journal of Geographical Systems*, in press.

Xie, Y. (1996) "A generalized model for cellular urban dynamics," *Geographical Analysis*, 28, 350–373.

Xiong, D. and Marble, D. F. (1996) "Strategies for real-time spatial analysis using massively parallel  D computers: An application to urban traffic flow analysis," *International Journal of Geographical Information Systems*, 10, 769–789.

Xu, Y. (1999) "Development of transport telematics in Europe," *Proceedings of the International Workshop on Geographic Information Systems for Transportation (GIS-T) and Intelligent Transportation Systems (ITS)*, Joint Laboratory for GeoInformation Science of the Chinese Academy of Sciences and the Chinese University of Hong Kong, China, and the National Center for Geographic Information and Analysis, U.S.A.

Yang, H., Akiyama, T., and Sasaki, T. (1998) "Estimation of time-varying origin-destination flows from traffic counts: A neural network approach," *Mathematical and Computer Modelling*, 27, 323–334.

Yang, H. and Qiao, F. (1998) "Neural network approach to classification of traffic flow states," *Journal of Transportation Engineering*, 124, 521–525.

Yearsley, C. M. and Worboys, M. F. (1995) "A deductive model of planar spatio-temporal objects," in P. Fisher (ed.) *Innovations in GIS 2*, London: Taylor and Francis, 43–51.

You, J., Nedović-Budić, Z., and Kim, T. J. (1997a) "A GIS-based traffic analysis zone design: Implementation and evaluation," *Transportation Planning and Technology*, 21, 69–91.

You, J., Nedović-Budić, Z., and Kim, T. J. (1997b) "A GIS-based traffic analysis zone design: Technique," *Transportation Planning and Technology*, 21, 45–68.

Yu, S., Van Krevald, M., and Snoeyink, J. (1996) "Drainage queries in TINs: From local to global and back again," *Proceedings of the 7th International Symposium on Spatial Data Handling*, 13A.1–13A.14.

Yuan, M. (1997) "Use of knowledge acquisition to build wildfire representation in geographic information systems," *International Journal of Geographical Information Systems*, 11, 723–745.

Yuan, M. (1999) "Use of a three-domain representation to enhance GIS support for complex spatiotemporal queries," *Transactions in GIS*, 3, 137–159.

Yun, S.-Y., Namkoong, S., Rho, J.-H., Shin, S.-W., and Choi, J.-U. (1998) "A performance evaluation of neural network models in traffic volume forecasting," *Mathematical and Computer Modelling*, 27, 293–310.

Zagami, J. M., Parl, S. A., Bussgang, J. J., and Melillo, K. D. (1998) "Providing universal location services using a wireless E911 location network," *IEEE Communications Magazine*, 36(4), 66–71.

Zeiler, M. (1999) *Modeling Our World: The ESRI Guide to Geodatabase Design*, Redlands, CA: ESRI, Inc.

Zhan, F. B. (1997) "Three fastest shortest path algorithms on real road networks: Data structures and procedures," *Journal of Geographic Information and Decision Analysis*, 1, 69–82.

Zhan, F. B. and Noon, C. E. (1998) "Shortest path algorithms: An evaluation using real road networks," *Transportation Science*, 32, 65–73.

Zhao, F. and Shen, L. D. (1997) *GIS Applications For Public Transportation Management*, Research report, Lehman Center for Transportation Research, Florida International University, Miami.

Zhao, F., Wang, L., Elbadrawi, H., and Shen, L. D. (1997) "Temporal geographic information system and its application to transportation," *Trasportation Research Record*, 1593, 47–54.

Zhao, Y. (1997) *Vehicle Location and Navigation Systems*, Boston: Artech House.

Zografos, K. G. and Davis, C. F. (1989) "A multiobjective programming approach for routing hazardous materials," *Journal of Transportation Engineering*, 115, 661–673.

# Index

A* algorithm, 131, 154–155, 226, 371
accessibility measures, 271–276
   attraction accessibility, 272, 275
   relative accessibility, 260
   space-time accessibility, 271, 272–274, 276
   user-benefit accessibility, 275
accidents, 360–361
accident analysis, 361–364
ActiveX controls, 235, 236, 388. *See also*
   Component Object Model (COM)
address matching, 61, 79, 105, 117–119, 273, 283,
   333, 360–361, 369, 372, 391
aerial photography, 107
air quality models, 12, 231, 355–359
   airshed dispersion models, 357–359, 367
   dynamic models, 357–358
   MOBILE model, 356
   mobile source emissions, 355–356
   particle models, 359
algorithm, 9, 130
   definition of, 134
   efficiency of, 135–137
   exact, 130, 137, 156
   heuristic, 130, 137, 139, 153. *See also* flow
      capturing location problem (FCLP)
      heuristics, *p*-median problem heuristics,
      TSP heuristics, VRP heuristics
   properties of, 134–139

*See also* shortest path problem
anchor point, 72, 322
anchor section, 73
arcs,
   directed arcs, 20, 21, 54, 55, 60, 132, 173
   logical arcs, 394
   relational schema, 25
   transfer arcs, 55, 56–57, 68, 264
   undirected arcs, 132
   weights, 54, 57, 132, 133, 140, 159, 174
arrays,
   arc array, 143
   arc weight array, 143
   point array, 143
artificial neural network (ANN), 265–268,
   362–363
   feed-back network, 266, 268
   feed-forward network, 266, 267, 268, 363
   finite impulse response network, 268
   time-delayed recurrent network, 268
automated highway system (AHS), 297, 302,
   303, 306, 310, 311
automatic vehicle location (AVL), 96, 304, 305
Avenue programming language, 235

bid rent curve, 279
Big-O notation, 136
binary large object (BLOB), 31

Index    449

Boolean logical operators, 218, 220–221, 222, 366
buffer operation, 222–224, 274, 282, 284, 287, 367, 385, 386, 398

C programming language, 231, 232, 392
capacity constraints, 158
cardinality of set, 174
carriers, 381, 397
cellular automata (CA), 198, 231, 270, 280, 355
cellular geography, 280
Census Feature Class Codes (CFCC), 105
central place theory, 280
choice set, 270, 272, 275
client-server architecture, 293, 335, 388
　heavy server/thin client, 335
　light server/thick client, 336
cluster analysis, 250
collaborative spatial decision support system (CSDSS), 293–294, 342. *See also* spatial decision support system (SDSS)
common lines problem, 262
compensatory decision rules, 286
complexity analysis, 136–137
　asymptotic complexity classes, 137, 146
　average-case analysis, 136
　exponential time, 137
　polynomial time, 137, 158
　time complexity, 136, 147, 148, 149, 150, 185
　worst-case analysis, 136, 147, 148, 149, 156, 348, 349, 351
Component Object Model (COM), 235, 293, 392
component GIS, 232, 233, 236
componentware, 7, 11, 235–236
Composite Theme Grid (CTG), 107
computational intelligence (CI) methods, 165, 168, 265
computer-aided drafting and design (CADD), 282, 290, 343
concurrency control, 36, 40
confirmatory spatial data analysis, 216–217
congestion pricing, 196
constrained optimization problem, 159, 262, 282, 284, 378
　constraints, 159
　decision variables, 159
　feasible region, 159
　for departure-based dynamic user optimal (DUO) equilibrium, 191–192
　for flow capturing location problem (FCLP), 210
　for land use/transportation (LUT) modeling, 279
　for maximal covering problem, 208
　for maximum flow problem, 177
　for minimum cost flow problem (MCFP), 173–174
　for $p$-centers problem, 209
　for $p$-medium problem, 200–201
　for preventive inspection model (PIM), 211
　for set covering problem, 207–208
　for the transportation problem, 176
　for traveling salesman problem (TSP), 159–160
　for user optimal (OU) equilibrium, 183–184
　for vehicle routing problem (VRP), 165–166
　objective function, 159, 195, 199, 202, 228
coordinate systems, 113
　British National Grid System, 91–93
　Cartesian, 89–90, 238
　geographic, 89
　plane, 89–93
　polar, 89
　State Plane Coordinate System (SPCS), 90–91, 324
　U.S. Military Grid Reference System, 90
　Universal Transverse Mercator (UTM), 90, 106, 324
corridor design problem, 284–285, 289–290, 343
corridor location problem, 155, 284, 285–289, 342
cost functions,
　arc flow cost function, 60, 178, 180–181, 183, 195, 357
　density-based, 195
　generalized cost function, 55, 262
cost polygons, 156
cost surface, 155
coupling GIS with transportation analysis and modeling, 229–232, 280, 281, 292–293, 378
cross classification analysis, 254, 261
cross elasticity, 253
customization, 232–233
cut-and-fill, 106

data accuracy. *See* data quality
data conversion, raster-vector, 119–120, 123
data dictionary, 48, 112
data domain, 15, 23, 31, 33
data errors, 342, 348, 349, 361, 378
　handling of, 124–125
　natural variations, 124
　quality assurance/quality control (QA/QC) procedures, 124, 125
　sources of, 122, 343
　topological, 120
　*See also* data quality, fuzzy boundary, linear referencing
data exchange, 112, 113, 114. *See also* standards
data integration, 9, 85, 111–112, 114–120, 270, 271, 357, 377, 388, 390

data mining, 15, 49, 238
data models, 8, 14
  conceptual, 17, 18–23, 35, 41, 322
  definition of, 16
  distributed, 41, 53
  interoperable, 41, 53, 78
  lane-based network, 79–81, 131, 217, 236
  logical, 17, 23, 35, 41, 59, 133
  matrix, 384
  navigable, 78–79, 80, 81, 236
  physical, 17, 133
  raster, 14, 41, 43, 237, 347
  semantic, 17
  three-dimensional, 81–82
  topological, 121, 217, 225
  vector, 14, 41, 43, 237
  *See also* dynamic segmentation, georelational data model, intelligent transportation systems (ITS), linear referencing, object-oriented (OO), relational data model
data products,
  private sector, 110
  public sector, 104–109
data quality, 120–125
  accuracy, 83, 120
  attribute accuracy, 113, 121
  completeness, 113, 121
  consistency, 113, 121
  precision, 120
  lineage, 48, 121
  positional accuracy 93, 106, 113, 120, 121
  *See also* metadata
data redundancy, 16, 23, 30
data sharing, 38, 82–83, 112–113
data sources, 9, 48
data standards, 112–114. *See also* standards
data structures, 143–150, 203–205
  approximate bucket strategy, 148
  binary tree, 149
  bucket, 147, 148
  deque, 145, 146
  distance strings, 203
  double bucket strategy, 148
  forward star structure (FSS), 143, 378
  heap, 147, 149–150
  linked list, 147, 150
  list, 145
  overflow bag strategy, 148
  priority queue, 147
  queue, 145, 146
  stack, 145, 146
  two-queue, 146
data uncertainty, 343
data warehouse, 8, 15, 49–52

database,
  database management system (DBMS), 36, 41, 42, 50
  design, 8, 15, 39–41
  distributed database, 8, 38–41, 53, 82–84
  extensible database, 219
  indexing, 17
  integrity rules, 16
  requirements, 36
  replication, 39
  schema, 17
  transaction, 36, 39, 40
  transparency, 40
datum, 62, 70, 72, 76, 86–88, 113, 324
  for ITS location referencing, 321–324, 330, 332
dead reckoning, 333
deadhead, 226, 398
decision support system, 214, 230, 377. *See also* spatial decision support system (SDSS)
Delauney triangulation, 156, 352. *See also* triangulated irregular network (TIN)
depot, 131, 158, 165
deregulation, 381, 382, 398
descriptive analysis, 383
destination choice, 258, 259
Digital Elevation Model (DEM), 99, 106, 289, 350
digital imagery, 119–120
Digital Line Graphs (DLG), 105–106
digital terrain model (DTM), 106, 121, 289, 351, 352, 357
  extracting hydrological features from, 347–350
  *See also* Digital Elevation Model (DEM)
disaggregate modeling, 236. *See also* travel demand analysis
discrete choice model, 261. *See also* travel demand analysis
distance measuring instruments (DMI), 95, 331
distribution centers (DCs), 384, 386, 389, 390, 393, 395
drainage network, 348, 349
dual, 228, 393
Dual Independent Map Encoding (DIME) files, 104
dynamic binding, 38
dynamic link libraries, 378
dynamic programming, 167, 368
dynamic segmentation, 65–68, 79, 123, 283
  data model, 53, 128, 131, 217, 236, 360
  events, 62, 65–68, 72–76, 221, 331, 360
  route, 65, 68, 123, 217, 221
  route system, 65, 68, 79, 123
dynamic user optimal (DUO) problem,
  convergent dynamic algorithm (CDA), 192, 193, 194

dynamic traffic assignment (DTA), 192–193
　ideal DUO, 189
　predictive DUO, 189
　reactive DUO, 189

ecological fallacy, 261
electronic commerce, 381
ellipsoid, 86
emergency 911 services, 369–370, 372
emergency management, 368–369
emergency vehicle routing 152, 369–372
encapsulation, 33, 34. *See also* object-oriented (OO)
enterprise resource planning (ERP), 387–388
entity,
　definition of, 18
　subtype of, 20, 34
　supertype of, 20, 34
　transportation, 15–16
entity-relationship (E-R) model, 18–20
entropy-maximizing, 258, 279
environmental impacts, 12, 341, 355, 357
　of transportation facilities, 342
　of transportation systems, 342
epsilon band, 121, 344
Euclidean distance, 155, 169, 170, 212, 225, 373
European GISDATA program, 216
European Road Telematics Implementation Coordination Organization (ERTICO), 297, 298
European Umbrella Organization for Geographic Information (EUROGI), 114
evacuation planning, 12, 374–378
　out-of-kilter algorithm, 374
event-driven programming, 234
event tracking, 388
event trees, 366
extended entity-relationship (EER) model, 18, 20
exploratory data analysis (EDA), 49, 50, 216, 238, 271

facility location problems, 10, 173, 199–200, 228–229, 367, 368, 369, 372–374, 380, 390, 394, 400
　flow-intercepting location problems, 209–211
　maximal covering location problem (MCLP), 206, 207, 208–209, 372
　maximum attendance facility location problem, 206, 395
　maximum availability location problem (MALP), 373
　$p$-centers (minmax) problem, 209, 395
　$p$-median problem with maximum distance constraints, 205, 395
　$p$-median with powered distance, 206, 395
　probabilistic set covering location problem (PSCLP), 373
　set covering location problem (SCLP), 205, 206, 207–209, 372, 398
　types of, 199
　unified linear model (ULM), 205–207
　*See also* flow capturing location problem (FCLP) heuristics, $p$-median problem, $p$-median problem heuristics
false origin, 90
feasible activity space, 271
file transfer protocol (FTP), 41, 335
fixed-length segmentation, 65
flow capturing location problem (FCLP) heuristics,
　cannibalizing, 210
　noncannibalizing, 211
flow interception, 10, 103
flow sampling, 95, 102–103
　sample dependency, 103
fly-through, 281
fuzzy boundary, 122, 123, 344–345, 370
fuzzy logic, 265
fuzzy set theory, 100, 343–345, 368
fuzzy tolerance, 124

genetic algorithms, 261, 265, 267, 289
geocoding, 111, 333, 360, 384, 385. *See also* address matching
geographic data capture system (GDCS), 94–95
Geographic Data Description Directory (GDDD), 109
Geographic Information Retrieval and Analysis System (GIRAS), 107
geographic information science for transportation (GISci-T), 6–7, 247
geographical information systems for transportation (GIS-T),
　applications, 3, 247
　data sources, 103–111
　definition, 3
　demand for, 4–5
Geographic Information Systems International Group (GISIG), 110–111
geographic potential, 260
geographic process, 126
geographic scale. *See* map scale
geographic visualization. *See* visualization
geographically weighted regression (GWR), 255
geoid, 86
GEOLINUS, 48, 121
Georeferencing, 49, 112
georelational data model, 17, 30–31, 394
GeoVRML, 244. *See also* virtual reality modeling language (VRML)

Global Positioning System (GPS), 9, 49, 78, 95–96, 238, 260, 265, 270, 283, 284, 306, 318, 321, 322, 325, 328, 331, 332, 333, 360, 369, 370, 371, 387, 388
goal programming, 368
graph theory, 8, 54, 61, 131, 378
  vertices, 54, 131, 199. *See also* nodes
  edges, 54, 131. *See also* arcs
  directed graph, 131, 132, 178
  face, 54
  planar graph, 54, 112, 131
  nonplanar graph, 54, 131
  undirected graph, 131, 132
  weighted graph, 132
graphic user interface (GUI), 112, 232, 233, 235, 240, 377
gravity model. *See* spatial interaction (SI) models
greedy approach, 139

hazards, 231, 341, 360, 364
hazardous material (HazMat), 12, 210, 341, 365–368, 369, 372
  risk estimation, 365–366
  routing, 366–368
heteroscedasticity, 254
historical data, 50, 364, 369. *See also* spatiotemporal
hydrological impacts, 345
hydrological modeling, 367
  finite difference methods (FDM), 346, 347
  finite element methods (FEM), 346, 347, 358
  flow routing, 349
  GIS-based, 350
  stochastic models, 347
  taxonomy, 346
  watershed models, 347
hypertext markup language (HTML), 244
hypertext transfer protocol (HTTP), 335

impedance mismatch, 33
in-vehicle navigation systems, 12, 111, 297, 298, 304, 332–334
inertial navigation systems, 95
inheritance, 20, 33, 34. *See also* object-oriented (OO)
input-output model, 279
integer programming (IP), 378
integrated highway information systems (IHIS), 82–84
Intelligent Transportation Society of America (ITS America), 297, 301
intelligent transportation systems (ITS), 8, 11–12, 49, 64, 152, 155, 182, 238, 247, 371, 375, 389
  applications, 82, 218, 237, 271, 303–307, 328, 332
  architecture in Europe, 316–317
  architecture in Japan, 312–316
  architecture in the United States, 307–312
  communication systems, 298, 302, 307, 310–311, 318, 320
  data model, 53, 77–82
  datum, 321–324
  deployment, 82, 199, 302, 354
  description and objectives of, 295–296
  development in Europe, 297–301
  development in Japan, 296–297
  development in the United States, 301–302
  flow simulation models, 196
  integrating GIS and ITS, 332–339
  locational referencing for, 320–332
  object-oriented approaches, 36–37
  requirements for geographic information, 318–320
  system optimal equilibrium, 186
  user services, 297, 308, 312–313, 318–319, 322, 326, 327, 334, 336, 354, 380
intelligent vehicle highway systems (IVHS) 83, 296, 301. *See also* intelligent transportation systems (ITS)
Intermodal Surface Transportation Efficiency Act (ISTEA), 105, 301
intermodal transport, 299, 302, 393–394, 398
Internet, 293, 304, 320, 334–339, 387. *See also* World Wide Web (WWW)
interoperability, 10–11, 39, 41–44, 69, 83, 232, 326
  for ITS, 302, 309, 311–312, 316–317, 319, 321, 322, 323, 332
interoperable databases, 82–84, 237
interpolation,
  areal, 9, 114–116, 128, 216, 282, 287
  elevation, 353
  linear, 117, 123
  over space and time, 355, 358
  spatial, 358, 367, 385
intervisibility, 351. *See also* viewshed analysis.
isochrones, 272, 274
isomorphic, 54

Java applet, 336, 339
join operation, 28, 29, 35. *See also* spatial join operation

knowledge-based (KB), 265, 363, 366, 374, 377
  expert system (KBES), 101, 231–232, 363
  GIS, 231, 363–364
knowledge discovery, 49

land allocation model, 231
land use/land cover, 99, 106–107
land use/transportation (LUT) modeling, 227, 229, 282, 354–355

mathematical programming, 278–279
microsimulation models, 280–281, 355
multisector models, 279
  Lowry model, 278
  urban economic models, 279
land use/transportation systems, 5, 11, 247, 248, 295, 368
Landsat Thematic Mapper (TM), 45, 98
lattice, 157, 347, 353, 358
least cost path, 155, 156, 182, 288. *See also* shortest path problem
line of sight problems, 352
linear programs (LP), 177, 193, 201, 207, 209, 228, 262, 393, 394. *See also* constrained optimization problem
linear referencing,
  enterprise linear referencing systems (LRS) data models, 36, 69–77, 123, 322
  errors, 74, 123–124
  linear referencing methods (LRM), 62–64, 72
  linear referencing profile (LRP), 331–332
  linear referencing systems (LRS), 62, 123, 221, 360
  pathologies, 76–77, 81
  unified linear referencing system, 324
linear regression, 126, 254–255, 261
link,
  line-haul, 56
  transfer, 56
location-allocation problem, 201, 228, 230–231, 373, 395. *See also* p-median problem
location referencing, 51–52, 318, 320–330
  Location Referencing Message Specification (LRMS), 324–325, 332
location theory, 400
logistics, 12, 158, 172, 369, 379
  continuous move problem, 397–398
  cross-dock costs, 384, 385, 396
  definition of, 382
  delivery time windows, 391, 392
  inventory carrying cost, 385, 396
  just-in-time (JIT), 380, 396
  third-party logistics (3PL), 381, 397, 398
  three-tier systems, 176
  time-based measures, 396
logit models, 187, 267, 275
  independence of irrelevant alternatives (IIA) property, 187, 261
  multinomial logit (MNL) model, 256, 258, 259, 260, 261, 264
  nested logit (NL) model, 188, 259–260, 261, 264, 265, 269

macro languages, 232, 233–234, 280, 293, 378, 392
Manhattan distance, 155, 169, 170
map algebra, 280

map cube, 52
map generalization, 122, 127, 326, 331
map matching, 78, 333
map pattern effect, 260
map projections, 88, 90–91, 113, 123
map scale, 14, 51, 93, 326, 331
marginal cost, 186
mathematical plane, 131
mathematical programming, 201, 393, 398
matrices,
  adjacency matrix, 132
  arc-path incident matrix, 133, 179, 191
  cost matrix, 133, 165
  incident matrix, 132, 133
  matrix algebra operations, 228
  misclassification matrix, 121, 122
  node-arc incident matrix, 132
  origin-destination matrix, 103, 193, 227, 256, 257, 356. *See also* origin-destination (O-D) flows
maximum flow problem, 176, 228, 392
maximum likelihood (ML), 260–261, 267
mean center, 360
metadata, 8, 15, 30, 32, 44, 47–49, 51, 108, 109, 112–114, 245. *See also* standards
microeconomic theory, 178, 196, 264
microsimulation models, 252, 375, 377. *See also* land use/transportation (LUT) modeling, simulation
milepost, 62–64, 123, 236
minimum cost flow problem (MCFP), 173–178, 228, 392
  feasible solutions property, 174, 176
  integer solutions property, 174
  the transportation problem, 174–176, 228, 392
  transshipment problem, 176, 392
mixed integer program, 211
model-base management system (MBMS), 231
modifiable areal unit problem (MAUP), 9, 102, 125–127, 212, 216, 250, 281
  boundary effect, 102, 216
  scale effects, 126, 216
  zoning effects, 126
multicollinearity, 254
multicriteria decision making (MCDM), 285–287, 292, 293
multilevel analysis, 261
multimodal network, 68, 264, 274, 393–394
Multi-purpose European Ground-Related Information Network (MEGRIN), 109
multipurpose trips, 269
multistop trips, 269

National Digital Orthophoto Program (NDOP), 107
National Transportation Atlas (NTA), 105
nearest neighbor, 362

neighbors,
  queen's case, 157, 353
  rook's case, 157, 353
network,
  acyclic, 133
  cycle, 133
  diameter of a, 155
  Hamiltonian cycle, 134, 137, 158
  Hamiltonian path, 134
  logical, 8, 15–16, 394
  mathematical representation of, 131–134
  nonplanar, 54, 132, 236, 322
  path, 133, 178
  physical, 10, 15–16, 394
  planar, 54, 58, 59, 132, 217, 236
  properties of, 131
  subnetwork, 133–134
  tour, 134, 158, 365
  tree, 134
  walk, 133
network aggregation, 127–129, 182
network analysis, 72, 109, 130, 132
network assignment methods, 192, 234, 270, 271, 356
  all-or-nothing assignment, 184
  capacity restraint method, 184, 210
  incremental assignment method, 184
  public transit assignment problem, 262
  *See also* network flow equilibrium
network autocorrelation, 103, 362
network conflation, 9, 116, 128
network connectivity, 80, 130, 133–134, 136, 152, 217
  degree of a node, 133
network distance, 225
network flow equilibrium, 181–196
  behavioral assumptions, 182, 187
  behavioral critique, 195–197
  convex combinations method, 184–185, 193, 264
  dynamic user optimal (DUO), 188–195, 228, 262. *See also* dynamic user optimal (DUO) problem
  equilibrium, concepts of, 161, 178, 189, 195–196
  equivalent optimization (EO) strategy, 163, 195
  method of successive averages (MSA), 188
  stochastic user optimal (SUO), 187–188, 262, 264, 375
  system optimal (SO), 185–186, 262
  user optimal (UO), 182–185, 193, 194, 228, 252, 262, 264, 279
network flow problems, 10, 172, 228, 268, 374
  congested, 172, 178–181
  uncongested, 172, 173–178
network routing problems, 380
  arc routing problem (ARP), 226
  fleet routing problems, 130
  traveling salesman problem (TSP), 10, 131, 137, 158–165, 166, 391. *See also* optimization-based approaches, TSP heuristics
  vehicle routing problem (VRP), 10, 131, 158, 165–170, 226, 391. *See also* optimization-based approaches, VRP heuristics
node search rules,
  best-first (BF), 141, 144, 146, 147, 148, 149, 154
  breadth-first search, 145, 155
  depth-first search, 145
  first-in first-out (FIFO), 145, 146, 147, 193
  last-in first-out (LIFO), 145, 146, 147
  hybrid FIFO-LIFO, 145, 146
nodes 20, 21, 25, 54, 55, 73, 132, 199, 324
  demand node, 173, 177
  from node, 60
  logical node, 394
  supply node, 173, 177
  to node, 60
  transfer node, 393
  transshipment node, 173, 177
node-arc-area (NAA) model, 20–23, 25, 59
node-arc representation, 53, 55–56, 198
  for public transit network, 56–57
  data model, 8, 11, 59–61, 127, 215, 217, 400
  weaknesses of, 61–62
noncompensatory decision rules, 287
nondeterministic polynomial time (NP) problems, 161
  NP-complete, 160–161, 166, 201
  NP-hard, 166, 207, 209, 210
North American Datum (NAD), 76, 87–88, 104, 106, 321. *See also* datum
NULL, 24

objects, 33, 41, 234
object behavior, 34, 235
object class, 37
  aggregation, 36
  association, 37
  hierarchy, 34–35, 41, 44
object events, 235
object identity, 32, 33, 34–35
object linking and embedding (OLE), 235, 293, 392. *See also* ActiveX controls
object-message paradigm, 34
object methods, 234, 235
object modeling technique (OMT), 36, 72
object-oriented (OO),
  abstract data type, 31, 33, 219

Index  455

advantages and disadvantages of object orientation, 36–38
analysis, 33, 35–36
applications, 282, 313, 377
concepts, 33–35
data models, 33, 81
database management systems (OODBMS), 33, 234
design methodologies (OODM), 234
interfaces, 33, 231
programming (OOP), 33, 234–235, 293
object properties, 234, 235
online analytical processing (OLAP), 51
Open Geodata Model (OGM), 42, 43
Open GIS Consortium (OpenGIS or OGIS), 31, 42–44, 83, 232
optimization-based approaches, 158, 161–163
  branch and bound (BB), 162, 167, 201, 208, 209, 211, 372
  brute force methods, 160
  combinatorial optimization, 162, 174
  cutting plane, 162, 167, 208, 209
  decomposition, 201
  integer optimization, 162, 174
  Lagrangian relaxation, 201
  *See also* constrained optimization problem, network routing problems, *p*-median problem, shortest path problem
ordinary least squares (OLS), 254
origin-destination (O-D) flows, 53, 193, 195, 258, 267, 375. *See also* matrices
orthophotos, 102, 107–108
outlier, 121
overlay operations, 282, 286, 287, 357, 367, 385, 386, 398
  dynamic segmentation overlay, 221–222. *See also* dynamic segmentation
  map overlay, 123, 219–222, 225
  topological overlay, 72, 220
overriding, 33, 34. *See also* object-oriented (OO)

*p*-median problem, 200–203, 373, 390
  enumeration solution procedure, 201
  graph-theoretic solution procedure, 201
  heuristic solution procedures, 201–203, 229. *See also* *p*-median problem heuristics
  mathematical programming solution procedures, 201
  statement and characteristics, 200–201
  *See also* facility location problems, *p*-median problem heuristics
*p*-median problem heuristics,
  add algorithm, 202, 395
  alternating algorithm, 202
  drop algorithm, 202, 395
  global/regional interchange algorithm (GRIA), 202, 229, 396
  interchange algorithm, 202–203, 229, 395–396
  *See also* facility location problems, *p*-median problem
parallel processing, 10, 131, 152–154, 168, 352, 371
  data parallelism, 153, 154, 352
  multiple instruction multiple data stream (MIMD), 153
  nonoverlapping partitions, 153, 154
  overlapping partitions, 153
  single instruction multiple data stream (SIMD), 153
  task parallelism, 153
partial linearization algorithm, 264
partitioning geographic space, 39
partitioning problems, 378
path, 130
  complex, 53, 236–237
  shortest. *See* shortest path problem
  *See also* network
Pathfinder towards the European Topographic Information Template (PETIT), 109
personal digital assistants (PDA), 270
pixel, 98
planar embedding, 22
point-based spatial statistical methods, 361–362
point-in-polygon, 169
Poisson regression, 115
polyhedral surfaces, 157
polylines, 22, 25
potential path area, 273
predictive modeling, 367
prescriptive analysis, 367, 383, 389–390
probit models, 187–188, 259
procedural programming, 234

quadtree, 157
query operations, 383, 392
  Boolean query, 218
  spatial query, 32, 219, 385, 386
  *See also* structured query language (SQL)
queuing, 193, 195, 210, 359, 374, 375, 377

random utility theory/models, 256, 261, 264, 275, 358
rangefinders, 95
real-time flow monitors, 265
real-time navigation, 236. *See also* in-vehicle navigation systems
rectification, 107
refraction, 156
regular square grid (RSG), 106, 157, 288, 289, 343, 347, 348, 349, 350, 351, 353, 359

relational algebra, 23, 28–30
  property of closure, 29
  operations, 29
relational data model, 23–28, 65, 234, 384
  alternative key, 24
  extensions to, 31
  foreign key, 25, 80, 384
  functional dependency, 23, 25–28, 68, 384, 394
  integrity constraints, 24, 31, 48
  normalization, 23, 25–28
  primary key, 24, 25, 60, 384
  relation, 23, 384
  tuple 23, 45
relational database management system (RDBMS), 24, 31, 45, 219, 227
relationships,
  one-to-many, 8, 16, 53–54, 61–62, 65, 79, 369
  one-to-one, 65
  spatial, 32
  topological, 61, 104, 106, 121, 326. See also topology
remote sensing, 95, 97–102, 115, 238, 281, 343, 360
  active systems, 97
  electromagnetic spectrum, 98
  extracting transportation information, 100–101
  filtering techniques, 101
  geosynchronous, 98
  image correction/registration, 100
  image parallax, 102
  light detection and ranging (LIDAR), 97–98
  passive systems, 97
  photogrammetry, 101
  radio detection and ranging (RADAR), 97–98
  scene size, 98
  spatial resolution, 98
  spectral resolution, 98
  SPOT, 98
  stereopairs, 102
  supervised classification, 100
  temporal resolution, 98
  unsupervised classification, 100
resolution, 97–98, 107, 347, 352, 356, 358, 366. See also remote sensing
right-of-way (ROW), 284–285, 342
road transport informatics (RTI), 298
root mean square (RMS) error, 121
route choice, 139, 198, 267, 304
rubbersheet transformation, 116, 326, 327
rule-based reasoning, 269, 271

scenarios, 240, 247, 271, 288, 294, 350, 354, 355, 366, 377, 378, 384, 389
schematic heterogeneity, 41, 84

Seamless Administrative Boundaries of Europe (SABE), 109, 110
semantic, 42, 44, 54, 76
semantic heterogeneity, 41, 84
set theoretic operations, 29. See also Boolean logical operators
shadow price, 228, 393, 394
shippers, 381, 397
shortest path problem, 333, 365, 369, 371
  algorithms, 9, 130, 136, 139, 140–143
  gateway shortest path problem, 288
  implementation of, 130, 143–152
  iterative penalty method, 288
  labeling method, 140, 143, 144
  network storage structure, 143
  properties of, 139–140
  selection rule/node processing structure, 141, 143, 144. See also node search rules
  shortest path tree (SPT), 134, 140, 184, 185, 194, 225–226. See also network
  through networks, 140–143
  through surface, 131, 155–158, 287–288, 349, 353
  See also data structures
simplex method, 177–178, 393
  network simplex method, 177
  transportation simplex method, 177–178
simulation, 172, 262, 366, 374–375
  for activity-based travel demand modeling, 270
  cell transmission model, 198
  DRACULA, 198
  DYNASMART, 197
  INTEGRATION, 198
  macroscopic models, 197, 374
  mesoscopic models, 197, 376
  microscopic models, 197, 231, 375
  Monte Carlo, 188, 345, 359
  NETFLO, 198
  NETSIM, 198
  network flow, 196, 197–199
  TRANSIMS, 196, 198
  TRANSYT, 197
SmallTalk programming language, 235
space and time utility, 381, 382
space-time constraints, 270, 271, 274
space-time networks, 194–195
space-time path, 218
space-time prism, 272–274
spanning tree, 134, 140, 152, 167, 177. See also shortest path problem
spatial aggregation, 85, 125, 212–213
  types of error, 212
spatial analysis, 10, 215, 216
spatial autocorrelation, 126, 250–251, 254, 255, 256, 349, 360, 362

Index 457

spatial autoregression, 255, 362
spatial compactness, 249, 250, 251
spatial contiguity, 249, 250, 252
Spatial Database Engine (SDE), 379
spatial decision support system (SDSS), 11, 231, 240, 293–294, 342, 354, 367, 390
spatial dependency, 6, 11, 247, 254–255, 261, 262, 362
spatial heterogeneity, 6, 11, 247, 254, 255, 256, 261
spatial interaction (SI) models, 126, 226–228, 256–261, 278, 279, 373, 375
   competing destinations, 260
   doubly-constrained, 257, 258, 260, 264, 267
   singly-constrained, 257, 258, 260
   unconstrained, 257, 258, 260, 267
spatial join operation, 224–225
spatial reasoning, 238
spatial sampling, 102–103, 216
spatial search operation, 223. *See also* buffer operation
spatiotemporal, 15, 43, 50, 238, 244, 270, 358
   changes, 53, 236, 237
   concepts, 45–46
   constraints, 272
   data models, 44, 46–47, 237
   *See also* visualization
standard deviation distance, 360
standards,
   Content Standards for Digital Geospatial Metadata, 48, 113–114, 125
   European, 114
   Federal Information Processing Stanadrds (FIPS), 104
   for ITS communications, 321
   National Map Accuracy Standard, 106, 108
   Spatial Data Transfer Standard (SDTS), 83, 106, 112–113, 121
   *See also* Open GIS Consortium (OpenGIS or OGIS)
static binding, 38
stochastic modeling, 270
structural programming, 234
structured query language (SQL), 30, 31, 32, 38, 218–219
suitability mapping, 286, 288
supply chain, 5, 381, 382, 384, 387, 389, 399, 400
sustainable development, 353–355
   transportation control measures (TCM), 354, 355
syntactic heterogeneity, 41, 84
synthetic aperture radar (SAR), 98

telematics, 298
TELNET, 41
temporal events, 45, 47. *See also* spatiotemporal

terrain modeling, 106
tessellation, 156, 198, 347, 358, 359
thematic layer, 39
three-dimensional (3-D) GIS, 281, 282
time, value of, 181, 185
time budget, 269, 273
time geography, 272
Topological Vector Profile (TVF), 112
Topologically Integrated Geographic Encoding and Referencing (TIGER) files, 104–105, 116, 217, 326
topology 53, 58, 230, 328, 329
   network, 76, 79, 112, 118, 128, 187, 268, 333, 362
   topological consistency, 21, 61
   *See also* relationships
traffic analysis zone (TAZ), 57, 102, 114, 125
   centroid, 57
   connector arcs, 57
   design of, 126, 127, 248–252
traffic flow, 194
traffic load, 194
TRANPLAN, 253. *See also* travel demand analysis
Transfer Control Protocol/Internet Protocol (TCP/IP), 335
transport telematics, 299, 337. *See also* intelligent transportation systems (ITS)
Transportation Equity Act for the 21$^{st}$ Century (TEA-21), 302, 306, 311
Transportation Network Profile (TNP), 112
transportation operations/management centers, 82, 303, 310, 326, 333
transportation planning, 11, 83, 196, 248, 290–292, 389
travel behavior, 252
travel demand analysis, 117, 125, 127, 227, 248, 252–263, 268, 271, 276, 354–355, 357
   activity-based approach, 218, 253, 268–271, 389
   aggregate, 248, 254, 256, 261, 264, 268
   computational intelligence approach, 253
   derived demand, 268
   disaggregate, 11, 252, 254, 256, 260, 261, 264
   equilibrium approach, 253, 263–265
   market concepts, 195–196
   modal split (MS), 252, 261–262, 267
   network assignment (NA), 252, 262
   sequential (four-step) approach, 231, 253–254, 262–263
   trip generation (TG), 252, 254–256, 356
   trip distribution (TD), 227, 252, 256–261, 267, 356, 375
   *See also* network flow equilibrium
travel diary, 117, 269, 270, 271

traveling salesman problem (TSP) heuristics, 161, 163–165
  construction procedures, 163–164, 391
  improvement procedures, 163, 164, 391
traversal, 73
triangulated irregular network (TIN), 106, 156, 288, 289, 347, 348, 349, 351, 352, 353
turn impedance, 55, 128
turn table, 8, 54, 60–61, 80, 217, 218, 226, 236

U.K. Ordnance Survey, 108
U.S. Federal Geographic Data Committee (FGDC), 108, 113
U.S. National Center for Geographic Information and Analysis (NCGIA), 215
U.S. National Geospatial Data Clearinghouse, 108, 113
U.S. National Spatial Data Infrastructure (NSDI), 108, 113, 322
urban transportation modeling system (UTMS), 253. *See also* travel demand analysis
utility maximizing, 269, 270

variable-length segmentation. *See* dynamic segmentation
variational inequality (VI), 189–190, 262, 264
vectorization algorithms, 120
Vehicle, Road and Traffic Information Society (VERTIS), 297
vehicle routing problems (VRP) heuristics, 167–168, 226, 367
  cluster first-route second strategy, 167, 392
  construction procedures, 167
  improvement procedures, 167
  route first-cluster second strategy, 167, 392
  set-partitioning procedures, 168
versioning, 36, 45
video log, 240. *See also* visualization
viewshed analysis, 350–353
  front-to-back algorithm, 352
  fuzzy viewshed, 353
  multiresolution viewshed algorithms, 352
  ray-tracing strategy, 351, 353
virtual reality (VR), 11, 241, 243–245, 291. *See also* visualization
virtual reality modeling language (VRML), 242
visibility location problems, 352
Visual Basic programming language, 235, 236
Visual Basic for Applications (VBA) programming language, 235, 236
Visual C++ programming language, 235, 236
visualization, 238–243, 247, 271, 363, 367, 390, 392, 394, 395, 399
  animation, 240, 241, 242, 245, 281
  dynamic/temporal, 241–242. *See also* spatiotemporal
  geographic visualization (GVis), 11, 101, 215, 238–240, 242, 289, 293–294, 343, 359, 374, 377, 383
  integration with GIS, 242–243
  scientific visualization, 11, 238–239, 240, 242
  visual communication, 239
  visual realism, 240–241
  visual thinking, 239
virtual network (vnet), 68–69

World Geodetic System (WGS), 88, 106, 321
World Wide Web (WWW), 5, 12, 104, 281, 283, 335, 337, 388

CPSIA information can be obtained at www.ICGtesting.com
Printed in the USA
BVOW07*0338191114

375706BV00003B/49/P